American Institute of Chemical Engineers

Particle Technology Forum

Presentations at the
2007 AIChE Annual Meeting

November 4-9, 2007
Salt Lake City, Utah, USA

Printed from e-media with permission by:

Curran Associates, Inc.
57 Morehouse Lane
Red Hook, NY 12571
www.proceedings.com

ISBN: 978-1-60560-001-7

Some format issues inherent in the e-media version may also appear in this print version.

ISBN: 978-1-60560-001-7

American Institute of Chemical Engineers
Proceedings
Three Park Avenue
New York, NY 10016-5991
Phone: 212-591-8100

www.aiche.org

American Institute of Chemical Engineers

Particle Technology Forum

TABLE OF CONTENTS

21a Dynamics, Packing Porosity, And Conformation Variations Of Granular Chains

Amy Shen, Department of Mechanical and Aerospace Engineering, Washington University in St. Louis, Campus Box 1185, 1 Brookings Drive, St.Louis, MO 63130

We report results concerning the dynamical behavior of collections of beaded chains rotated in a two dimensional cylinder. During our experiments, we observe a variety of complex spatio-temporal patterns. These patterns depend on the chain length, the size of the system, and the rotation rate of the cylinder. We report the porosity of the granular chains under flow, the conformation of chains, and the end-to-end distance of chains in comparison to the polymer melt system.

21b Mixing Effects On Heat Transfer In Rotating Tumblers

Isabel Figueroa, Watson L. Vargas, and Joseph J. McCarthy. University of Pittsburgh, 1249 Benedum Hall, University of Pittsburgh, Pittsburgh, PA 15261

Slow granular flows play an important role in industries ranging from food to pharmaceuticals to ceramics, and this subject has received much recent attention in the literature. In contrast, heat transfer in even the simplest particle flows is poorly understood. In granular flows, conductivity depends strongly on the micro-structure of the bulk material: therefore, the particle mixing/segregation (i.e., "convective" heat transfer) will dramatically impact the total heat flow in the granular material. In this work, a multi-scale, multi-physics modeling technique – Thermal Particle Dynamics (TPD) – is used to examine how transient heat transfer and particle mixing interplay in rotating tumblers. We study the effects of tumbler cross section shape, and operation parameters – rotation rate and tumbler filling level – in the heating rate of the granular material. We present/compare temperature profiles and mixing patterns for equivalent granular systems. Our TPD simulation results are used to demonstrate that the rate of mixing is critical to the heat transfer rate in the granular material.

Surface flow of granular materials in non-circular rotating cylinders

D. V. N. Prasad and D. V. Khakhar*

Department of Chemical Engineering,

Indian Institute of Technology - Bombay, Powai, Mumbai 400076, India

Abstract

An experimental study of the flow of glass beads in quasi-2d half-filled cylinders with different cross-sectional shapes (square, star, circle with 2 wedges or 4 wedges) rotated about their axes is carried out using flow visualization. The flow of particles in such geometries is time-periodic and the length and thickness of the flowing layer shrink and expand with rotation of the cylinder. The surface angle increases and decreases with the layer length. The shape of the layer in different geometries, however, is qualitatively similar to that for a circular cylinder. A depth-averaged flow model to predict the time-varying layer thickness profile is presented, along with a perturbation solution in terms of a small parameter, k, which is the ratio of the layer thickness at midpoint to the half-length of the layer at the cross-section orientation when the length is minimum. The $O(k)$ perturbation solution and the full theory both predict that the scaled layer thickness varies periodically; the deviations are proportional to the rate of change of the length. The perturbation solution gives results close to those from numerical solution except at cylinder orientations when the length of the flowing layer changes sharply. Measured variation of the scaled midlayer thickness with time for all geometries is well predicted by the theory.

*Electronic address: `khakhar@iitb.ac.in`

I INTRODUCTION

Mixing of granular materials in rotating cylinders is widely encountered in chemical, metallurgical, pharmaceutical and fertilizer industries. The mixing in the transverse direction is several times faster than that in the axial direction and plays a major role in the formation of a uniform end product. The effects of particle size, fill fraction of cylinder and rotational speed on transverse mixing have been studied both experimentally and theoretically [1–3].

In a circular cylinder, the flow of particles is time-independent and the streamlines are time-invariant. There is no radial flow of material in this case and hence convective mixing is slow. When the cross-section of the cylinder is changed to any non-circular shape, the stream function becomes time-dependent [4] and the streamlines at different times intersect. This condition is an indicator of chaotic advection [5].

Khakhar et al. [4] considered rotating cylinders with elliptical and square cross-sections by means of theory and experiment. The continuum model of Khakhar et al. [2] with the assumption that the flowing layer profile is self-similar was found to describe the flow reasonably well. Computed Poincaré sections showed chaotic advection, and the predicted mixing patterns for tracers matched experimental observations. The mixing was found to be faster in non-circular cross-sections as compared to the circular cross-section. More detailed theoretical studies, including the effect of fraction filling was carried by Hill et al. [6].

In non-circular cross-sections, the segregated structures at equilibrium are time-periodic. They are sensitive to the shape and fill level of the geometry [7]. The segregated structures are a result of dynamic competition between the chaotic mixing and segregation, which leads to a periodic variation of the equilibrium patterns. The segregated structures in the polygonal shaped rotating containers resemble the Poincaré sections of the geometry [8] and the rotational orientation of these structures is given by the direction of the eigen vectors at the hyperbolic points observed in the Poincaré maps [9].

The flow of granular materials in non-circular cylinder cross-sections has not been studied before in detail. Though a self-similar layer thickness profile assumption [4] is applicable at very low cylinder rotations, the variation of layer thickness profile at higher rotational speeds and for different particle sizes has not been studied systematically.

II EXPERIMENTAL

Flow visualization experiments are performed to study the layer thickness profiles in non-circular geometries at four different rotational speeds for two particle sizes. The study is limited to non-cohesive granular materials filled to 50% of the cylinder volume. A circular and four non-circular cylinder geometries are chosen for the study - square, star, a circular geometry with 2 and 4 triangular inserts (referred to as W2 and W4) - and are shown in Figure 1. The geometric parameters of the cross-sections are shown in Table I. Broadly all the geometries are similar; however, the differences lie in the number of corners each geometry has and the ratio of L_{max} to L. W2 is rotationally symmetric to 180° whereas other geometries are symmetric with respect to 90°.

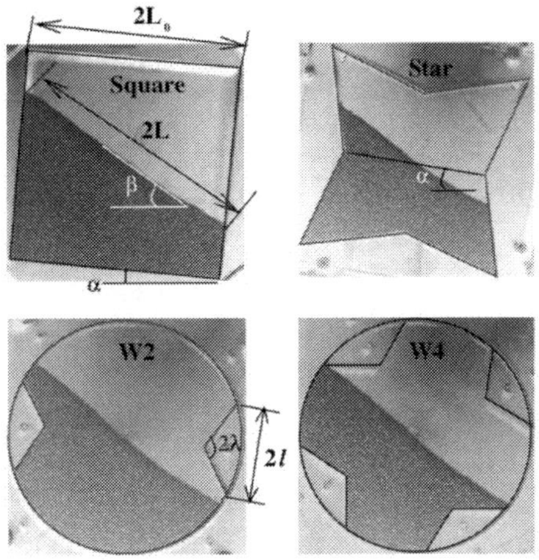

FIG. 1: Various non-circular geometries considered for experiments. $2L_0$ is the minimum length of the top free surface in all geometries.

III RESULTS AND DISCUSSION

The length and thickness of the flowing layer shrink and expand with rotation of a non-circular cylinder. The shape of the layer in different geometries, however, is qualitatively similar to that for a circular cylinder. The surface angle increases and decreases with the layer length. The length of the free surface is not sensitive to the variation of the free surface angle and is well-predicted based on the geometrical considerations using a constant value of the surface angle.

TABLE I: Geometric parameters in different cross-sections.

Shape	L_0 (mm)	λ (deg)	l (mm)	L_{max}/L_0	No. of corners
Circle	160.0	-	-	1.00	0
Square	141.5	-	-	1.41	4
Star	119.0	73	165	2.00	8
W2	110.0	60	64	1.45	6
W4	110.0	60	64	1.45	12

A depth-averaged flow model is obtained, with an assumption of a constant shear rate in the flowing layer, to predict the time-varying layer thickness profile. A perturbation solution is also obtained in terms of a small parameter k, which is the ratio of the layer thickness at midpoint to the half-length of the layer at the cross-section orientation when the length is minimum. The perturbation solution for the scaled layer thickness profile, ϕ, is as below:

$$\phi(\eta,\theta) = \left(1-\eta^2\right)^{1/2} - kf(\theta)\left(\frac{\cos^{-1}|\eta|}{2\phi_0} - \frac{|\eta|}{2}\right). \qquad (1)$$

where, η is dimensionless distance along the free surface with its origin at the midlayer length. θ is the orientation angle of the geometry and $f(\theta) = (1/L)dL/d\theta$ is the specific rate of change of free surface length. The perturbation solution gives results close to those from numerical solution of the unsteady mass balance equation except at cylinder orientations when the length of the flowing layer changes sharply.

The zeroth order solution of the above equation predicts that the layer profiles scaled with $L(t)$ at different rotational angles are identical. The measured layer thickness profiles averaged over different orientations of non-circular mixers match reasonably well with the theory, but the variation of the profiles about the mean value are larger for the non-circular cylinders compared to the circle. The $O(k)$ perturbation solution and the full theory both predict that the scaled layer thickness varies periodically; the deviations are proportional to the rate of change of the length.

A detailed comparison between experimental results and theory is made in terms of the scaled midlayer thickness ($\phi(0,\theta)$) and is shown in Figure 2. The midlayer thickness in square and star varies in 3 different phases: a slow decrease between $\theta = 0° - 45°$, a rapid increase between 45° and 60° and a slow decrease between 60° and 90°. The different phases observed in the experiments can be explained using eq. (1). When θ is between 0° - 45°, $dL/d\theta$ is positive and its magnitude increases as it approaches the corner of the mixer (where length of the layer is maximum). As a result, the second term in eq. (1) is negative and the midlayer thickness decreases monotonically. When the corner of the cylinder is reached, the transient term changes sign resulting a sharp increase in the midlayer thickness. In the final phase where $dL/d\theta$ is negative, this term is positive but reduces in magnitude with rotation. Thus the midlayer thickness decreases. The computational results for the unsteady-state mass balance equation are close to the experiments in all the three phases. The perturbation theory gives good predictions everywhere except near the corner of the square where a discontinuous change in f occurs. Square has discontinuity at $\theta = 45°$, star has at $\theta = 45°$ and 90°, W2 has at $\theta \approx 23°$ and W4 has discontinuity at $\theta = 0°, 22°, 62°$ and 90° (see Figure 2) and hence the variation of the midlayer thickness in W4 is discontinuous a number of times. However, in both W2 and W4 geometries, when $dL/d\theta = 0$ the layer thickness is constant. The deviations of the model predictions at higher rotational speeds indicate the violation of the constant shear rate assumption.

The computed Poincaré maps obtained using the modified theory are different from those obtained by assuming a self-similar layer profile [4] and are compared in Figure 3. The regular regions are asymmetric in the modified theory. The motion of the particles inside the regular island zones is also chaotic.

IV CONCLUSIONS

A continuum model based on the depth averaged continuity equation assuming the shear rate to be constant is developed to describe the flow in non-circular cylinders. The model predicts a symmetric layer thickness profile for all cases. A perturbation solution is obtained considering an expansion in terms of the small parameter, k. The perturbation solution is in reasonable agreement with the numerical solution of the full model at all orientations except near corners of the geometry when there is a discontinuous change in $dL/d\theta$. The perturbation solution to $O(k^0)$ indicates that the layer thickness profiles scaled with $L(t)$ are the same at different cylinder orientations. This is in agreement with the model proposed by Khakhar et al. [4]. To $O(k)$ the theory indicates that the changing layer length results in additional source term which produces a periodically varying scaled layer thickness. For

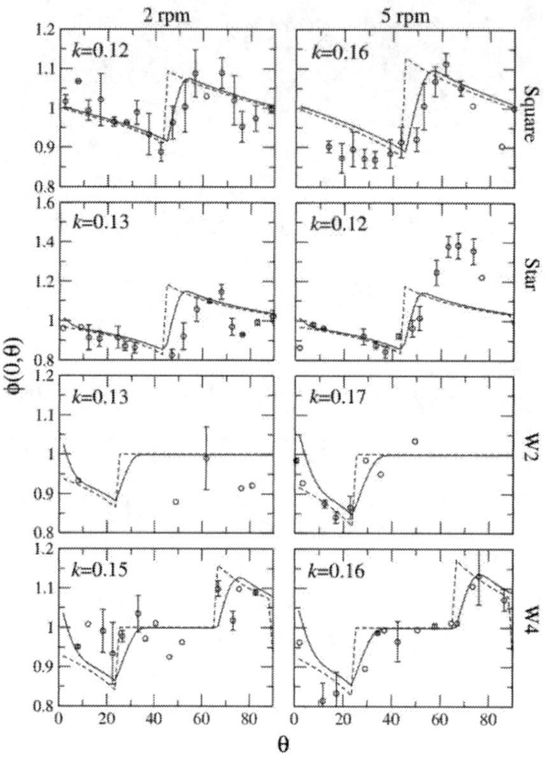

FIG. 2: Variation of scaled midlayer thickness with time in different cylinder cross-sections rotated at 2 rpm (left column) and 5 rpm (right column) for 3 mm glass beads. Solid lines are results from numerical solution of depth-averaged mass balance equation and dashed lines are predictions of the perturbation theory (eq.1).

orientations where the length is increasing ($dL/d\theta > 0$), the source term is negative resulting in a lower scaled thickness and for orientations for which $dL/d\theta < 0$, the reverse is true.

Detailed comparisons of the measured midlayer thickness and predictions of the full model and perturbation solution are presented. The variation of the midlayer thickness is significant ($\pm 10\%$ about the $O(k^0)$ value). The full model gives good predictions for all geometries, particularly at low rotational speeds.

The results presented here indicate that the approach of using depth average equations for the description of surface granular flows is robust and is useful for periodically steady flows as well. The analysis may be extended in a straight forward way to cylinders with

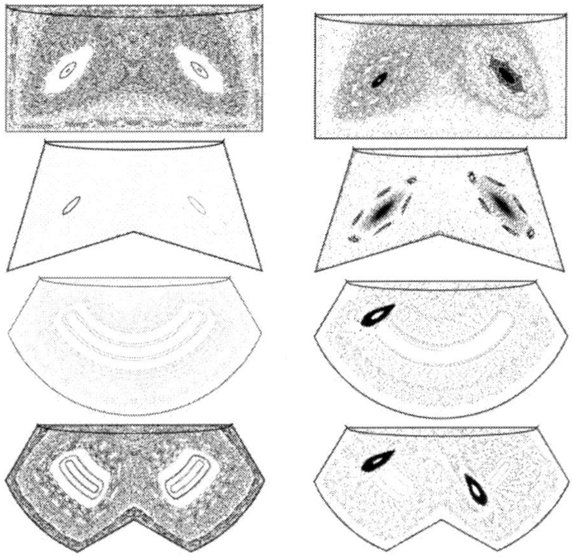

FIG. 3: Comparison of computer generated Poincaré sections using the model by Khakhar et al. [4] (left column) and modified theory (right column).

volume filled by particles different from 50%, as in [6].

---

[1] G. Metcalfe, T. Shinbrot, J. J. McCarthy, and J. M. Ottino, Nature **374**, 39 (1995).

[2] D. V. Khakhar, J. J. McCarthy, T. Shinbrot, and J. M. Ottino, Phys. Fluids **9**, 31 (1997).

[3] D. R. V. Puyvelde, B. R. Young, M. A. Wilson, and S. J. Schmidt, The Canadian Journal of Chemical Engineering **78**, 635 (1999).

[4] D. V. Khakhar, J. J. McCarthy, J. F. Gilchrist, and J. M. Ottino, Chaos **9**, 195 (1999).

[5] J. M. Ottino, *The Kinematics of Mixing: Stretching, Chaos and Transport* (Cambridge University Press, Cambridge, 1989).

[6] K. M. Hill, J. F. Gilchrist, J. M. Ottino, D. V. Khakhar, and J. J. McCarthy, Int. J. of Bifurcation and Chaos **9**, 1467 (1999).

[7] K. M. Hill, D. V. Khakhar, J. F. Gilchrist, J. J. McCarthy, and J. M. Ottino, Proc. Natl. Acad. Sci. **96**, 11701 (1999).

[8] S. E. Cisar, P. B. Umbanhowar, and J. M. Ottino, Phys. Rev. E **74**, 051305 (2006).

[9] S. W. Meier, S. E. Cisar, R. M. Lueptow, and J. M. Ottino, Phys. Rev. E **74**, 031310 (2006).

21d Discrete Element Simulation Of The Mixing Process In A Pancoater

Atul Dubey[1], Bodhisattwa Chaudhuri[2], M. Silvina Tomassone[1], and Fernando J. Muzzio[1]. (1) Chemical and Biochemical Engineering, Rutgers University, 98 Brett Road, Piscataway, NJ 08854, (2) Pharmaceutical Sciences, University of Connecticut, 69 North Eagleville Road, Storrs, CT 06269

Computer Simulations using Discrete Element Method have been performed to simulate the flow and mixing of tablets in a pancoater. A comparison between the mixing performance using front-back and side-by-side loading was studied. Differently colored tablets of 6mm diameter were loaded in the pancoater in side-side and front-back loading. The simulations were performed using a system of granular material of same mechanical and physical characteristics as that of a tablet. Five sets of spiral shaped ribbon baffles were incorporated in the pancoater in enhance the mixing performance. In order to maintain high level of computational performance, the baffles were made of particles of appropriate sizes and they remained glued and perform solid body rotation with the wall of the pancoater. The degree of uniformity of the mixture was observed in multiple locations of the pancoater by performing sampling. It was observed that in the side-by-side loading, the mixing performance was significantly better than the front-back loading. A uniform mix was produced in as little as 2 revolutions at a rotation rate of 15rpm in the case of side-by-side loading. This amount of mixing was not obtainable in the front-back loading even after 15 revolutions.

21e Electrostatics And The Flow Of Granular Materials

Keirnan R. LaMarche, Shejal K. Shah, Troy Shinbrot, and Benjamin J. Glasser. Department of Chemical and Biochemical Engineering, Rutgers University, 98 Brett Rd., Piscataway, NJ 08854

The effects of electrostatics can have a large impact on granular flows. These effects can cause segregation and agglomeration and have an important impact on many industrial processes. The charging of grains also leads to dangerous electric discharges and dust explosions. Many industries spend large sums of money to prevent or control electrostatic charging. The pharmaceutical industry is especially susceptible because it requires that small amounts of drug be mixed with large amounts of excipient and the unpredictable nature of electrostatics can cause segregation and product non-uniformities.

How charge is accumulated in the flow of granular materials through a stainless steel cylinder was analyzed. It has been found that the net charge accumulated by the grains was dependent on the surface area of the cylinder initially covered by particles. In mixtures that were both homogeneous and heterogeneous, only those grains in contact with the walls of the cylinder became charged. The preferred charging of one component in a mixture of granular materials based on composition or size could result in segregation, producing a change in product concentration. In bi-disperse mixtures, the smaller particles were observed to segregate toward the walls; only these smaller particles charged, even in a mixture of 50% larger particles by volume. These experiments show the possibility for the segregation of smaller particles due to electrostatic effects.

We also examined the effects of non-uniform electric fields on uncharged grains. A large non-uniform electric field can polarize particles and cause them to agglomerate or adhere to surfaces. The amount of materials that adhered to the surface of a rod in a strongly non-uniform field were measured. These forces acting on the flow of granular materials, both down an inclined plane and from a hopper, created agglomerates, which had a major impact on the behavior of the flows.

Dynamics of Surface Flow During Heap Formation

Subhashis Sarkar[1] *and D. V. Khakhar*[2,*]
Department of Chemical Engineering
Indian Institute of Technology-Bombay Powai, Mumbai, 400076, India
E-mail: [1]*subhashis@iitb.ac.in,* [2]*khakhar@iitb.ac.in**

Introduction

Surface flows comprise a shallow layer of material flowing over a static region of the same material. These flows are important in nature in phenomena such as lava flow, avalanches (Hutter, 1997), sedimentation at river bed (Bounhoure et al., 2002), sand dunes (Andreotti et al., 2002b,a) as well as in industrial processes like transportation of material, storage in silos (Nedderman, 1992) and mixing in rotary kilns. The flow rate is the only independent parameter in such flows and the interface between the flowing layer and static region depends on the flow determining the surface angle and the depth of flowing layer.

Steady state surface flows have been studied by means of experiments in heaps and rotating cylinders. Within the surface layer the velocity profile is found to be linear (Bonamy et al., 2002; Taberlet et al., 2003; Orpe and Khakhar, 2004; Jain et al., 2002) with an exponential decay near the static region (Komatsu et al., 2001; Bonamy et al., 2002; Taberlet et al., 2003; Orpe and Khakhar, 2004; Jain et al., 2002). Shear rate is found to be constant in some studies (Bonamy et al., 2002; Khakhar et al., 2001) while it is found to vary by Orpe and Khakhar (2004). Transient behavior of surface flow have not been studied much. du Pont et al. (2005) looked at the instantaneous velocity profile within the flowing layer during avalanches in a rotating cylinder. The velocity profile is found to be purely exponential over the depth. The profiles are found to be self-similar with a decay length independent of time.

To model surface flows, depth averaged conservation equations defined over the surface layer have been widely used in literature (Bouchad et al., 1994; Boutreux et al., 1998; Douady et al., 1999; Khakhar et al., 2001). The system is not a closed one as there is interchange of material between the flowing and the static regime. The rate of erosion of the bed or deposition on the bed is dependent on interchange velocity (Γ) and is an important factor for surface flows. The interchange velocity predicted by different models based on various assumptions are discussed below. Experimental determination of Γ during transient flow can serve as a robust tool to examine the models.

Here we present an experimental study of transient flow during heap formation in a quasi 2D system. The objective of the work is to gain an understanding of the rate of deposition and erosion in surface flows. Using video photography and image analysis technique we have determined the spatio-temporal variation of macroscopic properties like height of the free surface and interface, layer thickness, interchange velocity, mean velocity and local flux. Some of the existing theoretical models are compared with the experimental results.

Experiments

Experiments are carried out in a rectangular bin as shown in figure 1. The bin is divided into two parts by an inclined metal plate. The upper part acts as a hopper and the lower part acts as a sink where the heap forms. An auxiliary hopper is attached at the top to ensure constant supply of material during the experiment. The front and the back walls are made of glass to facilitate imaging and are separated by 1 cm. Both the hoppers are filled with 1 mm SS316 balls and are allowed to flow through the opening with gap k. The flow rate is varied by changing the size of the gap k. The process of heap formation is imaged (captured area is shown by a dotted rectangle in figure 1) with a video camera at a rate of 25 frames per second using a shutter speed $1/100^{th}$ s, until the flow becomes steady. Captured images have a resolution of 1280 x 1024 pixels and each pixel corresponds to $\sim 2\mu$m. Steady state mass flow rate ($\dot{m} = W/\Delta t$) is measured by collecting the material (W) at the exit for sufficient amount of time (Δt) during steady flow. The apparatus is earthed to minimize any effect of static charge that might develop due to inter-particle and particle-wall friction. Experiments are repeated for five different flow rates ranging from 35 g/s - 160 g/s, and for each flow rate six sets of data are taken.

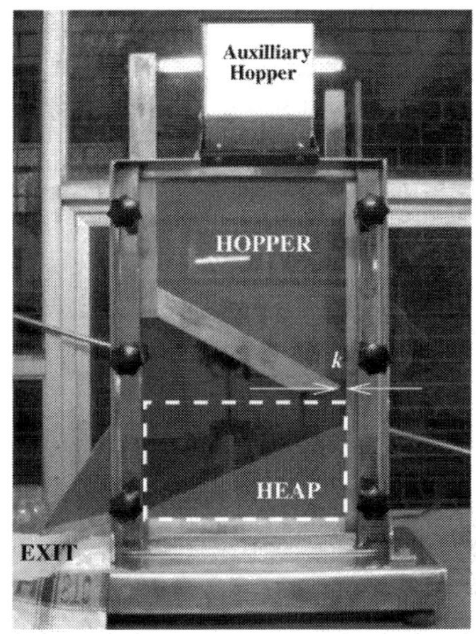

Figure 1: Experimental setup used

Figure 2(a) shows a captured image. The image analysis technique involves detection of the free surface and the static-flowing interface in every image. The flowing layer is detected by subtracting the gray scale value of each pixels for two consecutive images. The difference in gray scale value is

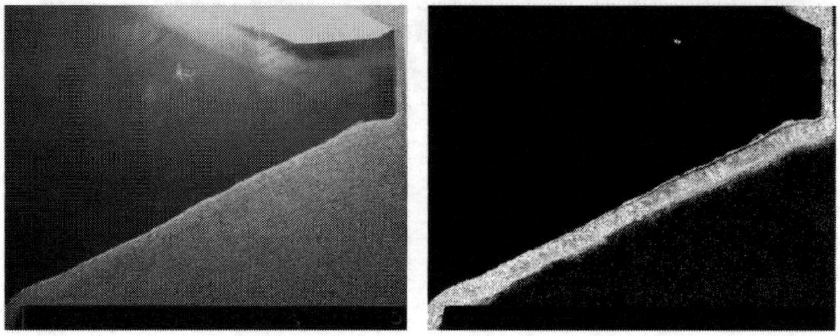

(a) (b)

Figure 2: (a) Typical captured image (b) Detected surface layer

significant for flowing region and almost zero for static region. The pixels are made white if difference is higher than a threshold value and black otherwise. Thus the flowing layer appears as a white region on a black background as shown in figure 2(b). To detect the free surface we search downwards from top for white pixels at all x-positions. The first white pixel encountered having sufficient neighboring white pixels is considered to be a point at the free surface for corresponding x-positions. Similarly the interface is determined by a upward search from bottom of the image. The detected free surface and interface are found to be linear and used to determine parameters like layer thickness, interface angle, interchange velocity from geometry of the system. Steady state volumetric flow rate is found from mass flow rate assuming a constant density within flowing layer. The volumetric flux and average velocity at each position is determined using mass balance equation.

Theory

A number of theoretical models have been proposed to describe granular surface flows. All the models present a macroscopic view of the process and are able to predict the layer thickness, height of the free surface and the interface. Interchange of material between the flowing layer and the static region (rate of interchange of material per unit area is called interchange velocity, Γ) is allowed in all the models and determines the rate of erosion of the heap or deposition on the heap.

Some of the models (Bouchad et al., 1994; Boutreux et al., 1998) uses a mass conservation equation (eqn. 1) for flowing particles along with an expression for interchange of material based on physical arguments.

$$\frac{\partial \delta}{\partial t} = \Gamma + v\frac{\partial \delta}{\partial x} + \text{diffusion terms} \tag{1}$$

Here δ is the layer thickness, v is mean velocity of the particles and Γ is the interchange velocity. Bouchad et al. (1994), argued Γ to be proportional to layer thickness assuming a constant mean velocity as

$$\Gamma = \gamma\delta(\beta_s - \beta) \tag{2}$$

where β_s is a characteristic angle, β is the local interface angle and γ is a constant. Boutreux et al. (1998) argued that the above expression is true only for thin layers and for thick layer he proposed

$$\Gamma = v_{up}(\beta_n - \beta) \tag{3}$$

where β_n is called neutral angle, angle at which there is no deposition or erosion and v_{up} is a characteristic velocity constant.

A more detailed study involves depth averaged mass (eqn. 4) and momentum balance (eqn. 5) equations for flowing layer (Douady et al., 1999; Khakhar et al., 2001).

$$\frac{\partial}{\partial t}(\delta\langle\rho\rangle) + \frac{\partial}{\partial x}(\delta\langle\rho v_x\rangle) = (\rho v_y)|_{y=0} \tag{4}$$

$$\frac{\partial}{\partial t}(\delta\langle\rho v_x\rangle) + \frac{\partial}{\partial x}(\delta\langle\rho v_x^2\rangle) = \frac{\partial}{\partial x}(\delta\langle\tau_{xx}\rangle) + \tau_{xy}|_{y=0} + (\rho v_x v_y)|_{y=0} + \langle\rho\rangle g\delta sin\beta \tag{5}$$

where $\langle\phi\rangle = (1/\delta)\int_0^\delta \phi\, dy$ denotes average across layer. The models are different in forms of stress constitutive equations used. Using the assumption of constant shear rate and linear velocity profile, Douady et al. (1999) found the expression for interchange velocity as

$$\Gamma = \frac{g\, sin(\beta_m - \beta)}{\dot{\gamma} cos\, \beta_m} \tag{6}$$

Here β_m is characteristic friction angle which depends on velocity, and shear rate $\dot{\gamma} = 2u/\delta$ where u is the mean velocity. Taking shear stress as a combination of collisional stress and Coulomb stress along with a condition of continuity of shear stress at the interface Khakhar et al. (2001) found similar expression for interchange velocity but with shear rate

$$\dot{\gamma} = \left[\frac{g\, cos\beta\, sin(\beta_m - \beta_s)}{c\, d\, cos\beta_m\beta_s}\right]^{1/2} \tag{7}$$

The models are compared using the experimental results.

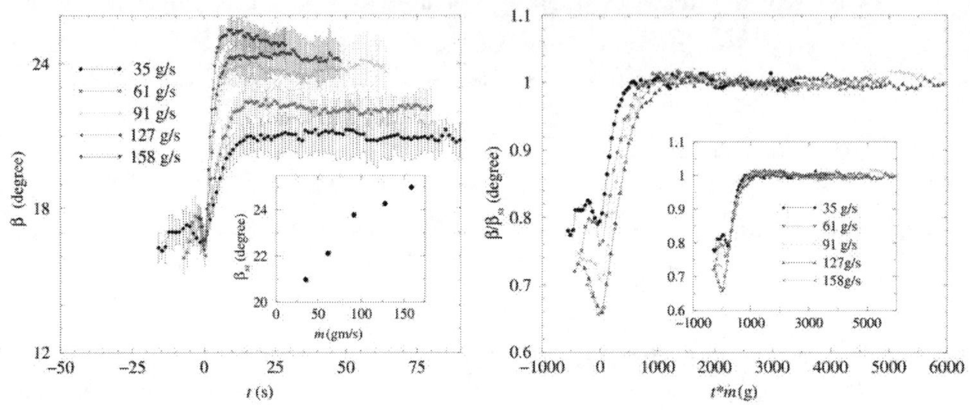

(a) (b)

Figure 3: (a) Variation of interface angle (β) with time for different flow rates; Inset shows variation of steady-state value of interface angle (β_{ss}) with mass flow rate ($\dot{m}$) (b) Variation of dimensionless angle of repose (β/β_{ss}) with time multiplied by mass flow rate. Inset shows that, on shifting of some of the plots along $x-$axis, all the plots collapse on a single plot.

Results and Discussions

Figure 3(a) shows the variation of interface angle (β) with time (t) for different mass flow rates ($\dot{m}$). For all the cases, data corresponds to a position sufficiently away from the entrance and exit (14-cm away from exit) to minimize any entrance and exit effects that might be present. Time zero corresponds to start of the outflow. The error bar shows the standard deviation over six data sets. Before the outflow begins, β is almost constant, at the point when outflow starts, β is minimum and after that there is a sudden increase of β. Eventually β becomes constant indicating steady flow. The steady state interface angle β_{ss} increases with mass flow rate (inset figure 3a). Figure 3b shows that when β is non-denationalized with β_{ss} and plotted against $\dot{m}t$, all the plots collapse on a single line which indicates $1/\dot{m}$ is proportional to the time scale of the process.

The variation of system parameters like layer thickness (δ), interchange velocity (Γ), local flow rate (q), average velocity (u) and shear rate ($\dot{\gamma}$) with time (t) for different mass flow rates are shown in figure 4. All the parameters show a sudden change immediately after outflow begins. The increase in mass flow rate is accommodated by increase in layer thickness (δ) and average velocity (u). The initial deposition rate (Γ) is high. The deposition rate decreases slowly till the time of outflow and rapidly after outflow. Zero Γ value signifies steady state.

Comparing deposition rate after outflow at different position, it is observed that rate of decrease of Γ is faster near the exit leading to a rapid increase in β. Shear rate ($\dot{\gamma}$), found assuming a linear velocity profile within the layer is not constant and increases with mass flow rate.

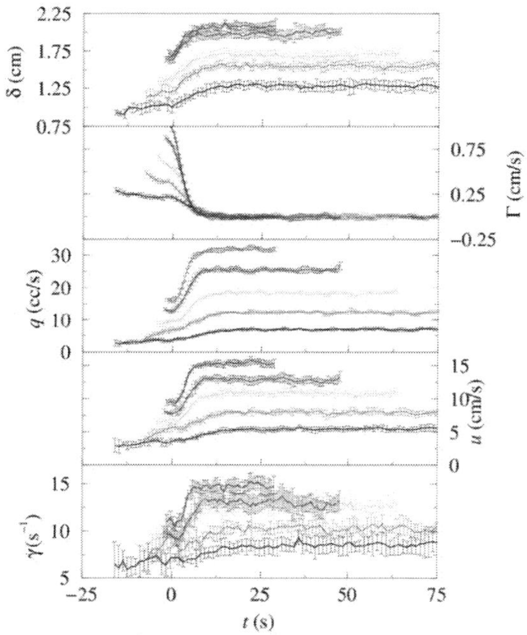

Figure 4: Variation of δ, Γ, q, u, and $\dot{\gamma}$ with time for different mass flow rates. Corresponding flow rates are 35, 61, 91, 127 and 158 g/s as we go upwards for each panel.

Cumulative weight (W_Q) of the exit material is measured throughout the process using a weighing balance kept at the exit. The mass flow rate at any time is determined by numerical differentiation of W_Q vs time data. Volumetric flow rate is determined assuming a constant density of the flowing layer. A comparison of the variation of exit flow rate during the process obtained using two independent method (image analysis and direct measurements at exit) is shown in figure 5. The match is good within experimental error. This validates our assumption of constant density (taken equal to the bulk density) within the flowing layer and indicates the accuracy of the analysis technique.

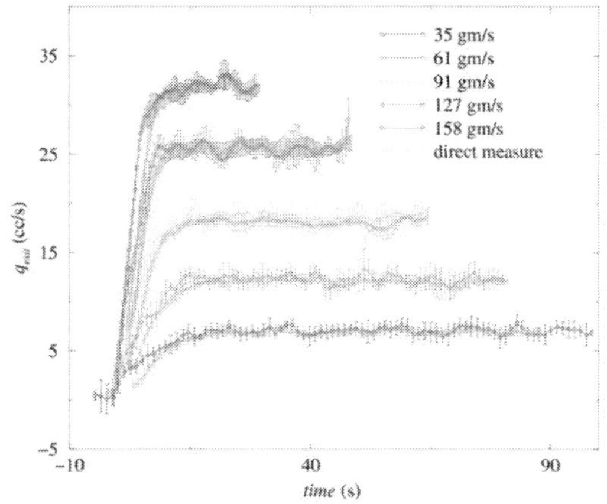

Figure 5: Variation of exit flow rate q_{exit} with time (t) for different mass flow rates using image analysis technique and through direct measurements.

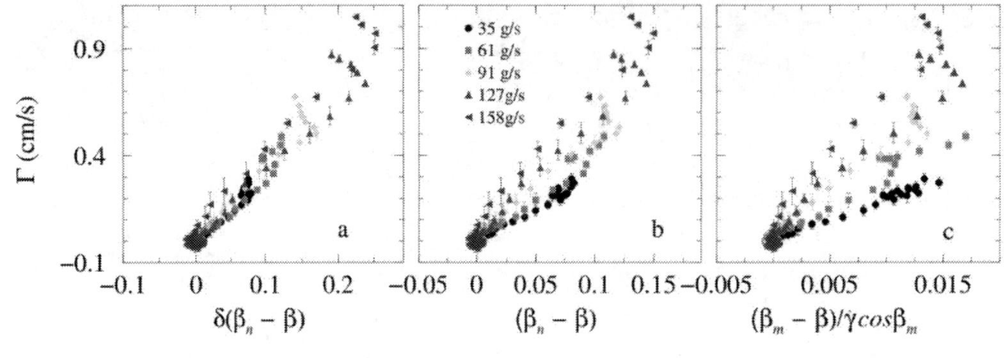

Figure 6: Comparison of interchange velocity (Γ) as predicted by different theories for different flow rates. (a) BCRE (b) BrdG (c) DAD and KOAO model. All the data corresponds to a position sufficiently away from the exit and entrance (x = 14 cm).

Figure 6 shows a comparison of Γ, as predicted by various models for different mass flow rate. Here β_s, β_n, β_m, different characteristic angle of repose as defined in different models, all are taken equal to the steady state angle of repose (β_{ss}). The plots are not linear and data for different mass flow rates does not superimpose for any of the model. Thus, none of the model is good enough to describe the experimental data satisfactorily. Among the three models BCRE model describes the data best. One possible reason is some of the assumptions used in the models are not true for our system, like velocity and shear rate is not constant as assumed in BCRE and BRdG model respectively. Also wall effect has not been considered in any of the models which is found to be significant by Taberlet et al. (2003) and Jop et al. (2005).

Conclusions

We have studied the dynamics of surface flow during heap formation in a quasi-2D system. Using video photography and image analysis we have detected the flowing layer which enable us to determine the variation of interface angle, layer thickness, average velocity, local flux, shear rate and interchange velocity with time. We found a sudden change in all the parameters studied immediately after outflow starts. The results are qualitatively similar for the range of flow rates studied. Four models (BCRE, BrdG, KOAO and DAD) are compared against the experimental results. Although none of the models predict the experimental data satisfactorily, among these models the BCRE model proposed by Bouchad et al. (1994) describes the experimental data best.

References

Andreotti, B., P. Claudin, and S. Douady, "Selection of dune shapes and velocities part 1: A two-dimensional modelling," *Eur. Phys. J. B*, **28** (2002a).

Andreotti, B., P. Claudin, and S. Douady, "Selection of dune shapes and velocities part 1: Dynamics of sand wind and barchans," *Eur. Phys. J. B*, **28**, 312–339 (2002b).

Bonamy, D., F. Daviaud, and L. Laurent, "Experimental study of granular surface flows via a fast camera: A continuous description," *Phys. Fluids*, **14**, 1666–1673 (2002).

Bouchad, J. P., M. Cates, J. R. Prakash, and S. F. Edwards, "A model for the dynamics of sandpile surface," *J. Phys. Paris I,*, **4**, 1383–1409 (1994).

Bounhoure, C., Y. Brunet, and A. Merlen, "Hydrodynamic study of sedimentation flows and avalanches," *Powder Technol.*, **125**, 306–312 (2002).

Boutreux, T., E. Raphaël, and P. G. de Gennes, "Surface flows of granular materials: A modified picture for thick avalanches," *Phys. Rev. E*, **58**, 4692–4700 (1998).

Douady, S., B. Andreotti, and A. Daerr, "On granular surface equations," *Eur. Phys. J. B*, **11**, 131–142 (1999).

du Pont, S. C., R. Fischer, P. Gondret, B. Perrin, and M. Rabaud, "Instantaneous velocity profiles during granular avalanches," *Phys. Rev. Lett.*, **94**, 048003 (2005).

Hutter, K., "Snow and glacier rheology," *Appl. Rheol.*, **7**, 267–277 (1997).

Jain, N., J. M. Ottino, and R. M. Lueptow, "An experimental study of the flowing granular layer in a rotating tumbler," *Phys. Fluids*, **14**, 572–582 (2002).

Jop, P., Y. Forterre, and O. Pouliquen, "Crucial role of side walls in granular surface flows: consequences for the rheology," *J. Fluid Mech.*, **541**, 167–192 (2005).

Khakhar, D. V., A. V. Orpe, P. Andresen, and J. M. Ottino, "Surface flow of granular materials: model and experiments in heap formation," *J. Fluid Mech.*, **441**, 255–264 (2001).

Komatsu, T. S., S. Inagaki, N. nakagawa, and S. Nasuno, "Creep motion in a granular pile exhibiting steady surface flow," *Phys. Rev. Lett.*, **86**, 1757–1760 (2001).

Nedderman, R. M. Cambridge University Press (1992).

Orpe, A. V. and D. V. Khakhar, "Solid-fluid transition in a granular shear flow," *Phys. Rev. Lett.*, **93(6)**, 068001 (2004).

Taberlet, N., P. Richard, A. Valance, W. Losert, J. M. Pasini, J. T. Jenkins, and R. Delannay, "Superstable granular heap in a thin channel," *Phys. Rev. Lett.*, **91**, 264301 (2003).

21g Assessing The Discharge Performance Of Hoppers Using The Discrete Element Method

William R. Ketterhagen[1], Jennifer S. Curtis[2], Carl R. Wassgren[3], and Bruno C. Hancock[1]. (1) Pfizer Global Research and Development, Pfizer Inc., Groton, CT 06340, (2) University of Florida, Chemical Engineering Department, P.O. Box 116005, Gainesville, FL 32611-6005, (3) Mechanical Engineering, Purdue University, 575 Stadium Mall Drive, West Lafayette, IN 47907-2088

The performance of hoppers has been fairly well studied both experimentally and theoretically. In this work, the discrete element method (DEM) is used to assess hopper performance and the results are compared to widely-used empirical or theoretical relations. These include the Beverloo correlation for hopper discharge rate and the Jenike hopper design charts that specify the necessary material properties and hopper wall angle required to obtain mass flow. The flow of frictional, non-cohesive, bidisperse granular mixtures through hoppers of various wall angles is modeled. In each case, the performance is assessed through measurement of the discharge rate and the extent of segregation during discharge, as well as characterization of the flow pattern as mass-flow, funnel-flow, or some intermediate. The flow pattern is characterized in a quantitative manner using the material velocity profile. The discharge rate results are in excellent agreement with those predicted by the Beverloo correlation. Similarly, the flow characterization results are in good agreement with the Jenike design charts. These favorable results indicate that DEM can not only reproduce the current estimates of hopper performance, but also provide additional insight into the flow – such as the internal granular structure – that may be difficult to obtain otherwise.

21h Triboelectric Charging In Fluidized Bed Systems Under Vacuum

Keith Forward, Dept. of Chemical Engineering, Case Western Reserve University, Cleveland, OH 44106, R. Mohan Sankaran, Chemical Engineering, Case Western Reserve University, 10900 Euclid Avenue, Cleveland, OH 44106-7217, and Dan Lacks, Case Western Reserve University, Dept. of Chemical Engineering, Cleveland, OH 44106.

Triboelectric charging has a large impact in numerous areas of science and industry. Recently, several studies of granular insulator systems have shown bipolar triboelectric charging occurs, which is dependent on the size distribution of the particles. A bench-top fluidized bed apparatus has been constructed to investigate this phenomenon. To study under highly controlled conditions, the fluidized bed is operated under vacuum (approximating 50 torr) with a nitrogen flow on the order of one liter per minute. Particles composed of a range of insulating materials, including silica-based and polymer materials, are examined. The particles are in the size range of 100- 1000 microns, and both mono-disperse and polydisperse particle size distributions are examined. The magnitude and polarity of the charging is related to the type of material and the particle size distribution.

29a Effects Of Vibration And Electric Field On The Agglomerate Particulate Fluidization Of Nanoparticles

Daniel Lepek[1], M.A.S. Quintanilla[2], Jose-Manuel Valverde[2], Antonio Castellanos[2], Rajesh Dave[3], and Robert Pfeffer[4]. (1) Department of Chemical Engineering, New Jersey Institute of Technology, Newark, NJ 07102, (2) Faculty of Physics, University of Seville, Seville, 41012, Spain, (3) New Jersey Center for Engineered Particulates, New Jersey Institute of Technology, Newark, NJ 07102, (4) New Jersey Center for Engineered Particulates, New Jersey Institute of Technology, 138 Warren Street, YCEES 208, Newark, NJ 07102

Sieved silica agglomerates (< 500μm) and of primary particle size 12nm (Aerosil R974) are uniformly fluidized by nitrogen in a fluidlike and bubbleless highly expanded state. The bubbling regime, commonly observed for micron-size and larger particles above a critical gas velocity, is fully suppressed and the fluidized bed transits progressively to elutriation as the gas velocity is increased. The condition of nonbubbling fluidization has been related to the formation of highly porous, fractal structured agglomerates of several hundred microns in size, thus this type of fluidization has been termed as Agglomerate Particulate Fluidization (APF). Several studies have suggested that fluidized agglomerates of nanoparticles have a spherical shape with a fractal dimension D close to 2.5, in agreement with the diffusion-limited-agglomeration (DLA) model. The size and fractal structure of the agglomerates determines bed expansion as described by the modified Richardson-Zaki (R-Z) empirical equation. We have studied the effect on bed expansion of externally applied fields such as vertical sinusoidal vibrations and electric field. As seen in vibrofluidized beds (VFB) of both micron-sized and nano-sized particles, the agglomerate size is decreased and the bed expansion enhanced by the action of external vibration, without a relevant change of the fractal dimension and shape of the agglomerates. On the other hand, a recent study has shown a collapse of the fluidized bed upon application of an electric field, suggesting an increase of the agglomerate size. In this case, further agglomeration due to the attractive interaction between the induced electric dipoles in the direction of the field is expected, giving rise to anisotropic larger agglomerates. In our experimental work we apply vibration to the fluidized bed under the action of the electric field. Again vibration is expected to decrease agglomerate size, but the effect on the fractal dimension and shape of the agglomerates would depend on the relative strength between the van der Waals and dipole-dipole attractive forces, which is determined by the strength of the electric field. In the light of this simple view we will describe how bed expansion is affected by varying the relative intensity of the coupled vibrations and electric field.

29b Closures For Filtered Two-Fluid Models Of Gas-Particle Flows

Yesim Igci and S. Sundaresan. Chemical Engineering, Princeton University, A-214 Equad, Princeton, NJ 08544

It is well known that gas-particle flows exhibit large fluctuations in velocities and local suspension density. In riser flows, these fluctuations are associated with the random motion of the individual particles (typically characterized through the granular temperature) and with the chaotic motion of particle clusters, which are repeatedly formed and broken apart. These clusters occur over a wide range of length scales and their dynamics span a broad range of time scales.

The origin of these clusters is well understood, and two-fluid model equations are able to capture their existence in a robust manner; however, to resolve the clusters at all length scales, extremely fine spatial grids are necessary [1]. Due to computing limitations, the grid size used in simulating industrial scale gas-particle flows is invariably much larger than the length scales of the finer particle clusters. Such a coarse-grid simulation will clearly not resolve the structures which exist on sub-grid length scales. The need to account for the consequences of these unresolved structures through suitable sub-grid models is now well established [1–3].

The goal of our study is to construct filtered two-fluid models that average over small scale structures. In these filtered equations, the consequences of the flow structures occurring on a scale smaller than a chosen filter width appear through correlations for which closure relations should be derived or postulated. If constructed properly, the filtered equations should produce a solution with the same macroscopic features as the finely resolved kinetic theory model results; however, as the filtered equations place less stringent requirements on the grid resolution than the original two-fluid model equations, they would be easier to solve.

In the present study, we have shown that filtering of the statistical data obtained from highly resolved 2-D and 3-D simulations – employing a kinetic theory based two-fluid model for uniformly sized particles [1–4] implemented into the MFIX platform [5] – in a periodic domain is a fruitful way to develop closures for coarse-grid simulation of fluidized gas-particle flows. We have completed several highly resolved 2-D and 3-D simulations in various domain sizes with different spatial resolutions to extract filtered quantities for various filter sizes. Using the computational "data" generated through such simulations, we have extracted filtered drag coefficient and particle-phase stresses as functions of the local particle volume fraction and the size of the spatial averaging window (i.e. filter size).We found that both the filtered drag coefficient decreased systematically with increasing filter width both in 2-D and 3-D simulations, whereas the particle-phase stresses increased with increasing filter width. We have also found that these filtered statistics become less dependent on spatial resolution and essentially independent of the domain size for both 2-D and 3-D simulations provided that the filter size is appreciably smaller than the domain size and is much larger than the grid size.

In addition, we have studied the effects of various model parameters on these filtered quantities to expose the robustness of the region-averaged statistics against small changes in these parameters. We have found that the dimensionless parameters of the kinetic theory model have a weaker effect on the dimensionless filtered drag coefficient than a dimensionless group involving the filter size; so, the filtered closures can be modeled to a good accuracy simply as functions of the dimensionless filter size and particle volume fraction.

As a metric of the spatial inhomogeneity observed in the kinetic theory model simulations (at various times in the course of the simulation under statistical steady state conditions), we examined the particle volume fraction fluctuations. Here, we define two types of inhomogeneities: the external fluctuation in particle volume fraction $<\phi'^2>_{ext}$, where "external" refers to inhomogeneities on a scale larger than the filter size, and the internal fluctuation in particle volume fraction $<\phi'^2>_{int}$, where "internal" refers to the spatial inhomogeneities occurring on a scale smaller than the filter size. We have seen that $<\phi'^2>_{ext}$ decreases as the filter size increases as one would indeed expect, whereas $<\phi'^2>_{int}$ increases as the filter size increases and eventually reaches an asymptotic value. It seems reasonable to hypothesize that a filter (region) size where $<\phi'^2>_{ext}$ and $<\phi'^2>_{int}$ are comparable sets the lower threshold for a meaningful filter size. One can also argue that the filter size at which $<\phi'^2>_{int}$ begins to reach its plateau is a physically meaningful estimate (as most of the inhomogeneities occur on a scale smaller than this, as evidenced by the approach to the asymptotic value). According to these qualitative arguments, a rational choice of filter size may lie in the range of 8 – 16 dimensionless units for a 2-D system and between 2 - 4 dimensionless units for a 3-D system. [The filter size is made dimensionless using v_t^2/g and as a characteristic length, where v_t refers to the terminal settling velocity of a single particle.]

We have examined the stability of a uniformly fluidized gas-particle suspension through a 1-D linear stability analysis (LSA) of the filtered model equations. As expected the filtered model predicts that the state of uniform fluidization will give way to nonuniform time-dependent solution through a Hopf bifurcation. The wavenumber corresponding to the Hopf bifurcation point, k_{HB}, is a function of the filter size; in particular, the k_{HB} decreases as the filter size increases. We have also carried out test MFIX simulations of the filtered model equations with fairly high spatial resolutions and ascertained that as one increases the filter size the inhomogeneous structures become coarser and coarser. These findings are reassuring, as they show that filtering is indeed erasing the fine structure and only presenting coarser structures.

The details of these results will be described in the presentation.

References:

[1] Agrawal, K., Loezos, P. N., Syamlal, M. & Sundaresan, S., 2001 The role of meso-scale structures in

rapid gas-solid flows. J. Fluid Mech. 445, 151 – 185.

[2] Zhang, D. Z. & VanderHeyden, W. B. 2002 The effects of mesoscale structures on the macroscopic momentum equations for two-phase flows. Int. J. Multiphase Flow. 28, 805 – 822.

[3] Andrews, A.T. IV, Loezos, P. N. & Sundaresan, S. 2005 Coarse-grid Simulation of Gas-Particle Flows in Vertical Risers. Ind. Eng. Chem. Res., 44, 6022 – 6037 (2005).

[4] Gidaspow, D., 1994 Multiphase Flow and Fluidization, Academic Press, CA. 31-58, 197 – 238.

[5] Syamlal, M., Rogers, W. & O'Brien, T. J., 1993 MFIX Documentation, U.S. Department of Energy, Federal Energy Technology Center, Morgantown, WV.

29c Influence Of The Kinetic Theory Model On Segregation Of Gas-Solid Systems With Binary Mixtures

Janine E. Galvin[1], Jose Leboreiro[1], Christine M. Hrenya[1], and Sofiane Benyahia[2]. (1) Chemical & Biological Engineering, University of Colorado, Boulder, CO 80309-0424, (2) National Energy Technology Laboratory, Department of Energy, 3610 Collins Ferry rd., Morgantown, WV 26505

Multiphase flows in which the particulate phase consists of particles of differing size and/or density are common in industry. Flows involving such mixtures tend to segregate and this can lead to unexpected and undesired results. Traditionally, monodisperse kinetic theories with ad hoc modifications have been applied in polydisperse solids systems. In this work, a kinetic theory model developed specifically for polydisperse systems (Iddir and Arastoopour, AIChE J., 2005) is incorporated in MFIX. The impact of the polydisperse kinetic theory model on simulations predictions of gas-solid fluidized beds with a binary particulate mixture is evaluated. Specifically, predictions obtained with the new polydisperse model, the existing ad hoc model in MFIX, and with no kinetic theory model are compared against one another and against experimental data. The influence of various forms of the drag law, as applied to binary mixtures, is also explored. Two systems are considered, namely bubbling beds and risers, which are composed of binary mixtures that differ in size and/or density are considered. For the case of the bubbling bed, both low- (near minimum fluidization velocity) and high-velocity systems are examined. For the low-velocity systems, the predicted steady state axial segregation profile corresponding to the three different treatments of the solids are essentially equivalent with one exception. At high superficial gas velocities, beyond those employed in the experimental setup, differences in the steady state axial segregation profiles between the three different treatments are evident. An examination of the axial momentum equation indicates that body force, gas-solid drag, and the gas pressure drop are dominant terms in both the low- and high-velocity systems. Accordingly, the role of gas-solid drag on segregation will be explored further. The analysis also shows that kinetic theory has little role in the predictions at low-velocities, but at high-velocities the role of kinetic theory, in particular, the role of solid-solid drag due to the relative velocity between solids phases, becomes more significant. The impact of the polydisperse kinetic theory model is further assessed in the case of dilute riser flow. Previous experiments show that, in a binary mixture of powders, larger particles tend to accumulate near the walls of the riser. The polydisperse kinetic theory model captures this radial segregation behavior, whereas the ad hoc treatment does not. An analysis of the terms in the radial momentum equations shows that the terms in the solids-solids drag expression due to gradients in solids concentration and gradients in granular temperatures are the cause of this radial segregation of particles by size.

29d Ldv Measurements And Cfd Modeling Of Liquid-Solid Flow With Particle-Particle Collisions

Mark Pepple and Jennifer S. Curtis. Chemical Engineering, University of Florida, Gainesville, FL 32611

Operations involving solid-liquid flows play a significant role throughout many industries, including pharmaceutical manufacturing, mining, and oil refining. Unfortunately, the limited understanding of solid-liquid systems results in significant loss in productivity. The development of accurate models that predict the behavior of these solid-liquid systems is needed to improve their design, scale up, and optimization. Non-intrusive flow measurements, where multiple flow variables are measured simultaneously, are needed to develop and validate flow models.

The transition from flow dominated by particle collisions to flow dominated by fluid-particle shear was investigated. Laser Doppler velocimetry (LDV) was used to collect non-intrusive data of the mean and fluctuating velocities of both the solid and liquid components of a turbulent two-phase vertical flow. The velocity and solids concentration of the slurry were varied so that data extended from a collision dominated flow regime to a viscous dominated flow regime. Water and 1 mm soda lime (glass) particles were used and solids concentrations between 0 and 4 percent by volume were investigated over a range of flow velocities. A pilot-scale flow loop made of 3 inch stainless steel pipe, including a vertical test section 60 diameters in height was constructed. The loop was operated at flows of 200 to 600 gpm.

The data collected were compared with predictions from a two-phase flow model which includes a fluctuating energy balance for the particle assembly. The results show the transition from viscous dominated to collision dominated flow as a function of solids concentration and flow velocity. Additionally, the results indicate the importance of the magnitude of the particle impact velocity in the presence of a viscous fluid on accurate prediction of the fluctuating solids velocity.

Assessing The Influence Of The Particle Size Distribution On Fluidized Bed Hydrodynamics Using High-Throughput Experimentation

John Nijenhuis, Renske Beetstra, Ronald van Dijk, and J. Ruud van Ommen*
Delft University of Technology - DelftChemTech
Julianalaan 136, 2628 BL Delft, The Netherlands
*T: +3115 278 2133; F: +3115 278 5006; E: j.r.vanommen@tudelft.nl

ABSTRACT

The goal of the described project is to design mixtures of particles with optimal fluidization properties. Using high-throughput experimentation, a novel approach for hydrodynamic research, the relevant properties can be obtained in a limited period of time. This approach is demonstrated by assessing the influence of the particle size distribution on fluidized bed hydrodynamics of Geldart A and B powders. By manipulating the width of the particle size distribution of alumina powder the bubble diameter is reduced up to 40%. The addition of fines to a given particle size distribution decreases the bubble diameter up to 40%, whereas the addition of coarse particles hardly influence the bubble size.

INTRODUCTION

The two main reactor types for catalysed gas phase reactions are the packed bed and the fluidized bed. In packed beds, the catalyst particles are normally not smaller than 1 mm to prevent excessive pressure drop. This leads to mass transfer limitations within the particles. Fluidized beds do not suffer from this drawback: small particles (typically 20 – 400 μm) can be used, while the pressure drop stays low. It has been predicted that catalyst efficiency will increase strongly in the coming years because of the increased possibilities of molecular modelling, the use of high-throughput techniques in catalysis engineering, and the advances in synthesis methods (e.g., 1,2,3). This will make the mass transfer problems in packed beds even more severe and a shift from packed beds to fluidized beds and structured reactors based on thin catalyst layers can be expected in the process industry within the next decade. Although fluidized beds do not suffer from mass transfer limitations on the particle scale, they have mass transfer limitations on a larger scale: the transport from gas in the dilute phase (voids or bubbles) to the particles in the dense phase. Consequently, the conversion of reactant(s) in the dilute phase is much lower than in the dense phase (see Figure 1a). This leads to a low overall conversion, since for fluidized beds of small particles the dilute phase conversion dictates the overall conversion.

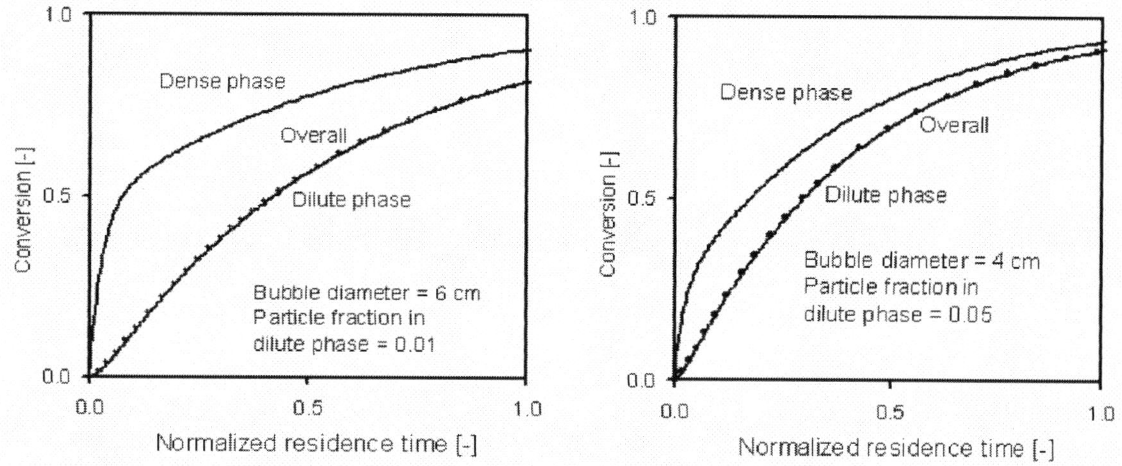

Figure 1: The conversion of reactant(s) in the dilute and dense phase for a fluidized bed with a bubble diameter of 6 cm and 4 cm, calculated with the Kunii and Levenspiel model (4).

A reduction of the bubble size and/or an increase of the particle content of the bubbles will lead to a much smaller difference in conversion between the two phases and a much higher overall conversion;

calculations with the simple Kunii and Levenspiel model (4) show this clearly (Figure 1b). Moreover, in the case of parallel and/or series reactions (of the most realistic situations) the narrower residence time distribution resulting from a lower concentration difference between the phases, leads to a higher selectivity for the desired product. In the past, our group has studied three different ways to introduce structure in fluidized beds in order to reduce the bubble size (5): oscillating the gas supply, distributing the gas supply over the height of the bed, and varying the interparticle forces using electric fields. In addition to these three approaches, we have recently started to study a fourth way to increase the efficiency of fluidized beds: optimising the particle size distribution (PSD) and other distributed particle properties. Compared to the other methods of improving fluidized bed performance, this approach has the advantage that, in principle, no modifications to the fluidized bed equipment are necessary (6).

It is a well-known fact in fluidization technology that the addition of fines (particles with a diameter < 45 mm (7)) improves the fluidization behaviour and leads to better mass transfer. The effect of fines has never been fully explained, although it has often been speculated that fines act as a kind of lubricant to lower the apparent viscosity of the dense phase, leading to smaller voids and more uniform gas-solid distribution. Sun and Grace (8) have shown by experiment that a wider particle size distribution leads to a higher conversion for ozone decomposition (see Figure 2). They suggested that this effect is due to a disproportionate amount of fines in the dilute phase (9).

The current practice is, however, that particles for fluidized beds (carriers for catalytic material) are optimised mainly on the scale of a single particle. Most attention is given to their pore size distribution such that a high surface area is achieved and that the active sites are easily accessible by the gaseous components.

Little attention is paid to the mass transfer from the gas in the dilute phase to the particles in the dense phase, which is essential to practical fluid bed operation. Therefore, this work is aimed at improving the conversion and selectivity of gas-solid fluidized reactors by designing mixtures of particles with optimal properties (size distribution, density, shape, elasticity) with the aid of high-throughput experimentation, a novel approach for hydrodynamic research.

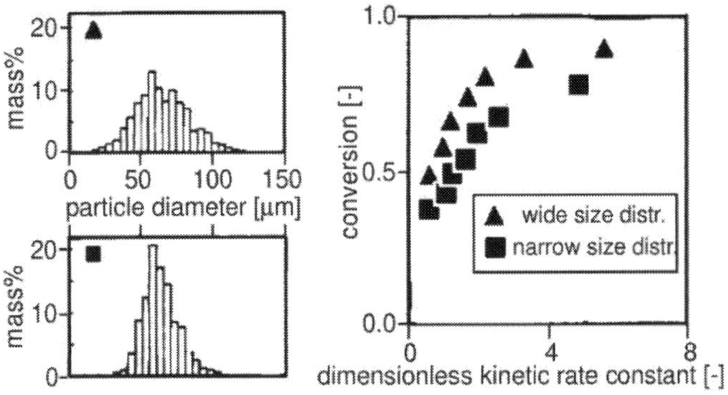

Figure 2: The effect of the particle size distribution on the decomposition of ozone. Adapted from Sun and Grace (7).

It is desirable to be able to obtain experimental data for all relevant variables over a wide range at a high resolution, to find trends and optimum values in fluidization characteristics. Furthermore, considerable variance in the measured data is expected due to the dynamic behaviour of the fluidized bed itself. Therefore, to obtain a significant ability to distinguish between the results and to find for instance an 'optimal' value for a property (e.g., fraction of fine particles of a given size), all main input variables need to be controlled and measured at the highest accuracy possible. Examples of these input variables are temperature, flow rate, pressure, relative humidity, bed mass and composition. This, in combination with the need of testing a large number of particle mixtures, leads to the need for high-throughput experimentation approach. An additional advantage is that no direct human interaction is needed due to the fully automated character of this high-throughput experimentation approach. Therefore, more dangerous materials could be tested safely under a wide range of conditions. Examples are nanoparticles and particles coated with nanomaterials (as, for example, used in pharmaceutics).

EXPERIMENTAL

To obtain quantitative information from large numbers of particle mixtures, an automated setup was constructed. The operation procedure of the set-up is schematically depicted in Figure 3. It is possible to load and empty the fluidized bed column with designed particle mixtures (of e.g. sand or catalyst carrier materials such as silica or alumina) fully automatically. Both the preparation of a desired particle size mixture and the loading of this mixture in the fluidized bed column can be handled by a Zymark XP Robot (see Figure 4). The column is emptied by blowing the particles out with a very high gas flow rate.

The experiments can be carried out in two industrially relevant fluidization regimes: bubbling fluidization and turbulent fluidization. In the experiments, pressure measurements are used to assess the hydrodynamics. We perform two types of pressure measurements. First, the pressure drop over parts of the bed is measured to determine the average bed density and the bed expansion, which is a measure for the total amount of gas in the bed. The pressure drops were measured with two Validyne DP15 pressure transducers. Second, high-frequency pressure fluctuations were measured in the plenum and 14 cm above the distributor plate with probes 10 cm long and an internal diameter of 4 mm. These dimensions guarantee an undisturbed transfer of the signal in the frequency range of interest (10). All probes were purged with air at 0.5 m/s to prevent blocking. The pressure fluctuations were measured with Kistler piezoelectric pressure sensors (type 7261). A recently developed spectral decomposition method (11) enables us to determine the bubble size from these pressure fluctuation measurements. This technique separates the part of the pressure time-series that is directly related to bubbles passing the pressure sensor, giving a measure for their diameter. Local transient voidage characteristics were measured by optical probes at 14 and 16 cm above the distributor plate

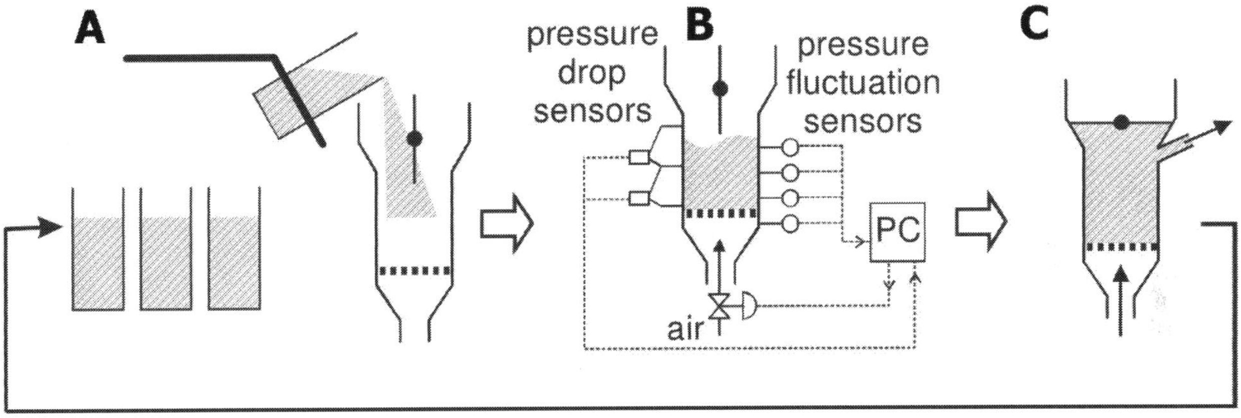

Figure 3: Schematic representation of the high-throughput screening procedure. (A) Filling the fluidized-bed column. (B) Carrying out the measurement program. (C) Emptying the column. After C, the whole procedure is automatically repeated with the next batch of particles.

To illustrate the novel concept of using high-throughput experimentation techniques to determine the effect of particle size and particle size distribution on the fluidization behaviour, three types of experiments were conducted with alumina powder:

- *Experiment A:* The effects of fines (i.e. particles < 45 μm) on the bubble size was examined by adding fines to a powder with a median particle size (D_{50}) of 70 μm. The fines weight fraction is varied in seven steps from 0 to 30 wt.%.
- *Experiment B:* The effect of broadening the PSD on the fluidization hydrodynamics is investigated by varying the standard deviation of the PSD in 7 steps from 40 to 188 μm, while D_{50} is kept constant at 165 μm.
- *Experiment C:* To determine the influence of the particle size, 10 fractions of particles with different D_{50} are added one by on (20 wt.%) to a base powder with a D_{50} of 70 μm. In other words, 10 bed masses are created of which 80 wt.% is alumina powder with a D_{50} of 70 μm and 20% is alumina powder with a D_{50} of 23, 49, 58, 69, 83, 98, 115, 138, 165 or 195 μm respectively.

All three types of experiments were carried out at superficial gas velocities of 6, 10, 18 and 24 cm/s. The experiments were conducted in a stainless steal column of 40 cm high with an internal diameter of 7.3 cm. To ensure constant conditions during experiments, the setup was operated in a temperature controlled box. Three mass flow controllers controlled the airflow through the fluidized bed with maximum superficial velocities of 0.8 cm/s, 8 cm/s and 80 cm/s respectively. All signals were low-pass filtered with a cut-off frequency of one half of the sample frequency, satisfying the Nyquist criterion. Subsequently, 16-bit analogue to digital conversion was applied at a sample frequency of 400 Hz. All experiments were performed at room temperature and with a constant relative humidity.

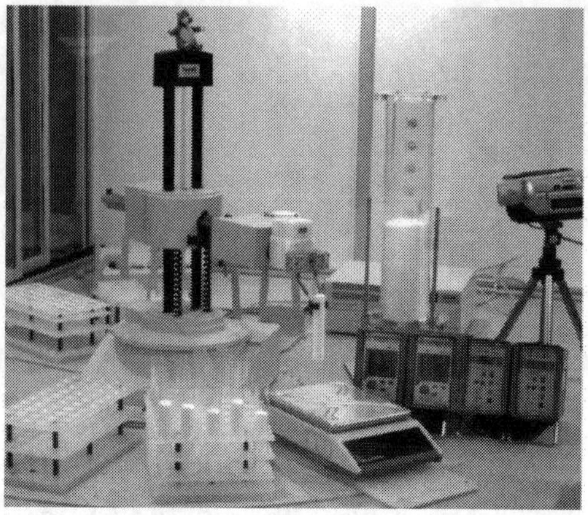

Figure 4: The Zymark XP Robot

RESULTS AND DISCUSSION

Experiment A

To investigate the effect of adding fines on the fluidization behaviour, the fines concentration was varied in seven steps from 0 to 30 wt.%. Figure 5 shows three of the resulting particle size distributions.

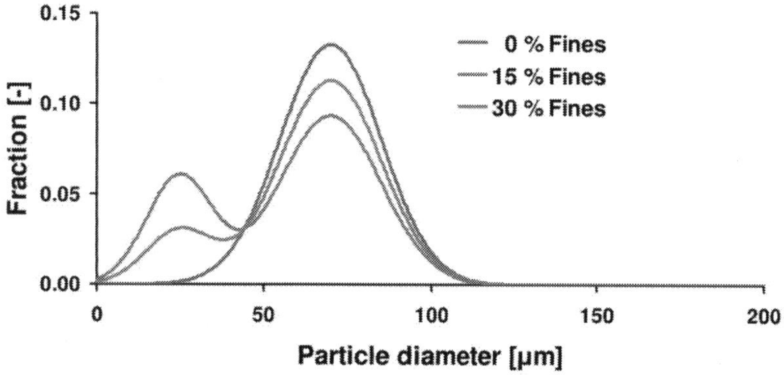

Figure 5: The experimental approach: fines are added to a powder with a D_{50} of 70 µm. The fines concentration is varied for seven bed masses from 0 to 30 wt.%.

In these experiments, the relative bubble size was measured by spectral decomposition of the measured pressure fluctuations and plotted as a function of the fines weight fraction. Figure 6 shows that the addition of fines causes the bubble size to decrease up to 40%. Furthermore, a strong dependence on the superficial gas velocities is shown.

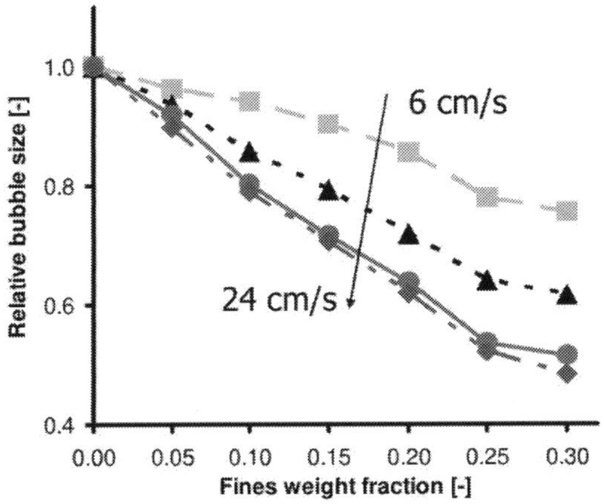

Figure 6: Relative bubble size based on pressure fluctuation measurement as a function of fines weight fraction at varying superficial gas velocities

Several authors describe that the conversion of a chemical reaction in a fluidized bed is increased at higher fines contents (7,8,12); it is often assumed that this is due to bubble size reduction. In our experiments we demonstrate the direct relation between the fines contents and this bubble size reduction, which is a first step in finding the fundamental mechanism behind this phenomenon.

Experiment B

Sun and Grace demonstrated in 1990 (8) that a wider particle size distribution results in higher conversions than obtained by narrow distributions. The bubble size reduction induced by changing the standard deviation of the PSD was measured as a function of the gas velocity. The PSD broadens with constant median particle size (D_{50} = 165 μm), by addition of smaller and larger particles. Spectral decomposition of the measured pressure fluctuations revealed that the bubble size decreases significantly while increasing the standard deviation of the PSD (see Figure 7). The effect becomes stronger at higher gas velocity. This is supported by the results obtained by the optical probes (see Figure 8): they are in rather good agreement with the results from the pressure fluctuations Although we observed that broadening the standard deviation of the PSD led to an increase in the minimum fluidization velocity, the decrease in relative bubble size is much larger than we could expect based on the decreased excess gas velocity (u_0-u_{mf}) only.

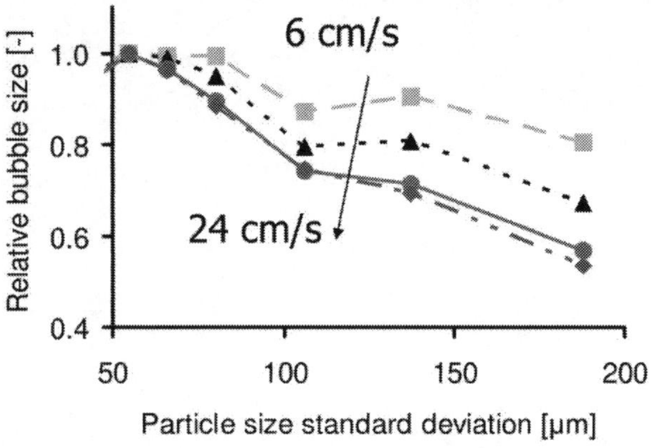

Figure 7: Average relative bubble size based on pressure fluctuation measurement as a function of the standard deviation of the PSD with a D_{50} of 165 μm, measured at four superficial gas velocities.

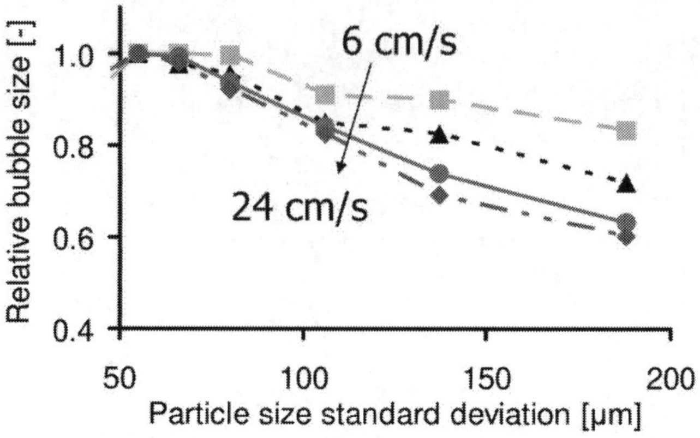

Figure 8: Average relative bubble size based on optical probe measurement as a function of the standard deviation of the PSD with a D_{50} of 165 μm, measured at four superficial gas velocities.

Experiment C

Experiment C was designed to reveal to influence on fluidization behaviour of adding fine or coarse particles to a base material. To investigate, 10 fractions of particles with different D_{50} (below and above 70 μm) are added one by on (20 wt.%) to the base powder with a D_{50} of 70 μm. Figure 9 shows the changes in bubble size from the base bed mass with a D_{50} of 70 μm as a function of the D_{50} of the added fraction.

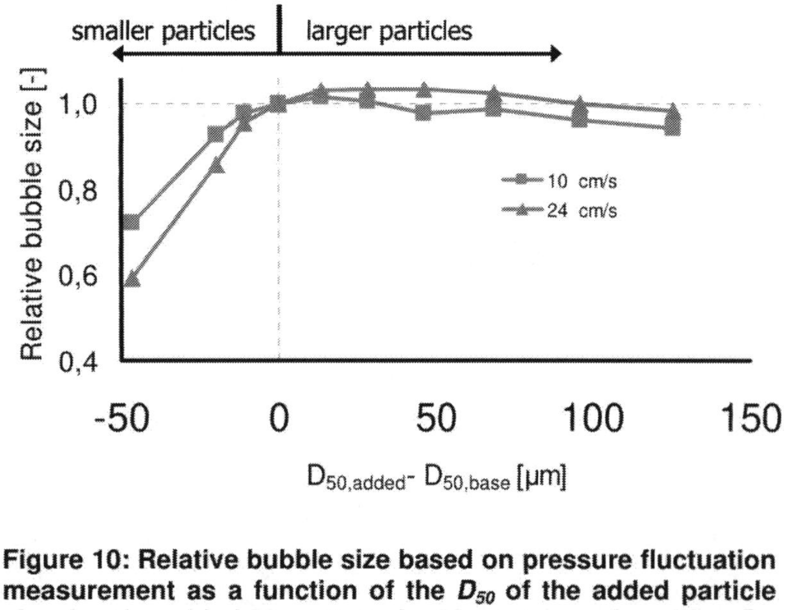

Figure 10: Relative bubble size based on pressure fluctuation measurement as a function of the D_{50} of the added particle size, i.e. the added 20 wt.% to the 80 wt.% powder with a D_{50} of 70 μm. The bed mass with a D_{50} of 70 μm is used as reference.

These results show that smaller particles affect the fluidization hydrodynamics to a much larger extent than larger particles at both superficial gas velocities. We are currently investigating the mechanisms behind the obtained results. Nevertheless, the current results clearly illustrate that it is very helpful to have proper tools to conduct a large set of experiments with changing PSD with high precision and in a limited period of time.

CONCLUSIONS AND OUTLOOK

The use and potential of a novel high-throughput experimentation test facility for fluidization research has been demonstrated by assessing the influence of the particle size distribution on fluidized bed hydrodynamics. With this approach, the relevant fluidization properties can be obtained of a wide range of particle size distributions in a limited period of time. By manipulating the width of the particle size distribution of alumina powder the bubble diameter is reduced up to 40%. The addition of fines to a given particle size distribution decreases the bubble diameter up to 40%, whereas the addition of coarse particles hardly influence the bubble size.

ACKNOWLEDGEMENTS

R. Monna is gratefully acknowledged for his skilful help in developing the setup. We want to thank the Dutch National Science Foundation, NWO, for their financial support.

NOTATION

D_{50}	Median particle diameter	[m]
u_{mf}	Minimum fluidization velocity	[m/s]
u_0	Fluidization velocity	[m/s]

REFERENCES

1. Broadbelt, L.J., Snurr, R.Q., 'Applications of molecular modeling in heterogeneous catalysis research', Applied Catalysis A-general 200, 23-46, 2000.
2. Senkan S, 'Combinatorial heterogeneous catalysis - A new path in an old field', Angewandte Chemie-International edition, 40 (2), 312-329, 2001.
3. Ying, J.Y., Mehnert, C.P., Wong, M.S., 'Synthesis and applications of supramolecular-templated mesoporous materials', Angewandte Chemie-International edition, 38 (1-2), 56-77, 1999.
4. Kunii, D., Levenspiel, O., 'Fluidization Engineering', 2nd Edition, Butterworth-Heinemann, Stoneham, Massachusetts, 289-292, 1991.
5. Coppens, M.-O., van Ommen, J.R., 'Structuring chaotic fluidized beds', Chem. Eng. J., 96, 117-124, 2003.
6. van Ommen, J.R., Nijenhuis, J., van den Bleek, C.M., Coppens, M.-O., 'Four ways to introduce structure in fluidized bed reactors', Ind. Eng. Chem. Res. 46, 4236-4244, 2007.
7. Yates, J.G., Newton, D., 'Fine particle effects in a fluidized-bed reactor', Chem. Eng. Sci., 41, 801-806, 1986.
8. Sun, G., Grace, J.R., 'The effect of particle size distribution on the performance of a catalytic fluidized bed reactor', Chem. Eng. Sci. 45, 2187-2194, 1990.
9. Sun, G., Grace, J.R., 'Experimental determination of particle dispersion in voids in a fluidized bed', Powder Technol. 80, 29-34, 1994.
10. van Ommen, J.R., Schouten, J.C., Vander Stappen, M.L.M., Van den Bleek, C.M., 'Response characteristics of probe-transducer systems for pressure measurements in gas-solid fluidized beds: How to prevent pitfalls in dynamic pressure measurements'. Powder Technol. 106, 199–218, 1999, Erratum, Powder Technol. 113, 217, 2000.
11. van der Schaaf, J., Schouten, J.C., Johnson, F., van den Bleek, C.M., 'Non-intrusive determination of bubble and slug length scales in fluidized beds by decomposition of the power spectral density of pressure time series', Int. J. of Multiphase Flow 28, 865-880, 2002.
12. Pell, M., Jordan, S.P., 'Effect of fines and velocity on fluid bed performance', AIChE symposium series, 262, 68-73, 1988.

29f Prediction Of The Minimum Bubbling Point

Stefano Brandani, School of Engineering and Electronics, University of Edinburgh, Sanderson Building, The King's Buildings, EH9 3JL, Edinburgh, United Kingdom, Parimanan Cherntongchai, Industrial Chemistry, Chiang Mai University, Chiang Mai, Thailand, and Kai Zhang, Chemical Engineering, China University of Petroleum, Beijing, Beijing, China.

We present the extension of the model of Brandani and Zhang (2006) to include the effect of voidage on the characteristic dimension. The constitutive equation for this model parameter is obtained through a series of well characterised bed collapse experiments (Cherntongchai and Brandani, 2005).

The equations needed to calculate the minimum bubbling point are obtained through a linear stability analysis. We compare the predictions to over 700 literature data points and discuss in detail the influence of pressure, temperature and the nature of the particles.

From this comprehensive comparison, it is also possible to establish that there is an upper maximum in the value that the minimum bubbling point of any fluid particle system can reach. We discuss how to formulate the problem of predicting this maximum and map this value as a function of the Archimedes number and density ratio for our model.

Brandani S. and Zhang K. Powder Technology, 2006, 163, 80-87.

Cherntongchai P. and Brandani S., Powder Technology, 2005, 151, 37-43.

Onset of Bubbling in Fluidized Beds of Silica Sand Particles: Determination from Pressure Drop Fluctuations

C. E. Davies[1], D. Krouse[2] and A Carroll[3]
[1] Institute of Technology and Engineering, Massey University, Private Bag 11 222, Palmerston North, New Zealand.
[2] Industrial Research Limited, P. O. Box 31-310, Lower Hutt, New Zealand.
[3] Carroll Consulting, P. O. Box 572, Taupo, New Zealand.

E-mail: C.Davies@massey.ac.nz

Batches of silica sand with diameters that place them in the vicinity of the Geldart A/B boundary have been used in experiments where the fluctuations in bed pressure drop have been measured over a range of superficial velocities. Plots were made of the standard deviation of pressure drop, σ, against superficial velocity; by inspection, it was apparent that the transition from values of σ that were zero or very close to it was rapid and took place over a very small velocity change. Thereafter, the increase in standard deviation was linear with increasing superficial velocity. Supporting evidence for this is provided by an analysis that uses the ratio of total variance to high frequency variance, $\{V_{TOT}/V_{HI}\}$. One consequence of (apparent) rapid initial growth in standard deviation is that linear extrapolation of the standard deviation plot to zero could underestimate the point at which bubbling actually begins.

By way of example, we consider data obtained with a mixture of 55% by weight "coarse" silica sand, and 45% by weight "very fine" silica sand; these base materials have the size distributions shown in Figure 1. The sand was fluidized with dry air in a cylindrical bed with an internal diameter of 80 mm; all data in this paper are for a sampling rate of 50 Hz. Details of the experimental system and method are given by Davies et al. [1].

Figure 2 is a plot showing the variation of the standard deviation of bed pressure drop with superficial velocity, U. It is apparent that the data divide loosely into two groups. Below a superficial velocity of about 0.03 m s^{-1}, σ is small and relatively constant; above ~0.03 m s^{-1} σ increases, apparently linearly, with increasing U. Extrapolation of σ to U=0, has been suggested as a method for estimating minimum fluidizing velocity [2] [3].

In the example given here, Figure 2, extrapolation of three data points, all apparently co-linear, gives an intercept for U=0, to the left of a single point, (0.0326, 3.29). Davies et al [4] have previously noted that the statistical parameter $\{V_{TOT}/V_{HI}\}$ has utility in the signalling transition to bubbling in a fluidized bed, where V_{TOT} and V_{HI} are respectively given by Equations (1) and (2):

$$V_{TOT} = \frac{\sum_{i=1-n}(x_i - \bar{x})^2}{(n-1)} \qquad (1)$$

$$V_{HI} = \frac{\sum_{i=2-n}(x_i - x_{i-1})^2}{(n-1)} \qquad (2)$$

where x is the measured variable, in this case, bed pressure drop, and n in the number of measurements.

To further examine the behaviour of $\{V_{TOT}/V_{HI}\}$ with increasing U, we have replotted Figure 2, with the associated values of $\{V_{TOT}/V_{HI}\}$ also shown; see Figure 3. There are two significant features to Figure 3. It is clear that $\{V_{TOT}/V_{HI}\}$ is approximately constant above a value of U between 0.03 m $s^{-1}$and 0.04 m s^{-1}; also $\{V_{TOT}/V_{HI}\}$ peaks at a value of U just under 0.02 m s^{-1} and then falls to a constant value. With reference to experimental observations of the system as U was being increased, the 3 constant values at U >~0.03 m s^{-1} are for a bubbling fluidized bed; it appears therefore that the value of $\{V_{TOT}/V_{HI}\}$ for a fully fluidized bed is independent of σ. In Figure 4, it is apparent that $\{V_{TOT}/V_{HI}\}$ for σ (0.0326, 3.29) is noticeably larger than the limiting values, and we speculate that this is evidence that this point is identified with an incomplete transition, and should indeed not be grouped with the linearly increasing σ of a fully fluidized bed. To further explore the transitions in this system, we have plotted normalized values of σ, $\{V_{TOT}/V_{HI}\}$, bed pressure drop, and bed height against U, as shown in Figure 4; each dependant variable plotted on the ordinate was normalized with respect to its maximum value.

Inspection of Figure 4 reveals that the onset of fluidization as signalled by the change in slope of the bed pressure drop occurs at a superficial velocity of approximately 0.014 m s^{-1}; at much the same superficial velocity (normalized) $\{V_{TOT}/V_{HI}\}$ peaks following a transition in bed height. There is also a hint of another transition in bed height at U slightly less than 0.04 m s^{-1}, much the same value suggested by the approach of $\{V_{TOT}/V_{HI}\}$ to a constant value where the bed is known to be fully fluidized.

We suggest that the onset of fluidization is associated with a peak in the plot of $\{V_{TOT}/V_{HI}\}$ versus U, and that the onset of bubbling is associated with $\{V_{TOT}/V_{HI}\}$ assuming an unchanging value as U increases. However, further experimental work with smaller increments in U is needed to validate these conjectures.

References

[1] Davies C.E., Carroll A. and Flemmer R.. *Particle size monitoring in a fluidized bed using pressure fluctuations*, Powder Technology, In press.

[2] Puncochar M., Drahos J., Cermak J., and Selucky K., Chem. Eng. Commun. 35 (1985) 81-86.

[3] Wilkinson D., Canad. Jnl. Chem. Eng. 73 (1995) 562 - 565.

[4] Davies C.E., Krouse D. and Carroll A. (2007). *Particle size estimation and monitoring in a bubbling fluidized bed using pressure fluctuation measurements*, Proceedings of the 12th International Conference on Fluidization, Eds. Bi, X., Berruti, F. and Pugsley, T., Engineering Conferences International, Vancouver Canada, May.

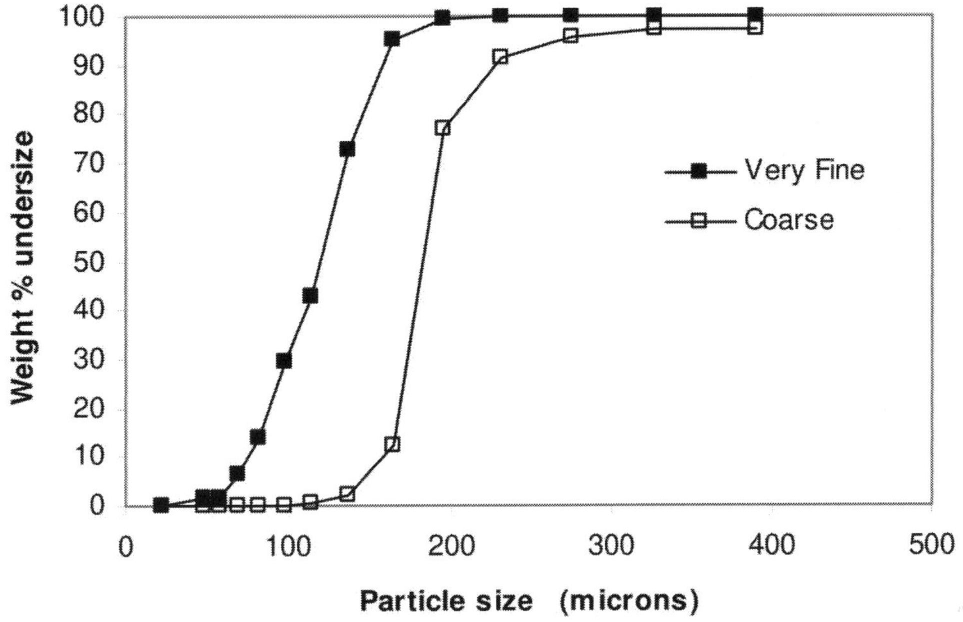

Figure 1 Size distributions of silica sand base materials.

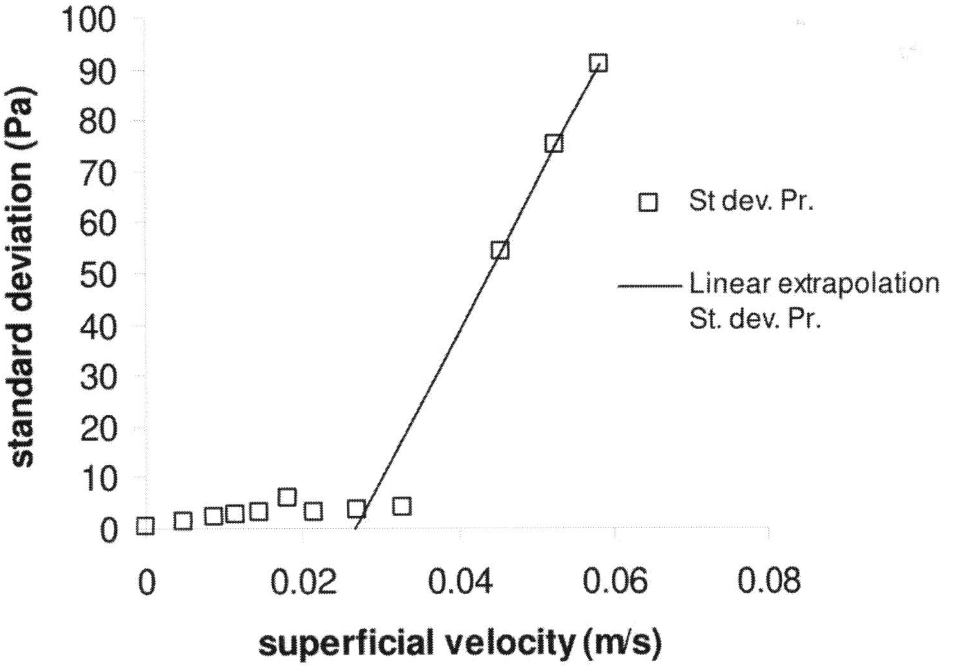

Figure 2 Standard deviation of bed pressure drop as function of superficial velocity.

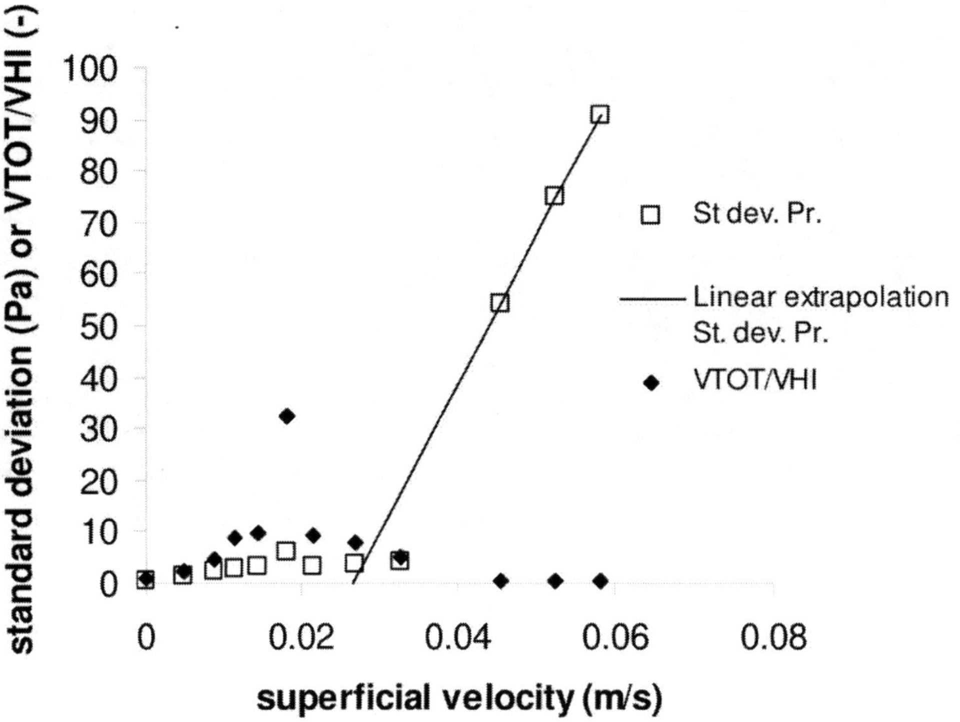

Figure 3 V_{TOT}/V_{HI} and standard deviation of bed pressure drop, as function of superficial velocity.

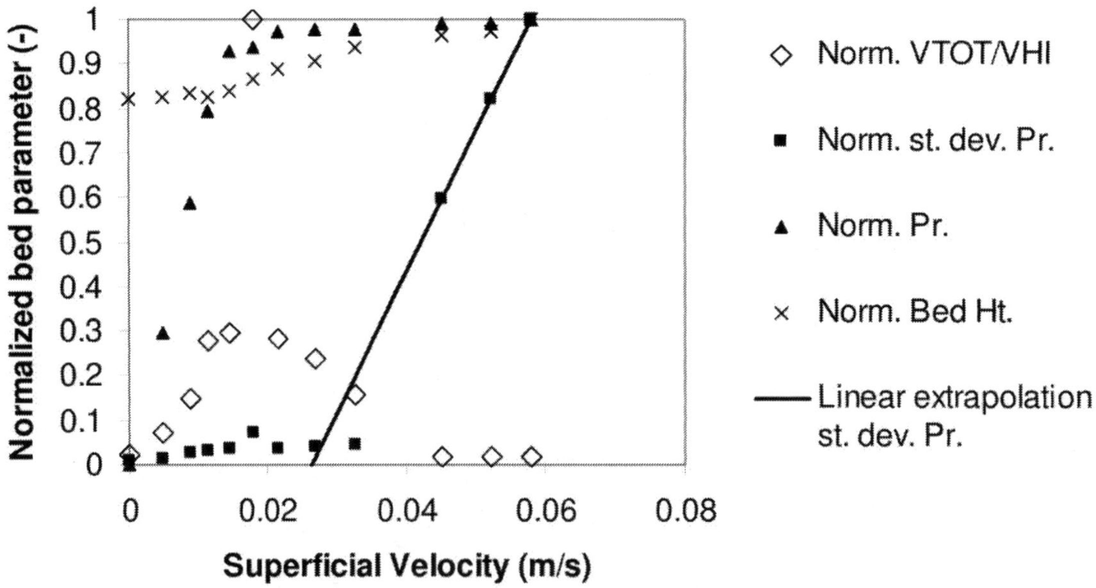

Figure 4 Normalized bed parameters as a function of superficial velocity.

33a Nanocomposites Of Inorganic Oxide Coated Gold Nanorods

David Walker, Department of Chemical Engineering, University of South Florida, 4202 E. Fowler Ave., Tampa, FL 33620 and Vinay K. Gupta, Chemical Engineering, University of South Florida, 4202 E. Fowler Ave., Tampa, FL 33620.

Combining rod-shaped metallic nanoparticles such as gold nanorods (GNR) with inorganic oxides can be used to tailor the functional properties of the composite nanomaterial. For example, a dielectric SiO2 coating allows one to use traditional means for immobilizing biomolecules on the oxide and exploit the optical absorbance of the GNR for biosensing. Coating GNRs with a semiconducting oxide such as titanium dioxide (TiO2) can have uses in catalysis and a material such as iron oxide (Fe3O4) can be used in magnetic devices. In this presentation, we will discuss our research on fabricating and characterizing these nanocomposites. Specifically, we will discuss strategies that facilitate coating of inorganic oxides by first modifying GNRs with various polyelectrolytes by a layer-by-layer (LBL) technique. The polyelectrolyte layers stabilize the GNRs during re-dispersion in an organic solvent and allow partial shielding of the surface charge of the gold nanorods. Subsequently, techniques such as hydrolysis of tetraethyl orthosilicate or titanium(IV) isopropoxide can be used to coat the GNR. The presentation will focus on the synthesis of GNRs, the LBL polyelectrolyte coating technique, and the controlled coating of various inorganic oxides. Supporting characterization of the composites using UV spectroscopy, TEM and dynamic light scattering (DLS) will also be presented.

33b Amphiphilic Block Copolymer Inverse Micelles For Delivery Of Hydrophilic Drugs

Andrew C. Miller[1], Paula T. Hammond[1], Robert E. Cohen[1], and Darrell J. Irvine[2]. (1) Chemical Engineering, Massachusetts Institute of Technology, 77 Massachusetts Ave., 66-357, Cambridge, MA 02139, (2) Biological Engineering, Materials Science and Engineering, Massachusetts Institute of Technology, 77 Massachusetts Ave., 8-425, Cambridge, MA 02139

Amphiphilic block copolymers have been the subject of significant recent research due to their ability to self-assemble on the nanometer length scale into a variety of morphologies, enabling a diverse range of applications. For many biological applications these systems are assembled in water, however, in our work we are studying inverse micelles assembled in an oil phase resulting in a hydrophobic corona and a hydrophilic core. Our work has focused on the bioresorbable polymer, poly(epsilon-caprolactone)-block-poly(2-vinylpyridine), as well as nonresorbable poly(styrene-b-2-vinylpyridine) and poly(styrene-b-polyacrylic acid). We have previously shown the ability cast these micelles on substrates to form quasi-hexagonal micellar arrays. These films undergo a rearrangement driven by the swelling of hydrophilic core domains when exposed to aqueous solution that opens ('cavitates') the micelles and exposes the hydrophilic domains to the solution. We have demonstrated the ability to load hydrophilic small molecules, polysaccharides, and proteins into the hydrophilic core of these micellar systems. When cast into films, the micelles release their contents upon exposure to aqueous solutions, giving them the potential to be used for film-based drug delivery. Alternatively, the oil phase containing the micelles can be placed in contact with an aqueous phase and the drug will be released into the aqueous phase. We have investigated the assembly of these polymer systems in biocompatible oil phases of interest for transdermal drug delivery; due to their hydrophobic exterior and the ability to encapsulate hydrophilic drugs, oil suspensions of these inverse micelle materials may provide improved transcutaneous delivery of compounds through the stratum corneum of skin.

33c Synthesis And Steric Stabilization Of Magnetite Nanoparticles With Block Copolymers: Effect Of Anchor Block Type

Richey M. Davis, Virginia Tech, Department of Chemical Engineering, Blacksburg, VA 24061 and Judy Riffle, Chemistry, Virginia Tech, 2102 Hahn Hall, Blacksburg, VA 24061.

Magnetic nanoparticles are of great interest in developing new drug delivery systems and in medical diagnostic tests. The stabilization of nanoparticles of magnetite coated with a biocompatible polymer is of particular interest. The adsorption of novel triblock copolymers with end blocks comprised of copolymers of ethylene oxide (EO) and anchor blocks comprised of: (i) a carboxylic acid/urethane center block, (ii) a tricarboxylate end group, and (iii) a cationic PEI end block was studied on particles of magnetite, Fe3O4 approximately 10 nm in diameter in aqueous media. Adsorption via the anchor blocks led to the formation of a brush layer that sterically stabilized the particles. The extension of the brush layer from the particle surface was modeled using a model based on star polymer molecules and the resulting steric contribution to the modified DLVO theory was sufficient to stabilize the particles in aqueous buffer for periods of weeks in some cases. Particle size of the bare magnetite was measured by electron microscopy while the size of the particles with adsorbed block copolymers was measured by dynamic light scattering. The particle surface area was measured by the BET method and the amount of bound polymer was measured by thermal gravimetric analysis. The zeta potential was measured with an electrophoretic dynamic light scattering instrument. The effect of an external magnetic field on the colloid stability of the particles was also accounted for in the DLVO theory. The block copolymer composition was varied and the resulting effect on particle stabilization will be discussed.

33d Turbulent Precipitation of Nanoparticles through Solvent Displacement

Federica Lince[1], Daniele Marchisio[1], Anna Jawor-Baczynska[2], Jan Sefcik[2], and Antonello Barresi[1]. (1) Scienza dei Materiali e Ingegneria Chimica, Politecnico di Torino, C.so Duca degli Abruzzi 24, Torino, 10129, Italy, (2) Chemical and Process Engineering, University of Strathclyde, James Weir Building, 75 Montrose Street, Glasgow, United Kingdom

Colloidal particulate systems have attracted the growing interest of the scientific community in the last decades due to their interesting properties as drug carriers for therapeutic applications. Drugs could be entrapped in polymeric matrices resulting in increased therapeutic benefits and minimized side effects. The particulate carrier must have a controlled Particles Size Distribution (PSD), in fact, in order to avoid negative interactions with the Reticulo-Endothelial System (RES) particle size must be smaller than 250 nm, guaranteeing an adequate circulation half-life, and a continuous and controlled drug release.

In polymeric nanoparticles the drug can be dissolved, entrapped, encapsulated or attached to the polymeric matrix, depending on the method of preparation, namely: emulsification-evaporation, salting-out, emulsification-diffusion and solvent displacement [1, 2]. Rapid solvent displacement precipitation can be also used to assemble therapeutic biomacromolecules such as proteins and nucleic acids onto the surface of suitable excipient particles [3, 4], such as sugars or amino acids, while preserving structural integrity and biological activity of biotherapeutics.

In this work we investigated formation of two classes of nanoparticles for drug delivery applications through solvent displacement (anti-solvent precipitation). Poly-ε-Capro-Lactone (PCL) nanoparticles for controlled drug release applications were prepared by using acetone and/water as solvent and anti-solvent, respectively. Nanoparticles of DL-valine, an amino acid, were prepared by using water and isopropanol as solvent and anti-solvent respectively.

This methodology for obtaining nanoparticles involves two main steps: dissolution of the carrier (polymer or excipient) and the drug into a solvent and rapid mixing of the obtained solution with the anti-solvent leading to spontaneous particle formation. The solvent displacement method offers several important advantages, such as reproducible carrier sizes in the nanometer range and the use of ingredients with low toxic potential. However, since particle formation is extremely rapid, mixing must be very fast, and therefore special continuous mixers must be used and in this work Confined Impinging Jets Reactors (CIJR) are employed [5, 6]. In CIJR the two solutions of solvent and anti-solvent mix in a very small volume in the centre of a cylindrical chamber, where due to collision and impingement of the two jets, turbulent kinetic energy is generated and then quickly dissipated, inducing very rapid mixing. In turns rapid mixing involves very high nucleation rates and therefore results in the production of very small particles.

The effect of the most important operating parameters such as initial solute concentration, mixing rate and solvent-to-antisolvent ratio on the final PSD was investigated, analyzed and rationalized. The mixing rate in CIJR is quantified by the flow rate with which the two solutions are fed into the mixer, and it is often expressed in terms of the Reynolds number in the feeding pipe. The flow rate in our experiments was varied between 3 and 120 ml/min resulting in Reynolds numbers ranging from 500 to 3000. Eventually the solvent/anti-solvent ratio was kept between 1 and 4.

Experimental results show that the effect of mixing on the PSD is crucial. We found that increasing the feeding rate of the solution of solvent and anti-solvent in the CIJR, and therefore improving mixing, nucleation of smaller particles is favored and the mean size is remarkably reduced. We also found that at constant Reynolds numbers a significant reduction of the mean particle size is detected increasing the anti-solvent to solvent ratio. This is probably caused by the fact that the carrier solubility is lower at higher anti-solvent concentrations, resulting in increased super-saturation levels and higher nucleation rates. Moreover there is also a dilution effect with increasing anti-solvent to solvent ratio, which can further stabilize the nanoparticle suspension, suppressing agglomeration and therefore reducing the final mean particle size.

Our experimental data clearly show that the controlling parameter for nanoparticle precipitation is solute supersaturation. This quantity is defined as the ratio between the actual carrier concentration in the solvent/anti-solvent mixture and the carrier solubility in the mixture (the equilibrium concentration experimentally determined in this work). Another very important parameter is the interfacial tension between the solid particles and the mixture, that changes dramatically with the mixture composition, and therefore changes with the anti-solvent to solvent ratio.

[1] Soppimath, K.S.; Aminabhavi, T.M.; Kulkarni, A.R.; Rudzinski, W.E., J. Contr. Rel., 70, (2001), 1-20.

[2] Horn, D., Rieger, J., Angew. Chem. Int. Ed., 40, (2001), 4330-4361.

[3] Kreiner, M.; Moore, B.D.; Parker, M.C. Chem. Comm., (2001), 1096-1097.

[4] Kreiner, M.; Fuglevand, G.; Moore, B.D.; Parker, M.C. Chem. Comm., (2005), 2675-2676.

[5] Fessi, H., Puisiex, F., Devissaguet, J.P., Int. J. Pharm. , 55, (1989), R1–R4.

[6] Johnson, B.K. ; Prud'Homme, R.K., AIChE J., 49, (2003), 2264-2282.

Control of Elastomer Cluster Structure by Tuning of Particle Surface Characteristics

Cornelius Gauer, Hua Wu, Marco Lattuada and Massimo Morbidelli

Institute for Chemical and Bioengineering, Department of Chemistry and Applied Biosciences, ETH Zurich, 8093 Zurich, Switzerland

Abstract

Clusters of elastomer particles can be susceptible to densification from sintering. This process is governed by surface properties of the primary particles and contrary to clusters of rigid particles where structural universality has been found. In order to investigate the cluster sintering (or coalescence) process the aggregation kinetics of industrial elastomer latices has been studied in the diffusion limited cluster aggregation (DLCA) regime in quiescent fluid. The aggregation extent was followed by particle sizing via dynamic light scattering (DLS). Simulation of the aggregation using the kinetic approach (population balances) allows an interpretation of the light scattering result in terms of cluster structure, which is described by the cluster fractal dimension – the key parameter in the simulation for fitting the average hydrodynamic radius $\langle R_h \rangle$ from the DLS experiments. Additionally the cluster structure has been probed by small angle static light scattering (SASLS) and cryogenic electron microscopy. Depending on the surface properties of the particles with respect to latex stability two limiting cluster structures can be found. Bare elastomer particles, stabilized with surfactants (mobile charges) only, form dense, sphere-like clusters due to very strong sintering of primary particles. On the other hand sintering is practically absent if the particles carry a sufficiently high number of fixed charges. This observation is explained by the permission of polymer chain inter-diffusion due to physical particle-particle contact or its prevention by a barrier.

Keywords

Aggregation, Diffusion Limited Cluster Aggregation (DLCA), Cluster Structure, Fractal Clusters, Dense Clusters, Sintering, Soft Particles, Nanoparticle Dispersion

1 Introduction

Processing of particulate systems is a typical intermediate step from material synthesis towards technological application. Properties of the final product are essentially linked to particle size distribution (PSD) and structure of the processed solids. Industrial coagulation of solid-liquid dispersions generally comprises application of shear flow to control size and structure of evolving clusters by breakage and restructuring. Colloidal systems of rigid particles have been studied extensively. Comparably little is known about aggregation of soft particles which has application e.g. in coagulation of elastomer particles where problems of fouling or clogging are faced.

In order to address industrial problems related to cluster structure, quantitative models with physically sound parameters can be applied to understand observations from aggregation experiments. Kinetic models based on population balance equations (PBEs) are widely employed to study the physicochemical nature of aggregation processes. Information about the cluster structure can be extracted from the cluster fractal dimension, D_f, a key parameter of kinetic models. The typical structure of aggregates is a random arrangement of primary particles characterized by a self similar fractal scaling[1]. Universal cluster structure has been found for clusters of rigid particles independent of their chemical nature[2]. Basic studies often use quiescent fluid to grow aggregates in order to eliminate breakage and shear influence on cluster structure. It is known from experiments and simulations that clusters of rigid particles have an open structure. This translates into D_f close to 1.8 for clusters growing without interaction barrier, i.e., under diffusion controlled conditions[3,4,5] and into D_f close to 2.1 for cluster growth with a limited interaction barrier[4,5,6], which leads to a little denser conformation. However, it was found that dense, sphere-like clusters can result from aggregation of soft elastomer particles[7].

It is expected that the involved cluster sintering is greatly governed by the surface properties of respective primary particles. To identify the features that control the so called sintering or coalescence process, we compare two latices with identical polymer properties but different surface characteristics. The first latex contains bare particles that are stabilized with carboxylic surfactants, whereas particles of the second latex carry fixed charges from chain end groups. Influence of the charge state is studied employing sodium chloride (NaCl) and nitric acid (HNO_3) as coagulants. Diffusion limited cluster aggregation (DLCA) in quiescent fluid allows of the most reliable application of the kinetic model since the cluster fractal dimension D_f is basically the only free parameter. Accordingly, cluster structure information can be extracted from the value of D_f that leads to the best simulation of the cluster growth in terms of size.

2 Experiments & Methodology

Two latices of elastomer particles used in the experiments have similar polymer composition, with primary particle radii R_0 in the 100 nm range. According to information from the supplier (Solvay Solexis), the first latex is stabilized only by carboxylic surfactants. The dissociated end groups of the adsorbed surfactants are considered as mobile charges with respect to their position at the particle surface. Therefore, this latex is labeled as mobile carboxyl charge (mcc)-latex in the following. The second latex is stabilized only by charged end groups of polymer chains, which are considered as fixed non-carboxylic charges (fncc) as declared by the supplier.

Aggregation experiments were conducted at a temperature about 50 K above the polymer's glass transition temperature T_g. This condition establishes a rubber-like state of the material characterized by relatively high mobility of polymer chains[8]. The pre-diluted latex was destabilized by addition of coagulant at a concentration that is well above the critical coagulant concentration for fast aggregation, i.e., within DLCA regime. In order to moderate cluster growth to a traceable rate but to ensure reliable detection by light scattering instruments the final solid volume fraction of latex ranged between $\phi = 10^{-5}$ and 10^{-4}. Aggregation within the DLCA regime was confirmed by recurring equal time evolutions of the measured average hydrodynamic radius $\langle R_h \rangle$ at different coagulant concentrations. Data sets from at least three experiments were combined to an average set for each coagulant (NaCl, HNO_3) and latex type (fncc- and mcc-latex). Evolution of $\langle R_h \rangle$ was recorded on-line by means of dynamic light scattering (DLS) at 90 degree and laser wavelength $532\,\text{nm}$, equivalent to a scattering wave vector $q = 2.23 \times 10^{-2}\ \text{nm}^{-1}$. The average hydrodynamic radius $\langle R_h \rangle$ is extracted from the intensity autocorrelation function of the scattered light employing the method of cumulants[9]. Photo detector and the BI-200SM goniometer are from Brookhaven Instruments Corp., the ventus LP 532 from Laser Quantum provided the source of light. Information about the cluster structure was gained from small angle static light scattering (SASLS) of aggregating latices. Within the limits of the power law region, where sufficiently large aggregates exhibit a well developed fractal scaling, the slope of log-log plot of scattered light intensity $I(q)$ versus q can be interpreted as cluster fractal dimension D_f [10]. The SASLS instrument was a Mastersizer 2000 from Malvern Instruments Ltd.

Reconstruction of the experimentally obtained $\langle R_h \rangle$ from the simulated cluster mass distribution (CMD) was used as a tool to extract the structure of evolving clusters. For a given CMD the average hydrodynamic radius can be expressed as[2,11]

$$\langle R_h \rangle = \frac{\sum_i i^2 N_i S_i(q)}{\sum_i i^2 N_i S_i(q) R_{h,i,eff}^{-1}} \qquad (1)$$

which is an average of the effective hydrodynamic cluster radii $R_{h,i,eff}$ weighed by the relative cluster mass i, number of clusters N_i and the cluster structure factor $S_i(q)$. For fractal aggregates $S_i(q)$ is approximated through the Fisher-Burford relation[10]. In case of condensed sphere-like aggregates with D_f close to 3 $S_i(q)$ essentially becomes the form factor of the respective sphere which is calculated according to Lorenz-Mie theory [12]. Access to the CMD evolution in time $N_i(t)$ is gained by solving population balance equations (PBEs) [13] for cluster-cluster aggregation which is generally understood as a second order process[14].

$$\frac{dN_i(t)}{dt} = -\sum_{j=1}^{\infty} K_{i,j}\, N_i(t)\, N_j(t) + \frac{1}{2}\sum_{j=1}^{i-1} K_{i-j,j}\, N_{i-j}(t)\, N_j(t) \qquad (2)$$

The matrix of rate constants (so called kernel) $K_{i,j}$ includes all physical information of the simulated system. In DLCA condition it reduces to[15,16]

$$K_{i,j} = \frac{K_B}{W}\frac{1}{4}\left(i^{-1/D_f} + j^{-1/D_f}\right)\left(i^{1/D_f} + j^{1/D_f}\right) \qquad (3)$$

The kernel accounts for diffusive motion and collision cross section of approaching clusters and is actually an extension of the aggregation rate of identical spherical particles driven purely by diffusion $K_B = 8kT/(3\mu)$ as derived from von Smolukchowski[14]. The aggregation rate due to unrestricted Brownian motion is determined only by thermal energy kT and dynamic viscosity of the continuous phase μ. Deviations are lumped into the Fuchs stability ratio W [17] which relates the effective aggregation rate of primary particles $K_{1,1}$ to the ideal, diffusion limited clusters approach $W = K_B/K_{1,1}$. Even though introduced to account for a reduced sticking probability due to an interaction barrier in reaction limited cluster aggregation (RLCA) it may serve in DLCA to adjust $K_{1,1}$ to small resistances e.g. hydrodynamic perturbations which can reduce $K_{1,1}$ by a factor up to 2 compared to K_B[15,18]. Practically, W is adjusted to reproduce $\langle R_h \rangle$ at the aggregation onset whereas D_f needs to take a value that enables to fit the whole evolution of $\langle R_h \rangle$. It can be convenient to eliminate specific effects like initial particle concentration $N_{1,0}$ and deviations from ideal Brownian coagulation that are lumped into the W by introduction of a dimensionless time defined as[11]

$$\tau = t K_B N_{1,0} W^{-1} \tag{4}$$

Plotting of experimental as well as simulation results in dimensionless form will highlight general features by a collapse of data to a single line or so called master-curve. Such master-curves can be understood as a proof of universal nature of the investigated process with respect to cluster structure[2,3,6] and kinetics[5]. This procedure is very beneficial to demonstrate structural differences arising from an aggregation process.

3 Results & Discussion

Evolution of average hydrodynamic radii follows the typical power law behavior of DLCA kinetics regardless of latex and coagulant type. Let us now discuss the observations for the fncc-latex in detail. The best fit of the kinetic model (Eqs. (2) and (3)) is achieved for fncc-latex coagulation if D_f and W values are respectively 1.6 and 2.0 in the case of using HNO_3 as the coagulant, and 1.6 and 2.1 in the case of using NaCl as the coagulant. It is clear that the influence of electrolyte is negligible which can be consulted conveniently in Figure 1. Respecting the limits of DLCA, both parameters have reasonable quantities even though $D_f = 1.6$ is a little smaller than the usually obtained D_f values that are close to 1.8[3,4,5,10]. This deviation might arise from some considerable polydispersity of the fncc-latex, as well as influence from sedimentation that might contribute especially when the relative size $\langle R_h \rangle / R_0$ exceeds 5 (corresponding to cluster diameter of about $1\,\mu m$). However, the fractal nature of the fncc-latex clusters and their rather open structure is confirmed from kinetically obtained D_f. A second proof is given by the D_f values extracted from the average structure factor of sufficiently large aggregates by means of SASLS, which range between 1.6 and 1.7. Using $D_f = 1.7$ and lowering W a good reproduction of kinetic data could be achieved as well but the quality of the fit is reduced in this case. Nevertheless, the main conclusion is that very open clusters of fractal nature result from aggregation of particles that carry fixed-charges at their surface (fncc-latex) despite the rubbery nature of the polymer.

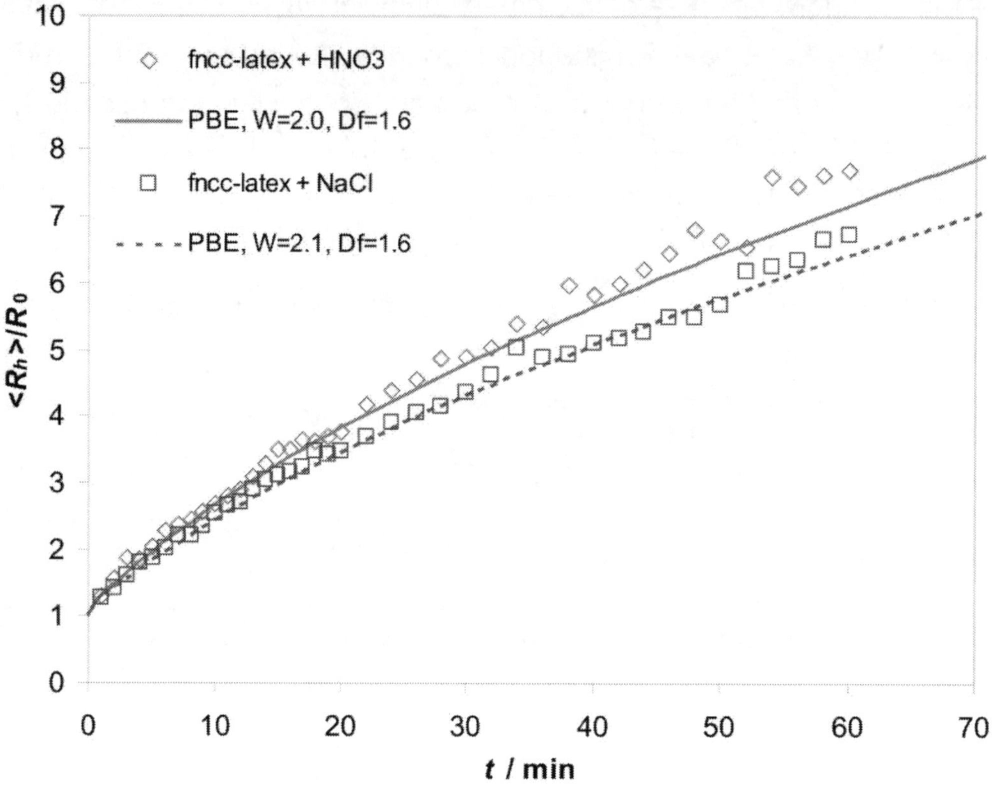

Figure 1: Evolution of size $\langle R_h \rangle / R_0$ for aggregation of the fncc-latex completely destabilized (DLCA) by HNO_3 (◇) and NaCl (□).

The mcc-latex behaves different in terms of aggregate structure. Proper simulation of size evolution requires to set $D_f = 3.0$, which suggests the existence of very dense, sphere-like clusters. The same information is directly accessible from plotting of average size $\langle R_h \rangle / R_0$ versus time on a double logarithmic grid. According to the linear dependence of average mass with time in DLCA[3,14] and the scaling relationship between mass and size[3,5,10]

$$i = m_i / m_1 \approx \left(R_{g,i} / R_0 \right)^{D_f} \qquad (5)$$

a scaling exponent of 1/3 corresponds to $D_f = 3.0$ because of the fixed relation $R_{g,i} / R_{h,i} = \sqrt{3/5}$ for spheres. Time evolutions of simulated and experimental data in Figure 2 show nicely the sensitivity of $\langle R_h \rangle$ to the CMD and spherical cluster structure (Eq. (1)), characterized by a hump at small times. From these considerations it is clear that particles of the mcc-latex undergo sintering after they have aggregated. The fusion process is obviously fast enough to allow the existence of only spherical aggregates within the studied framework. Influences from the coagulant type are limited to the W value which is similar to the fncc-latex.

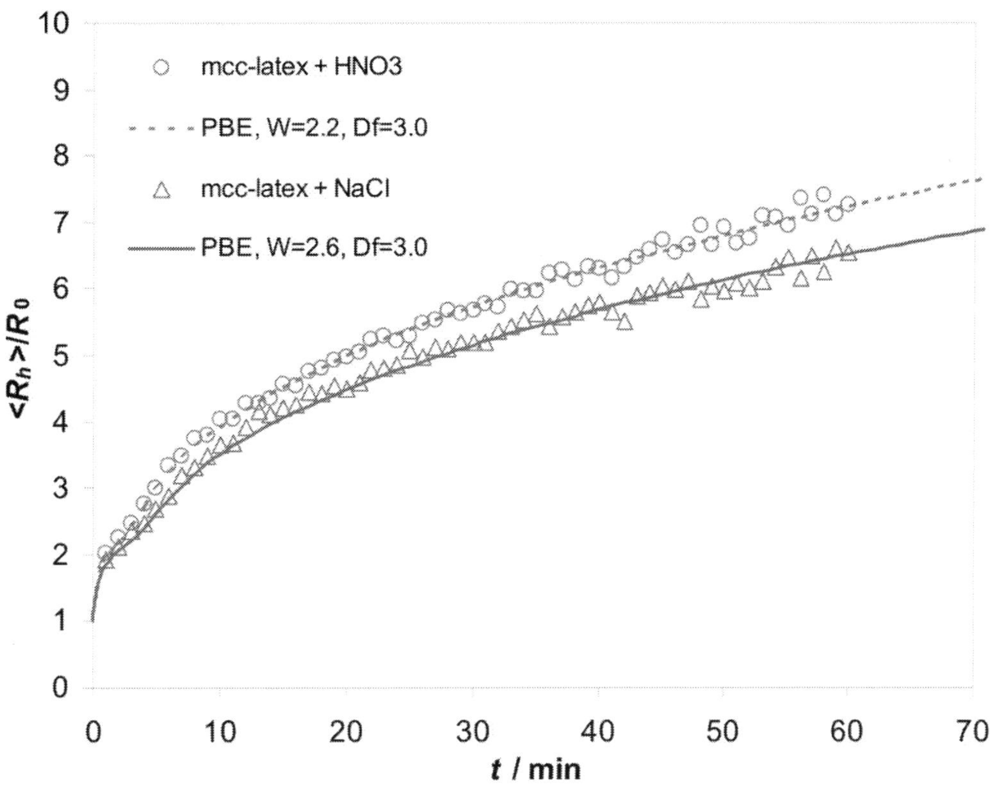

Figure 2: Evolution of size $\langle R_h \rangle / R_0$ for aggregation of the mcc-latex completely destabilized (DLCA) by HNO3 (○) and NaCl (△).

Let us now compare both latexes by eliminating all influences that are not based on the cluster structure. By plotting in dimensionless time τ the aggregation kinetics reduces to master-curves that are determined by the cluster structure only. As can be seen from Figure 3 each latex follows its own characteristic master-curve. The question has to be addressed why open, fractal aggregates result from fncc-latex aggregation and what enables particle fusion in the case of mcc-latex. Remembering that both latexes are of equal polymer material the sintering process must be controlled by the surface properties of the latex particles. Prerequisite for cluster sintering is certainly the high mobility of polymer chains. It seems reasonable that a real physical contact between the polymer phases must be established to allow fusion of particles from polymer chain inter-diffusion. Bare particles of mcc-latex obviously generate a sufficiently large contact area if they are attracted by Van-der-Waals forces. It has to be assumed that surfactants are squeezed out from the contact region no matter of their charge state. In case of the fncc-latex sufficient surface contact that would enable polymer chain inter-diffusion cannot be established. Supposedly the fixed chain ends act as barrier to prevent physical surface contact of particles in close proximity. Since the charges of the chain ends are not of carboxylic nature it cannot be resolved whether the fixed chain ends act as a steric or electrostatic barrier.

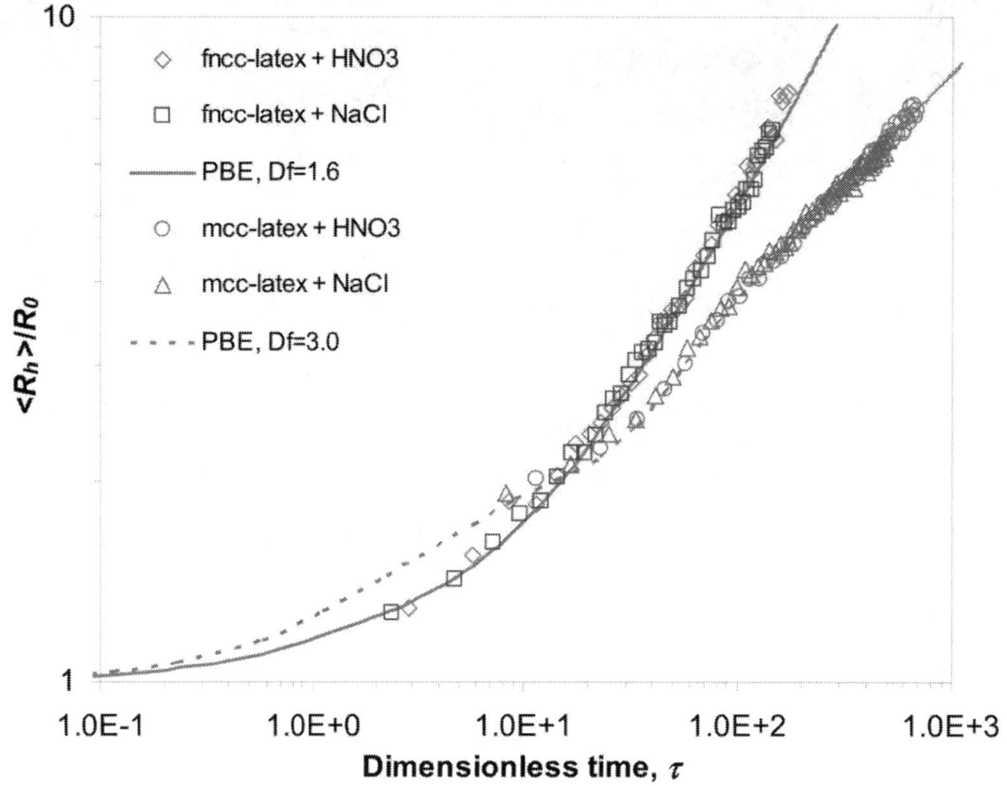

Figure 3: Master-curves resulting from plot of normalized size $\langle R_h \rangle / R_0$ versus dimensionless time $\tau = N_{1,0} K_B t / W$; (○,△) mcc-latex; (◇, □) fncc-latex.

4 Conclusion

Sintering of clusters that originate from particle aggregation can be studied by interpretation of the aggregation process using a kinetic model. Results from aggregation of elastomer particles show that sintering of corresponding clusters is controlled by surface properties of primary particles. It is suggested that bare particles of surfactant stabilized (mcc-) latex coalesce due to polymer chain inter-diffusion at the particle-particle contact area which is established by Van-der-Waals attraction and exclusion of surfactant. Sintering is absent when particles carry fixed charges from polymer chain end groups that are likely to form a barrier against physical particle-particle contact. Discrimination between a steric or electrostatic barrier remains an open issue since charge elimination of the respective (fncc-) latex e.g. by acid destabilization was not feasible. Preliminary electron micrographs of frozen latices confirm the kinetic observations. Light scattering is a fast tool and can provide sufficiently detailed and reliable information. The presented results and methodology are valuable for processing soft particulate systems e.g. elastomer latices. Tuning of surface properties might allow reduction of cluster sintering if handling problems occur. On the other hand sintering effects might be desired to increase cluster density which facilitates removal of water or by products.

5 Acknowledgement

The authors are thankful to Roger Wepf from Electron Microscopy Center of ETH Zurich

(EMEZ) for advancing freezing techniques that should promote cryogenic-SEM in future. Solvay Solexis is appreciated for latex supply and collaboration. Financial support from the Swiss National Science Foundation (Grant No. 200020-113805/1) is gratefully acknowledged.

6 References

1. Meakin, P. Fractal Aggregates. Advances in Colloid and Interface Science. 1988; 28:249-331.
2. Lin, M. Y.; Lindsay, H. M.; Weitz, D. A.; Ball, R. C.; Klein, R; Meakin, P. Universality in Colloid Aggregation. Nature. 1989; 339: 360-362.
3. Lin, M. Y.; Lindsay, H. M.; Weitz, D. A.; Klein, R; Ball, R.C.; Meakin, P., Universal Diffusion-limited Colloid Aggregation. Journal of Physics-Condensed Matter. 1990; 2:3093-3113.
4. Odriozola G.; Tirado-Miranda M.; Schmitt A.; Martinez Lopez F.; Callejas-Fernandez J.; Martinez-Garcia R.; Hidalgo-Alvarez R. A Light Scattering Study of the Transition Region between Diffusion- and Reaction-Limited Cluster Aggregation. Journal of Colloid Interface Science. 2001; 240:90–96.
5. Lattuada, M.; Wu, H.; A Simple Model for the Structure of Fractal Aggregates. Journal of Colloid and Interface Science. 2003; 268:106-120.
6. Lin, M. Y.; Lindsay, H. M.; Weitz, D. A.; Ball, R. C.; Klein, R; Meakin, P. Universal reaction-limited colloid aggregation. Phys.Rev .A 41 (1990) 2005–2020.
7. Gauer, C.; Jia, Z.; Wu, H.; Morbidelli, M.; Aggregation Kinetics of Soft Polymer Particles. Proceedings: PARTEC: 2007.
8. Cowie, J. M. G. Polymers: Chemistry & Physics of Modern Materials. Glasgow: Blackie: 1991.
9. Lindner, P.; Zemb, T. Neutrons, X-rays and Light: Scattering Methods Applied to Soft Condensed Matter. Amsterdam: Elsevier: 2002.
10. Sorensen, C. M. Light Scattering by Fractal Aggregates: A Review. Aerosol Science and Technology. 2001; 35:648-687.
11. Lattuada, M.; Sandkühler, P.; Wu, H.; Sefcik, J.; Morbidelli. M. Aggregation Kinetics of Polymer Colloids in Reaction Limited Regime: Experiments and Simulations. Advances in Colloid and Interface Science. 2003; 103:33-56.
12. Wiscombe, W. J. Mie Scattering Calculations: Advances in Technique and Fast, Vector-Speed Computer Codes. NCAR/TN-140+STR: National Center for Atmospheric Research: Boulder, Colorado: 1979.
13. Ramkrishna, D. Population Balances. San Diego: Academic Press: 2000.
14. von Smoluchowski, M. Versuch einer mathematischen Theorie der Koagulationskinetik kolloider Lösungen. Zeitschrift für Physikalische Chemie. 1917; 92:129-168.
15. Odriozola G.; Schmitt A.; Callejas-Fernandez J.; Martinez-Garcia R.; Hidalgo-Alvarez R. Dynamic Scaling Concepts Applied to Numerical Solutions of Smoluchowski's Rate Equation. Journal of Chemical Physics 1999; 111:7657–7667.
16. Sandkühler, P.; Sefcik, J.; Lattuada, M.; Wu, H.; Morbidelli, M. Modeling Structure Effects on Aggregation Kinetics in Colloidal Dispersions. AIChE Journal. 2003; 49:1542-1555.
17. Fuchs, N. Über die Stabilität und Aufladung der Aerosole. Z. Phys. 1934; 89:736–743.

18. Spielman, L. A. Viscous Interactions in Brownian Coagulation. Journal of Colloid Interface Science. 1970; 33:562–571.

Effect of pH and sulphate on the morphology of γ-AlOOH

Taobo He, Lan Xiang*, Shenlin Zhu
Department of Chemical Engineering, Tsinghua University, Beijing 100084, P. R. China

Abstract

The influence of pH and sulphate on the morphology of γ-AlOOH was investigated in the hydrothermal treatment of $Al(OH)_3$ formed by mixing $AlCl_3$ and NH_4OH solutions. Results show that the initial solution pH plays a crucial role in the morphology of the final γ-AlOOH products: the acidic conditions are favored toward the formation of nanorods and the basic conditions toward the formation of nanoflakes. The one dimensional growth of γ-AlOOH nanorods was enhanced by adding Na_2SO_4 under acidic conditions and the aspect ratio of γ-AlOOH nanorods increased from about 12 to 100 as the initial Na_2SO_4 concentration increased from 0.0536 to 0.286 mol/L. Thermodynamic calculation indicated that $Al(OH)_4^-$ is the predominant aluminum ion under alkaline conditions and positive aluminum ions ($Al(OH)_2^-$ and $AlOH^{2-}$) are the main ions under acidic conditions; the percentage of $Al(OH)_2^-$ to the total aluminum ions increases with increasing the initial Na_2SO_4 concentration.

Keywords: pH; sulphate; γ-AlOOH; morphology

1. Introduction

The synthesis of inorganic nanocrystals of controlled shape and size has drawn much attention because the shape and size of inorganic nanocrystals have great influence on their physical properties [1,2]. During the past decade, there has been an increasing number of reports on the synthesis of novel nanocrystals such as rods, cubes, prisms, disks, and tubes and some of the growth parameters (such as surface energy, growth rate, capping molecures, and reaction temperature) can be critical in determining the behavior of the ultimate nanocrystals. Understanding the shape-guiding process of nanocrystals is technologically important in the shape-and size-controlled synthesis of nanocrystals and will make it possible to program the system to yield the nanomaterials with a desired shape and size.

Recently, the preparation of γ-AlOOH nanocrystals has been a great interest due to their broad applications in ceramics, composites, catalysts, absorbents and flame retardation. γ-AlOOH nanocrystals, especially 1D nanocrystals, have been synthesized via the hydrothermal method. It was reported that nanotubes, nanorods, nanowires and nanofibers were synthesized by using sodium amide, alkali salt, aluminum hydrous, or aluminum isopropoxide [3-8]. However, the growth mechanism of γ-AlOOH was seldom studied and the factors influencing the morphology are not well understood.

In this study, a facile hydrothermal method was developed to synthesize γ-AlOOH nanorods using $AlCl_3$ and NH_4OH as the reactants and Na_2SO_4 as the morphology-directing agent. The influence of pH and sulphate on the morphology of γ-AlOOH nanocrystals was investigated. The corresponding reaction mechanism was studied via the thermodynamic equilibrium calculation of $Al(OH)_3$ aqueous system and $Al(OH)_3$-Na_2SO_4 aqueous system, respectively.

2. Experimental

In a typical synthetic procedure for γ-AlOOH nanocrystals, 15.0 mL of 3.0 mol·L^{-1} NH_4OH solution was added dropwise into 15.0 mL of 1.0 mol·L^{-1} $AlCl_3$ solution at room temperature. The precipitate was centrifuged and washed with distilled water. The process was repeated until no chloride was detected by the 10wt% silver nitrate solution. The precipitate was then dispersed in

* Corresponding author. Tel.: +86 10 62788984; fax: +86 10 62772051
E-mail address: xianglan@mail.tsinghua.edu.cn

distilled water and the pH was adjusted by the addition of 1.5 $mol{\cdot}L^{-1}$ NaOH solution or 1.5 $mol{\cdot}L^{-1}$ HCl solution. The as-received slurry and a certain amount of 0.75 $mol{\cdot}L^{-1}$ Na_2SO_4 solution were transferred into a Teflon-lined autoclave with an inner volume of 70 mL. Then, the distilled water was added to keep the total volume as 35.0 mL. The autoclave was subsequently heated (5 $°C{\cdot}min^{-1}$) to 240 °C and kept in isothermal condition for 16.0 hours. After hydrothermal reaction, the autoclave was cooled down to room temperature naturally and the precipitate was centrifuged, washed with distilled water and dried at 105 °C for 24.0 hours.

X-ray diffraction (XRD) measurements were performed on an X-ray powder diffractometer (D/max2500, Rigaku, Japan) using Cu Kα radiation (λ = 1.54178 Å). The morphology was examined by a high-resolution transmission electron microscopy (HRTEM, JEM-2010, JEOL, Japan) equipped with a selected area electron diffraction (SAED).

3. Results and discussion

3.1 Effect of initial solution pH

Fig. 1 shows the XRD patterns of the samples prepared at pH=4.0 (a) and pH=9.5 (b) at 240°C for 16 h. The essentially same XRD patterns were obtained when the sample was prepared under different initial solution pH. All of the reflection peaks are readily indexed as γ-AlOOH phase (space group: Cmcm, JCPDS: 74-1895). No characteristic peaks of impurity phases are present in Fig.1, indicating the high purity of the final products. It is worth pointing that the XRD patterns also indicate that there is a large difference from each other in the relative intensities based on (020), (021) and (002) peaks for two samples, indicating the possibility of different preferential orientation growth under different pH conditions, which was confirmed by the following TEM observation.

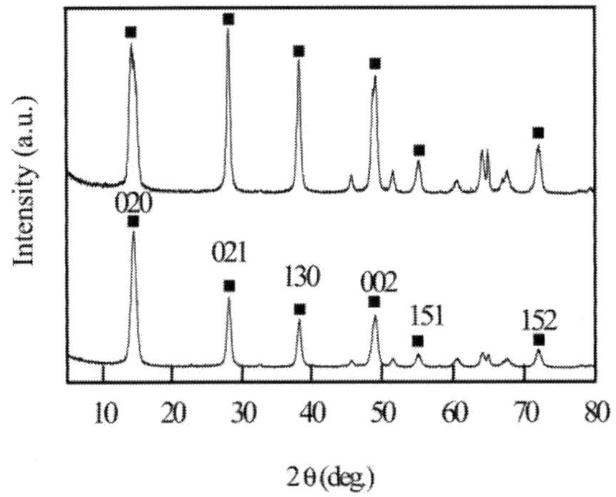

Fig.1. XRD patterns of the hydrothermal products in the absence of Na_2SO_4 pH of initial solution: a-4.0, b-9.5.

Fig. 2 shows the TEM images of the γ-AlOOH samples prepared at pH=4.0 and pH=9.5, respectively. At pH=4.0, the TEM image with low magnification reveals that the γ-AlOOH samples was composed of nanorods with the lengths ranging from tens to hundreds of nanometers. According to the TEM image in Fig. 2b, the γ-AlOOH nanorods had an average diameter of 10 nm and the surfaces of the nanorods were very coarse. The TEM images of the sample obtained under basic conditions (pH=9.5) (Fig. 2c) demonstrates the sample consists of many nanoflakes with the width of tens of nanometers. The SAED pattern recorded from the [020] zone axis was composed of high symmetrical dotted lattice, indicating the perfect crystallinity of the nanoflakes.

From the TEM images in Fig. 2, a conclusion can be drawn that the pH of the solution plays a crucial role in determining the morphology of the final γ-AlOOH nanocrystals: the acidic conditions are favorable for the formation of 1-D nanostructures and the basic conditions for the formation of the 2 D nanostructure.

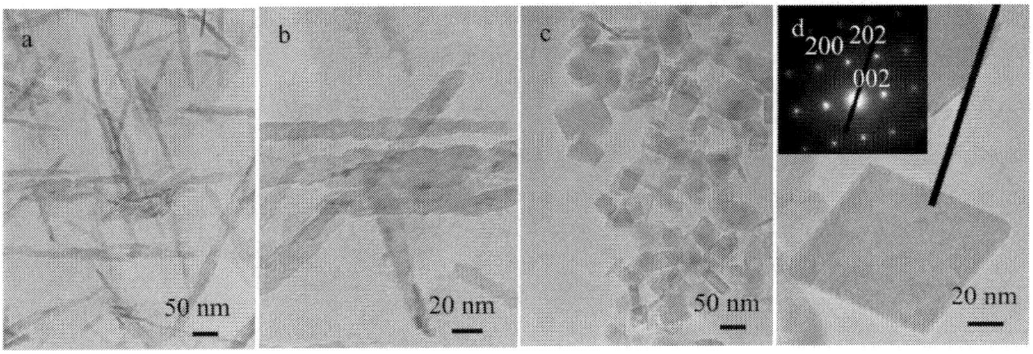

Fig.2. Morphology of the hydrothermal products. Initial solution pH: (a-b)-4.0, (c-d)-9.5; temperature (℃): 240; time (h): 16.0.

3.2 Effect of Na_2SO_4

Fig. 3 illustrates the XRD patterns of the final products prepared at different initial Na_2SO_4 concentration. Compared with the XRD patterns of the final product prepared in the absence of Na_2SO_4, Fig. 3 shows that another crystalline phase indexed to be $NaAl_3(SO_4)_2(OH)_6$ occurred in the final products at the initial concentration of 0.1608 and the amount of $NaAl_3(SO_4)_2(OH)_6$ in the final products increased with increasing the initial Na_2SO_4 concentration.

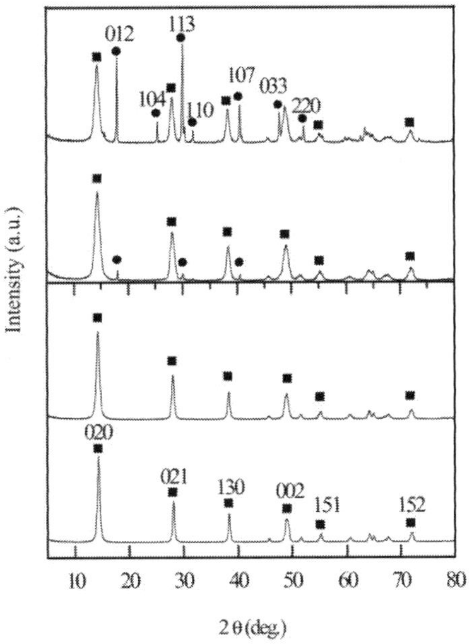

Fig.3. XRD patterns of the products at pH = 4.0. Initial Na_2SO_4 concentration (mol/L): a-0.0536, b-0.1072, c-0.1608, d-0.286; temperature (℃): 240; time (h): 16.0.

■ :γ-AlOOH, •: $NaAl_3(SO_4)_2(OH)_6$.

Fig. 4 shows the morphologies of γ-AlOOH obtained at different initial Na_2SO_4 concentration. γ-AlOOH nanorods with well-define shape were obtained in the presence of Na_2SO_4. With the increase of the initial Na_2SO_4 from 0.0536 to 0.286, the length of the nanorods increased from about 300 nm to about 1.0 μm and the diameter decreased from about 25 nm to about 10 nm, leading to the increase of the aspect ratio from about 12 to around 100.

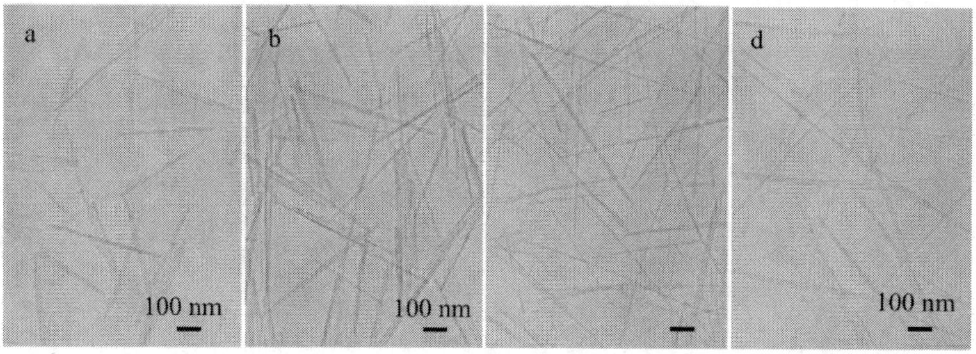

Fig.4. Morphology of γ-AlOOH at pH = 4.0. Initial Na_2SO_4 concentration (mol/L): a-0.0536, b-0.1072, c-0.1608, d-0.286; temperature (°C): 240; time (h): 16.0.

▪ : γ-AlOOH, • : $NaAl_3(SO_4)_2(OH)_6$.

3.3 Equilibrium of $Al(OH)_3$ aqueous system at elevated temperature

For hydrothermal synthesis of metal hydroxide salts, the growth mechanism is supposed to be the traditional dissolution-recrystallization process [9] and pH condition is quite important for controlling the morphology. In order to understand the effect of pH condition on the morphology of γ-AlOOH, the equilibrium of $Al(OH)_3$ aqueous system was investigated at elevated temperature.

The following equilibrium reactions exists in the $Al(OH)_3$ aqueous system:

$$H^+ + OH^- = H_2O \qquad (1)$$

$$Al^{3+} + OH^- = AlOH^{2+} \qquad (2)$$

$$Al^{3+} + 2OH^- = Al(OH)_2^+ \qquad (3)$$

$$Al^{3+} + 3OH^- = Al(OH)_3 \qquad (4)$$

$$Al^{3+} + 4OH^- = Al(OH)^{4-} \qquad (5)$$

The variation of the equilibrium constants of reactions (1)-(5) with temperature was listed in Fig. 5. An equilibrium calculation was carried out to investigate the behavior of soluble ions using HSC software. Fig. 6 shows that the solution pH decreases drastically with increasing temperature and the solution pH changes from 4.0 to 1.42 and from 9.5 to 5.62, respectively. Fig. 7 shows the speciation of aqueous aluminum species in equilibrium with $Al(OH)_3$ at 240°C. It is clearly indicated that aluminum species distribution is quite sensitive to the pH conditions. At pH = 5.62, the predominant aluminum specie is $Al(OH)_4^-$; however, the main aluminum species are $AlOH^{2+}$ and $Al(OH)_2^+$ at pH = 1.42. Thus, it can be inferred that $Al(OH)_4^-$ is favorable for the formation of γ-AlOOH nanoflakes and positively charged aluminum ions favor the formation of γ-AlOOH nanorods.

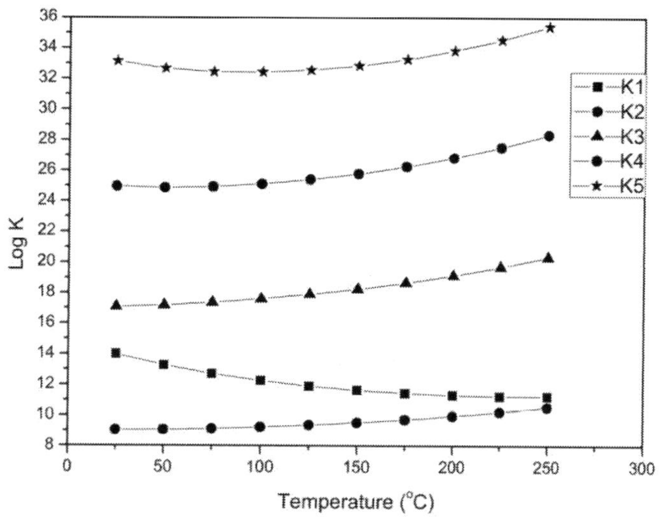

Fig. 5. Equilibrium constants of reactions (1)-(5)

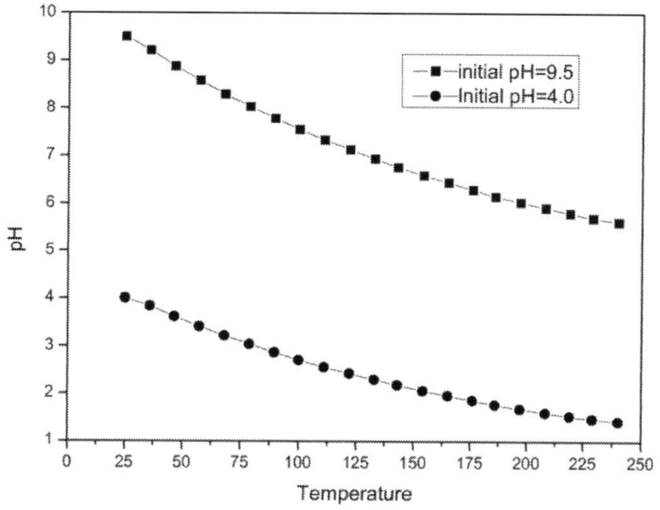

Fig. 6. Variation of the solution pH with temperature

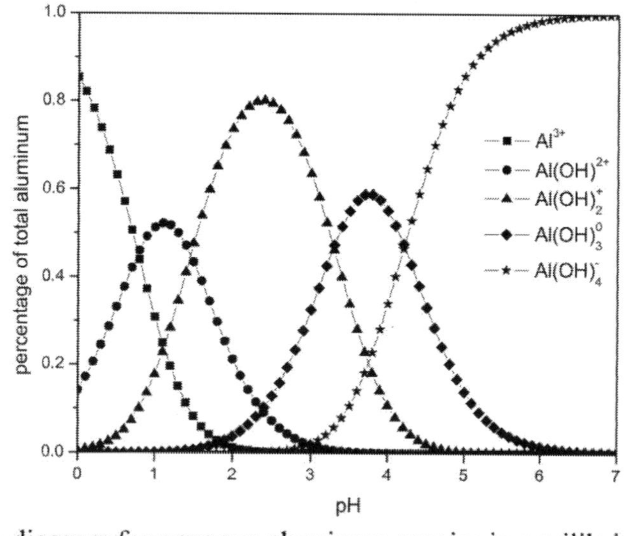

Fig. 7. Speciation diagram for aqueous aluminum species in equilibrium with $Al(OH)_3$.
Temperature (℃): 240.

3.4 Effect of SO_4^{2-} on the equilibrium

The following equilibrium reactions existed in the $Al(OH)_3$-Na_2SO_4 aqueous system:

$$H^+ + OH^- = H_2O$$
$$Al^{3+} + OH^- = Al(OH)^{2+}$$
$$Al^{3+} + 2OH^- = Al(OH)_2^+$$
$$Al^{3+} + 3OH^- = Al(OH)_3$$
$$Al^{3+} + 4OH^- = Al(OH)^{4-}$$
$$H^+ + HSO_4^- = H_2SO_4$$
$$H^+ + SO_4^{2-} = HSO_4^-$$
$$2Al^{3+} + 3SO_4^{2-} = Al_2(SO_4)_3^0$$
$$Na^+ + 3Al^{3+} + 2SO_4^{2-} + 6OH^- = NaAl_3(SO_4)_2(OH)_6$$

Fig. 7 shows the aluminum species' distribution for the $Al(OH)_3$-Na_2SO_4 aqueous system as a function of the initial concentration of Na_2SO_4 when the initial solution pH = 4.0. It shows that aluminum species' distribution varies with the initial concentration of Na_2SO_4. $AlOH^{2+}$ and $Al(OH)_2^+$ are the two main aluminum species in the range of the initial concentration of Na_2SO_4 investigated. Furthermore, the percentage of $AlOH^{2+}$ decreases, while that of $Al(OH)_2^-$ increases with increasing of the initial concentration of Na_2SO_4, which may be the cause for the precipitation of $NaAl_3(SO_4)_2(OH)_6$ for the formation of $Al(OH)_2^-$ was considered to be the critical step in the precipitation of alunite[10]. Our experiment has shown that the aspect ratio of γ-AlOOH increases with the initial concentration of Na_2SO_4, which may connect with the increase of the percentage of $Al(OH)_2^-$ in the solution.

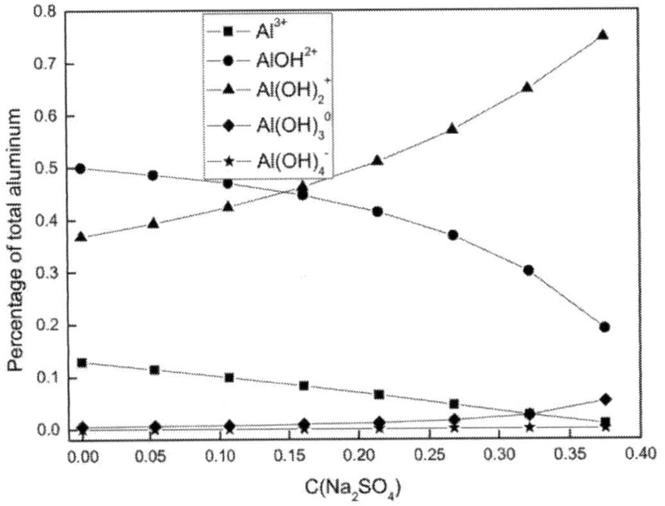

Fig. 7. Speciation diagram for aqueous aluminum species as a function of the initial concentration of Na_2SO_4. Temperature (℃): 240.

4. Conclusions

The initial solution pH plays a crucial role in determining the morphology of the final γ-AlOOH nanocrystals: the acidic conditions are favorable for the formation of 1-D nanostructures and the basic conditions for the formation of the 2 D nanostructure. Thermodynamic calculation indicates that $Al(OH)_4^-$ is favorable for the formation of γ-AlOOH nanoflakes and $Al(OH)_2^-$ favor the formation of γ-AlOOH nanorods. The 1 D growth of γ-AlOOH under acidic condition can be enhanced by adding Na_2SO_4 due to the increase of the percentage of $Al(OH)_2^-$ in the solution by the precipitation of $NaAl_3(SO_4)_2(OH)_6$.

Acknowledgment

This work was supported by the National Natural Science Foundation of China (No.50574051)

References

[1] Alivisatos AP. Science 1996;271:933.
[2] El-saved MA. Acc Chem Res 2001;34:257.
[3] Hou HW, Xie Y, Yang Q, Guo QX, Tan CR. Nanotechnology 2005;16:741.
[4] Tang B, He JC, Zhuo LH, Wang GL, Shi ZQ, Dong YB. Eur J Chem 2005;4366
[5] Zhang J et al. J Phys Chem B 2006;110:21680
[6] Zhang J et al. J Phys Chem B 2006;110:14249
[7] Wierenga A, Philipse AP, Lekkerkerker HNW. Langmuir 1998;14:55.
[8] Zhu HY, Riches JD, Barry JC. Chem Mater 2002;14:2086.
[9] Gong XY, Nie Z, Qian M, Liu J, Pederson LA, Hobbs DT, McDufie G.. Ind Eng Chem Res 2003; 42: 2163
[10] Hartig C, Brand P, Bohmhammel K. Neue Hutte 1990;35:205

SDS-PEG Clusters Templated Synthesis of Au Nanoparticles

Yun Fang*, Chunrong Wang

School of Chemical and Material Engineering, Southern Yangtze University, Wuxi, Jiangsu 214122, China

Abstract : We demonstrate a simple wet chemical approach to synthesize Au nanoplates and nanorings. The synthesis was templated by a soft template composed of sodium dodecyl sulfate (SDS)- polyethyleneglycol (PEG) clusters. In the process, $H[AuCl_4]$ was self-reduced by PEG to form Au quantum dots (AQD), the produced AQD self-assembled to hexagon-shaped gold nanoplates, and then turned into nanorings.

Keywords: gold; nanostructures; cluster; soft template; self-assembly

1. Introduction

A variety of methods have been applied to fabricate Au nanospheres(Han *et al.*, 1999; Jana *et al.*, 2001),nanorods(Yu *et al.*, 1997; Gai *et al.*, 2002; Chang *et al.*,1999; Nikoobakht *et al.*, 2003), nanowires(Pei *et al.*, 2004; Lisiecki *et al.*, 1996), nanodisks(Simakin *et al.*, 2001), nanocubes(Huang *et al.*, 2006) and nanoplates(Ibano *et al.*, 2003; Kim, *et al.*, 2004; Sun *et al.*,2004; Ah *et al.*, 2005; Wang *et al.*,2005; Li *et al.*,2006; Umar *et al.*,2006; Sun *et al.*,2005.) in virtue of their applications in catalysis process, photonics and biology(Daniel *et al.*,2004). Although various morphologies of Au nanostructures have been obtained, the production of nanoplates or nanorings has still met limited success(Sun *et al.*, 2003; Jiang *et al.*, 2004; Metraux *et al.*, 2003).SDS has been used as an additives in the synthesis of Au nanoparticles because of its surface activity, so do polymers(Porel *et al.*, 2005; Sun *et al.*, 2005). SDS has a weak protection to the Au nanoparticles (Deng *et al.*, 2005; Kuo *et al.*, 2004; Tang *et al.*,2001), therefore it was used together with some bigger molecules(Mandeep *et al.*,2007) or polymers (Longenberger *et al.*, 1995)to form a network, which has been applied as the soft template and provides better protection to the produced nanoparticles. However, there is still lack of research on using both surfactant and polymer to synthsize nanoparticles (Bakshi *et al.*,2006).

Previously, we studied the micelle behaviour of aqueous solution of SDS-PEG(Fang *et al.*,2004; Fang *et al.*,2006; Jiang *et al.*,2000), in which SDS micelle formed when the SDS concentration c_s reached the first critical concentration (c_1), and addition of inorganic salts can decrease c_1(Huang *et al.*,2005). Subsquently when c_s was above c_1, SDS and PEG could form the "necklace-like" soft clusters with the diameter about 10 nm. In the "necklace-like" soft clusters, PEG chains surrounded the hydrophilic head of SDS, as well as the first and second adjacent carbon atoms of the hydrophobic chain of SDS, which was proofed with NMR analysis(Fang *et al.*,2006). Therefore in this study we aim to use the SDS-PEG clusters as the single soft template to synthesize Au nanoparticles. In this simple wet chemical approach, $H[AuCl_4]$ was self-reduced by polyethyleneglycol (PEG), (SDS)-PEG clusters were served as a soft template to form Au quantum dots (AQD), and no hard template or extra reducer was needed.

2. Experimental procedures

$H[AuCl_4]$ and PEG 20000 were from China Medicine Shanghai Chemical Reagent Corporation. SDS was from Acros Organics. All reagents were used as received without further purification. Deionized water was used through out the experiment. In a typical experiment, an aqueous solution of 5 mM $H[AuCl_4]$ (1 mL) was added to the solution (19 mL) containing PEG 20000 and SDS. The volume was adjusted to 25 mL with water, and the final concentration for each component was 0.2 mM $H[AuCl_4]$, 5 mM PEG 20000 and 3 mM SDS (above c_1), respectively. Experiments were conducted at 25±0.5 °C in dark to avoid photoreactions of the Au complex. Yellow precipitate was observed 1 day later, and was collected

* Corresponding author. Tel: 86-510-89880600; Fax: 86-510-85917920;E-mail: yunfang@126.com

after centrifugation, washing several times with water, then resuspended in water. The resulting suspension was used for characterization.

The samples for TEM and SEM recording were prepared by doping the suspension on carbon-coated copper grids and copper, respectively. The samples were then allowed to dry at room temperature in a fume hood. The TEM images and electron diffraction patterns were taken using a Hitachi H-600 electron microscope operated at 100 kV. SEM images were obtained using FEI Quanta-200 operated at 15 kV. XRD was carried out with D/Max 2500 X-ray diffractometer using Cu (40 kV, 200 mA) radiation and the samples were prepared as thin layers on glass slides. The UV/Vis spectra were recorded on TU-1901 spectrophotometer.

3. Results and discussion

The morphology of the nanoplates prepared was characterized by SEM as shown in figure1.

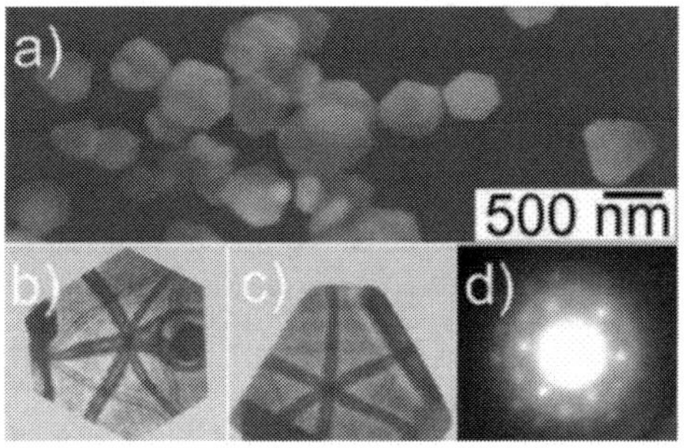

Figure 1. Au nanocrystals obtained from $H[AuCl_4]$ (0.2 mM)-SDS (3 mM)-PEG 20000 (5 mM) mixture (a) SEM image measured 2 days after preparation; (b) and (c) Typical TEM images; (d) Corresponding electron diffraction pattern of a single hexagonal Au nanoplate from (b) placed flatly on a TEM grid.

Figure 1a clearly indicates that these nanoplates have one-dimensional diameter (~500 nm) and are mainly hexagonal. Some of them have grown to equilateral hexagon and the others are evolving from truncated triangle to hexagon. The TEM images in Figure 1b and c show the morphologies of two typical nanoplates from Figure 1a. Three dark bands intersect in the central parts of the nanoplates and the six ends of these bands point at the middle of each side of a hexagon respectively, which denotes the median of each side. It is interesting that there are another three dim crewels enclosing an equilateral triangle on the surface of every nanoplate, the three sides of the triangle are vertically crossed with the above medians alternately, and the three vertices of the triangle intersect with the residual medians simultaneously. To our knowledge, there has been no report on such fine dark texture in Au nanostructures. Figure 1d shows selected area electron diffraction (SAED) pattern obtained by focusing the electron beam on a nanoplate lying on the TEM grid. The SAED pattern reveals that stronger hexagonal symmetry diffraction spot pattern was generated, demonstrating that the Au nanoplate is single crystal.

The crystalline character of the nanoplates (Maye *et al.*, 2000)were further confirmed by the XRD proof. The XRD pattern in Figure 2 shows sharp peaks corresponding to the (111), (200), (220) and (311) diffraction peaks of metallic Au, suggesting that the Au nanoplate has a preferential growth direction along the Au (111) plane(Sun *et al.*,2004).

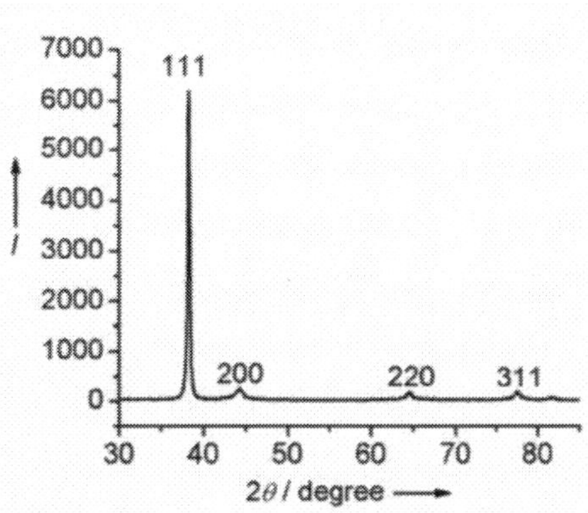

Figure 2. XRD pattern of the resulting Au nanoplates deposited on a glass slide. Nanoplates were obtained from $H[AuCl_4]$ (0.2 mM)-SDS (3 mM)-PEG 20000 (5 mM) mixture measured 2 days after preparation.

It is well-known that the surface plasma resonance (SPR) bands of Au nanostructures strongly depend on the sizes, shapes and aggregation of the nanoparticles(Link *et al.*,1999; Sosa *et al.*, 2003; Jana *et al.*,2001;Yu *et al.*,1997). When Au nanoplates formed, SPR band was red-shifted from 520 nm, the normal band(Sun *et al.*, 2004; Ah *et al.*, 2005; Wang *et al.*, 2005).

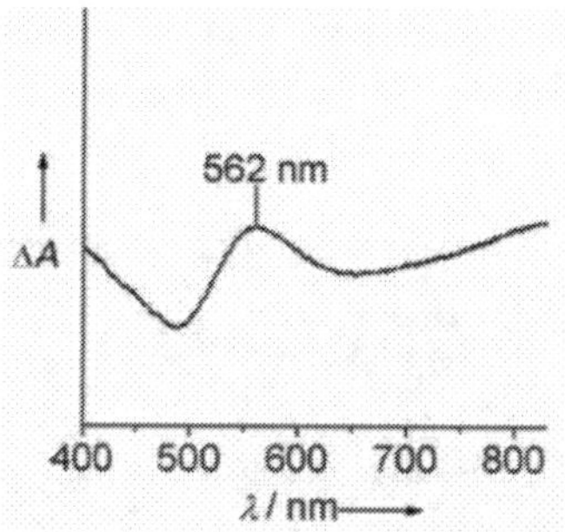

Figure 3. UV/Vis spectrum of the Au nanoplates measured 2 days after preparation from $H[AuCl_4]$ (0.2 mM)-SDS (3 mM)-PEG 20000 (5 mM) mixture.

Figure 3 shows a strong absorption band above 800 nm near IR region, which suggests there were some irregular nanostructures in solution. This is another evidence for the formation of anisotropic Au particles(Sun *et al.*, 2004; Ah *et al.*, 2005; Wang *et al.*, 2005). Besides, SPR band at 562 nm arises from small Au particles(Kuo *et al.*, 2004; Kuo *et al.*, 2005).

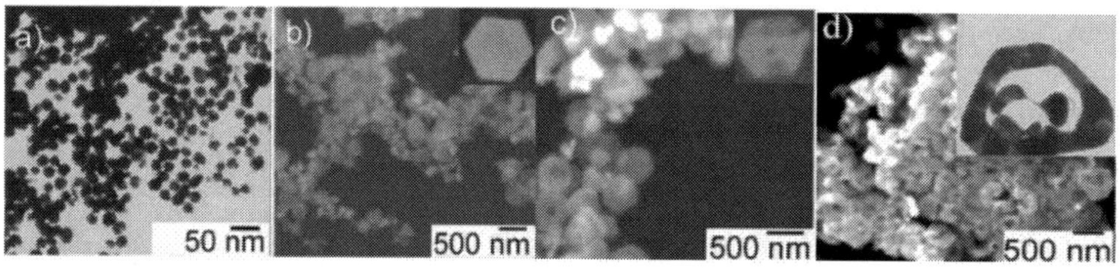

Figure 4. Images of Au nanostructrues obtained from $H[AuCl_4]$ (0.2 mM)-SDS (3 mM)-PEG 20000 (5 mM) solution. (a) TEM measured 6 h later, (b) SEM measured 1 day later, (c) SEM measured 3 days later, (d) SEM measured 10 days later.

Most strikingly, the aging time was found influencing the structures of Au nanoparticles dramatically. The Au nanospheres primarily produced in SDS-PEG solution, quickly turned into nanoplates and then became nanorings as shown in figure 4. Many AQD (~20 nm) were generated in the solution after 6 h (figure 4a), then polygonal planar Au nanostructures (~100 nm) such as hexagonal or truncated triangular Au nanoplates appeared after 1 day (figure 4b); as the reaction proceeded, the number of small Au nanospheres decreased with the increase of hexagon or truncated triangle Au nanoplates after 3 days (figure 4c) and finally melted to nanorings 10 days after (figure 4d). Qi *et al* (Qi *et al.*, 2002)prepared calcium carbonate disks using the mixed micelles of SDS and PEO-b-PMAA as template, which led the calcium carbonate disks to be changed to hollow disks after aged in the mother solution in 1 week.

4. Conclusions

In conclusion, soft template composed of SDS-PEG clusters provides a simple and controllable wet chemical approach to prepare Au nanoparticles. In the process, AQD self-assembles to sub-structures of Au quantum wells and then break down into Au quantum wires at ambient temperature and atmosphere. These Au nanostructures are homogeneous in size and dispersed well.

Acknowledgement

We are grateful for financial support from National Natural Science Foundation of China (20371021).

Reference

Ah, C. S.; Yun, Y. J.; Park, H. J.; Kim, W.; Ha, D. H.; Yun, W. S., Size-Controlled Synthesis of Machinable Single Crystalline Gold Nanoplates. *Chem. Mater.* **2005**, 17, 5558-5561.

Chang, S.; Shih, C.; Chen, C.; Lai, W.; Wang, C. R., The Shape Transition of Gold Nanorods. *Langmuir* **1999**, 15, 701-709.

Daniel. M; Astruc. D., Gold nanoparticles: assembly, supramolecular chemical, quantum-size-related properties, and applications toward biology, catalysis, and nanotechnology *Chem. Rev.* **2004**, 104, 293-346.

Fang, Y.; Cai, K.; Zhong, L-Y.; Liu, X-F.; He, M-S., Specific Saturation Capacity of Clusterization in Cluster of $ROSO_3Na$ With Water- soluble Nonionic Macromolecules. *Chem. J. Chinese Universities* **2004**, 25, 888-891.

Fang, Y.; Liu, X-F.; Xia, Y-M.; Zhong, L-Y., Characterization of Clusters Interaction Sites at Cluster of Sodium Dodecyl Sulfate and Water Soluble Nonionic Macromolecules Investigated by ^1H NMR, ^{13}C NMR and 2D NMR. *Chem. J. Chinese Universities* **2006**, 27, 731-734.

Gai, P. L.; Harmer, M. A., Surface Atomic Defect Structures and Growth of Gold Nanorods. *Nano Lett.* **2002**, 2, 771-774.

Han, M. Y.; Quek, C. H.; Huang, W.; Chew, C. H.; Gan, L. M., A simple and effective chemical route for the preparation of uniform nonaqueous gold colloids. *Chem. Mater.* **1999**, 11, 1144-1147.

Huang. C.; Wang. Y.;. Chiu. P.;. Shih. M.; Meen, T., Electrochemical synthesis of gold nanocubes *Mater. Lett.* **2006**, 60, 1896-1900.

Huang, S-Y.; Fang, Y.; Chen, F-B.; Fu, S-S.; Ren G-Q., Characteristics of Bound Micelle in the Ternary Mixing System of Alkali Metal Sulfate-SDS-PEG. *Chemical Journal of Chinese Universities* **2005**, 26, 1881-1885.

Ibano, D.; Yokota, Y.; Tominaga, T., Preparation of gold nanoplates protected by an anionic phospholipid. *Chem. Lett.* **2003**, 32, 574-575.

Jana, N. R.; Gearheart, L.; Murphy, C. J., Seeding Growth for Size Control of 5-40 nm Diameter Gold Nanoparticles. *Langmuir* **2001**, 17, 6782-6786.

Jana, N. R.; Gearheart, L.; Murphy, C. J., Seed-mediated growth approach for shape-controlled synthesis of spheroidal and rod-like gold nanoparticles using a surfactant template. *Adv. Mater.* **2001**, 13, 1389-1393.

Jiang, L.; Xu, S.; Zhu, J.; Zhang, J.; Zhu, J.; Chen, H., Ultrasonic-Assisted Synthesis of Monodisperse Single-Crystalline Silver Nanoplates and Gold Nanorings. *Inorg. Chem.* **2004**, 43, 5877-5883.

Jiang, W. H.; Han, S. J., Viscosity of Nonionic Polymer/Anionic Surfactant Complexes in Water. *J. Colloid and Interface Science* **2000**, 229, 1-5.

Kim, J.; Cha, S.; Shin, K.; Jho, J. Y.; Lee, J., Preparation of gold nanowires and nanosheets in bulk block copolymer phases under mild conditions. *Adv. Mater.* **2004**, 16, 459-464.

Kuo, C.; Chiang, T.; Chen, L.; Huang, M. H., Synthesis of Highly Faceted Pentagonal- and Hexagonal-Shaped Gold Nanoparticles with Controlled Sizes by Sodium Dodecyl Sulfate. *Langmuir* **2004**, 20, 7820-7824.

Kuo, C.; Huang, M. H., synthesis of branched gold nanocrystals by a seeding growth approach. *Langmuir* **2005**, 21, 2012-2016.

Li, C.; Cai, W.; Cao, B.; Sun, F.; Li, Y.; Kan, C.; Zhang, L., Zhang, L., Mass synthesis of large, single-crystal Au nanosheets based on a polyol process. *Adv. Funct. Mater.* **2006**, 16, 83-90.

Qi, L.; Ma, J., Biomimetic Morphogenesis of Calcium Carbonate in Mixed Solutions of Surfactants and Double-Hydrophilic Block Copolymers. *Advanced Materials* **2002**, 14, (4), 300-303.

Link, S.; El-Sayed, M. A., Spectral Properties and Relaxation Dynamics of Surface Plasmon Electronic Oscillations in Gold and Silver Nanodots and Nanorods. *J. Phys. Chem. B* **1999**, 103, 8410-8426.

Lisiecki, I.; Billoudet, F.; Pileni, M. P. Control of the Shape and the Size of Copper Metallic Particles. *J. Phys. Chem.* **1996**, 100, 4160-4166.

Longenberger, L.; Mills, G., Formation of Metal Particles in Aqueous Solutions by Reactions of Metal Complexes with Polymers. *J. Phys. Chem.* **1995**, 99, 475-478.

Maye, M. M.; Zheng, W.; Leibowitz, F. L.; Ly, N. K.; Zhong, C. J., Heating-Induced Evolution of Thiolate-Encapsulated Gold Nanoparticles: A Strategy for Size and Shape Manipulations. *Langmuir* **2000**, 16, 490-497.

Metraux, G. S.; Cao, Y. C.;. Jin, R; Mirkin, C. A., Triangular Nanoframes Made of Gold and Silver. *Nano Lett.* **2003**, 3, 519-522.

Nikoobakht, B.; El-Sayed, M. A., Preparation and Growth Mechanism of Gold Nanorods (NRs) Using Seed-Mediated Growth Method. *Chem. Mater.* **2003**, 15, 1957-1962.

Pei, L.; Mori, K.; Adachi, M., Formation Process of Two-Dimensional Networked Gold Nanowires by Citrate Reduction of $AuCl_4^-$ and the Shape Stabilization. *Langmuir* **2004**, 20, 7837-7843.

Porel, S.; Singh, S.; Radhakrishnan, T. P., Polygonal gold nanoplates in a polymer matrix. *Chem. Commun.* **2005**, 2387-2389.

Simakin, A. V.; Voronov, V. V.; Shafeev, G. A.; Brayner, R.; Bozon-Verduraz, F., Nanodisks of Au and Ag produced by laser ablation in liquid environment. *Chem. Phys. Lett.* **2001**, 348, 182-186..

Sosa I. O.; Noguez C.; Barrera R. G., Optical Properties of Metal Nanoparticles with Arbitrary Shapes. *J. Phys. Chem. B* **2003**, 107, 6269-6275.

Sun, X.; Dong, S.; Wang, E., High-yield synthesis of large single-crystalline gold nanoplates through a polyamine process. *langmuir* **2005**, 21, 4710-4712.

Sun, X.; Dong, S.; Wang, E., Large-scale synthesis of micrometer-scale single-crystalline Au plates of nanometer thickness by a wet-chemical route. *Angew Chem Int Ed Engl* **2004**, 46, 6360-6363.

Sun, Y.; Xia, Y., Triangular Nanoplates of Silver: Synthesis, Characterization, and Their Use as Sacrificial Templates in Generating Triangular Nanorings of Gold. *Adv. Mater.* **2003**, 15, 695-699.

Umar, A. A.; Oyama, M., Formation of Gold Nanoplates on Indium Tin Oxide Surface: Two-Dimensional Crystal Growth from Gold Nanoseed Particles in the Presence of Poly(vinylpyrrolidone). *Crystal Growth & Design* **2006**, 6, 818-821.

Wang, L.; Chen, X.; Zhan, J.; Chai, Y.; Yang, C.; Xu, L.; Zhuang. W.; Jing. B., Synthesis of gold nano- and microplates in hexagonal liquid crystals. *J. Phys. Chem. B* **2005**, 109, 3189-3194.

Yener, D. O.; Sindel. J.; Randall, C. A.; Adair, J. H., Synthesis of Nanosized Silver Platelets in Octylamine-Water Bilayer Systems. *Langmuir* **2002**, 18, 8692-8699.

Yu, Y.; Chang, S.; Lee, C.; Wang, C. R. C., Gold Nanorods: Electrochemical Synthesis and Optical Properties. *J. Phys. Chem. B* **1997**, 101, 6661-6664.

Dynamical Tests on Fiber Optic Data Taken from the Riser Section of a Circulating Fluidized Bed

Emily M. Taylor, Christopher P. Guenther, Ronald W. Breault, US Department of Energy, National Energy Technology Laboratory, Morgantown, WV 26507, USA

Abstract

Dynamical tests have been applied to fiber optic data taken from a cold-flow circulating fluidized bed to characterize flow conditions, identify three time and/or length scales (macro, meso, and micro), and understand the contribution these scales have on the raw data. The characteristic variable analyzed is the raw voltage signal obtained from a fiber-optic probe taken at various axial and radial positions under different loading conditions so that different flow regimes could be attained. These experiments were carried out with the bed material of 812 μm cork particles. The characterization was accomplished through analysis of the distribution of the signal through the third and fourth moments of skewness and excess kurtosis. A generalization of the autocorrelation function known as the average mutual information function was analyzed by examining the function's first minimum, identifying the point at which successive elements are no longer correlated. Further characterization was accomplished through the correlation dimension, a measure of the complexity of the attractor. Lastly, the amount of disorder of the system is described by a Kolmogorov-type entropy estimate. All six aforementioned tests were also implemented on ten levels of detail coefficients resulting from a discrete wavelet transformation of the same signal as used above. Through this analysis it is possible to identify and describe micro (particle level), meso (clustering or turbulence level), and macro (physical or dimensional level) length scales even though some literature considers these scales inseparable [6]. This investigation also used detail wavelet coefficients in conjunction with ANOVA analysis to show which scales have the most impact on the raw signal resulting from local hydrodynamic conditions.

Introduction

Deterministic chaos theory has recently been used as a tool to analyze the dynamics within gas-solid systems, such as fluidized beds. This theory has been used for the purpose of examining scale-up effects [8, 12], identifying regime changes [3, 17, 18], characterizing fluidization [5], and determining how changes in conditions affect dynamics [4, 10].

Three scales can be identified within a fluidized bed: the macroscale representing the physical or global aspects of the unit, the mesoscale expressing the clusters and their interactions, and microscale of individual particle behavior. Bai et al.[1] and Ji et al. [11] examined the three scales by observing three separate time series collected from three separate probes. A single time series decomposed through wavelet decomposition into series of detail and approximation coefficient was utilized to identify the three scales by Zhao and Yang [20] and Wu et al. [16]. This method of decomposition allows information from different frequency components to be analyzed separately.

A majority of the current work have used deterministic chaos or wavelets techniques to analyze signals from a bubbling bed where the regime transitions and scales are well-defined. The objective of this work is to extend and add upon the characteristic techniques and regime and scale identification to a circulating fluidized bed, where these transitions are more subtle.

It was concluded by Zhao and Yang [20] and implemented by many others that characterization of a fluidized bed cannot be accomplished through a single test or approach. Therefore, multiple tests are being used in the current work on both the raw and wavelet data. In general, the wavelet data served to examine each scale independently. Combine this with the examination of the raw data and based on trends and ANOVA results, it can be concluded which scale most contributes to the changes seen within the raw data.

Experimental

The experimental unit is a cold flow circulating fluidized bed located at National Energy Technology Laboratory (NETL) in Morgantown, WV. It consists of a riser 15.3 m high and 0.3 m diameter and a 0.25 m diameter standpipe. The transfer leg is a loopseal with a 0.23 m diameter allowing the solids to enter the riser from a side port 0.27 m above the distributor. Cork with an average size of 812 μm was the test bed material. Solids circulation rate (Ms) and gas velocity (Ug) was varied independently; studied values are shown in table 1.

Table 1
Operating Conditions

Ug (m/s)	Ms (kg/h)	Load Ratio
4.6	907.2	0.62
5.5	2721.6	1.55
4.6	2721.6	1.86
3.8	2721.6	2.25
4.6	4535.9	3.10

Data was also obtained for three separate axial locations and five radial positions. Axial locations were chosen in an effort to characterize the acceleration zone (3.9m), the fully developed zone (8.3m), and deceleration zone (12.1m). Radial positions were chosen as the midpoint of five equal area circles within the riser. These positions will be referred to as P0, P1, P2, P3, and P4 and represent distances of 0.12, 0.07, 0.05, 0.03, and 0.01 m from the wall, respectively.

The data was acquired using a multi-fiber optical reflective probe developed by Vector Scientific instruments. It consists of two fiber bundles located on the same vertical line. Each bundle is composed of 150 randomly distributed of each transmitting and receiving fibers. A light emitting diode (LED) transmits light through the transmitting half of the fibers, where upon hitting particle(s) in the riser will be reflected back to the probe. The intensity of the reflected light depends on the concentration, size distribution, composition, and shape of the particles. The sampling rate was12.5 kHz and data was collected for 61.536 sec.

Calculations

Various statistics can be used to describe the distribution of a dataset. Higher order moments of skewness and excess kurtosis are of interest. Skewness is a measure of the asymmetry of the distribution and excess kurtosis quantifies the peakedness. As a point of reference, a Gaussian distribution has skewness and kurtosis values of zero.

Also of interest is the amount of dependence the signal has with itself, or amount of memory a signal possesses. The AMI function is a general measure of dependence that can detect both linear and non-linear correlations. Its first minimum represents the point at which successive elements are no longer correlated.

The two chaotic statistics used here are the correlation dimension which is a dimensionless number that represents the complexity of the signal, and the correlation entropy which is a measure of the predictability. Both values are calculated from the reconstructed attractor most commonly computed through Takens' delay embedding theorem [15]. The correlation dimension can be interpreted as the number of degrees of freedom that dominate the dynamic behavior of the system. Entropy can be interpreted as a quantitative measure of the amount of information needed to predict the value of an observation within a certain precision.

Statistics will also be computed for the series of coefficients resulting from a discrete wavelet transformation (DWT). This process involves passing the signal *f(x)* through a low and high pass filter multiple times to examine the different levels of detail of *f(x)* independently. The output of the low pass filter known as the approximation coefficients, a^j, form a smoother and smoother version of the raw signal. The detail lost by going to the next higher level is captured in the detail coefficients, d^j, output of the high-pass filter. Each filter and level of decomposition represents a different frequency band, given by

$$d^j : \left[2^{-(j+1)} f_s, 2^{-j} f_s\right] \qquad a^j : \left[0, 2^{-(j+1)} f_s\right] \quad (1)$$

where f_s represents the sampling frequency. Therefore, a reconstruction of the original signal is accomplished with the summation of the detail coefficients at all levels and the highest level of approximation. In this work the detail coefficients up to level $j = 10$ are considered. The Daubechies mother wavelet was utilized here for its simplicity, compact support, perfect reconstruction, orthogonality, highest number of vanishing moments, *L*, for a given support width and popularity within the literature [2, 4, 8, 9, 16, 20].

Scale Identification

A ten level DWT was performed on the raw signal to examine the different frequency bands and determine which scale, micro, meso, or macro, they represent. It is desired to determine which phenomena within the CFB these bands are able to characterize.

Figure 1 presents the trend of the first minimum at increasing scales for three randomly chosen combinations of independent variables. At the first half of the scales the first minimum value remains low. Once the sixth level is attained the amount of memory within the decomposed signal begins to significantly increase and steadily increases to the maximum scale of d10. This trend can be explained by the occurrence of a change within the CFB system will not be absorbed by the macro scale as quickly as the micro or meso scales. A greater number of time increments will then be necessary before elements are no longer correlated at the macroscale, leading to a larger value for the first minimum. Therefore, the examination of the first minimum of the AMI function indicates a boundary between the macro and meso scales with d6-d10 being identified as the macro scale.

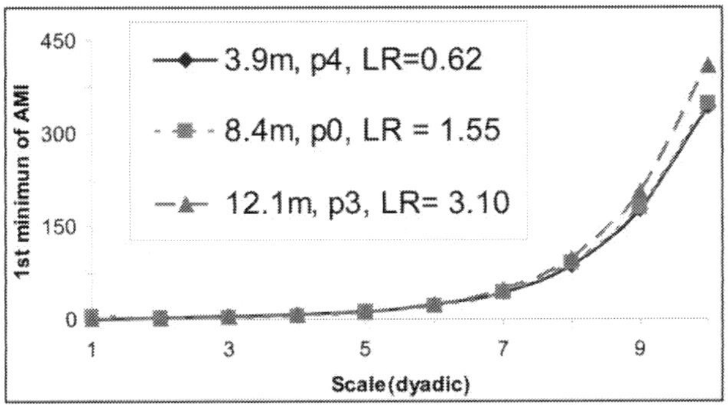

Figure 1-First minimum of the AMI function of the detail coefficient series (d1-d10) for three representative data sets

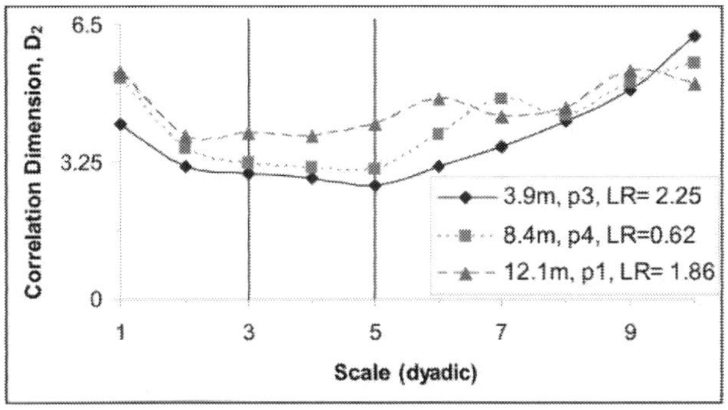

Figure 2- Correlation dimension, D_2, of the detail coefficient series (d1-d10) for three representative data sets.

A plot of the correlation dimension as a function of the dyadic scale for three representative time series can be seen in figure 2. For the lowest two scales, the dimension value decreases for all series. At the third detail level, the estimates level off and do not vary until d6 where the values begin to increase and continue to do so until d10. It has already been determined that the macro scale begins at d6, but the first shift in trends indicate that d1 and d2 represent the micro scale, or individual particle interactions, and d3-d5 characterizes the meso scale, or clusters. This is consistent with the view of larger changes in the high frequency fluctuations due to the random movement of particles which would correspond to higher dimensionality.

With a sampling frequency of 12.5 kHz and equation 1, it can be concluded the frequency band which corresponds to the mesoscale is (195 Hz, 1563 Hz). This implies that the clustering is occurring at time scales between 0.00064 s and 0.00512 s. Guenther and Breault [9] found clusters to occur within similar time scales for cork bed material.

Results and Discussion

The characterization of flow regimes, phenomena within scales, and the dominant contributing scale to the raw signal can be accomplished with the examination of a combination of tests on both the raw voltage signal and detail wavelet coefficients. Many plots will be presented as a group and will follow the order of the circles shown in figure 3. Here, changes in the solids circulation rate (Ms) are along the abscissa and the riser gas velocity (Ug) along the ordinate. The relative size of the circles in figure 3 represents load ratio. A dilute upflow regime was seen in the case with the least amount of loading and smallest Ms value. For Ms = 2721.6 kg/hr, the larger two values of gas velocity exhibit behavior within the core annular flow regime, and the lowest Ug value operates in the fast fluidization regime. The point with the largest amount of loading and the highest Ms can demonstrate characteristics within the fast fluidization and core annular regimes. The lines in figure 3 are shown to classify the regime in which the points studied in this work are operating and do not represent boundaries for the flow regimes of values not plotted here. These divisions were suggested by the solids velocity in [21] and riser pressure profiles in [13]. Since the heaviest loaded case has been known to show behavior within both core annular and fast fluidization, the boundary can be represented by either the solid or dashed line.

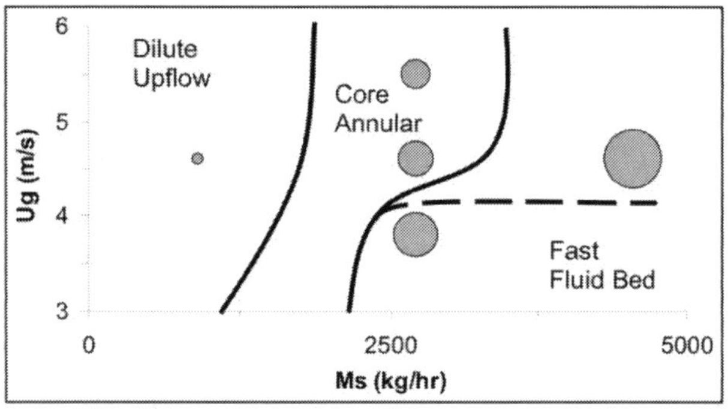

Figure 3- The structure of most groups of graphs in this section.

For the raw voltage signal, the first minimum of the AMI function is plotted against radial position at varying axial locations in figure 4. The trend for most cases is a steady increase from the center (p0) to the wall (p4). Since it takes more time steps for elements to become uncorrelated, this implies that there is more memory in the raw data at the wall. This increase becomes more pronounced as the load ratio increases. Dilute upflow has the least amount of memory overall. The discrepancy between the lower elevation (3.9 m) and the other axial locations is largest while operating at the highest two load ratios. A characterization of the fast fluidization is the "S" shaped pressure profile which is indicative of and exhibits "classical Yerushalmi [19] behavior" in the dense region at the bottom of the riser. Figure 4 indicates that the two heaviest loaded cases are within the fast fluidization regime. The two highest load ratios also show that fast fluidization and core-annular can exist simultaneously in the riser at two different elevations which explains the ambiguity of the dashed line in figure 3. By examining the incremental differential pressure profiles, this transition is not as obvious, providing an advantage of the AMI over incremental pressure measurements. If the correlation dimension, D_2, is plotted in the same manner, it can be seen that the complexity of the attractor is also greatest at the wall of the riser. The transition between fast fluid and core annular is also present, but not as well-defined.

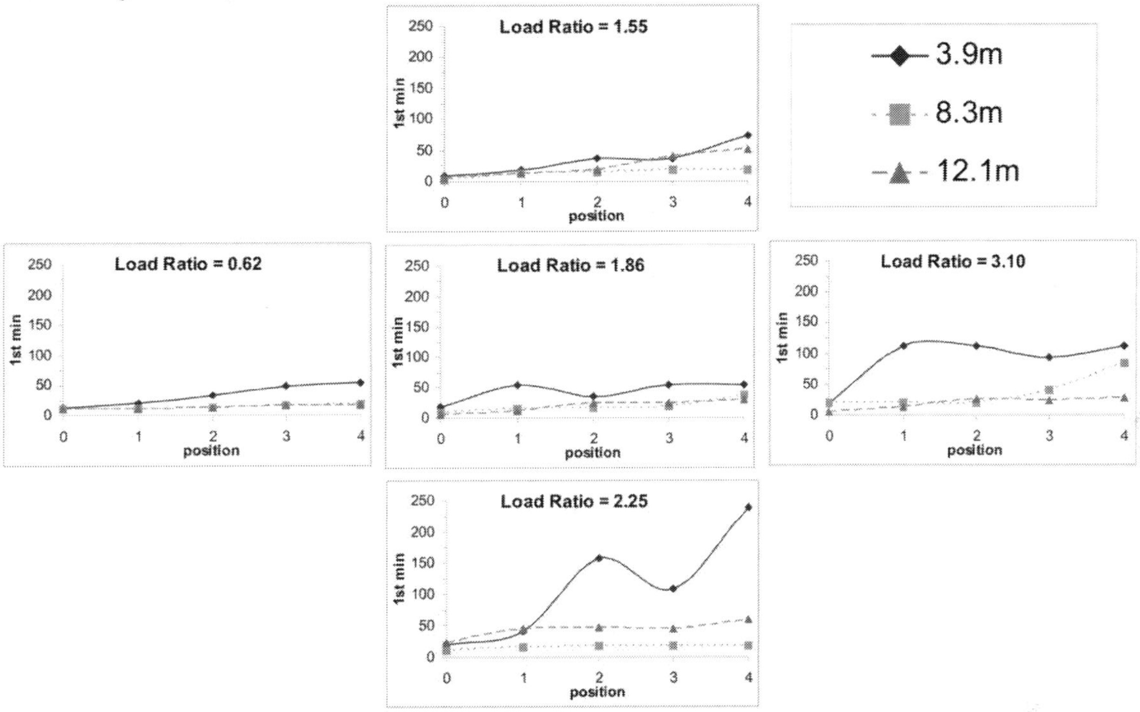

Figure 4- 1st min of the AMI function versus radial position for the raw voltage data at three axial locations.

Figure 5 characterizes the distribution of the raw signal with excess kurtosis by plotting it against the radial position at each axial location with a different graph for each load ratio. A similar analysis was also performed for the skewness, but because of space limitations not shown here. None of the distributions are statistically equivalent to the Gaussian distribution with skewness and excess kurtosis values of 0 using the standard error of skewness of $SES = \sqrt{\frac{6}{n}}$ and kurtosis of $SEK = \sqrt{\frac{24}{n}}$ defined in [14]. All cases are positively skewed i.e. have a longer tail in the positive direction because the fiber optic probe measures the intensity of reflected light and this intensity is low due to the frequently passing voids. A majority of the distributions are also leptokurtic (kurtosis > 0) which can be seen in figure 5. At the heavier loaded cases, more platykurtic (kurtosis < 0) distributions occur. The trend is for the distribution to approach Gaussian closer to the wall in relation to both the skewness and kurtosis. One exception to this trend is the dilute upflow case (LR = 0.62) which remains consistently positively skewed and leptokurtic across the radius. This boundary for dilute upflow is also present in the skewness plots; however, not as pronounced as that seen for the kurtosis.

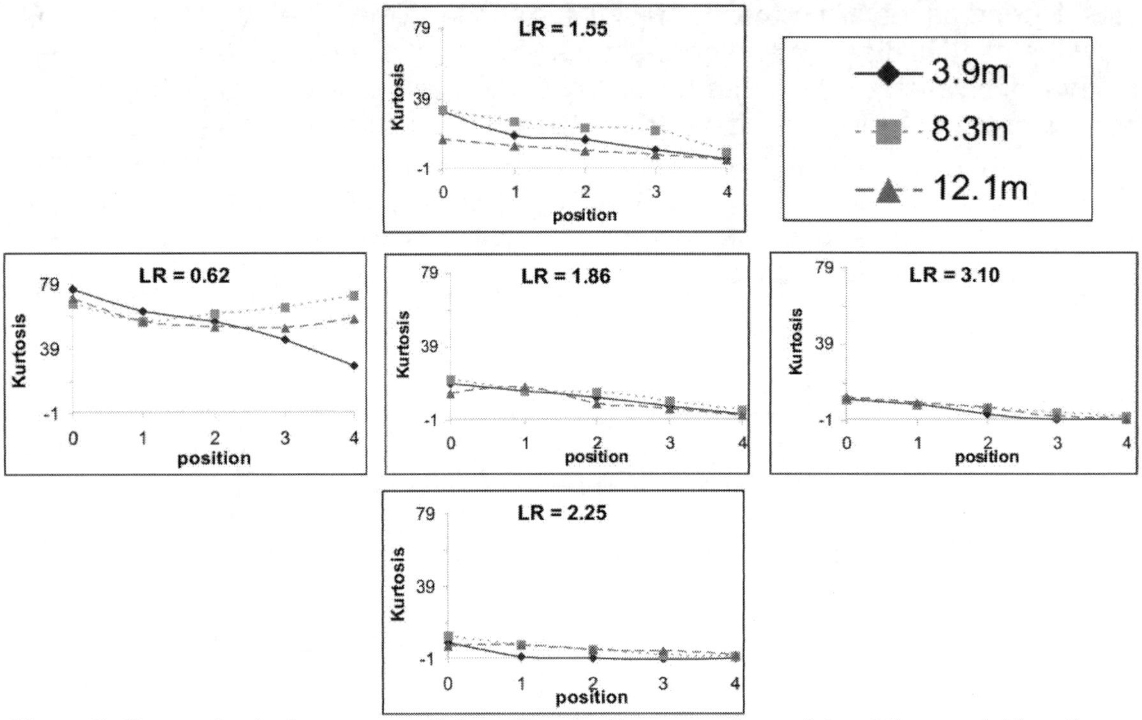

Figure 5- Excess kurtosis versus radial position for the raw voltage data at three axial locations.

Since the detail coefficients that relate to micro, meso, and macro scales have been identified, it is possible to analyze each scale independently by averaging over the corresponding detail levels (i.e. d1 and d2 for micro, d3-d5 for meso, and d6-d10 for macro). The result of doing this for the excess kurtosis for at 8.3m are plotted against the radial position at varying scales with a separate plot for each load ratio in figure 6. High kurtosis values of detail coefficients can indicate intermittency caused by high frequency fluctuations from particle movements [2]. The fact that the microscale is always the most intermittent is further indication that this scale is capturing particle behavior. For the micro and meso scales, as the load ratio increases the kurtosis value decreases with a considerable decrease going from the lowest load ratio to all others indicating that detail wavelet coefficients at these scales can be used to identify a boundary for dilute upflow. To further quantify these results a separate ANOVA was performed at each scale for the data at all three elevations with table 2 indicating which independent variables were significant. The value in parenthesis of each heading indicates the adjusted R^2 value, a measure of the amount of variability explained by the model. Since the

Table 2

ANOVA for response variable of excess kurtosis.

Scale (Adjusted R^2)	micro (.42)	meso (.70)	macro (.14)	raw (.95)
Axial	x			
Ug		x	x	x
Ms	x	x	x	x
Radial	x	x		x
Axial*Ug				
Axial*Ms				
Axial*Radial				x
Ug*Radial				
Ms*Radial			x	
Axial^2				x
Ug^2				
Ms^2	x	x		x
Radial^2				

*x indicates significant at a 0.01 level of significance.

macroscale did not pick up on the regime transition and the model is such a poor fit (R_a^2 = 0.14), nothing can be concluded about the macroscale from examination of the kurtosis. In comparing the significant terms of the ANOVAs in table 2 of each scale to that of the raw, it can be seen that the scale that has the greatest influence on the excess kurtosis of the raw data is the mesoscale.

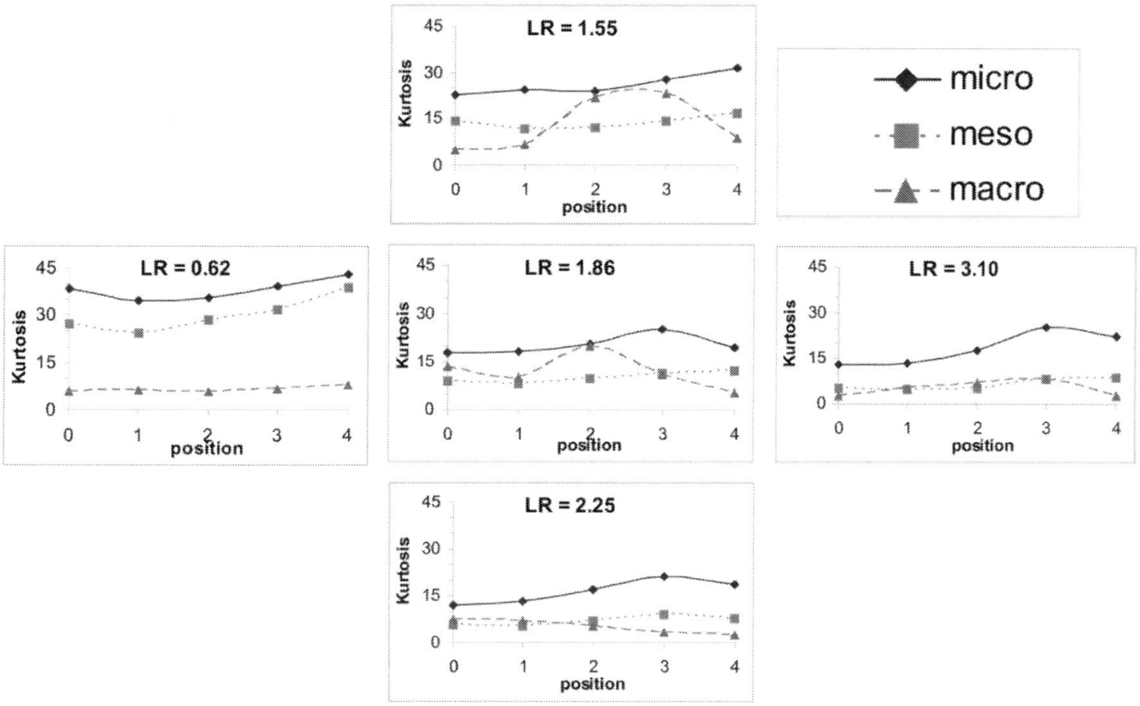

Figure 6 - Average excess kurtosis values versus radial position at the 8.3m elevation for each scale identified with wavelet analysis.

Table 3
ANOVA for response variable of correlation dimension.

Scale (Adjusted R^2)	micro (.19)	meso (.75)	macro (.18)	raw (.80)
Axial	x	x		
Ug		x		x
Ms		x		x
Radial		x	x	x
Axial*Ug				
Axial*Ms			x	
Axial*Radial				x
Ug*Radial			x	
Ms*Radial				
Axial^2				
Ug^2		x		x
Ms^2				
Radial^2	x	x	x	x

*x indicates significant at 0.01 level of significance

Correlation entropy, h_2, as a function of radial position can be seen for each scale at the 8.3m elevation in figure 7. For almost every set of points, the mesoscale has the lowest entropy value. In terms of all other independent variables, there is no noticeable trend from these graphs. This implies that the meso scale has the lowest rate of information loss, or the most memory. For the raw data, it was shown in figure 4 that the first minimum of the AMI function was greatest at the wall and thus the amount of memory increased towards this wall. This implies that the mesoscale has more influence at the wall than at the center. Furthermore, from table 3 it can be concluded that since the model with the response variable of correlation dimension for the raw data most closely agrees with the mesoscale, it is this scale that is dominating the dynamics of the raw signal.

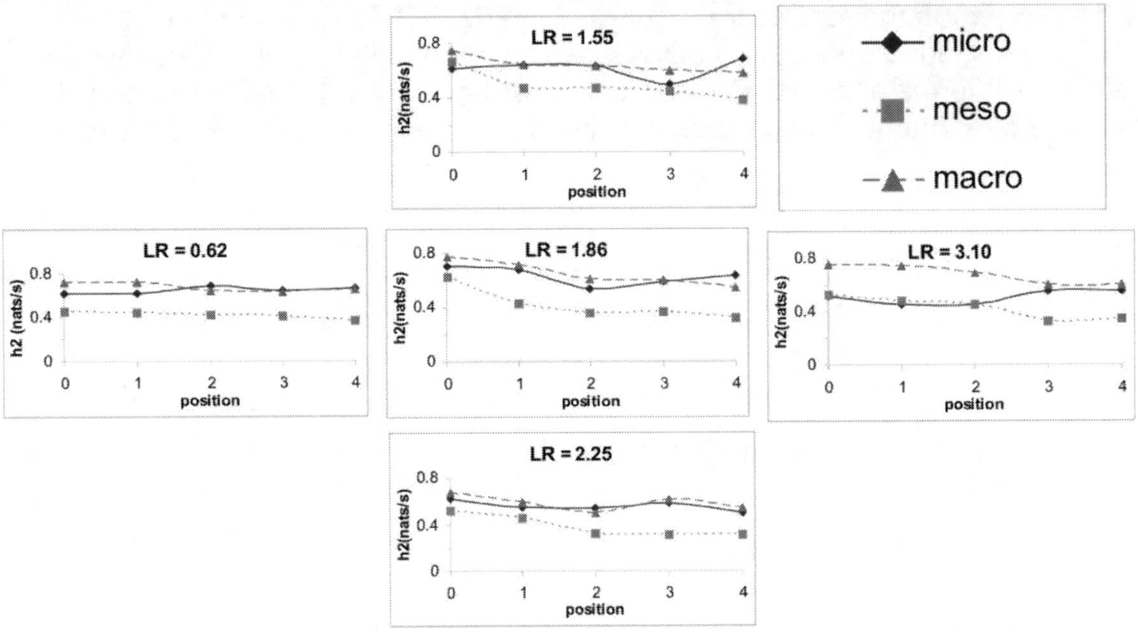

Figure 7- Average correlation entropy versus radial position at the 8.3m elevation for each scale identified with wavelet analysis.

Summary

It has been shown that the raw voltage signal obtained from a fiber optic vector probe has more memory, a more complex attractor, and is more closely related to the Gaussian distribution at the wall of the riser of a CFB when compared to data at the center. These trends become more pronounced for increased load ratio values. Through the analysis of the first minimum of the AMI function, correlation dimension, skewness, and excess kurtosis of the raw signal, boundaries for the flow regimes of dilute upflow to core annular and core annular to fast fluid bed can be identified.

The wavelet decomposition of the raw signal allows for the examination of the micro, meso, and macro scales independently. From this it can be seen that the mesoscale has the greatest influence on the intermittency and dynamics of the raw signal. Also, the influence of this scale is greater at the wall than at the center of the riser.

Acknowledgements

This research was supported in part by an appointment to the National Energy Technology Laboratory Research Participation Program, sponsored by the US Department of Energy and administered by the Oak Ridge Institute for Science and Education.

References

1. Bai, D., H. T. Bi, and J. R. Grace, (1997a). Chaotic Behavior of Fluidized Beds Based on Pressure and Voidage Fluctuations. American Institute of Chemical Engineers Journal, 43, 1357-1361.
2. Briens, L. A., and N. Ellis, (2005). Hydrodynamics of Three-Phase Fluidized Bed Systems Examined by Statistical, Fractal, Chaos and Wavelet Analysis Methods. Chemical Engineering Science, 60, 6094-6106.
3. Cabrejos, F. J., and G. E. Klinzing, (1995). Characterization of Dilute Gas-Solids Flow using the Rescaled Range Analysis. Powder Technology, 84, 139-156.

4. Ellis, N., L. A. Briens, J. R. Grace, H. T. Bi, and C. J. Lim, (2003). Characterization of Dynamic Behavior in Gas-Solid Turbulent Fluidized Bed Using Chaos and Wavelet Analyses. Chemical Engineering Science, 96, 105-116.
5. Fuller, T. A., T. J. Flynn, C. S. Daw, and J. S. Halow, (1993). Interpretation of Pilot-Scale, Fluidized Bed Behavior using Chaotic Time Series Analysis. Proceedings of the 12th International Conference on Fluidized Bed Combustion (pp. 141-153).
6. Goldhirsh, I. (1999). Scales and Kinetics of Granular Flows. Chaos, 9, 659-672.
7. Grassberger, P., and I. Procaccia, (1983). Characterization of Strange Attractors. Physical Review Letters, 50, 346-349.
8. Gu, L.-L., and J.-X. Zhu. Study on Flow Microstructure and Scale up Effect in Circulating Fluidized Bed Risers Using Solids Concentration Signals. Powder Technology, manuscript submitted for publication.
9. Guenther, C., and R. Breault, (2007). Wavelet Analysis to Characterize Cluster Dynamics in a Circulating Fluidized Bed. Powder Technology, 173, 163-173.
10. Hay, J. M., B. H. Nelson, C. L. Briens, and M. A. Bergougnou, (1995). The Calculation of the Characteristics of a Chaotic Attractor in a Gas-Solid Fluidized Bed. Chemical Engineering Science, 50, 373-380.
11. Ji, H., H. Ohara, K. Kuramoto, A. Tsutsumi, K. Yoshida, and T. Hirama, (2000). Nonlinear Dynamics of Gas-Solid Circulating Fluidized-Bed System. Chemical Engineering Science, 55, 403-410.
12. Li, J., and M. Kwauk, (2003). Exploring Complex Systems in Chemical Engineering-The Multi-Scale Methodology. Chemical Engineering Science., 58, 521-535.
13. Monazam, E. R., L. J. Shadle, J. S. Mei, and J. Spenik, (2005). Identification and Characteristics of Different Flow Regimes in a Circulating Fluidized Bed. Powder Technology, 155, 17-25.
14. Tabachnick, B.G., and L. S. Fidell, (1996). Using Multivariate Statistics (3rd ed). New York: Harper Collins.
15. Takens, F. (1981). Detecting strange attractors in Turbulence. In D. A. Rand, & L.-S. Young (Eds.), Lecture notes in mathematics, Dynamical Systems and Turbulence, 898, 366-381. Berlin: Springer.
16. Wu, B., A. Kantzas, C. T. Bellehumeur, Z. He, and S. Kryuchkov, (2007). Multiresolution Analysis of Pressure Fluctuations in a Gas-Solid Fluidized Bed: Application to Glass Beads and Polyethylene Powder Systems. Chemical Engineering Science, 131, 23-33.
17. van den Bleek, C. M., M. O. Coppens, and J. C. Schouten, (2002). Application of Chaos Analysis to Multiphase Reactors. Chemical Engineering Science, 57, 4763-4778.
18. van den Bleek, C. M., and J. C. Schouten, (1993). Deterministic Chaos: A New Tool in Fluidized Bed Design and Operation. The Chemical Engineering Journal, 53, 75-87.
19. Yerushalmi, J., D. H. Turner, and A. M. Squires (1976). The Fast Fluidized Bed. Industrial and Engineering Chemistry Process Design and Development, 15, 47-53.
20. Zhao, G. B., and Y. R. Yang, (2003). Multiscale Resolution of Fluidized-Bed Pressure Fluctuations. American Institute of Chemical Engineers Journal, 49, 869-882.
21. Zinn, A., E. Monazam, J. Spenik, J. Ludlow, (2006). REM Laser Probe Contract Results and Summary. DOE/NETL-2006/1229.

Distinctions between Circulating-turbulent Fluidized Beds, Turbulent Fluidized Beds and High-Density Circulating Fluidized Beds

Xiaobo Qi, Haiyan Zhu and Jesse Zhu
Department of Chemical and Biochemical Engineering
The University of Western Ontario, London, Ontario, Canada, N6A 5B9

Abstract: Flow characteristics in a novel circulating-turbulent fluidized bed (C-TFB) was further investigated. The present work focuses on the hydrodynamic distinctions between the flow regime of turbulent fluidization, fast fluidization, dense suspension upflow and newly proposed circulating-turbulent fluidization. The experimental results show that the circulating-turbulent fluidization is a new flow regime, independent of turbulent fluidization, fast fluidization and dense suspension upflow. As a result, the gas-solid flow in the C-TFB has its inherent characteristics, different from those in CFB, TFB and HDCFB reactors.

Introduction

Gas-solid fluidized beds have been widely applied in various industrial processes involving gas-solid contact including combustion, gasification, catalytic cracking, calcinations, drying, etc, due to their excellent gas-solid contacting, mixing and heat/mass transfer characteristics [1-4]. Good understanding of gas/solids distribution and movement in fluidized beds is essential for successful design and operation of fluidized bed reactors. It is well known that gas-solid fluidized beds may operate in the following flow regimes: particulate fluidization, bubbling (slugging) fluidization, turbulent fluidization, fast fluidization, dense suspension upflow and pneumatic transport. While each of these flow regimes has applications in industrial practice, most key commercial gas-solid fluidized bed reactors operate in the flow regime of turbulent fluidization, fast fluidization and dense suspension upflow due to their favorable gas-solid contacting, mixing and transfer characteristics [5]. Despite their advantages such as the high throughput per unit reactor volume and independent gas/solids flow rate control in circulating fluidized beds (CFBs), the vigorous gas-solid contacting, favorable bed-to-surface heat transfer, and high solids concentration (typically 25-35% by volume) in turbulent fluidized beds (TFBs), and the high solids concentration and no backmixing in the wall region in the high-density circulating fluidized beds (HDCFBs), some inherent shortcomings of the extremely non-uniform axial and radial gas/solids distribution and low solids concentration (usually less than 10%) in CFBs, the intense axial solids backmixing near the wall in TFBs, and the significant gradient in radial solids concentration profile in HDCFBs limit or hinder further improving these reactor performances with better gas-solid contacting efficiency and higher conversion per unit volume [5-10]. The non-uniform axial and radial flow structure and the relatively low solids concentration result in a key disadvantage, the serious gas by-passing through the core dilute region and extensive backmixing of solids in the wall region, consequently leading to low gas-solid contact efficiency and poor selectivity of chemical reactions [11]. Although the gas and solids phases in TFBs have much more uniform radial distributions than in CFBs, serious backmixing of both gas and solids phases leads to a broad residence time distribution of the two phases, resulting in poor chemical reaction selectivity in TFBs. Due to high superficial gas velocity (usually 6-10 m/s), the non-uniform radial gas and solids distribution, disadvantage of CFBs, to some extent, exists in HDCFBs as well [8], consequently causing a non-uniform residence time of the gas and solids phase.

Taking all favorable and unfavorable characteristics of CFBs, TFBs and HDCFBs into consideration, Zhu and Zhu [12] proposed a novel circulating-turbulent fluidized bed (C-TFB), which operates in a special mode with low superficial gas velocities and high solids circulation rates, resulting in a highly dense suspension and uniform axial flow structure. Their C-TFB design and operation is developed especially to overcome those disadvantages encountered in CFB, TFB and HDCFB reactors, while integrating the advantages of them, such as the high solids concentration, excellent gas-solid contacting and mixing, as well as largely reduced axial solids back-mixing with high solids circulation rates. The preliminary study of Zhu and Zhu [12] has shown that the high density gas-solid suspension, having solids volume concentration 25-30%, could be maintained throughout the entire C-TFB and that the C-TFB also exhibited an almost homogeneous axial flow structure with no net downflow of solids across the whole column section under all operating conditions. Based on the preliminary experimental results in the C-TFB, Zhu and Zhu [13] compared the macro hydrodynamics in the C-TFB, TFB and CFB. Their comparison has shown that flow behaviors in the C-TFB are different from both TFB and the bottom region of a high flux CFB and that the C-TFB maintains the advantages of TFB and HDCFB but overcome their disadvantages.

However, deeply investigation of the newly proposed circulating-turbulent fluidization is of considerable academic interest, helping to understand the circulating-turbulent fluidization regime and to develop a theoretical model for design and scale-up purposes. Also, the flow regime demarcation between turbulent, fast fluidization, dense suspension upflow and the newly proposed circulating-turbulent fluidization needs to be clarified to design and operate the possible circulating turbulent fluidized bed reactors. To this end, the present work focuses on the micro hydrodynamic distinctions between the flow regime of turbulent fluidization, fast fluidization, dense suspension upflow and circulating-turbulent fluidization. A novel multi-functional optical fiber probe was used to measure the dynamic local solids concentration and particle velocity simultaneously in a circulating-turbulent fluidized bed, a conventional turbulent fluidized bed and a high-density circulating fluidized bed riser.

Experimental

All experiments were conducted in a combined fluidized bed system, schematically shown in Figure 1. The system includes two fluidized beds, the right-hand fluidized bed serves as the newly proposed circulating-turbulent fluidized bed and a conventional turbulent fluidized bed, and the left-hand fluidized bed serves as a conventional circulating fluidized bed and a high-density circulating fluidized bed. A common downcomer with a diameter of 0.3 m and a height of 8.0 m was used for solids storage and return for both fluidized beds. Total solids inventory in the downcomer was about 420 kg, equivalent to a height of approximately 6.0 m. This high solids level ensures high pressure head in the downcomer and enables high solids circulation rates and high suspension density in the two fluidized beds. The C-TFB column shown on the right consists of two parts: 1) a C-TFB column at the bottom, with i.d. of 0.1 m and a 3.6 m height; 2) an upper section above the C-TFB column with a diameter of 0.2 m and a total height of 6.4 m for quick solids discharging. Secondary air supply added at the bottom of the discharging section can entrain particles upward as soon as possible. The secondary air velocity was kept about 5.0 m/s. The air flow rates were independently controlled in each column to allow the flexible operation over a wide range of gas velocities and solids circulation rates. For the operation in the conventional TFB, the static bed height was kept at 1.8 m. The conventional CFB and high-density CFB operation were conducted in the column on the left, with inner diameter of 0.076 m and a height of 10 m. More details about the fluidized bed system can be found in Zhu and Zhu [12].

In order to minimize the electrostatics found in the two columns, the whole fluidized bed system was electrically grounded. At the same time, a small stream of steam was introduced into the main air pipeline to humidify the fluidization air to a relative humidity of 50-60%. This has been shown to be very effective to control the electrostatic effects and the capillary forces [14].

The fluidization gas used in this study was air at ambient temperature and pressure, supplied by a Roots-type blower. Particles used in this work were FCC catalysts (d_p=65 µm, ρ_p=1780 kg/m^3). A series of orifice plates were employed to measure the gas flow rates. A solids circulation rate measuring device is located at the top of the downcomer sectioning the column into two halves with a central vertical plate and two half butterfly valves fixed at the top and the bottom of the two-half section. By appropriately flipping over the two valves from one side to the other, solids circulated through the system can be accumulated on one side of the measuring section for a given time period and then the solids circulation rate can be obtained.

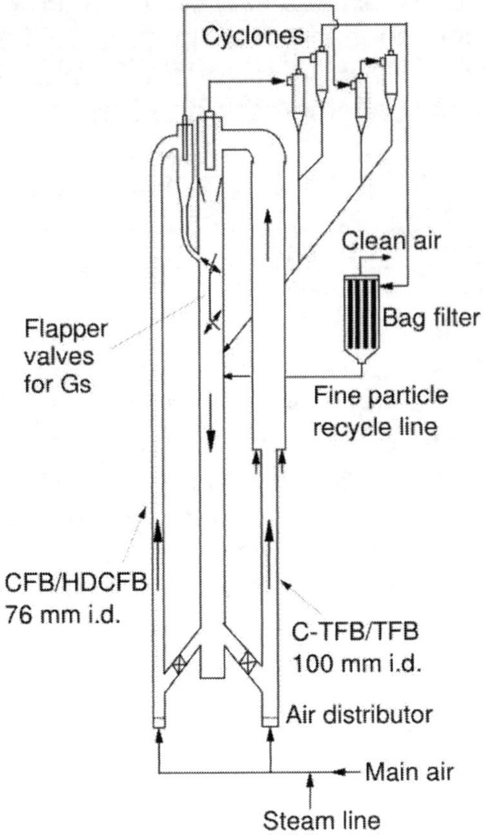

Figure 1. Experimental setup

Solids concentration and particle velocity were measured by an multi-functional optical fiber probe (Model PV5, newly developed by the Institute of Process Engineering, Chinese Academy of Sciences, Beijing, China) at 11 radial positions (r/R=0.00, 0.16, 0.38, 0.50, 0.59, 0.67, 0.74, 0.81, 0.87, 0.92 and 0.98) on two axial levels (z=0.8 and 1.5 m) for all four types of fluidized beds. This novel probe system is capable of measuring instantaneous local solids concentration and particle velocity simultaneously, and then the local instantaneous solids flux can thus be calculated from them. At each measurement location, raw data were sampled with a frequency of 50 kHz and a sampling time of 30 second, and all measurements were repeated five times. More details about the multi-functional optical fiber probe can be found in Zhu and Zhu [12].

Results and Discussion

Instantaneous Solids Concentration

Instantaneous local solids concentration and its fluctuations reflect the dynamic behavior of the interaction between gas and solids phase and then influence the gas-solid contacting efficiency and heat/mass transfer properties in fluidized beds [15]. To compare the turbulence of local gas-solid suspension flow in C-TFBs, TFBs and HDCFBs, Figure 2 plots instantaneous solids concentration traces obtained by the multi-functional optical fiber probe under typical turbulent, circulating-turbulent and high-density circulating fluidization conditions, respectively. Clearly, the transient solids concentrations in the HDCFB are much lower than those in the C-TFB and TFB except the radial position nearest the column wall and therefore the gas-solid flow dynamics in the HDCFB are quite distinct from those in the C-TFB and TFB. In the HDCFB, the most transient solids concentrations in the most area of the section ($r/R < 0.98$) are of low value, indicating that dilute gas-solid suspension dominate the most space and time in the HDCFB, but there are some higher solids concentrations peaks, which reflect the formation of clusters. The fluctuations of the solids concentration increases from the center to the wall. Moreover, in the column center, there is only dilute phase. With increasing radial position, more and more relatively dense clusters are formed, and the solids fluctuation becomes more vigorous.

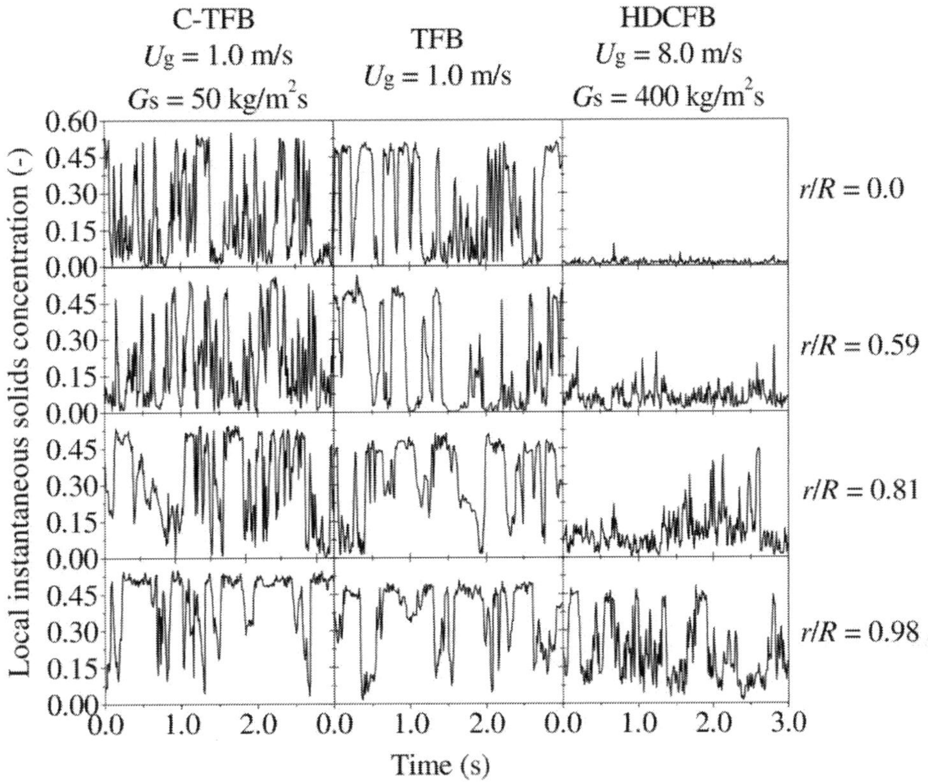

Figure 2. Instantaneous solids concentration in the C-TFB, TFB and HDCFB (z=0.8m)

Comparing the solids concentration traces in the C-TFB and TFB, one can find that the time series of instantaneous solids concentration in the C-TFB and TFB are somewhat similar but there are also some noticeable differences. In the central region of the column, the solids concentration is high and accompanied by high-amplitude fluctuations, different from the characteristics of the instantaneous solids concentrations in the same region of the HDCFB. Further outwards towards the column wall, the fluctuations in local solids concentration tend to be weaker and less frequent. And, the transient solids concentration and its fluctuations show that two phases clearly exist in the gas-solid flow of TFB. One phase could be termed as void/dilute phase with low solids concentration; the other phase is dense/cluster phase with high solids concentration.

Due to intenser turbulence and more frequent fluctuations in the local solids concentration in the C-TFB than those in the TFB, the distinction between the dense and dilute phases in the C-TFB is diffuse or even lost completely. The continuous distribution of transient solids concentration indicates that the dense phase in the C-TFB contained more gas than in the TFB, and more particles entered the void/dilute phase, and the clusters have a continuous variation in solids concentration, resulting in intenser fluctuations and stronger turbulence and better gas/solids contacting quality and thus reduced gas by-passing, which occurs in the core region of the HDCFB risers. This excellent contacting and mixing characteristic between the solids and gas phases at almost all radial positions is a great advantage of C-TFBs over TFBs, and HDCFBs.

To further compare the complexity of the gas-solid flow in the C-TFB and TFB, another parameter is used to characterize instantaneous solids concentration time series signals, cycle frequency, defined as the number of times per second that an instantaneous solids concentration crosses its time-averaged value. Higher cycle frequencies correspond to higher Kolmogorov entropies and thus to more complex dynamic gas-solid flow system in the fluidized beds [16]. Figure 3 shows the cycle frequencies in term of solids concentration in the C-TFB and TFB. In the near wall region (r/R > 0.81), the cycle frequencies in the C-TFB and TFB are very close, suggesting there is no marked difference between the turbulence of gas-solid suspension in the wall region of C-TFB and TFB. The cycle frequencies in the TFB change little with the radial position. However, the cycle frequencies in the C-TFB are seen to be a strong function of the radial position, with a maximum value of about 225 Hz at the column center and decreasing to about 50 Hz at the wall, indicating that the closer to the column center, the intenser the turbulence of gas-solid suspension. This suggests that the dynamic flow behaviors in the C-TFB are quite different from those in the TFB.

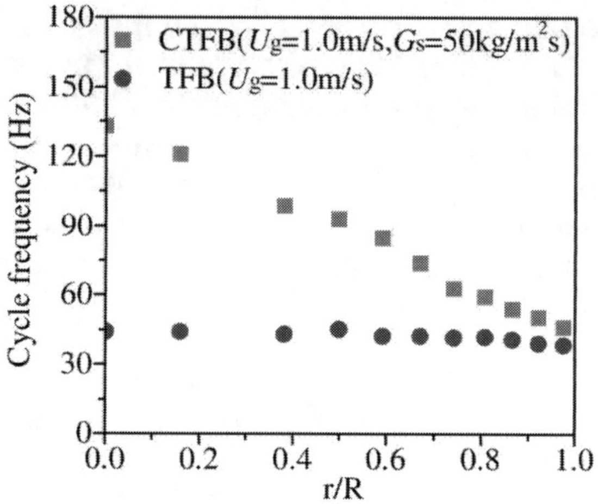

Figure 3. Radial profiles of cycle frequency in C-TFB and TFB on the same axial level (z=0.8m)

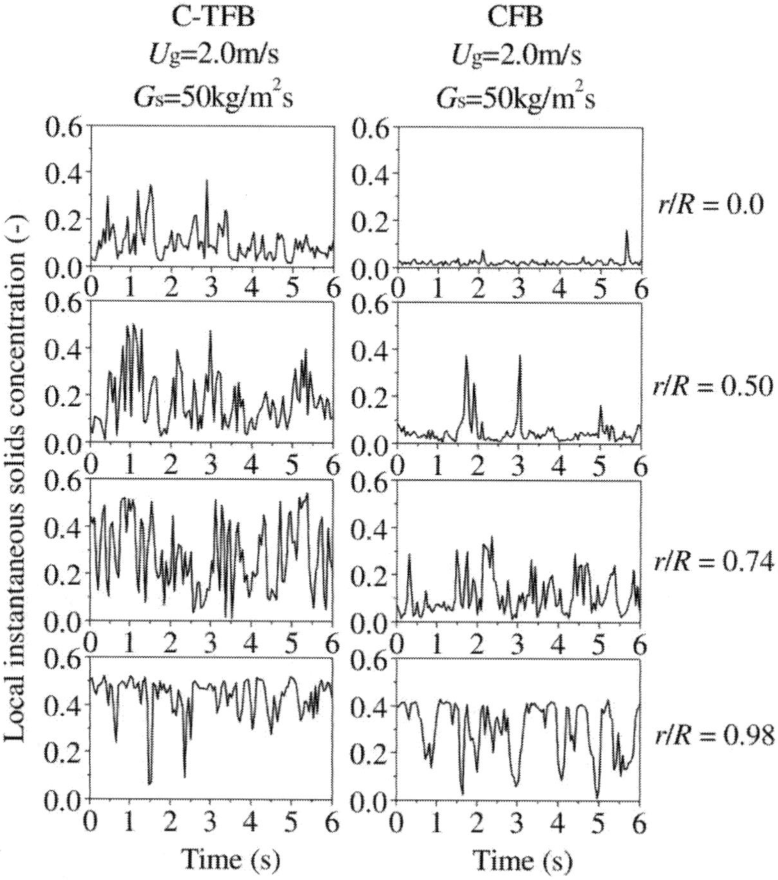

Figure 4. Instantaneous solids concentration in C-TFB and CFB (z=1.5m)

Figure 4 compares the instantaneous local solids concentrations in the CFB and C-TFB under the same operating condition. Clearly, an evident distinction can be observed between the C-TFB and TFB although the operating condition is the same. The broad and continuous spectrum of transient solids concentration in the C-TFB suggests that the gas and solids phases have a greatly improved mixing, and the clusters have a continuous variation in solids concentration. This is a key advantage of C-TFBs over CFBs.

Time-Mean Solids Concentration

Typical radial profiles of time-mean solids concentration on the same axial level of the C-TFB, TFB and HDCFB are presented in Figures 5(a) and 6(a). In the four fluidized beds, the time-mean solids concentrations all increase monotonically from the column center to the wall, but the radial solids concentration distribution in the C-TFB and TFB are much more uniform than that in the HDCFB and CFB. Obviously, the solids concentration and its radial distribution in the C-TFB are closer to those of the conventional TFB, different from those sharply demarcated 'core-annulus' flow structure with dilute core region and dense annulus region in the CFB and HDCFB. Since high solids volume concentration significantly increases interactions between particles, the contacting efficiencies between particles, and gas and solids phase in the center region of the C-TFB may be much higher than those in the CFB and HDCFB, leading to much higher heat/mass transfer rate in the center region of the C-TFB and the then more uniform radial distribution of contacting efficiency. At the same time, the high solids concentration in the C-TFB would also results in a higher effective viscosity of gas-solids suspension [8], thereby, imposing more shear on the descending particles and reducing the tendency for the solids downflow. As a result, there is no net solids back-mixing in the C-TFB [12], which lead to a reduction in the extent of axial dispersion of gas and solids.

Segregations of gas phase from solid phase can be reflected quantitatively by the intermittency index of gas-solid flow in term of the local solids concentration, which would be zero for perfect core-annulus flow and one for perfect cluster flow [17]. To compare the extent of the segregation between the gas and solids phase in the different flow regimes, radial profiles of intermittency index on the same axial level of the C-TFB, TFB, CFB and HDCFB are plotted in Figures 5(b) and 6(b). The comparison of the intermittency index profiles in the C-TFB, TFB, CFB and HDCFB shows some similarity. In general, all the radial profiles of intermittency index increase with radial position to a maximum at an intermediate radial position around r/R =0.6~0.8 and decrease thereafter toward the wall. A possible explanation could be that lower solids concentration in the center region and higher solids concentration in near wall region induce less fluctuations and turbulence (as shown in Figures 9 and 10) and then result in greater phase segregation. In spite of the above similarities, quite a difference happens in the intermittency index profiles between the C-TFB, TFB and CFB and HDCFB. First, the intermittency indexes in the C-TFB and TFB are much higher than those in the CFB and HDCFB across the whole column section, indicating that the extents of the segregation of the gas phase from the solids phase in the CFB and HDCFB are much greater than those in the C-TFB and TFB. Second, compared with C-TFB and TFB, the intermittency indexes in the core region of the CFB and HDCFB are much lower and thus the radial profiles of intermittency index in the CFB and HDCFB are more non-uniform than those in the C-TFB and TFB. The extremely non-uniform distribution of intermittency index in the CFB and HDCFB may be due to the significant gradient in radial distribution of local gas velocity, as high as 10 m/s in the center and as low as 0.6 m/s at the wall, coupled with dilute solids concentration in the core region and much denser solids concentration in the wall region. This serious radial separation of the gas phase from the solids phase in the CFB and HDCFB would result in the worse gas-solid interactions [18]. However, for the dense conditions, such as in the C-TFB and TFB, the relative low superficial gas velocities lead to a more uniform distribution of local gas velocity across the column section, and therefore a more uniform solids concentration, particle velocity and solids flux [12]. This hydrodynamic characteristic in the C-TFB would consequently improve reactor performance, such as more uniform gas and solids contact efficiency in radial direction and residence time distribution of gas and solids phase. Furthermore, it can also be noted from Figure 5(b) that the intermittency index and its radial profiles in the C-TFB and TFB are almost the same, suggesting an inherent similarity between these two fluidized beds.

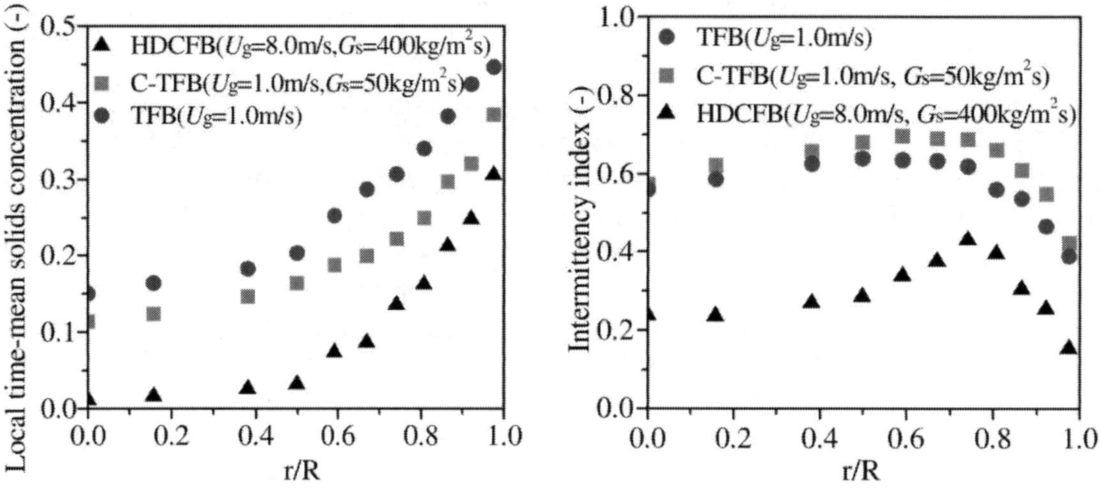

(a) (b)

Figure 5. Radial profiles of (a) time-mean solids concentration and (b) intermittency index in C-TFB, TFB and HDCFB (z=0.8m)

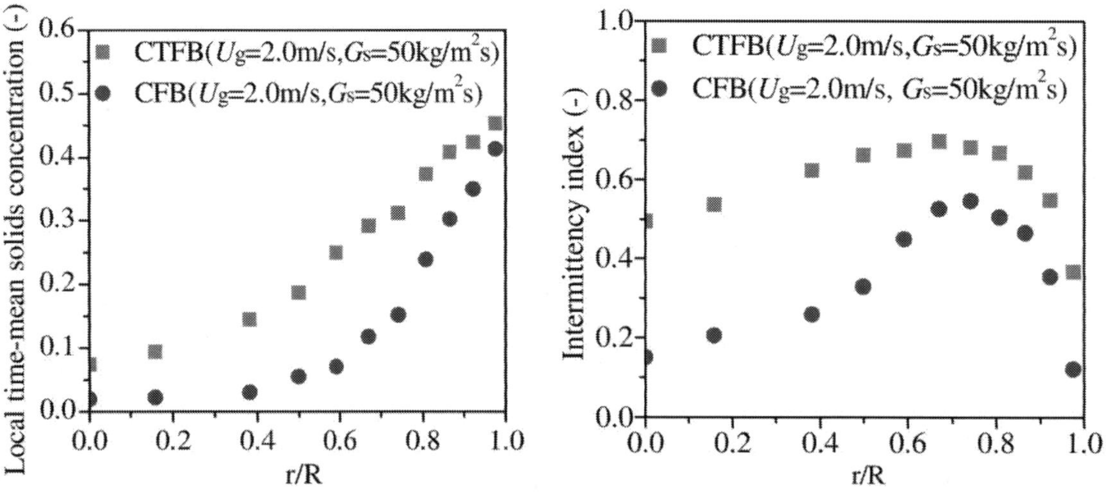

(a) (b)

Figure 6. Radial profiles of (a) time-mean solids concentration and (b) intermittency index in C-TFB and CFB (z=1.5m)

Relationship between Instantaneous Solids Concentration and Particle Velocity

Simultaneous measurements of instantaneous local solids concentrations and particle velocities could provide more detailed dynamic information on gas-solid suspension flow in fluidized beds. To investigate the relationship between the local instantaneous solids concentrations and particle velocities, Figure 7 plots the instantaneous particle velocities against the instantaneous solids concentrations determined simultaneously by our novel multi-functional optical probe at the same radial and axial position in the C-TFB, TFB and HDCFB. Clearly, a strong correlation can be observed for the gas-solid flow in the HDCFB, with higher solids concentrations usually corresponding to lower even negative velocities, while dilute gas-solid suspension mostly correspond to higher upflow particle velocities. Such a strong dependence suggests that the particle movements are mainly determined by gas-particle interaction in the HDCFB, where the particle-particle interactions (collisions) are weaker than the gas-particle interactions [19, 20]. However, for the gas-solid flow in the C-TFB and TFB, the instantaneous particle velocities and the corresponding solids concentrations appear to be uncorrelated. As shown in Figure 7, both solids concentrations and particle velocities all have a very wide distribution in the C-TFB and TFB. No matter what is the solids concentration, the particle velocities all vary from negative to positive values, indicating a random flow structure with intense fluctuations of solids concentrations and particle velocities in the C-TFB and TFB. Even for a very dilute gas-solid suspension flow, the particles still have a wide distribution of velocities. This suggests that in the dense conditions, such as in the C-TFB and TFB, particles motion is mainly controlled by the particle-particle interactions, while the gas-particle interactions have insignificant influence on the particle movements.

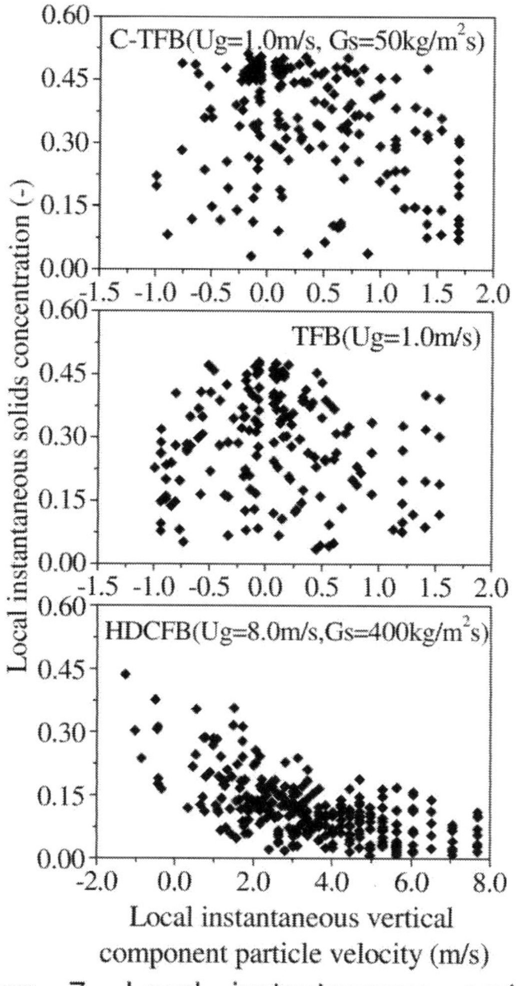

Figure 7. Local instantaneous particle velocity vs. concentration in C-TFB, TFB and HDCFB (z=0.8m, r/R=0.81)

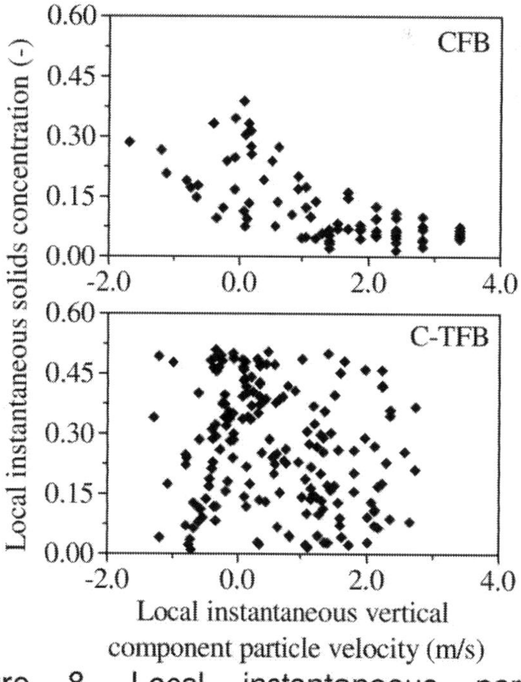

Figure 8. Local instantaneous particle velocity vs. concentration in C-TFB and CFB (U_g=2.0m/s, G_s=50kg/m^{2}s; z=1.5m; r/R=0.81)

Figure 8 displays the instantaneous particle velocities against the instantaneous solids concentrations determined simultaneously at the same radial and axial position in the C-TFB and CFB under the same operating condition. Similarly, the particle movements in the CFB are dominated by the gas-particle interactions, while for the gas-solid flow in the C-TFB, the particle-particle interactions predominated over the gas-particle interactions.

In fact, the same phenomena can also be observed on all other radial positions except the one nearest the column wall, due to the significant effects of the column wall on the gas-solid suspension flow. As a result, it could be concluded that the movements of particles in the CFB and HDCFB are controlled by the gas-particle interactions, the particle-particle interactions (collisions), however, dominate the movements of particles in the C-TFB and TFB. This further suggests that the characteristics of gas-solid flow in the C-TFB are essentially close to those in the TFB in this respect, but totally different from those in the CFB and HDCFB. Consequently, the flow regime of circulating-turbulent fluidization is a new flow regime, different from the turbulent fluidization, fast fluidization and dense suspension upflow.

Solids Flux Deviation

Fluctuations in local instantaneous solids flux reflect turbulence of local gas-solid suspension flow dynamics. Non-isokinetic sampling probes have been used to measure the radial distribution of vertical solids fluxes in CFBs by many researchers [21, 22]. But only time-averaged values can be obtained by this probe. Our newly developed multi-functional optical fiber probe is capable of measuring instantaneous solids fluxes and their fluctuations. The standard deviation of solids flux quantitatively reflects the turbulent intensity. To this end, Figures 9 and 10 compare the radial profiles the standard deviation of solids flux in the C-TFB, TFB, HDCFB and CFB on the same axial level, respectively. It can been seen from Figure 9 that the radial profiles of the solids flux standard deviation in the C-TFB, TFB and HDCFB all reach a maximum at some intermediate radial location between the column center and the wall, but the solids flux standard deviations and the radial positions to reach their maxima differ greatly in the C-TFB, TFB, and HDCFB. As shown in Figure 9, the solids flux standard deviations in the center region of the HDCFB are much lower than those in the C-TFB and TFB, while in the wall region, the solids flux standard deviations in the center region of the HDCFB are rather higher that those in the C-TFB and TFB. Consequently, the radial solids flux

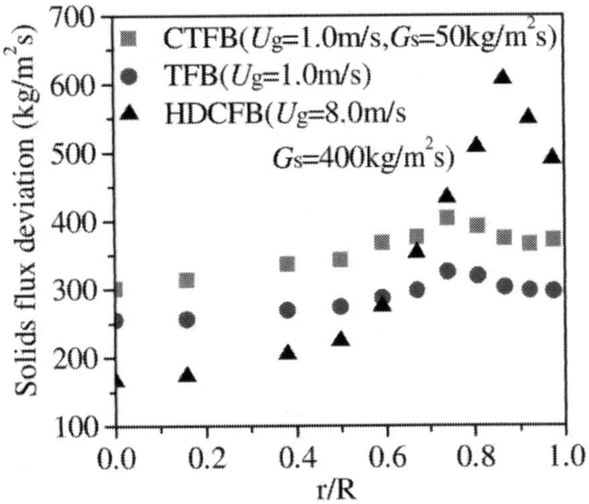

Figure 9. Solids flux deviation in C-TFB, TFB and HDCFB (*z*=0.8m)

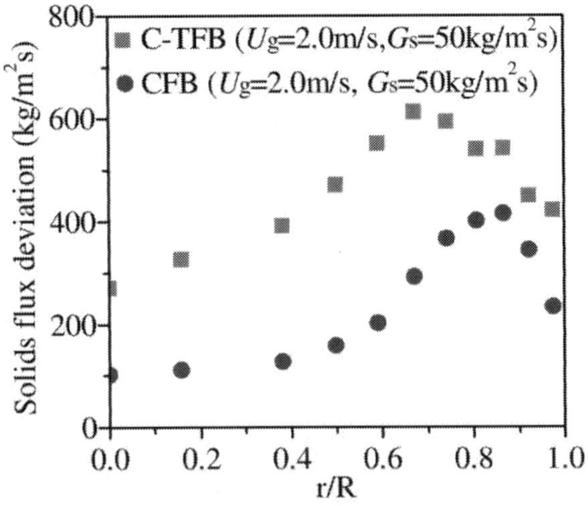

Figure 10. Solids flux fluctuation in C-TFB and CFB (*z*=1.5m)

standard deviation profile in the HDCFB is more non-uniform that those in the C-TFB and TFB, unfavorable for uniform gas/solids contacting and heat/mass transfer in radial direction. Furthermore, within the cross-section of the column, the solids flux standard deviations in the C-TFB are higher than those in the TFB, indicating intenser turbulence of gas-solid flow dynamics in the C-TFB.

As shown in Figure 10, under the same operating condition, the solids flux standard deviations in the C-TFB are remarkably higher than those in the CFB; furthermore, the radial distribution of the solids flux standard deviations in the C-TFB is also more uniform than that in the CFB. This indicates that the turbulence of gas-solid suspension flow in the C-TFB is much intenser than that in the CFB due to the higher solids concentration and then significantly increased particle-particle interactions in the C-TFB. The strong fluctuations in the C-TFB result in excellent contacting efficiency and heat/mass transfer.

Conclusions

This work presents further investigation into the flow characteristics in the novel C-TFB. The present work focuses on the characteristic distinctions between the flow regime of turbulent fluidization, fast fluidization, dense suspension upflow and newly proposed circulating-turbulent fluidization. The experimental results show that the circulating-turbulent fluidization is a new flow regime, independent of turbulent fluidization, fast fluidization and dense suspension upflow. As a result, the flow behavior in the C-TFB has its inherent characteristics, different from those in the CFB, TFB and HDCFB reactors. Preliminary study has shown that this novel C-TFB reactor is an attractive new reactor with high conversions and/or selectivities for rapid heat/mass transfer and good mixing of solids.

References

1. Kunii, D., and Levenspiel, O., (1991), Fluidization Engineering, Butterworth-Heinmann, London.
2. Grace, J.R. and Bi, H.T. (1997), "Introduction to circulating fluidized beds", Chapter 1 in *Circulating Fluidized Beds* (Grace, J.R. Avidan, A.A., Knowlton, T.M., eds.), Chapman & Hall, London, pp. 1-18.
3. Grace, J.R., Bi, H.T. and Golriz, M., (2003), "Circulating Fluidized Beds", Chapter 19 in *Handbook of Fluidization and Fluid-Particle Systems* (Yang, W.C. ed.), Marcel Dekker, New York, pp. 485-544.
4. Johnsson, F., (2007), "Fluidized bed combustion for clean energy", in: *Fluidization XII* (Bi, H.-T., Berruti, F. and Pugsley, T., eds.), Engineering Foundation, Vancouver, pp. 47-62.
5. Grace, J.R., (2000), "Reflections on turbulent fluidization and dense suspension upflow", *Powder Technology*, 113, 242-248.
6. Lim, K.S., Zhu, J.X., Grace, J.R., (1995), "Hydrodynamics of gas-solid fluidization", *International Journal of Multiphase Flow*, 21, 141-193.
7. Bi, H.T., Ellis, N., Abba, I. A., Grace, J.R., (2000), "A state-of-the-art review of gas-solid turbulent fluidization", *Chemical Engineering Science*, 55, 4789-4825.

8. Grace, J.R., Issangya, A.S., Bai, D.R., Bi, H.T., Zhu, J.X., (1999), "Situating the high-density circulating fluidized bed", *AIChE Journal*, 45, 2108-2116.
9. Du, B., Warsito, W., Fan, L.S., (2003), "Bed nonhomogeneity in turbulent gas-solid fluidization", *AIChE Journal*, 49, 1109-1126.
10. Zhu, J.X., (2005), "Circulating fluidized beds - recent developments and research requirements in the near future", in: *CFB VIII* (Cen, K., ed.), International Academic Publishers, Hangzhou, pp. 41-55.
11. Dry, R.J., Christensen, I.N., White, C.C., (1987), "Gas-solids contact efficiency in a High-velocity fluidised bed", *Powder Technology*, 52, 243-250.
12. Zhu, H.Y. and Zhu, J.X., (2007), "Gas-solids flow structures in a novel circulating-turbulent fluidized bed", submitted to *AIChE Journal*.
13. Zhu, H.Y. and Zhu, J.X., (2007b), "Comparative study of flow structures in a circulating-turbulent fluidized bed", submitted to *Chemical Engineering Science*.
14. Park, A., Bi, H.T. and Grace, J.R., (2002), "Reduction of electrostatic charges in fluidized beds", *Chemical Engineering Science*, 57, 153-162.
15. Cui, H.P., Mostoufi, N., Chaouki, J., (2000), "Characterization of dynamic gas-solid distribution in fluidized beds", *Chemical Engineering Journal*, 79, 133-143.
16. van der Stappen, M.L.M., Schouten, J.C., van den Bleek, C.M., (1993), "Deterministic chaos analysis of the dynamical behavior of slugging and bubbling fluidized beds", In: *Proceedings of the 12th Fluidized Bed Combustion*; ASME, New York, pp. 129-140.
17. Brereton, C.M.H. and Grace, J.R., (1993), "Microstructural aspects of the behavior of circulating fluidized beds", *Chemical Engineering Science*, 43, 2565-2572.
18. Lin, Q., Wei, F., Jin, Y., (2001), "Transient density signal analysis and two-phase micro-structure flow in gas-solids fluidization", *Chemical Engineering Science*, 56, 2179-2189.
19. Mathiesen, V., Solberg, T., Hjertager, B.H., (2000), "An experimental and computational study of multiphase flow behavior in a circulating fluidized bed", *International Journal of Multiphase Flow*, 26, 387-419.
20. Li, J. and Kuipers, A.M., (2007), "Effect of competition between particle–particle and gas–particle interactions on flow patterns in dense gas-fluidized beds", *Chemical Engineering Science*, 62, 3429-3442.
21. Herb, B., Dou, K., Tuzla K. and Chen J.C., (1992), "Solid mass fluxes in circulating fluidized beds", *Powder Technology*, 70, 197-205.
22. Wei, F., Lu, F., Jin, Y. and Yu Z.Q., (1997), "Mass flux profiles in a high density circulating fluidized bed", *Powder Technology*, 91, 189-195.

74c Experimental And Computational Study Of T- And L-Outlet Effects In A Dilute Circulating Fluidized Bed Riser

Gorik Van engelandt, Total, Antwerp, B2000, Belgium, **Juray De Wilde**, Materials and Process Engineering, UCL, Place Sainte Barbe, 2, Louvain-la-Neuve, B - 1348, Belgium, Geraldine J. Heynderickx, Laboratorium voor Petrochemische Techniek, University of Ghent, Krijgslaan 281(S5), Ghent, B-9000, Belgium, and Guy B. Marin, Ghent University, Krijgslaan 281, S5, Gent, 9000, Belgium.

Outlet effects induced by an L-outlet and T-outlets with different extension heights and outlet surfaces are studied experimentally and computationally. Experiments are carried out in a cold flow riser. The mean and fluctuating particle velocities are measured using a 3D Laser Doppler Anemometer (LDA). The riser has a diameter of 0.1m and a height of 8.765 m and is operated in the dilute regime with a superficial gas velocity of 2.65-7.43 m/s and a solids flux of 3.0 kg/m2/s. A T-outlet configuration induces recirculation by vortex formation in the extension part of the riser above the outlet, resulting in steep velocity gradients and off-centre maxima in the velocity field. The vortex has a 3D nature, but recirculates the flow along the wall opposite of the outlet, inducing reflux down to about 0.1m upstream of the outlet. High Root Mean Square (RMS) fluctuating particle velocities are observed at the vortex boundaries. The vortex is affected by the extension height of the T-outlet and stretches with growing extension. The use of an L-outlet significantly reduces vortex formation. With decreasing outlet surface area, the over-all solids hold-up in an abrupt T-outlet increases and the recirculation zone at the opposite of the outlet broadens. The experimental measurements are used to validate an Eulerian-Eulerian and Kinetic Theory of Granular Flow (KTGF) based gas-solid flow model. In general, the 3D simulation results are in agreement with the experimentally observed phenomena, that is 3D vortex formation is simulated, but the exact length, shape and position of the vortex are not always accurately predicted.

A Model for Cluster Size in Risers **74d**

Duvvuri Subbarao
Department of Chemical Engineering
Univerisiti Teknologi Petronas
Bandar Seri Iskander, Tronoh 31750, Malaysia

Abstract

Slip velocities for gas solid flow in risers are higher than the terminal velocity of particles. This is generally attributed to aggregation of particles as clusters and gas as voids/bubbles. Clusters and voids form, coalesce and break and are in a state of dynamic equilibrium. A model is proposed to estimate size of clusters considering that "ratio of cluster volume to void volume is equal to the ratio of cluster to void volumetric fractions in the bed". The model predictions compare well with the available experimental data.

Introduction

In processes such as fluid catalytic cracking, combustion of coal, calcination of lime, gas and particles are brought into contact at high gas velocities with particle entrainment. To sustain gas solid contact in such beds, known as risers, it is essential to introduce particles along with the gas to the bottom of the bed. Modeling the performance of these vital processes can help in their rational design and further development. Relationship between particle holdup along the riser height as a function of gas and solid velocities is the key hydrodynamic information needed to develop performance models.

Experimental investigations of Yerushalmi et al (1976), Yerushalmi and Cankurt (1979) on particle volume fraction as a function of gas velocity and particle flux in risers revealed that slip velocity between gas and particles is greater than the terminal velocity of the particles. Based on voidage fluctuations, they attributed this to particles moving as aggregates (clusters). Emerging techniques such as capacitance probes, laser optic probes, LDA, PIV and analysis of fluctuations have made it possible to get information on cluster size. Harris et al (2002) reviewed various correlations that appeared in the literature for cluster size and proposed another correlation for cluster voidage and size as

$$D_c = \frac{(1-\varepsilon)}{40.8 - 94.5\,(1-\varepsilon)} \qquad 1$$

This equation does not reflect the dependence of cluster size on properties of gas and particles as well as column diameter.

Subbarao (1986) conceptualized that particles get together to move as clusters while gas gets together to develop in to voids. Horio and Kuroki (1994) observed the existence of clusters and voids in lean beds using laser sheet technique. Assuming ratio of cluster volume to void volume is equal to the ratio of cluster to void volumetric flow rates, Subbarao (1986) proposed an equation for cluster diameter in terms of void size as

$$D_c = \left[\frac{W}{\rho_p(1-\varepsilon_c)u_o}\right]^{1/3} D_v \qquad 2$$

The voids were imagined to have originated in a bubbling bed (in the vicinity of the grid) and erupted into the freeboard with a velocity of

$$u_{vr} = 0.71(g\, D_v)^{1/2} \qquad 3$$

In high velocity fluidized beds, the void size was assumed to be equal to the expected maximum stable bubble size (for which the rise velocity is equal to the terminal velocity of the particles)

$$D_v = \frac{2u_t^2}{g} \qquad \textit{for } D_v \le \frac{D_t}{4} \qquad 4$$

For high velocity fluidization with coarse particles and/or small diameter columns, void size is restricted to the column diameter and rise velocity is restricted to slug rise velocity. Slug rise velocity depends on column diameter as

$$u_{sr} = 0.35\,(g\, D_t)^{1/2} \qquad 5$$

Specifying the diameter of such voids as

$$D_v = \frac{D_t}{4} \qquad \textit{for } D_v > \frac{D_t}{4} \qquad 6$$

will ensure void rise velocity to be restricted to slug rise velocity.

Assuming that the terminal velocity of cluster is the slip velocity, particle slip velocity in lean beds was obtained as

$$\left[\frac{u-v}{u_t}\right]^2 = \frac{8(1-\varepsilon_c)}{3\,c_{Dc}}\frac{(\rho_p-\rho)}{\rho}\left(\frac{W}{\rho_p u_o(1-\varepsilon_c)}\right)^{1/3} \qquad 7$$

The assumption in the model of Subbarao (1986) ("ratio of cluster volume to void volume is equal to the ratio of cluster to void volumetric flow rates") implicitly assumed pseudo homogeneous flow; instead, "the ratio of cluster volume to void volume is equal to the ratio of cluster to void volumetric fractions in the bed" could be a more realistic description.

2. A New Model for Cluster Size

a) Structure of fluid Particle flow

A visualization of the structure of fluid particle aggregates in a riser is shown in fig. 1. Particles are assumed to move as clusters. A cluster is considered to be a deformable impermeable packet of particles. Gas flows around the cluster forming a wake, free of particles, behind the cluster. Fluid velocity profile in the wake deforms the cluster by fluid drag as an inverted paraboloid. This can be a description to explain the shape of clusters as observed by Horio and Kuroki (1994).

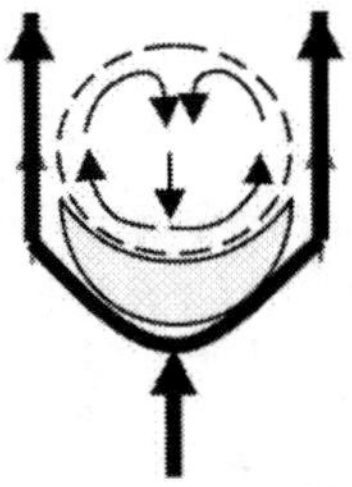

Fig.1 Structure of fluid particle aggregation in risers

The wakes behind clusters in a riser experience buoyancy due to the higher density of dispersed phase and develop a rise velocity similar to bubbles in a bubbling bed. Then clusters can be viewed as wakes behind rising bubbles. Bubbles can coalesce and clusters also can coalesce. Bigger bubbles develop higher rise velocities and internal circulation velocities to cause bubble breakup by particle entrainment from clusters in the form of “streamers”. This process leads to dynamic equilibrium resulting in bubbles of stable size dictated by fluid drag and gravitational forces.

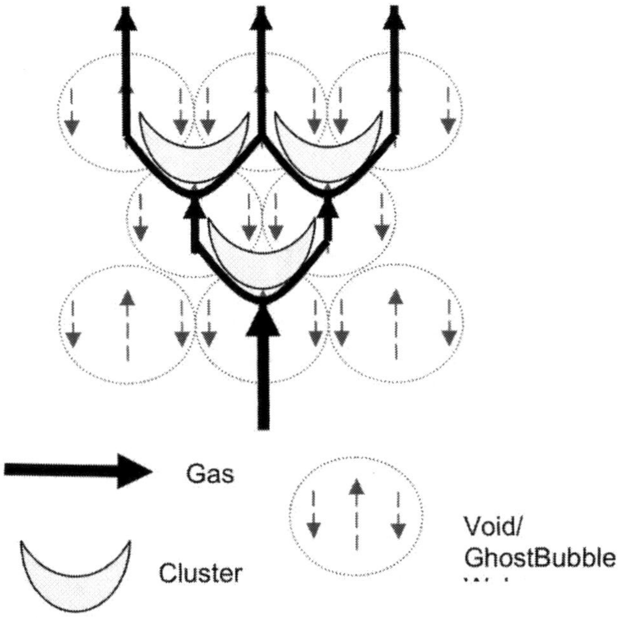

Fig.2 Gas-cluster flow structure in a riser

b) Model Equations

It is proposed that ratio of cluster volume to void volume is equal to the ratio of cluster to void volumetric fractions in the bed.

$$\frac{D_c^{\,3}}{D_v^{\,3}} = \left[\frac{\delta_c}{1-\delta_c}\right] \qquad 8$$

Recognizing that cluster size can not be smaller than particles size, cluster size can be written as

$$D_c = \left[\frac{\delta_c}{1-\delta_c}\right]^{1/3} D_v + D_p \qquad 9$$

Considering that in high velocity fluidization regimes all the particles move as clusters with a voidage of ε_c

$$1-\varepsilon = \delta_c(1-\varepsilon_c) \qquad 10$$

$$D_c = \left[\frac{1-\varepsilon}{\varepsilon-\varepsilon_c}\right]^{1/3} D_v + D_p \qquad 11$$

The asymptotic equations for void size, eq. 4 and 6, can be combined to obtain void size as

$$\frac{1}{D_v} = \frac{1}{2u_t^2/g} + \frac{1}{D_t/4} \qquad 12$$

This can be rewritten as

$$D_v = \frac{2u_t^{\,2}}{g}\left(1+\frac{u_t^{\,2}}{u_{sr}^{\,2}}\right)^{-1} \qquad 13$$

With this cluster size can be obtained as

$$D_c = \left[\frac{1-\varepsilon}{\varepsilon-\varepsilon_c}\right]^{1/3} \frac{2u_t^{\,2}}{g}\left(1+\frac{u_t^{\,2}}{u_{sr}^{\,2}}\right)^{-1} + D_p \qquad 14$$

Distinctive existence of clusters is possible for

$$\frac{\delta_c}{1-\delta_c} \leq 1 \qquad 15$$

beyond which clusters coalesce excessively to form a continuous dense phase in which voids can move as bubbles.

The terminal velocity of a cluster can be expressed as

$$(u-v)^2 = \frac{4}{3\,c_{Dc}}\frac{D_c g(\rho_c-\rho)}{\rho} \qquad 16$$

Assuming that the terminal velocity of cluster is the slip velocity in lean beds,

$$\left[\frac{u-v}{u_t}\right]^2 = \frac{8(1-\varepsilon_c)}{3\,c_{Dc}}\frac{(\rho_p-\rho)}{\rho}\left[\frac{1-\varepsilon}{\varepsilon-\varepsilon_c}\right]^{1/3}\left(1+\frac{u_t^2}{u_{sr}^2}\right)^{-1} + \frac{c_{Ds}}{c_{Dc}}(1-\varepsilon_c) \qquad 17$$

For not so lean beds, following Richardson-Zaki equation for hindered settling, slip velocity for clusters in presence of other clusters can be written as

$$\left[\frac{u-v}{u_t}\right] = \left\{\frac{8(1-\varepsilon_c)}{3\,c_{Dc}}\frac{(\rho_p-\rho)}{\rho}\left[\frac{1-\varepsilon}{\varepsilon-\varepsilon_c}\right]^{1/3}\left(1+\frac{u_t^2}{u_{sr}^2}\right)^{-1} + \frac{c_{Ds}}{c_{Dc}}(1-\varepsilon_c)\right\}^{1/2}(1-\delta_c)^{1.3} \qquad 18$$

In terms of voidage, this equation can be written as

$$\left[\frac{u-v}{u_t}\right] = \left\{\frac{8(1-\varepsilon_c)}{3\,c_{Dc}}\frac{(\rho_p-\rho)}{\rho}\left[\frac{1-\varepsilon}{\varepsilon-\varepsilon_c}\right]^{1/3}\left[1+\frac{u_t^2}{u_{sr}^2}\right]^{-1} + \frac{c_{Ds}}{c_{Dc}}(1-\varepsilon_c)\right\}^{1/2}\left(\frac{\varepsilon-\varepsilon_c}{1-\varepsilon_c}\right)^{1.3} \qquad 19$$

The parameter $\frac{c_{Ds}}{c_{Dc}}(1-\varepsilon_c)$ can be in the range of 0.1 to 10 and assigning a value of 1 demonstrates the essential features of the model.

Results and Discussion

During the process of coalescence and break up the cluster can take different shapes with different voidages. To be in line with the conceptual frame work of aggregative fluidization, voidage of such clusters, ε_c , is expected to be at ε_{mf} and is taken as equal to 0.5. Flow of gas in excess of incipient fluidization through such clusters can cause formation of bubbles in clusters leading to their destruction.

Diameter of Cluster - Comparison with experimental observations

Equation 14 summarizes the present model to estimate diameter of clusters in gas solid flows through risers considering the properties of gas, particles as well as column diameter. This equation is compared with the available experimental observations.

a) Zou et al (1994) investigated fluidization of FCC particles (D_p = 54 μm, ρ_p=930 kg/m^3 , D_v = 1.29 mm) at gas velocities ranging between 1.3 to 5.5 m/s at solid feeding rates in the range of 9.3 to 64.7 kg/s in a 9 cm diameter risers. In lean phase systems, the clusters are smaller than the diameter of void D_v (1.29 mm). The experimental observations on cluster size are shown in Fig.3 along with the present model and the correlation of Harris et al (2002). Present model compares remarkably well with the experimental observations.

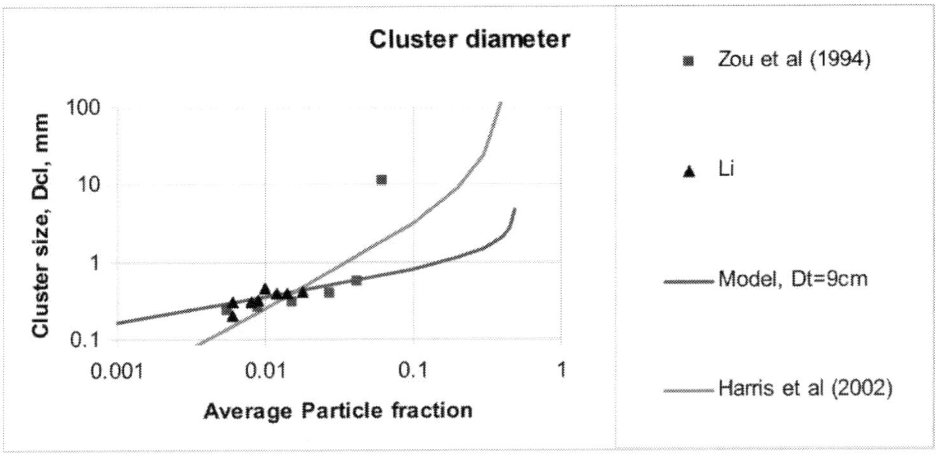

Fig.3 Comparison of the present model with data of Zou et al (1994)

b) Pandey et al. [2004] reported experimental observatons on diameter of clusters using a backscatter laser Doppler velocimeter (LDV) system. in a cold flow circulating fluidized bed (CFCFB) riser operating in the lean phase regime at the National Energy Technology Laboratory (NETL), in Morgantown, West Virginia. Guenther and Breault (2007) also studied size of clusters in the same facility using fiber optic probe through application of discrete wavelet analysis. The riser was of 0.305 m diameter and 15.45 m tall. They used cork particles (D_p = 812 microns, and ρ_p = 189 kg/m^3 , D_v= 50 mm) at gas velocities of 3.6 to 5.5 m/s and particle mass flux rates in the range of 0.35 to 1.75 kg/m^2/s. Experimentally measured cluster sizes along with model predictions are presented in Fig.4. The data of Pandey et al (2004) are very well explained by the model of Subbarao (1986).

However, observations of Guenther and Breault (2007) are higher compared to the predictions of Subbarao (1986). Considering that cluster terminal velocity is the slip velocity between gas and cluster flows, particle volume fractions are estimated to obtain diameter of cluster by the present model equations. The results are presented in Fig.4 and present model predicts the observations of Guenther and Breault (2007) reasonably well; correlation of Harris et al (2002) also is included and the comparison is very poor.

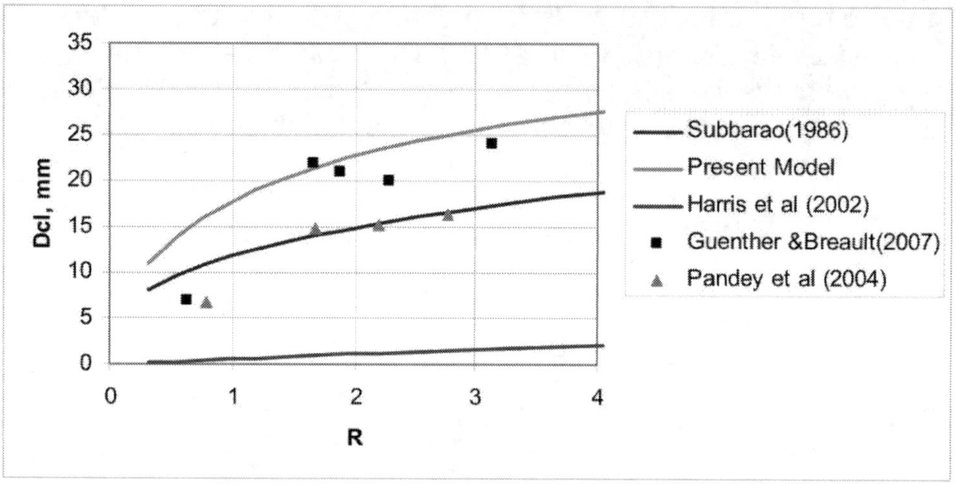

Fig.4 Comparison of the model with data of Pandey et al [2004] and Guenther and Breault [2007]

c) Horio et al (1988) investigated fluidization of FCC particles (D_p = 60 μm, ρ_p=1000 kg/m3, D_v=4.8 mm) at gas velocities ranging between 1.17 to 1.29 m/s at solid feeding rates of around 11 kg/s in 5 cm diameter risers. The bed is likely to be operating in turbulent bed regime. Horio et al (1994) investigated fluidization of FCC particles (D_p = 61 μm, ρ_p=1780 kg/m3, D_v=7 mm) at gas velocities ranging between 0.15 to 1.3 m/s at solid feeding rates less than 16.5 kg/s in 20 cm diameter risers. The bed is likely to be operating in bubbling/turbulent bed regime and the measurements are in the free board region. Clusters in this regimes are expected to be larger than in lean phase transport regime. Experimentally observed cluster sizes are larger compared to the present model as well as the correlation of Harris et al (2002).

d) Wei et al (1993, 1995) investigated fluidization of FCC particles (D_p = 54 μm, ρ_p=1389 kg/m3, D_v= 2.87 mm) at gas velocities ranging between 1.2 to 8.5 m/s at solid feeding rates in the range of 18 to 215 kg/s in 18.6 cm diameter risers. Reported data appears to be from low gas velocity – high solid feed rates operation as bed particle volume fractions are in the range of 0.15 to 0.4.

Slip Ratio – Sensitivity of the Model

a) Effect of fluid properties

Slip ratio is a vital indicator of nature of fluidization. Results of simulation of the present model, eq.19, on slip ratio as a function of particle holdup for sand particles (D_p = 100 μm and ρ_p = 2650 kg/m^3) flowing with air or water through a riser of 10 cm diameter are shown in Fig.5. It can be seen that the slip ratio for air-sand system is much higher than that of water-sand system. Present model predicts that slip ratio for water sand system also can be slightly greater than 1 for particle volume fractions smaller than 0.15. Experimental observations of Ying et al (2002) support this conclusion. For comparison, slip ratios by the correlations of Richardson Zaki (1952)

$$\frac{u-v}{u_t} = \varepsilon^{n-1} \qquad 20$$

and Matsen (1982)

$$\frac{u-v}{u_t} = \left(\frac{u_b}{u_t}\right)\left(\frac{\varepsilon-\varepsilon_{mf}}{\varepsilon\,(1-\varepsilon)}\right) + \frac{u_{mf}}{u_t\,\varepsilon} \qquad \text{for dense phase} \qquad 21$$

$$\frac{u-v}{u_t} = 10.8\,(1-\varepsilon)^{0.293} \qquad for \quad (1-\varepsilon) > 0.0003$$

$$\frac{u-v}{u_t} = 1 \qquad for \quad (1-\varepsilon) < 0.0003 \qquad 22$$

as a function of particle holdup are included in Fig.5. Slip ratios by the correlation of Richardson and Zaki (1952) is always smaller than 1. It should be noted that correlation of Matsen (1982) does not reflect the effect of particle and gas properties and column diameter.

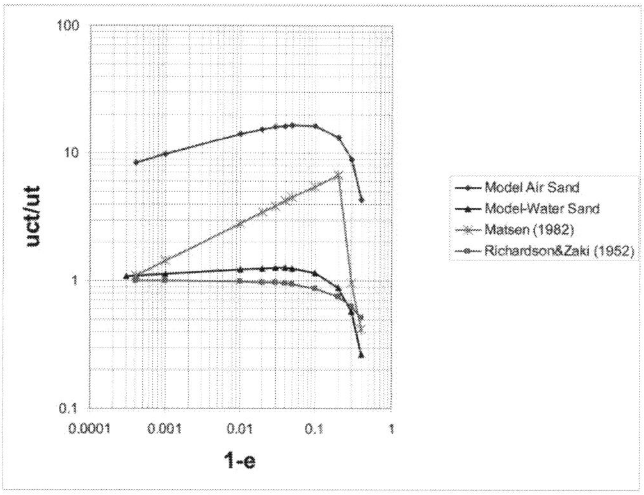

Fig.5 Effect of fluid properties on Slip Velocity

Concluding remarks

A model for fluid-particles flow in a riser at high fluid velocities is developed considering that the bed consists of cluster and void phases; particle are assumed to move as clusters by drag of fluid flowing through void phase.

Assuming ratio of cluster to void volume to be equal to ratio of cluster phase to void phase volume fractions in the column, diameter of cluster is estimated in terms of diameter of void. Diameter of the void is assumed to be given by the equation for the diameter of stable bubble size in a bubbling fluidized bed. Estimates from the model equation (eq.24) compare well with the literature data.

The correlation of Harris et al (2002) can be rearranged as

$$D_{cl} = \frac{\left[\dfrac{(1-\varepsilon)}{0.432}\right]}{94.4\left[1-\dfrac{(1-\varepsilon)}{0.432}\right]}$$

Comparing this with eq.9 in a generalized form

$$D_c = \left[\frac{\left(\dfrac{1-\varepsilon}{1-\varepsilon_c} \right)}{1-\left(\dfrac{1-\varepsilon}{1-\varepsilon_c} \right)} \right]^{1/m} D_v + D_p$$

it can be seen that the correlation of Harris et al (2002) corresponds to cluster voidage ε_c of 0.568 and "m" of 1; Diameter of void D_v is around 0.01 m (10 mm). It is to be noted that void size for most of the materials used in the study of risers is around 10 mm with the exception of Pandey et al (2004) and Guenther and Breault (2007).

Ratio of slip velocity in the risers to terminal velocity of the particles, slip ratio, is an important parameter. Present model predicts that slip ratio can be much greater than 1 for gas-particle flows while for liquid-particle flows it can be slightly more than 1 for low particle volume fractions. This is inline with most of the experimental observations. This also indicates the inadequacy of Richardson Zaki correlation even for liquid solid systems at low particle volume fractions. In view of this, effective drag coefficients/ momentum exchange coefficients used in CFD studies should not be based on Richardson Zaki correlation.

Nomenclature

c_{Dc}	Cluster Drag Coefficient
c_{Ds}	Single particle Drag coefficient
D_c	Diameter of Cluster, m
D_p	Diameter of particle, m
D_t	Column Diameter, m
D_v	Diameter of the void, m
g	Acceleration due to gravity, m/s^2
u	Actual gas velocity, m/s
u_o	Superficial gas velocity, m/s
u_{sr}	Slug Rise Velocity, m/s
u_t	Terminal velocity of a single particle, m/s
u_{vr}	Void rise velocity, m/s
v	Actual velocity of particles, m/s
W	Solid feed flux, $kg/m^2/s$

Greek Letters

δ_c	Cluster fraction
ε	Bed Voidage
ε_c	Cluster voidage

(1-ε)	Average bed particle fraction
μ	Viscosity of gas, kg/m/s
ρ_f	Density of gas, kg/m^3
ρ_p	Particle density, kg/m^3

References

1. J.Yerushalmi, M.J.Gluckman, R.A.Graff, S.Dobner and A.M.Squires in D.L.Kearns (Ed) Fluidization Technology, vol2, Hemisphere Publications Carp. Washington, 1976, p437
2. J.Yerushalmi and N.T.Cankurt, Powder Technology, 24, 1979, 187
3. J.F.Richardson and W.N.Zaki, Trans.Inst.Chem.Eng. 35, 1954 35
4. J.M.Matsen, Powder Technology, 32 (1982) 21-33
5. D.Subbarao. Powder Technology,(1986)
6. M. Horio, H. Kuroki, Chem. Eng. Sci. 49 (1994) 2413–2421
7. M Harris, A.T., Davidson, J.F. and Thorpe, R.B. (2002).127, 128
8. Horio, K. Morishita, O. Tachibana, N. Murata, in P. Basu, J.F. Large (Eds.), Circulating Fluidised Bed Technology II, Pergamon, Oxford,1988, pp. 147–154.
9. F. Wei, G. Yang, Y. Jin, Z. Yu, Can. J. Chem. Eng. 73 (1995) 650– 655.
10. F. Wei, Y. Jin, Z. Yu,, in: A. Avidan (Ed.), Circulating Fluidised Bed Technology IV, AIChE New York, 1993, pp. 588– 593.
11. B. Zou, H. Li, Y. Xia, X. Ma, Powder Technol. 78 (1994) 173– 178.
12. P. Pandey, R. Turton, P. Yue, L.J. Shadle, , Industrial and Engineering Chemistry Research 43 (2004) 5582–5592
13. Chris Guenther, Ronald Breault, Powder Technology 173 (2007) 163–173– 143.
14. Ying Zheng1, Jing-Xu Zhu, Narenderpal S. Marwaha, Amarjeet S. Bassi, Chemical Engineering Journal 88 (2002) 141–150

Abstract ID#: 98733
Password: 949257
Title: A Model For Size Of Clusters In Risers
Session Selection: 03B00 Circulating Fluidized Beds
Preferred Presentation Format: Oral
Lead Presenter's Email: subbarao_duvvuri@yahoo.com

74e The Transport Disengaging Height in Fluidised Beds for Fine and Coarse Particles Measured by Positron Particle Tracking

Manon Van de Velden[1], Xianfeng Fan[2], Andy Ingram[3], David J. Parker[2], and **Jan Baejens**[1]. (1) Department of Chemical Engineering, The University of Birmingham, Birmingham, B15 2TT, United Kingdom, (2) School of Physics and Astronomy, University of Birmingham, Edgbaston, Birmingham, B15 2TT, United Kingdom, (3) Department of Chemical Engineering, University of Birmingham, Edgbaston, Birmingahm, United Kingdom

In fluidised bed applications, bubbles bursting at the surface of the bed will project particles into the freeboard, the space between the bed surface and the gas off-take. Depending on their terminal velocities and the gas velocity, solids are carried up the column to various heights. The larger particles fall back and the smallest are evacuated with the off gas through the off-take tube; consequently the solids loading (commonly expressed as kg solids/m≥ gas) will decline with height. The region within which the solids loading falls is called the Transport Disengaging Height (TDH) and this term TDH is applied to two related but different circumstances: coarse particles having a terminal velocity larger than the superficial gas velocity are nevertheless flung upwards by the bursting bubbles at the bed surface and then fall back. Above the height they reach, only fines i.e. particles with terminal velocities smaller than the gas velocity are found. The concentration (or hold-up) of fines decreases with height and eventually reaches a constant value. The height which the coarse particles reach is called the splash height or TDH(C). The height above which the fines hold-up changes little or not at all will be called TDH(F). It is clear that the definition of a particle as coarse or fine will depend upon its terminal velocity and the superficial gas velocity in the freeboard.

The TDH was determined experimentally from positron emission particle tracking in bubbling fluidised beds of 8.6 and 16 cm I.D. respectively. The bulk bed material was sand of 120 and 250 μm. 18F-tracer particles were used. These tracers included sand (120, 250, 375, 580 μm), •-alumina of 150 μm and coal of 920 μm. The static bed height was between 30 and 40 cm. The operating velocity varied from 0.05 to 0.5 m/s. The detection camera covered approx. 0.6 x 0.6 m^2 and started detecting the tracer from about 20 cm above the distributor level. Collected data determine the particle velocity when ejected by a bubble, the particle profile in the freeboard and the maximum height reached (TDH) for particles with a terminal velocity in excess of the operating gas velocity in the freeboard. Illustration of particle tracking profiles will be given, clearly showing that the velocity of ejection exceeds the bubble rise velocity, as calculated from common equations..

The experimental results will thereafter be used (i) to assess the accuracy of existing empirical correlations for the TDH, (ii) to develop a fundamental-principle based empirical equation, and (iii) to develop a model that will predict the trajectories of ejected particles from a balance of forces.

To illustrate the use of both empirical equations and model approach, TDH predictions are made for various superficial gas velocities and/or diameters of the fluidised bed. At low superficial gas velocities, the TDH is determined by the ejection of coarse particles, at higher velocities entrained fines fix the TDH.

*Corresponding author: Professor Jan Baejens; e-mail: J.Baejens@bham.ac.uk

74f Design And Construction Of A Cold Flow Circulating Fluidized Bed Gasifier: A Technical Overview

Richard D. Lusk, Dave C. Swalm School of Chemical Engineering, 330 Swalm Chemical Engineering Building, Mississippi State University, MS 39762 and Mark Bricka, Dave C. Swalm School of Chemical Engineering, Mississippi State University, MS 9595, Mississippi State, MS 39762.

The gasification of biomass has continued to receive increasing attention as energy prices continue to climb and more focus is placed on alternative energy sources. The use of circulating fluidized beds (a natural expansion to fluidized beds) look to be a promising means of gasifying biomass. The design and construction of a small scale cold flow model circulating fluidized bed will be examined to characterize the flow properties of the bed material and the fluidizing medium. This paper also looks at how tar formation, particulate content, and syngas composition are affected by various factors including variations in temperature, fluidizing medium, oxygen to carbon ratio, and equivalent ratio in a circulating fluidized bed gasifier.

74g Predictions Of Cyclone Wear In A Large Industrial Unit Under High Loadings

Ken Williams, CPFD Software, LLC, 10899 Montgomery Blvd. NE, Suite B, Albuquerque, NM 87111 and Soundar Ramchandran, AN, Solutia, Inc, Chocolate Bayou, TX 77511.

Extending the performance of large industrial cyclones in a high-load environment is an important goal for many reactors systems. Soultia operates such cyclones in its acrylonitrile reactors, and have extensive operating data on wall wear and performance. Simulations made with the commercial engineering software Barracuda have been compared to wear details (photographs) in several regions of the cyclone, inlet, backwall, vortex tube. Trade studies have lead to design decisions that will improve cyclone lifetime, and aid with particle (catalyst) attrition.

Web Page: www.barracuda-cpfd.com

80a A New Kinetic-Theory-Based Model For Granular Flows With Size And/or Density Differences

Vicente Garzo, Department of Physics, University of Extremadura, Badajoz, Spain, James W. Dufty, Department of Physics, University of Florida, Gainesville, FL 32611, and **Christine M. Hrenya**, Chemical and Biological Engineering, University of Colorado, Boulder, CO 80309-0424.

Industrial and natural systems involving the rapid flow of solid particulates often are characterized by particles with differences in size and/or density. Previous kinetic-theory-based models have been developed for such polydisperse systems, though they typically involve limiting assumptions. For example, an equipartition-of-energy assumption has been applied often, though previous experiments and simulations have shown that non-equipartition is not only prevalent, but also plays a role in species segregation for certain systems. In the current effort, a rigorous Chapman Enskog expansion is applied to the revised Enskog kinetic equation for a system of N species. This expansion is carried out about the homogeneous cooling state, and thus the resulting constitutive equations are applicable to a wide range of inelasticities. Furthermore, because the time scale associated with individual species is of the kinetic (fast) order, momentum and granular energy balances are only required for the mixture, not for individual species. The corresponding computational effort associated with solving the closed set of equations is expected to be lower than previous theories.

80b Shear Flow Of Assemblies Of Cohesive Granular Materials

Lee R. Aarons and S. Sundaresan. Chemical Engineering, Princeton University, Princeton, NJ 08544

It is well known that assemblies of granular materials behave differently under different flow conditions. Under rapid flow conditions, the particles interact with each other predominantly through binary collisions, while at slow shear rates the interaction is dominated by enduring contacts. Campbell [1] carried out discrete element method (DEM) simulations of sheared assemblies of cohesionless particles in periodic domains and under constant volume conditions and presented a map of the different regimes of flow. Aarons & Sundaresan [2] performed similar simulations for cohesive granular materials and showed how the stresses varied with particle volume fraction, shear rate and particle properties including the strength of the cohesive interaction. This study revealed that inter-particle attractive forces expand the range of particle volume fractions and shear rates over which the quasi-static flow regime is observed. Furthermore, this study identified how the stresses in this regime should be scaled and showed that the apparent coefficient of friction varied systematically with particle volume fraction.

In practical devices, flows of granular materials rarely occur under constant volume conditions, and compaction and dilation do occur locally in response to the stresses. This and the fact that laboratory rheological experiments are often performed under constant applied overburden levels have prompted researchers to investigate regime maps under conditions of constant applied normal stress. The discrete element method (DEM) simulations performed by Campbell [3,4] revealed that the flow regime map for assemblies of cohesionless particles sheared at constant applied normal stress differed from the flow regime map he previously constructed for such assemblies sheared at constant volume [1].

In the present study, we have analyzed shear flow of cohesive granular materials under constant applied normal stress and compared the results with those obtained under constant volume conditions.

We consider the plane shear flow of particle assemblies in periodic domains, employing Lees – Edwards boundary conditions [5] across the faces. Simulations were performed for different strengths of cohesion, shear rates, and applied normal stresses. From each simulation, we have extracted the average volume fraction, apparent coefficient of friction, and coordination number. We have also extracted results on the statistics of the temporal fluctuations in the stresses.

Our simulations reveal that in a time-averaged sense shearing under constant applied stress turns out to be identical to shearing at constant volume. That is, the volume fraction – shear rate – stress relation is the same for both types of shearing. The probability distribution functions of the shear stress fluctuations under constant volume and constant applied normal stress conditions exhibit very similar behavior, becoming narrower as the strength of cohesion increases. In the inertial regime, the stress fluctuations show a relatively weak dependence on cohesion.

The apparent coefficient of friction for the cohesive assemblies varied systematically with volume fraction. Over the entire range of particle volume fractions studied, the apparent coefficient decreases with increasing volume fraction.

[1] C. S. Campbell, Granular Shear Flows at the Elastic Limit, J. Fluid Mech. 465 (2002) 261-291.

[2] L. Aarons, S. Sundaresan, Shear flow of assemblies of cohesive and non-cohesive granular materials, Powder Technol. 169 (1) (2006) 10-21.

[3] C. S. Campbell, Stress-controlled elastic granular shear flows, J. of Fluid Mech. 539 (2005) 279-297.

[4] C. S. Campbell, Granular material flows – An overview, Powder Technol. 162 (3) (2006) 208-229.

[5] A. W. Lees, S. F. Edwards, The computer study of transport processes under extreme conditions, Physics C: Solid State Phys. 5 (1972) 1921-1929.

80c Constitutive Modeling Of Particle-Phase Stress For Non-Spherical Particles

Benjamin D. James and Jennifer S. Curtis. Chemical Engineering, University of Florida, Chemical Engineering Department, P.O. Box 116005, Gainesville, FL 32611-6005

Virtually all solid handling operations involve particles that are non-spherical in shape. However, most fundamental studies of granular material undertaken to date have involved spherical particles. Hence, there is a current significant disconnect between the model particles which are used in fundamental research studies and the characteristics of real particles dealt with in industry. While industrial practitioners comprehend fully that the influence of particle shape on particle flow behavior is significant, the role of particle shape in flowing granular systems is not understood.

Three-dimensional DEM simulations in shear flow were conducted to understand the influence of various specific features of particle shape - such as aspect ratio or roughness - on granular stress. Various non-spherical particle shapes are formed from linking and/or overlapping spheres. Granular stress at dilute particle concentrations is shown to be a function of the projected area of the non-spherical particle. At dense particle concentrations, the granular stress is strongly influenced by the collision rate and the average impact velocity. The higher the particle aspect ratio, the higher the collision rate and the lower the average impact velocity, resulting in an increase in the granular stress at high concentrations. Decreasing the coefficient of elasticity decreases the granular temperature and in turn decreases the kinetic stress and decreases the impact velocity which decreases the collisional stress. However, the decrease in elasticity and the increase in aspect ratio leads to greater particle packing and, therefore, a greater collision rate at higher concentrations resulting in similar stress values for high and low elasticity. Increasing friction leads to a decrease in the granular temperature and in turn a decrease in the kinetic stress and the average impact velocity, but there is a much more significant increase in the collision rate at higher concentrations when the aspect ratio is increased leading to a rather large increase in the granular stress.

80d Simulation Of Shear Cells: Quantitative Validation Of Dem

Joseph J. McCarthy[1], C. Fred Higgs Jr.[2], and Venkat Jasti[2]. (1) Chemical and Petroleum Engineering, University of Pittsburgh, 1238 Benedum Hall, Pittsburgh, PA 15261, (2) Mechanical Engineering, Carnegie Mellon University, Pittsburgh, PA 15213

Particle Dynamics (alternately called the Discrete Element Method or DEM) is often used as the ``gold standard" for comparison to continuum-level theories and/or course-grained models of granular material flows due to its derivation from first-principal constructs, like contact mechanics. Despite its prevalence, the method is most often validated against experiment in only qualitative ways -- comparison of mixing rates, gross features of concentration profiles, etc. -- for exactly the reason it has found its popularity; detailed experimental measurements are difficult and often expensive. In this paper, we outline work aimed at using detailed, particle-level experimental measurements to quantitatively validate Particle Dynamics simulations. Specifically, we examine the flow in a horizontally-aligned annular shear cell. Measurements are performed using digital particle tracking velocimetry (DPTV) so that the velocity, granular temperature, and solids fractions profiles may be extracted. Computationally, we attempt to match the experimental measurements as closely as possible and study the impact of a variety of contact mechanics-inspired force laws as well as perform sensitivity analysis on materials properties employed.

80e Particle Dynamics Simulations To Elucidate The Electrostatic Charging Of Granular Materials

Nathan Duff and Dan Lacks. Chemical Engineering, Case Western Reserve University, Dept. of Chemical Engineering, Cleveland, OH 44106

As granular materials flow, the particle contacts can cause electrostatic charges to develop on the particles. This electrostatic charging is often significant in industrial processes such as fluidized beds, pneumatic conveying, and spray deposition, and pharmaceutical dispersal processes such as dry powder inhalers. Despite the widespread significance of this phenomenon, the origins of the electrostatic charging are not well known. In particular, it is not clear why charging occurs when all particles are composed of the same material, which would seem to preclue any driving force for charge transfer during particle contact.

In our work, particle dynamics simulations are carried out to address the origins of this electrostatic charging. The simulations test our hypothesis that the electrostatic charging in single-component granular materials can be explained in terms of simple population balance ideas. The simulations are based on hard-sphere particle dynamics coupled with an ad hoc model for the collision-induced transfer of trapped electrical charges. In this model, trapped electrons are randomly placed on the surface of the particles, and these electrons are transferred by a collision if the trapped electron is within a cutoff distance from the point of collision. The simulation results show that these simple ideas alone lead to electrostatic charging that increases in magnitude with the breadth of the particle size distribution, which has also been observed in experimental studies. Furthermore, the simulations show that smaller particles tend to charge negatively and larger particles tend to charge positively, as found in experimental and field studies.

80f An Approach For The Parallelization Of The Discrete Element Code

Atul Dubey, M. Silvina Tomassone, and Fernando J. Muzzio. Chemical and Biochemical Engineering, Rutgers University, 98 Brett Road, Piscataway, NJ 08854

A multi-stage approach for parallelization of a Discrete Element Methid (DEM) code has been developed. This algorithm consists of three stages with the first stage as the one that will reduce the simulation time the most. The multiple processor cluster has been divided among one lead processor (parent) and the remaining child processors which can number from four to over a hundred, depending on the availability. The main processor manages the neighbor lists of all particles and updates it periodically. It also distributes the particles equally among the child processors which do computations on only the particles that they are assigned. The message passing interface (MPI) algorithm is used to perform these scatter (dividing of particles) and gather (collecting data from the child processors). In stage 1, the parallelization involves only the distribution of particles among the processors which perform the dynamics calculations on the particles assigned to them serially. The dynamics calculations which involve calculation of parameters such as new particle position, force, velocity, temperature, humidity etc. are not parallelized. In stage two, these calculations are also parallelized to enhance the overall performance of the code. In the third stage, the handling and updating of neighbor list is performed by a number of processors instead of one main processor, hence further reducing computational times.

80g Impact Of The Time Step In Dem Simulations On Granular Mixing Properties

Jocelyn Doucet, **Francois Bertrand**, and Jamal Chaouki. Department of Chemical Engineering, Ecole Polytechnique de Montreal, University of Montreal Campus, 2920 ch. de la Tour, Montreal, QC H3T1J7, Canada

The numerical error in fluid mixing simulations using CFD is due to the discretization of the problem, round-off in floating point operations and the use of time integration. As a matter of fact, it is known that the combination of these errors may lead to spurious mixing patterns even if the problem is a priori well stated. On the other hand, it appears that there is a current lack of a similar analysis in the field of granular mixing simulation, in particular with the discrete element method (DEM).

The objective of this work is to illustrate by means of two case studies the impact of the time integration error on the macroscopic flow patterns and mixing properties predicted with DEM, and gain insight into the importance of selecting an appropriate time step when simulating granular flow with this method. First, the effect of the time step on the accuracy of the flow of spherical particles through an orifice will be investigated. Next, the effect of the time step will be studied in the case of the mixing of bidisperse spherical particles in a cylindrical tumbler.

80h Van Der Waals Forces Between Micron-/nano-Scale Contaminants And Non-Homogenous/patterned Substrates- Implications In Photomask Cleaning

Ravi P. Jaiswal and Stephen P. Beaudoin. Chemical Engineering, Purdue University, 480 Stadium Mall Drive, FRNY Hall of Chemical Engineering, W. Lafayette, IN 47907

Micron-/nano-scale contaminants adhering to microelectronic devices with dimensions in the nanoscale regime are a challenge to cleaning processes used in existing and future technologies. Such devices/ patterned surfaces are generally found in the front end of the line (FEOL) of microelectronic manufacturing and on extreme ultraviolet (EUV) photomasks used in the existing and future manufacturing processes. Effective removal of these contaminants without damage to the underlying substrate requires knowledge of the adhesion force between the contaminant particle, which may have dimensions on the order of a few tens of nanometers, and the patterned substrate. In the current work, an experimental and a continuum-based theoretical study has been presented for such systems. Specifically, patterns with nanoscale dimensions were made on photomask substrates using a focused ion beam (FIB). Adhesion forces between silicon nitride particles ranging in size from the order the feature size to the micron-scale and the patterned substrates were measured for multiple particles. The observed forces were modeled by dividing the surface of the particle and substrate into single-nm wide cylindrical elements and a macroscopic approach based on pair-wise addition was used to calculate the adhesion force between the particle and the substrate, with good agreement.

Advances in Liquid-Feed Flame Spray Pyrolysis: Innovative approaches to nano-scale oxide materials.

Richard M. Laine,* J. Marchal, J. Azurdia, M. Kim, T. Hinklin, Contribution from the Dept. of Material Science and Engineering University of Michigan, Ann Arbor, MI 48109-2136

Introduction

Liquid-feed flame spray pyrolysis (LF-FSP) as described in this paper was developed and patented in our group some 15 years ago and is the basis of a number of papers from our group.[1-8] LF-FSP is also practiced very successfully by a number of other groups.[9-19] In the UM process, metal alkoxide and/or carboxylate precursor complexes are dissolved in alcohol solvents (typically ethanol) at concentrations of 1-20 wt % ceramic content but typically 5-10 wt. %. These solutions are aerosolized with oxygen at atmospheric pressures into a 1.5 m x 50 mm dia. quartz tube and ignited generating very turbulent flames with peak temperatures of 2000°C. Combustion is quenched very rapidly such that temperatures are 300-500°C at the exit point in the system.

The resulting soot from combustion can be trapped in electrostatic precipitators or in a bag house, see Figure **1**. The former is used to produce samples of 100-300 g/h and the latter system allows production of powders at up to 2 kg/h.

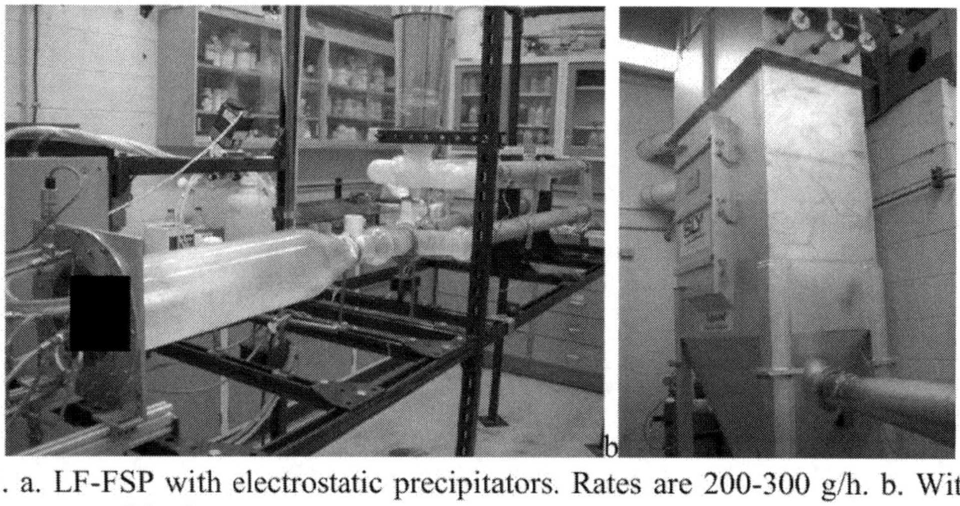

Figure 1. a. LF-FSP with electrostatic precipitators. Rates are 200-300 g/h. b. With the Bag-house rates ≤ 2 kg/h.

LF-FSP allows us to produce nanopowders that are often single phase and contain the exact composition of the original solution with control of stoichiometries down to ppm levels. Typical average particle sizes (APSs) are 10-100 nm with specific surface areas (SSAs) of 100 to 30 m^2/g. These particles are often unagglomerated single crystals and disperse very easily in water and alcoholic media.

Almost all of the nanopowders we obtain starting from metalloorganic precursors are kinetic products. Thus, we normally produce TiO_2 as the anatase phase and δ-Al_2O_3 as the alumina phase, despite flame temperatures that are high enough to produce rutile and α-Al_2O_3. We begin by discussing efforts to produce nano-Al_2O_3, which we used to benchmark the LF-FSP process.

Our efforts to benchmark LF-FSP began with studies on generating nano-Al_2O_3.[2] In these studies we compared alumatrane, $Al(OCH_2CH_2)_3N$ with $AlCl_3$ and $Al(NO_3)_3$ all of which are soluble in EtOH. Much to our surprise, with $AlCl_3$ and $Al(NO_3)_3$, we did not produce nano- only micron-size powders. Furthermore, from $AlCl_3$ roughly 50% of the powder produced still contained unreacted $AlCl_3$ as determined by TGA.[2] In addition, all of our instrumentation showed severe corrosion from co-production of HCl causing us to rapidly terminate these efforts. $Al(NO_3)_3$ provides hollow microspheres. In our system, it appears that the nitrates melt before they combust leading to a spray pyrolysis process instead for flame spray process. We believe that the difference is that the Cl and NO_3 ligands cannot oxidize whereas the triethanolamine ligand can that makes all the difference in these systems. More recently we have pursued the synthesis of mixed-metal oxide nanopowders.

Recent results

The rapid quench of more than 1000°C in < 100 ms provides access to unique nanopowders. For example we produce a hexagonal $Y_3Al_5O_{12}$ phase rather than the normal YAG phase.[4] An additional benefit of the Figure **1a** setup is that it allows us to combinatorially produce sets of nanopowders along such that we can make up to five 30 g samples of different compositions for example along the $(MO)_x(Al_2O_3)_{1-x}$ tieline of a set of materials to explore the kinetic phase diagrams as opposed to the thermodynamic phase diagrams. To this end we recently explored the MO-Al_2O_3 tie-lines for MgO, CoO, NiO, ZnO and MgO-Fe_2O_3 etc. These systems typically form a spinel phase at 1:1 MO:Al_2O_3. However, we can produce phase pure spinels in all of these systems at compositions not seen in their phase diagrams (Figure **2**). Furthermore these materials are

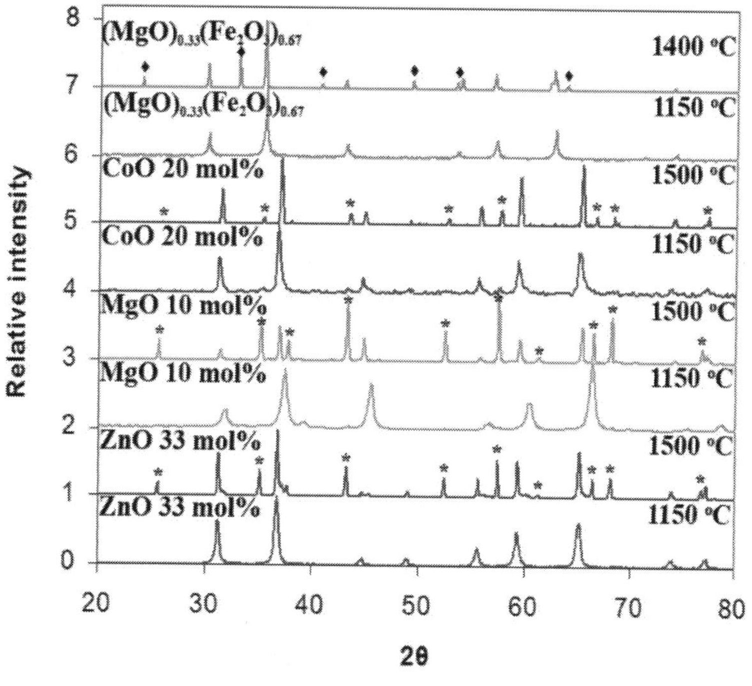

Figure 2. LF-FSP produced phase pure spinel nanopowders at compositions outside those found in their phase diagrams. These materials are stable to 1150°C.

are relatively stable undergoing transformation to their thermodynamic phase compositions only above 1150°C. Thus LF-FSP can provides access to sets of nanopowders of phase compositions hitherto unknown for a wide number materials. It is reasonable to expect that a number of these materials will have properties, e.g. catalytic activities, that are different from the mixed-phase thermodynamic systems made previously.

Still another aspect of LF-FSP that offers access to other novel materials was developed totally by accident. We recently made an effort to coat the kinetic nano-δ-alumina produced in our process with a small amount of chromium with the objective of making Cr doped alumina that could be sintered to fully dense and transparent ruby. Much to our surprise suspensions of nano-δ-alumina in EtOH when passed back through the LF-FSP system are suddenly transformed into nano-α-alumina, see Figure **3**.[7]

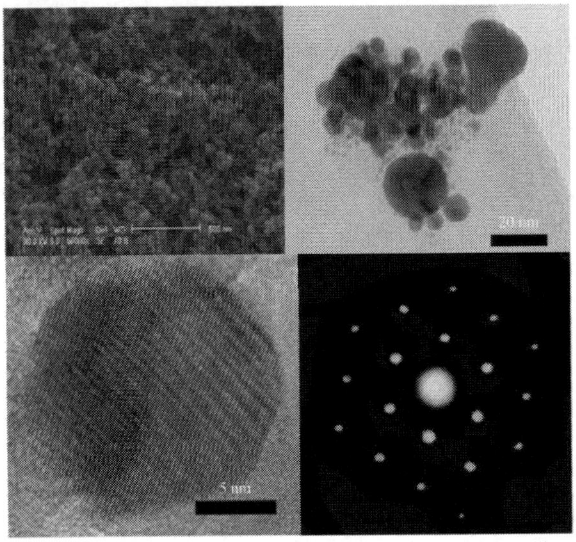

Figure 3. Electron microscopy of nano-α-Al_2O_3. From top left to bottom right: SEM, TEM, HRTEM, and electron diffraction of single particle showing hexagonal symmetry.[7]

Prior to our discovery of a route to nano-α-Al_2O_3 there were essentially no general ways to make this extremely important starting material which offers exceptional potential for making superior prosthetics for hip and knee implants, for making transparent alumina envelopes for sodium vapor lamps that are the primary component of almost all street lamps.

In addition to these efforts, we have recently developed routes to core-shell ceramics that offer extremely novel properties that will be discussed in the presentation.

Acknowledgements. The authors gratefully acknowledge support from the Air Force Office of Scientific Research (F49620-99-1-0158), the National Science Foundation (IGERT grant DGE-9972776 and DMR 9975542), and the Army Research Office (DAAD 19-99-1-0229), NASA Lewis (NLRC NCC3-381). Support from Maxit Corp. Oslo Norway, Defense Sciences Office (Singapore), and Tal Materials Inc.

References

1. Laine, R.M.; K. Waldner, C. Bickmore, D.R. Treadwell "Ultrafine powders by flame spray pyrolysis," U.S. patent 5,958,361, allowed Sept. 28, 1999.
2. Hinklin, T.; B. Toury, C. Gervais, F. Babonneau, J.J. Gislason, R.W. Morton, R.M. Laine "Liquid-Feed Flame Spray Pyrolytic Synthesis of Nanoalumina Powders," Chem. Mater. **16**, 21-30 (2004).
3. Kim, S.; J.J. Gislason, R.W. Morton, X. Pan, H. Sun, R.M. Laine, "Liquid-Feed Flame Spray Pyrolysis of Nanopowders in the Alumina-Titania System, Chem. Mater. **16**, 2336-43 (2004).
4. Laine, R.M.; J. Marchal, H.J. Sun, X.Q. Pan, "A new $Y_3Al_5O_{12}$ phase produced by liquid-feed flame spray pyrolysis (LF-FSP)," Adv. Mater. **17**, 830-33 (2005).
5. Azurdia, J.A.; J.C. Marchal, R.M. Laine, "Combinatorial processing of mixed-metal oxide nanopowders along the Co_3O_4-Al_2O_3 tie line using liquid-feed flame spray pyrolysis (LF-FSP)," J. Am. Ceram. Soc. **89**, 2749-56 (2006).
6. Baranwal, R.; M.P. Villar, R. Garcia and R.M. Laine, "Synthesis, Characterization, and Sintering Behavior of Nano-mullite Powder and Powder Compacts," J. Am. Ceram. Soc. **84**, 951- 61 (2001).
7. Laine, R.M.; J.C. Marchal, H.P. Sun, X.Q. Pan, "Nano-α-Al_2O_3 by liquid-feed flame spray pyrolysis (LF-FSP) of nano-transition aluminas," Nature Materials **5**, 710-2 (2006).
8. Laine, R.M.; J. Marchal, J. Azurdia, "Liquid Feed Flame Spray Modification of Nanoparticles," 264645000 (USPTO) issued Sept., 2006. Ser. No. 60/548,729 filed Feb. 27, 2004
9. Mueller, R.; R. Jossen, H.K. Kammler, S.E. Pratsinis, M.K. Akhtar, "Growth of zirconia particles made by flame spray pyrolysis", AIChE J., **50**, 3085-94 (2004).
10. Mueller, R.; R. Jossen, S. E. Pratsinis, M. Watson, M. K. Akhtar, "Zirconia Nanoparticles Made in Spray Flames at High Production Rates," J. Am. Ceram. Soc., **87,** 197-202 (2004).
11. Mädler, L.; W. J. Stark, S. E. Pratsinis, "Simultaneous deposition of Au nanoparticles during flame synthesis of TiO_2 and SiO_2," J. Mater. Chem., **13**, 2979-84, (2003).
12. Limaye, A.U.; J.J. Helble, "Effect of Precursor and Solvent on Morphology of Zirconia, Nanoparticles Produced by Combustion Aerosol Synthesis," J. Am. Ceram. Soc., **86,** 273-278 (2004).
13. Limaye, A.U.; J.J. Helble, "Morphological Control of Zirconia Nanoparticles through Combustion Aerosol Synthesis, J. Am. Ceram. Soc., **85,** 1127-1132 (2002).
14. Limaye, A.U.; J.J. Helble, "Secondary Atomization as a Mechanism for Controlling the Size of Ceramic Nanoparticles Produced by Combustion Aerosol Synthesis," J. Aerosol Sci. **35**, 599-610 (2004).
15. Srdi, V.V.; M. Winterer, H. Hahn, "Sintering Behavior of Nanocrystalline Zirconia Prepared by Chemical Vapor Synthesis," J. Am. Cer. Soc. **83**, 729-36, (2000).
16. Srdi, V.V.; M. Winterer, H. Hahn, "Sintering Behavior of Nanocrystalline Zirconia Doped with Alumina Prepared by Chemical Vapor Synthesis," J. Am. Cer. Soc. **83** 1853-60, (2000).
17. Glumac, N.G.; Y.-J. Chen, G. Skandan, B. Kear, "Scalable high-rate production of non-agglomerated nanopowders in low pressure flames, Mater. Lett. **34**, 148-53 (1998).

18. Glumac, N.G.; Y-J. Chen, G. Skandan, "Diagnostics and modeling of nanopowder synthesis in low pressure flames, J. Mater. Res. **13**, 2572-78 (1998).
19. Ulrich, G.D. "Flame Synthesis of Fine Particles," Chem. & Eng. News **62**, 22-29 (1984).

87b Opportunities And Limitations Of Flame Synthesis: Biomaterials, Metal And Alloys

Evagelos K. Athanassiou, Robert N. Grass, Fabian M. Koehler, and Wendelin J. Stark. Chemistry and Applied Biosciences, Swiss Federal Institute of Technology (ETH Zurich), Wolfgang-Pauli-Str. 10, ETH Hönggerberg, Zurich, 8093, Switzerland

Flame synthesis has been applied for several decades for the large-scale manufacturing of metal oxides such as silica and titania. Recent developments including flame spray pyrolysis have allowed the large scale production of biomaterials, carbonates and other salts. Controlled oxygen limitation during flame processing gives even access to non-noble metals, complex alloy nanoparticles and metal/ceramic hybrid materials.

We show how simple calculations based on solid state chemistry data may be applied together with gas phase thermodynamics to influence the product composition and to get a better insight in opportunities and limitations of flame processes. This detailed understanding provides a basis for chemical reaction engineering of flames and resulted in a variety of functional nanomaterials starting from nano-gypsum [1], calcium-phosphate[2] for bio applications, to non-noble metal[3-7] nanoparticles, complex alloys[8] and metallic / ceramic[9] nanocomposites for mechanical, magnetic or electronic applications. We demonstrate how such metal nanoparticles can be used for the preparation of bulk nanocrystalline materials, with an up to tripled Vickers hardness if compared to conventional metals.

Fig. 1: Scanning electron microscopy image of nano-NaCl (top left). Transmission electron micrograph of as prepared tricalcium phosphate nanoparticles (bottom left), which can be applied for biomedical applications such as medical implants (top right). Metallic bulk nano-bismuth pill exhibits conducting properties (bottom right).

References:

[1] N. Osterwalder, S. Loher, R. N. Grass, T.J. Brunner, L. K. Limbach, S. C. Halim, W. J. Stark, *J. Nanopart. Res.*, **2007**, 9, 275. [2] T.J. Brunner, M. Bohner, C. Dora, C. Gerber, W.J. Stark, *J. Biomed. Mater. Res. Part B*, **2007**, DOI: 10.1002/jbm.b.30809 [3] E. A. Athanassiou, R. N. Grass, W. J. Stark, *Nanotechnology*, **2006**, 17, 1668. [4] R. N. Grass, W. J. Stark, *J. Mater. Chem.*, **2006**, 16, 1825. [5] R.N. Grass, W.J. Stark, *J. Nanopart. Res.*, **2006**, 8(5), 729. [6] R.N. Grass, M. Dietiker, C. Solenthaler, R. Spolenak, W.J. Stark, *Nanotechnology*, **2007**, 18(3), 035703. [7] R.N. Grass, E.K. Athanassiou, W.J. Stark, *Angew. Chem. Int. Ed.*, **2007**, in print [8] E.K. Athanassiou, R.N. Grass, N. Osterwalder and W.J. Stark, *Chem. Mater*, **2007**, under review [9] R.N. Grass, T.F. Albrecht, F. Krumeich, W.J. Stark, *J. Mater. Chem.*, **2007**, 17, 1485.

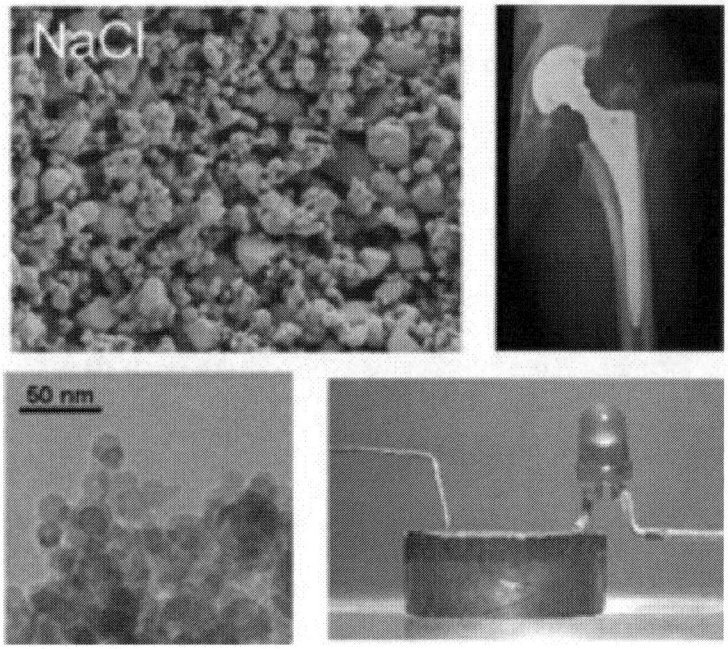
NaCl
50 nm

87c Effect of Dopants on the Phase Stabilization of Flame Made Zirconia Nanoparticles

Ranjan Pati[1], Hillary Sadoff[1], Osifo Akhuemonkhan[1], and **Sheryl Ehrman**[2]. (1) Chemical and Biomolecular Engineering, University of Maryland, 2113 Building 90, College Park, MD 20742, (2) University of Maryland College Park, 2113E Chemical & Nuclear Engineering Bldg., College Park, MD 20742

Zirconia (ZrO_2) occurs in three crystallographic forms- monoclinic, tetragonal and cubic with monoclinic being the most stable form among these three. Although cubic and tetragonal phases are not the thermodynamically favored at room temperature, these are stabilized by the addition of Y_2O_3, CaO, MgO, CeO_2, and CuO. The phase stabilization of ZrO_2 by these dopants is the basis of chemical and mechanical properties of zirconia based ceramics that are important for energy conversion related applications. In the present study, stabilized zirconia powders with different dopant (Y_2O_3, CaO, MgO, CeO_2, and CuO) concentrations are synthesized by a one step flame spray pyrolysis. In this process, aqueous solutions of precursors are atomized using a nebulizer. The atomized solution is then passed through the flame reactor. In the reactor, the precursors react in the solid phase, forming the final product. The particles are collected by thermophoresis onto a water-cooled surface. The materials are characterized by using FTIR spectroscopy, X-ray diffraction (XRD), high resolution transmission electron microscopy (HRTEM) and BET surface area. X-ray diffraction study shows the formation of cubic and tetragonal phases of the as prepared powders. BET surface areas of the as prepared samples are in the range of 10-80 m^2/g. The stabilized zirconia particles are roughly spherical in shape and size in the range of 30 - 100 nm depending on the stabilizing agents used. The tetragonal / cubic phase stabilization effect for the dopants were found to be, in order of decreasing ability to stabilize the cubic phase, Y_2O_3 > CaO > CeO_2 > MgO > CuO, whereas the ionic radii are Ca^{2+} > Y^{3+} >Ce^{4+} >Mg^{2+} > Cu^{2+}.

87d Spray Flame Pyrolysis For Synthesis Of Supported Gold Nano-Catalysts

Sachit Chopra and Gregory Beaucage. Department of Chemical and Materials Engineering, University of Cincinnati, Cincinnati, OH 45221-0012

Spray Flame Pyrolysis (SFP) is of great potential for mass production of nano-materials, especially oxides and supported metals. Flames offer a unique environment for nano-particle growth since they contain high degrees of super-saturation, extremely high temperatures and rapid quench conditions, locking in kinetic structures. Spray jets flames offer the most versatile and controllable of pyrolytic conditions for nano-particle formation. We report on the development of a SFP apparatus and synthesis of supported gold nano-particles on titania (Au/TiO_2). The synthesis of supported gold nano-particles with desired catalytic properties requires understanding of structural nucleation and growth in this complex "co-precipitation"- like reaction. This is possible due to recent advances in characterization of nano-particle nucleation, growth and aggregation in flames using in situ x-ray scattering. Anomalous scattering (ASAXS) is used to isolate the contribution due to gold from that of the support through contrast matching. USAXS, SAXS and XRD are used to determine the structure and morphology of the nano-particles produced. ICP-MS and other spectroscopic techniques are employed for the composition analysis of material produced. The reactivity of the supported Au/TiO_2 as a function of various synthesis parameters and the particle morphology is also studied.

87e Aerosol Synthesis Of Zns Nanoparticles

Hongwang Zhang, Sha Liu, and Mark T. Swihart. Chemical and Biological Engineering, University at Buffalo (SUNY), 303 Furnas Hall, Buffalo, NY 14260-4200

Zinc sulfide is a II-VI semiconductor with a large direct band gap (3.5 eV) in the near-UV region. It has found use in a variety of applications ranging from electronic devices to biology. Production of ZnS nanoparticles by spray pyrolysis has also been reported previously. However, to date, there are no reports on synthesizing ZnS NPs by spray pyrolysis using a single-source precursor. In this work, we demonstrate preparation of ZnS nanoparticles (NPs) by spray pyrolysis using inexpensive and commercially available zinc diethyldithiocarbamate as a single-source precursor. Solutions of this precursor in toluene are dispersed into fine droplets by an atomizer and then carried by nitrogen into a tube furnace where they evaporate fully and ZnS NPs nucleate from the vapor phase. The synthesized NPs were characterized by transmission electron microscopy (TEM), high-resolution transmission electron microscopy (HRTEM), photoluminescence (PL) spectroscopy, UV adsorption spectroscopy and photoluminescence excitation (PLE) spectroscopy. The NPs are range from 3 – 20 nm in diameter. The HRTEM analysis clearly shows the NPs are crystalline, as indicated by the lattice fringes and selective-area electron diffraction pattern (SAED). Hexane dispersions of them exhibit bright blue emission under UV illumination. The PL emission ranges from 300 nm to 500 nm. The UV adsorption spectrum shows peaks in the range of 340-380 nm. The reaction mechanism was investigated by the online measurement of particle size distributions using a scanning mobility particle spectrometer.

87f Synthesis And Functionalization Of Silicon Nanocrystals By Means Of Non-Thermal Plasmas

Lorenzo Mangolini, Chemical Engineering and Materials Science, University of Minnesota, Minneapolis, MN 55455 and Uwe Kortshagen, Mechanical Engineering, University of Minnesota, Minneapolis, MN 55455.

We present a gas-phase, non-thermal plasma approach for the synthesis of silicon quantum dots. Silicon is non-toxic to the human body, cheap, abundant, and compatible with the current micro-electronic technology, and thus presents many advantages with respect to other quantum dot materials, such as CdSe. The research in the field of silicon quantum dots has suffered from the lack of a synthesis technique capable of producing high-quality nanocrystals. Liquid-phase synthetic routes produce samples with a broad size distribution and with poor crystalline fraction. Gas-phase synthesis techniques are affected by fast particle agglomeration rates, which make it difficult to produce very small silicon crystals. Silane-containing non-thermal plasmas are well known to be prone to gas-phase nucleation of particles. Dust nucleated in non-thermal plasmas is well-known to acquire a negative charge, and the resulting Coulomb interaction between the particles greatly reduces the agglomeration rate. Moreover, the particle temperature can largely exceed the background gas temperature because of the recombination of ions and electrons at the particle surface and because of reactions with gas-phase radicals. This allows obtaining high-quality single-crystal particles. The reactor that we have developed consist of a 3/8" OD, 1/4" ID quartz tube with two ring electrodes mounted outside of the tube. An argon-silane-hydrogen mixture is fed into the system and a radiofrequency (RF) signal is applied to the electrodes, resulting in gas-breakdown and formation of a plasma. The system is a continuous flow-through reactor; particles are extracted from the plasma by aerodynamic drag and deposited on a filter downstream of the discharge. The mass yield of the system has been determined by measuring the mass of the reactor quartz tube and the mass of the filter before and after collection. From mass spectroscopy measurements we know that the reactor consumes completely the silane that is fed into the system. Half of the silicon mass introduced in the reactor as silane is converted into a film deposited onto the reactor wall and half is converted into particles. The measured mass yield can be as high as 50 mg/hour of material that has visible photoluminescence after oxidation in air. Extensive TEM analysis suggests that the process is capable of producing very small (<5 nm) crystallites, and it confirms that quantum confinement effects should be observed from the plasma-produced silicon nanocrystals. The quantum yield of the fluorescent particles, defined as the ratio of the number of photons emitted in the visible over the number of UV photons absorbed by the sample, has been measured using an integrating sphere and a USB spectrometer (Ocean Optics USB2000). A UV-LED emitting at 390 nm is used to excite the particle fluorescence. The hydrogen-termination of the plasma produced silicon nanocrystals does not prevent surface oxidation, and the quantum yield of the surface-oxidized silicon nanoparticles rarely exceeds 10%. It is in fact well known that the growth of a native oxide is associated with the formation of defects at the Si/SiOx interface. An alternative surface passivation technique is the grafting of organic molecules onto the hydrogen-terminated surface of the as-produced nanocrystals. Two different approaches have been explored for this purpose. The first approach is a liquid-phase thermal reaction.

The silicon particles have been collected on the usual stainless steel filter, then sonicated into a 5:1 mixture of mesitylene and 1-dodecene. The transfer from the system to the solution is done under air-exclusion by using a nitrogen purged glove bag mounted onto the system exhaust line. The liquid-phase reaction proceeds at the solvent boiling temperature and a clear colloidal solution of silicon nanoparticles is obtained in a few minutes. After hydrosilylation the particles can be dried and redispersed into non-polar solvents. The improved liquid-phase processibility of the material makes it very interesting for printed electronics applications. For liquid-phase hydrosilylated material we report quantum yield values exceeding 60% for samples with a peak emission wavelength around 800 nm. The second approach for the passivation of the silicon nanoparticle surface is a gas-phase approach. Gas-phase treatment avoids the use of solvents from the production scheme and allows a significant reduction in production time. Our approach to the problem is to use a second non-thermal plasma to provide the necessary reaction activation energy. The same molecule used for the liquid-phase passivation, 1-dodecene, is fed into a second-stage plasma reactor using a bubbler system. Particles produced in the first stage are dragged by the flow into the second stage where the in-flight hydrosilylation takes place. The in-flight grafted material is soluble in non-polar solvents without any further processing. When the particles are collected on the stainless steel filter they are heavily agglomerated, but the agglomerates dissolve into the solvent and freestanding particles are present in solution. FTIR data will be presented confirming that for gas-phase treated particles 1-dodecene has reacted with the particle surface, since the IR absorption spectrum from the in-flight hydrosilylated material practically overlaps with the IR absorption spectrum from the liquid-phase treated material. Unfortunately the quantum yield of the in-flight hydrosilylated material is typically <10%, and further annealing is necessary to improve the optical properties. The gas-phase reactor is undergoing further development to obtain silicon nanoparticles that are readily soluble after gas-phase production and that have optical properties as good as those of the liquid-phase reacted particles.

87g Molecular Dynamics Simulation Of Nanoparticle Fractal Aggregate Sintering: Developing A Simple Phenomenological Model

Takumi Hawa, University of Maryland, College Park, MD 20742 and **Michael R. Zachariah**, University of Maryland and NIST, 2125 Glenn L Martin Hall, College Park, MD 20742.

A simple modification to the Frenkel sintering law is developed for nanoparticle fractal aggregates, based on molecular dynamics (MD) simulations. The fractal aggregates investigated consist of up to 110 primary particles of silicon, with primary particles of 2.5 nm in diameter. Aggregates of Fractal dimension of 1 (wire), 1.9 (complex), and 3 (compact) were considered. Sintering of aggregates consists of three steps, a) reaction between particles to minimize surface defects, (b) sintering of multiple secondary branches to the primary branch, and (c) contraction of the primary branch. The sintering times normalized by the primary particle diameter showed a universal relationship that only depends on the number of particles in an aggregate and its fractal dimension. This result was found to be consistent with a continuum viscous flow mathematical model we developed. The results for the sintering of arbitrary fractal aggregates can be approximated with a power law modification of the Frenkel viscous flow equation, to include a dependence on the number of particles in a fractal aggregate and fractal dimension:

Web Page: www.enme.umd.edu/~mrz

172a Aspects of Latex Particle Size Control for Improved Water Blush Resistance

Natasha Starostina, Technology Center, Pacific Nanotechnology Inc., 17981 Sky Park Circle, Suite J, Irvine, CA 92614 and Scott Harvey, New Technology Solutions, Ashland Specialty Chemical Company, P.O. Box 2219, Colombus, OH 43216.

Latex films have a tendency to blush when exposed to water. The swelling of trapped hydrophilic material, resulting in pockets with different refractive indices is proposed as a possible model for blushing. A pressure sensitive latex was developed based on this model. The blush was improved by choice of surfactant and control of particle diameter. The resulting latex has an unusual particle growth behavior, during polymerization. The particle size distribution was characterized by dynamic light scattering (DLS), methods of electron microscopy (SEM/TEM) and with the help of atomic force microscopy (AFM). AFM data are in a good agreement with our growth model. A possible explanation, based on competitive growth of dissimilar particle diameters is proposed.

Web Page: www.nanoparticles.pacificnano.com

172b pH-Dependent Controlled Clustering of Polymer-Functionalized Gold and Magnetite Core-Shell and Janus Nanoparticles

Tatsushi Isojima, Chemical Engineering, Massachusetts Institute of Technology, 77 Massachusetts Avenue, Room 66-325, Cambridge, MA 02139 and T. Alan Hatton, Department of Chemical Engineering, Massachusetts Institute of Technology, 77 Massachusetts Ave, Cambridge, MA 02139.

In recent years, there has been increasing interest in the potential applications of functionalized nanoparticles, with a concomitant strong scientific focus on their fundamental properties. These functionalized nanoparticles, prominent among which are gold(Au) and magnetic (Fe_3O_4) nanoparticles, are applied in a wide range of fields, including drug delivery systems, diagnostics, gene analysis, proteomics, affinity purifications, quantum dots, and the like. There are undoubtedly still many gains to be made in the synthesis, functionalization, characterization and applications of these systems. Gold nanoparticles display electromagnetic resonance due to collective oscillations of their free electrons known as plasmons. For particles which are small compared to the wavelength of the incident light, plasmons are excited when light with an appropriate wavelength is shone on them. This results in a strong absorption at that wavelength. Because of plasmon behavior, different conditions of aggregation result in different colors; which helps us for a visual determination of the aggregation condition of particles. On the other hand, the magnetite nanoparticles can be controlled by an external magnetic field. It is very important to prepare nanoparticles with a well defined size distribution and with surface functionalization because of their ferrohydrodynamic behavior. In this report, we present polymer functionalized gold and magnetite nanoparticles. Two forms of functionalization have been employed which resulted in the core shell type and the Janus type nanoparticles.

Poly acrylic acid (PAA)-coated core-shell gold and magnetite nanoparticles were prepared by surface initiated atomic transfer radical polymerization (ATRP) (namely "grafting from method") as has been the recent trend in synthesizing core-shell type nanoparticles. ATRP is a well-known technique for creating a narrow polymer distribution which allows us to make core-shell nanoparticles with a precisely designed and high density polymer shell. The methods of attaching polymer on the surface of nanoparticles (namely "grafting onto method") tend to lead to a relatively low graft density because of the steric repulsions. In addition, PAA is a well-known and a very useful polymer because of its numerous carboxyl groups. We can control the aggregation and dispersion by changing pH because of the isoelectric point for PAA(pKa=4.5). Also these carboxylic groups on the surface help in the easy functionalization of the nanoparticles. We examined the PAA coated core shell structure of nanoparticles by TEM, TGA, FT-IR, GPC, DLS, UV/Visible spectroscopy (for gold particles) and Zeta potential measurement and their pH dependent behaviors by DLS, Zeta potential and UV/Visible spectroscopy (for gold particles) at pH values from 10 to 2.

Also, we have recently developed methods for the synthesis of a new set of functional nanoparticles with asymmetric surface properties "C Janus Particles. In previous studies, Janus particles have been prepared for particles on the order of 100s of nanometers, but never before at the 10 nm scale. Specifically, we have prepared nanoparticles that have a composite structure consisting of a nanocolloid and a soft material. The colloid size is approximately 5-10 nm and the functional polymer is on only one side of the colloid. This Janus particle can have a surfactant-like structure if the remaining colloid surface has different properties than the tail polymer. Consequently, Janus nanoparticles can self-assemble into micelle-like structures depending on medium conditions such as pH, temperature, concentration and additional salt. We characterized Janus particles by TEM, DLS, and Zeta Potential.

Preparation and characterization of the core-shell type of nanoparticles

Poly(acrylic acid) coated core-shell type of gold and magnetite nanoparticles were prepared by deprotection of poly(trimethylsilyl acrylate)(TMSA) which was polymerized on the gold nanoparticles by ATRP (by using the grafting from method). Gold nanoparticles were prepared in the presence of 11-mercaptoundecanol(MUD) and subsequently esterified with 2-bromoisobutyryl bromide (BIB) for ATRP initiation (Mandal et al.,NANO lett. 2, 3(2002)), and the following polymerization of TMSA occurred on the surface of gold particles. Magnetite nano particles were prepared following the organic synthesis route proposed by Sun et al.(JACS 126, 273(2004)). Ligand exchange reaction of galactic acid was done and subsequently esterified with BIB for ATRP initiation and the following polymerization of TMSA occurred on the surface of magnetite nanoparticles (Langmuir, 23, 2158 (2007)).

TEM images show the size of gold and magnetite to be about 10nm and 5nm respectively. We observed core-shell structure of both gold and magnetite nanoparticles with 2-3nm PAA shell. At pH greater than 4.5, the PAA coated nanoparticles were well-dispersed for at least 1 month. At acid condition smaller than pH 4.5, these particles aggregated due to a decrease in their negative zeta potential. The optical properties of the clustered PAA coated gold nanoparticles exhibit red shift, together with damping and broadening of the surface plasmon features at pH 2. We have investigated this clustering behavior by UV/visible spectroscopy and DLS as a function of time. These clusters slowly re-dispersed on changing the pH from 2 to 7.

Preparation and characterization of the Janus particle

Our Janus nanoparticles consist of either magnetite or gold crystals coated on one side with a pH-dependant polymer (PAA) that is charged at high pH but not at low pH, with the other side functionalized by a second polymer that is always charged, independent of the solution pH. Therefore, these Janus particles can be dispersed individually at high pH values, but can self assemble at low pH value. The first step in the preparation of Janus nanoparticels consists in preparing nanoparticles (magnetite and gold) with size approximately 5-10 nm, following which they are coated them poly acrylic acid (PAA), both of which have pKa values below neutral pH. In particular, these nano colloids have high negative surface charges of about -50mV at pH10 and are uncharged at a pH of 2. Next, large positively charged silica beads with diameters of approximately 700nm are prepared using a silane coupling agent with a trimethyl ammonium group. The positively charged silica beads are then dispersed in the nanoparticle suspension at high pH, and coated by a monolayer of nanoparticles through electrostatic adsorption. The surfaces of the nanoparticles facing the silica beads are protected from further functionalization while amino-end functionalized polystyrene sodium sulfonate (PSSNa Mw:90kD) are attached to the exposed surfaces of the nanoparticles through amidation chemistry. The recovery of the Janus particles thus synthesized is achieved by increasing the pH of the solution thereby inducing a charge reversal on silica bead surface. The pH dependent aggregation behaviors of both polymer-coated magnetic nanoparticles and Janus magnetic nanoparticles have been measured by dynamic light scattering (DLS). The results for the Janus nanoparticles show that self assembly does occur, and a stable dispersion of clusters of approximately 80 nm in diameter can be formed at pH 2. On the other hand, nanoparticles without the tail polymer become uncharged, aggregate and precipitate at pH values less than about 4, i.e., less than the pKa of the coating polymer. It is noted that the sizes of self-assembled Janus nanoparticle clusters change depending on their concentration, from a low value of 15 nm at low concentration (corresponding to individually dispersed nanoparticles) to a higher value of 95 nm at high concentration; this behavior is akin to that observed near the critical micelle concentration noted for conventional molecular and polymer micellar systems.

172c Application Of Functionalized Magnetic Nanoparticles In Separation Technology

Fabian M. Koehler, Robert N. Grass, Evagelos K. Athanassiou, and Wendelin J. Stark. Chemistry and Applied Biosciences, Swiss Federal Institute of Technology (ETH Zurich), Wolfgang-Pauli-Str. 10, ETH Hönggerberg, Zurich, 8093, Switzerland

Magnetic particles are routinely applied in biochemistry separations. Unfortunately their application is often limited due to high costs, poor stability and low binding capacities.

Here, we use carbon coated metallic nanoparticles prepared by reducing flame spray synthesis. The magnetic nanoparticles are highly air stable until 190°C. In a one step synthesis the carbon surface of the magnetic beads was functionalized with chlorine- or nitrogroups. Due to the strength of the generated carbon-carbon bond the links are stable under a wide variety of reaction conditions.

The derivatized material was separated within seconds from the reaction mixture using a permanent magnet. Coupling a molecule of interest to the functionalized particles generates an easily separate-able molecule for further treatments. The low costs and particle stability favor this preparation method and material for large-scale separation application of biological molecules such as proteins and antibodies.

The left figure shows the powder mass gain upon oxidation measured by thermogravimetry in air of pristine carbon coated cobalt nanoparticles (solid line) and chlorobenzene functionalized nanobeads (broken line). The scheme in the middle shows covalently functionalized metallic and magnetic nanoparticles as a platform for inexpensive magnetic separations in synthesis of biological molecules. The TEM picture on the right side presents carbon-coated cobalt particles.

References

[1] R. N. Grass, E. K. Athanassiou, W. J. Stark, *Angewandte Chemie* **2007**, in print.

[2] R. N. Grass, W. J. Stark, *J. Mater. Chem.* **2006**, 16, 1825.

[3] R. N. Grass, M. Dietiker, R. Spolenak, W. J. Stark, *Nanotechnology* **2007**, 18, 035703.

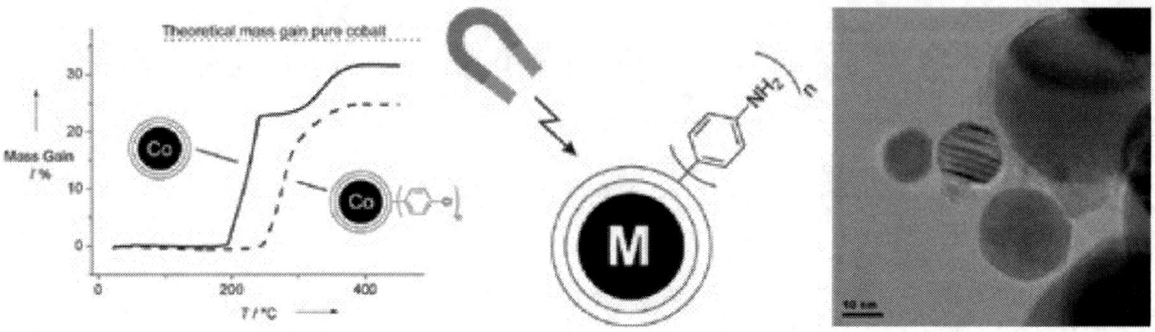
Theoretical mass gain pure cobalt
30
20
10
0
Mass Gain
/ %
Co
Co
0
200
400
T / °C
M
NH2
n
10 nm

172d Catalytic Magnetic Nanoparticles For Organophosphate Degradation

Huan Zhang, Lev Bromberg, Kenneth A. Smith, and T. Alan Hatton. Department of Chemical Engineering, Massachusetts Institute of Technology, 77 Massachusetts Avenue, Cambridge, MA 02139

Organophosphates (OP) represent a major class of pesticides as well as a number of chemical warfare agents. In addition to current remediation approaches, catalytic hydrolysis of these toxic environmental contaminants by recyclable particulates is emerging as a promising alternative. We have developed a catalytic system based on magnetic nanoparticles functionalized with appropriate reactive groups for catalytic decomposition of OP compounds; these particles can be recycled and reused repeatedly.

Magnetite (Fe_3O_4) nanoparticle clusters with an average size of approximate 100 nm were successfully functionalized by ƒÑ-nucleophilic groups, such as oximes and iodo-benzoates, through both water- and organic solvent-based reactions. These modified magnetite nanoparticle colloidal suspensions were electro-statically stabilized over a wide pH range. The particles significantly accelerated the hydrolysis reaction of a model compound, p-nitrophenyl acetate, and a real pesticide, paraoxon, and were readily recovered from the aqueous milieu for reuse by high-gradient magnetic separation methods with no loss of catalytic activity. A pseudo-phase model incorporating both surface adsorption of the substrate and the subsequent catalytic reaction by the functional groups on the particle surface captured well the experimentally observed trends.

172e "Nanoions": Fundamental Properties And Analytical Applications Of Charged Nanoparticles

Kyle J. M. Bishop, Chemical & Biological Engineering, Northwestern University, 2145 Sheridan Rd/TECH E136, Evanston, IL 60208 and Bartosz A. Grzybowski, Chemical Engineering, Northwestern University, 2145 Sheridan Rd/TECH E136, Evanston, IL 60208.

What happens when solutions containing objects bearing charges of opposite polarities are mixed? It depends on the scale and is thought to be fundamentally different in the molecular and the colloidal regimes. Whereas oppositely charged microparticles precipitate continuously over a wide range of relative particle concentrations, mixtures of oppositely-charged nanoparticles (NPs) precipitate sharply only at the point of NP electroneutrality. This behavior – reminiscent of threshold precipitation of inorganic ions – is specific to the nanoscale and is due to the formation of like-charged, "core-and-shell" clusters stabilized in solution by mutual electrostatic repulsions. The formation of these charged clusters is rationalized by deriving the electrostatic interaction potentials between NPs accounting for the adsorption/desorption of counterions from the NPs' surface. Interestingly, this model predicts that the attractive energy between oppositely charged NPs at contact is nearly twice the repulsive energy of like-charged NPs at the same distance. Monte Carlo simulations employing these potentials confirm the formation of "core-and-shell" aggregates and sharp precipitation at electroneutrality. Finally, we show how nanoparticle titrations based on these phenomena provide a uniquely accurate tool for determining charges tethered onto nanoscopic objects and for studying the thermodynamics of surface reactions at the nanoscale.

172f 1-Step Nanoparticle Coatings And Applications

Stoyan K. Smoukov[1], Kyle J. M. Bishop[1], Bartlomiej Kowalczyk[1], Alexander M. Kalsin[1], Anatoliy Pinchuk[2], Maciej Paszewski[1], Tim P. Gray[2], and Bartosz A. Grzybowski[1]. (1) Chemical & Biological Engineering, Northwestern University, 2145 Sheridan Rd/TECH E136, Evanston, IL 60208, (2) Chemistry and Chem. Engineering Departments, Northwestern University, 2145 Sheridan Rd., E136, Evanston, IL 60208

Facile 1-step functionalization of most hydrophilic surfaces with a dense monolayer of nanoparticles via a novel self-organization mechanism. These coatings display novel optical (plasmonic) properties, of interest both fundamentally and for sensor applications.

181a a Perspective on Andrew Jenike's Groundbreaking Work at the University of Utah Experimental Station

Jerry R. Johanson, Diamondback Technologies, Atascadero, CA 93422

to be completed at a later date

Jenike's Influence on the European Particle Technology Community

Prof. Gisle G. Enstad, Tel-Tek, dept. POSTEC, Kjoelnes ring, N-3918 Porsgrunn, Norway

Background

To most workers active in particle technology Salt Lake City is a well known name thanks to the pioneering work that was carried out here by Andrew Jenike and his co-workers in the period from 1952 to 1962 (1, 2, 3, 4). In this period the basic phenomena concerning flow in silos were revealed, and we got new concepts like mass flow and funnel flow, which changed our understanding of flow in silos fundamentally, and has become a part of our everyday vocabulary, like laminar and turbulent flow of fluids. In addition also basic flow properties of powders were investigated, and new concepts for describing these properties were developed, together with the Jenike tester, which is now in use all over the world to investigate flow properties of a countless number of different materials stored in silos. Finally, a complete method for safe design of silos for reliable flow and discharge was developed, known as the famous Jenike method. By the Jenike method it is possible to test the flow properties of powders, using the Jenike tester, and to calculate critical silo parameters, using the Jenike theory. Knowing the critical silo parameters, including critical hopper inclinations for mass flow and critical outlet widths for reliable gravity flow, for both mass and funnel flow, silos can be designed to give the desired mode of flow, mass or funnel flow and with sufficient outlet size to ensure reliable discharge by gravity alone.

The consultancy company Jenike & Johanson was founded right after the Jenike method was completed, and it was based on the application of this method for solving industrial silo problems, and designing new silos that would work reliably. The company was founded by Jenike and his closest co-worker during the pioneering period, Jerry Johanson, who made extensive contributions to the development both of the Jenike shear tester and the Jenike theory of flow in silos (5, 6). This was the first company that was solely based on the Jenike method, but the method was quickly picked up by other workers in the field, and was taken into use in the course of relatively few years more or less over the whole world, first in Europe and Australia, but soon also in Asia and Africa (7, 8, 9). A large number of research institutions involved in particle technology research are at present using the Jenike method for silo design and a countless number of silos have been built and designed by means of this method. The company Jenike & Johanson is still thriving long after Jenike retired and Johanson left. Johanson has founded his own company based on the Jenike method, but he has also developed further along his own lines, and is no longer limited only to the original Jenike method (10, 11).

The Jenike method in Europe

I joined the Chr. Michelsen Institute in Bergen in 1972, where I started to work with Rolf Eckhoff, who was the one who brought the science of powder technology to Norway. Jenike came to Norway in the late 1960-ies to visit William Bruff, who needed his assistance to solve a severe silo problem at Elkem Fiskaa. Compared to what had been tried so far, the solution Jenike came up with was both elegant and simple, and in contrast to the previous trials, it worked perfectly. Some problems emerged when the discharge capacity was increased over a critical limit, but together Jenike and Bruff came up with a simple solution where they increased the discharge capacity dramatically by blowing a small amount of air into the powder by means of an aquarium pump (12).

Bruff convinced Rolf Eckhoff at the Chr. Michelsen Institute that this was a new field he ought to get involved in, and a Jenike tester was purchased by him in 1969. With Rolf Eckhoff as his supervisor Per-Gunnar Leversen made his masters degree on the accuracy of the Jenike method, by carrying out experiments in both experimental silos and in an actual industrial silo installation at Norton in Lillesand in Norway, where it was particularly important to avoid segregation (13). Some over design was found, and I was then assigned the task to find the reason for the over design. One assumption in the original arching theory was that the pressure from the powder above an arch could be ignored, an assumption that would lead to over design. The idea was to make an amendment where the pressure on top of the arch was included in the analysis. Many years later this work led to my doctor's degree. There was some dispute whether the over design was real or not, especially Wright (14) came to a different conclusion, with more or less exact agreement between experimental and theoretical results.

In order to introduce the Jenike method to Norwegian and Scandinavian industry, a silo course was organized in the fall of 1972 in Geilo in the central part of Norway. Jenike was invited to give lectures about his discoveries at this course, and at this occasion I had the opportunity to meet him for the first time. At a small gathering the evening before the course he gave a small speech, and I remember he started by asking, what has gone wrong? We have known how to make silos for centuries, why do we have to ask for new expertise now? He gave a number of reasons why expertise was required, but I think he pointed at a problem we still have. Design of silos is regarded by many as something with long traditions, something simple that everybody knows how to do, although it is perhaps the most complicated building you can think of.

At this course I learned for the first time of mass flow, with complete movement of the whole silo content during discharge, and funnel flow, with only partial movement through a flow channel from the outlet to the top surface, where a funnel is formed by the material sliding along the top surface down to the flow channel. What practical consequences this had for the function of the silo, like the first in – first out principle for mass flow, and the first in – last out for funnel flow, made deep impressions. We also learned about one of the customs of Andrew Jenike, he always found some time for a mountain hike wherever he went in the world, and in Norway we have plenty of mountains. Per-Gunnar Leversen and I were the privileged two young guys who were chosen to accompany him for one day at the top of Hallingskarvet, jumping from boulder to boulder brought there thousands of years ago by the ice that once covered the whole of Norway and a large part of Europe. We two newcomers to the field tried to get so much information as possible from the pioneer, and I remember he said to me that if you want to join us, there are enough unsolved problems in this field that you can join us for eternity. Many years later I think I still can say the same thing to the young people who are now the newcomers to the field.

Having carried out his master's degree on the accuracy of the Jenike method, Leversen started to apply the method for silo design from the beginning of 1973 at the Chr. Michelsen Institute, and I started on my theoretical work to find the reason for the over design, with the aim to do some amendments that could give a higher accuracy. In this period we got an invitation from Jenike to attend a seminar he was giving in co-operation with Hans Rumpf in Karlsruhe, and it was decided that I should go to attend this seminar. Andrew's comment when he met me was that he thought I had a girlfriend in Karlsruhe, since I had travelled so far. But this was the only time I briefly met with Hans Rumpf, who also left us a legacy of important work on powder or particle technology, and I also had the opportunity to discuss my plans for a new stress and arching theory of mass flow silos with Andrew Jenike. In fact my ideas were based on a conceptual drawing in his Bull. 123 (4). Andrew agreed that it would be worth trying analytical solutions along these lines, a recognition that was of great importance for me. He also gave some hints on how to do it, and I went straight home to finish the mathematical analysis along the lines we had discussed. The results were first published in Chemical Engineering Science (15, 16), and later formed the basic contents of my doctoral dissertation.

The WPMPS of EFCE

The Working Party on the Mechanics of Particulate Solids (WPMPS) of the European Federation of Chemical Engineering (EFCE) was founded at the CHISA congress in Prague in 1968, and Rolf Eckhoff was one of the delegates of the founding meeting. This working party had as its main task to promote and co-ordinate research on the new scientific field of mechanics of particulate solids, or powder mechanics which it is also called, that had been initiated by Jenike here in Salt Lake City between 1952 and 1962. It has been one of the most active working parties of the EFCE, and was led by Jan Novosad from Prague for about 30 years. In this way it was also an important link between the scientific communities on both sides of the iron curtain, which was a reality until 1989. The WPMPS had many important members, who have contributed a lot to the field, and it was instrumental in the process of spreading the knowledge of the new theory, and taking it into use.

A number of Universities and research groups in Europe started very soon to use the Jenike method and continued the research that Jenike had initiated, and it will take too long to go into details of this. Only a few might be listed. In Germany Hans Rumpf at the University of Karlsruhe has already been mentioned, in the Netherlands Professor Rademacher at the University of Twente could be mentioned, in Great Britain professor John Williams at the University of Bradford, the Warren Spring Laboratory, the University of Loughborough, as well as the University of Cambridge were among the first to get involved in the Jenike method and his work on silo flow.

Jörg Schwedes from the group of Hans Rumpf ought to be mentioned. His work on testing of powders, where he first investigated the Jenike tester, and then developed the simple shear tester, which is far from simple, and compared its results with the results of the Jenike shear tester, is famous, and forms a mile stone in the scientific acceptance of the Jenike shear tester (17). Later on he went to Braunschweig were he taught powder technology, and created his own research group where he continued his work on investigation of flow properties of powders, after having developed a bi-axial tester for this purpose. Here not only strength properties of powders could be investigated, but also deformation as a response to external stresses. Also work on flow in silos, including the effect of inserts, has been investigated in Braunschweig.

I myself developed a bi-axial tester, inspired by the one in Braunschweig, but with a different technical solution. A simpler tester, the uniaxial tester has also been developed by my group in Norway, but these testers are not suitable as replacements of the Jenike shear tester. The only alternative so far seems to be the ring shear tester, where Dietmar Schulze, who came from the group of Schwedes, has marketed one version that has been sold in a considerable number. Also the Peschl tester, a torsion tester using the whole circular area, has been taken into use by many, and later also the Johanson Indecizers have been purchased by many. But still I think a majority of the people using the Jenike method, including myself, are using the original Jenike tester.

The Jenike tester has been a matter of concern for a long time, and the WPMPS undertook a considerable effort in order to find out how well it worked in the different laboratories where it was applied. A round Robin test where some 20 laboratories participated was initiated in 1978 by the WPMPS, who also led the work. In addition to members of the working party, a number of other people were also invited to take part in this work. A large number of identical samples of lime stone were prepared and distributed to the participating laboratories, which tested the samples and determined the flow function of the material. The results were far from encouraging. The spread was even worse than the most pessimistic participants had anticipated. To find out what could be the reason for the spread in the results, a new series of round tests were carried out, now with some specific guidelines on how to carry out the tests, and the spread in the second round was considerably reduced, but still bigger than what was considered acceptable. This clearly indicated that the major reason for the spread in the results was that the Jenike tester was not used in the same way by all the participating laboratories, and the work to develop a guideline on how to use the Jenike tester was initiated. This was a task that required a lot of work, and many meetings with long discussions. Many people in addition to the working party members were invited for these discussions, and Andrew Jenike often came, and initially he contributed substantially to the work, until he found we were moving too much in a circle, without sufficient progress from meeting to meeting. Finally, the whole working party spent nearly a whole week in Freising in Germany, where every sentence were discussed and spelled out before the final version of the Standard Shear Testing Technique (SSTT) was ready (18). This has now been accepted as the standard for using the Jenike shear tester, and is published by the Institution of Chemical Engineering in Great Britain. After some reworking by John Carson, it has now also become an American Standard.

As a continuation of the work on standardising the use of the Jenike tester, a reference material was developed by a group led by Brian Scarlett and sponsored by the Community Bureau of Reference of the European Union (BCR). In this case 6 well recognised laboratories in 6 European countries participated in the work. The same limestone as before was chosen as test material, but in this case we were determining four yield loci, and not the flow function. Still the spread of the results made it necessary to have a second round of tests, where it was emphasized that the SSTT had to be followed strictly, and also the limestone was modified in order to avoid possibilities of differences between the distributed samples. A considerable improvement was obtained in the second round, but still some outliers had to be removed before the spread just about was small enough for the material to be accepted by the BCR as a standard material, which is now known as CRM 116 (19). Samples of this material can still be purchased from EU together with the report with the standardized results of the Jenike tester.

In addition to the flow properties of the powder itself, also the wall friction is an important part of the parameters that need to be known before a silo can be designed. The WPMPS therefore also initiated round Robin tests on the part of the method developed to measure the wall yield locus. The outcome of the tests was a similar spread in results or perhaps even worse than for the previous work on yield loci and flow functions. Several rounds were carried out, and some reduction in the spread of results was achieved, but more consistent results would clearly have been desirable. The results were published by the chairman of this task, Gerard Haaker before he retired, but this work has not yet been concluded.

In addition to flow property testing, also work on the flow in the silo itself has been promoted by the WPMPS. Segregation is one task that has been carried out, and which has led to a hand book, mainly thanks to the efforts of Lyn Bates (20). Another task has looked into the use of different standards of how to calculate silo loads. Some results have been published by the chairman of this task, Harald Wilms.

It may seem a bit disappointing that so often big scatter in the results have emerged when comparing results from different laboratories. Clearly a lot of the scatter is caused by lack of skill, or different procedures at the various laboratories, but in my opinion this is not the whole explanation. There are still a number of factors we are not aware of that are of importance, and which have to be taken into account and controlled better than what is the case so far. Therefore a lot of research is still needed to understand fully the flow behaviour of particulate solids. One such factor that we already know about is anisotropy. We assume that the properties of the powder are isotropic, which means that they are the same in all directions. Jenike was well aware that this was a simplification. His philosophy was that there needed to be a considerable safety margin. Whenever he needed to make an approximation or an assumption, he made it on the safe side. The strength of a powder is therefore measured in the direction where it has its maximum, which for mass flow is also the correct direction, but not necessarily for funnel flow. Therefore an over design of his method is to be expected, and as the measured properties vary, sometimes the over design may be large, sometimes small. The main thing is that using the Jenike method gives safe design.

Preliminary investigations of anisotropy

In 1985 I went to Miami for an international conference on particle technology, and afterwards I went to Massachusetts where I visited Andrew Jenike. Here he took me on a mountain hike in the mountains where he often used to hike. He told me I had joined a famous club of people who had been hiking with him in these mountains, and he mentioned Otto Theimer, Jörg Schwedes and many more. I felt very honoured to be counted among these people. At that time he must have been around 70, whereas I was only 40. Still I had a hard time to follow him, as he was very fit. He told me how important it was for him to keep physically fit, in that way he could also keep mentally fit. He was busy writing a book summarizing, and elaborating on his most important findings. What he feared most was to get involved in something like a car accident that would prevent him from keeping physically fit. Then he might not be able to recover.

Later that year we organized the conference RelPowFlo I in Norway, and among many participants also Andrew Jenike came. After the conference he spent some time in Bergen, and in addition to hiking in at least two of the seven mountains surrounding Bergen, we were also able to discuss further work in the field of powder mechanics. I was able to show him an early version of our bi-axial tester, and quite a premature version of the uniaxial tester. One outcome of our discussions was that anisotropy needed more investigations, and this led to a student project with Fred Saraber from Twente University in the Netherlands later the same year. A very strong anisotropy was found for the powders we tested (21). Both the bi-axial tester and the Jenike tester were used. In the bi-axial tester the direction of the major principal stress could only be varied by 90°, whereas with the Jenike tester the whole tester was simply turned any angle between 0 and 180° from the pre-shear to the shear test. The shear strength was more and more reduced as the turning angle was increased to 180°, where the strength was reduced to only a fraction of what it was when it was measured in the same direction as the pre-shear.

Concluding remarks

When I heard about the car accident Andrew was involved in, I remembered his remark last time I met him. Sadly he was right in his predictions. His analytical mind had once again come up with the correct answer.

I am glad I had the opportunity to get to know him, and I am very impressed by, and very grateful for the pioneering work he did. I am only one of very many who have found the field he founded so interesting that I have spent most of my life to try to promote it further. Thousands of silos must have been built based on his method of silo design, and this may have improved the economy of industry by perhaps the order of billions of dollars every year worldwide, which was also his vision. He has made a deep impact on our understanding of how particulate solids behave during flow, and our understanding of silo flow has been changed for ever.

Fundamental phenomena like mass and funnel flow were bound to be discovered sooner or later by somebody who would be able to make the necessary observations, and would probably have been discovered by somebody else if Jenike had chosen another subject for his investigations. But Andrew Jenike was the one who made the discovery before anybody else, and formulated it in his own practical way, which initiated a completely new field of research, which will continue to challenge new generations of scientist for a long time still, perhaps for eternity, to use his own words. There must have been published thousands of papers in this field already, and there will probably be thousands more in the years to come.

His good friend William Bruff, who brought this science to Norway, and took great pleasure in contributing to and following the development, wrote a book in Norwegian explaining the findings of Jenike. After he had retired, he once commented that in spite of all the work that had been going on trying to find weaknesses and improvements, nothing substantially new had come up that could be added to the original pioneering work of Jenike and his co-workers. This proves the high quality of the work that was carried out here at Salt Lake City 50 years ago.

References

1. Jenike, Andrew W. (1954), "Flow of Solids in Bulk Handling Systems", and "Flow of Bulk Solids in Bins", Bul. 64 of the Utah Engineering Experiment Station, University of Utah.
2. Jenike, Andrew W., Elsey, P. J. and Wolley, R. H. (1960), "Flow Properties of Bulk Solids", Proceedings, ASTM, Vol. 6, pp. 1168 – 1181.
3. Jenike, Andrew W. (1961), "Gravity Flow of Bulk Solids", Bul. 108, Utah Engineering Experiment Station, University of Utah, Salt Lake City, Utah.
4. Jenike, Andrew W. (1964), "Storage and Flow of Solids", Bul. 123, Utah Engineering Experiment Station, University of Utah, Salt Lake City, Utah.
5. Johanson, Jerry R. (1964), "Stress and Velocity Fields in the Gravity Flow of Bulk Solids", Journal of Applied Mech., Series E, Vol. 86.
6. Johanson, Jerrey R. and Colijn, H. (1964), "New Design Criteria for Hoppers and Bins", Iron and Steel Engineer, Vol. XLI, pp. 85 – 104.
7. Walker, D. M. (1966), "An approximate Theory for Pressures and Arching in Hoppers", Chemical Engineering Science, Vol. 21, pp. 975 – 997.
8. Williams, John C. and Birks, Anthony H. (1967), "The Comparison of the Failure Measurements of Powders with Theory", Powder Technology, Vol. 1, pp. 199 – 206.

9. Molerus, Otto (1975), "Theory of Yield of Cohesive Powders", Powder Technology, Vol. 12, pp. 259 – 275.
10. Johanson, Jerry R. (1992), "The Johanson Indiciser System vs. the Jenike Shear Tester", Bulk Solids Handling, May, pp. 237 – 240.
11. Johanson, Jerry R. (1993), "Characterizing dry Particulate Solids for Systems Design", Proceedings, Reliable Flow of Particulate Solids II, Oslo, Norway.
12. Bruff, William and Jenike, Andrew W. (1967/68), "A Silo for ground Anthracite", Powder Technology, 1, pp. 252 – 256.
13. Eckhoff, Rolf and Leversen, Per-Gunnar (1974), Powder Technology 10, p. 51.
14. Wright, Harold (1972), "An Evaluation of the Jenike Bunker Design Method", Transactions of the ASME, Paper No. 72-MH-7.
15. Enstad, Gisle G. (1975), "On the Theory of Arching in Mass Flow Hoppers", Chemical Engineering Science, Vol. 30, pp. 1273 – 1283.
16. Enstad, Gisle G. (1977), "A Note on the Stresses and Dome Formation in axially symmetric Mass Flow Hoppers", Chemical Engineering Science, Vol. 32, pp. 337 – 339.
17. Schwedes, Jörg (1971), "Scherverhalten leicht verdichteter kohäsiver Schüttgüter", Dissertation, University of Karlsruhe.
18. The Institution of Chemical Engineers, „Standard Shear Testing Technique for Particulate Solids Using the Jenike Shear Cell, ISBN 0 85295 232 5.
19. Akers, Richard R. (1990), "The Certification of a Limestone Powder for Jenike Shear Testing – CRM 116", Loughborough University of Technology, UK, BCR/163/90 (Community Bureau of Reference)
20. Bates, Lyn (1997), "User Guide to Segregation", BMHB/DTI, UK.
21. Saraber, Fred, Enstad Gisle G. and Haaker, Gerard (1991), "Investigations on the Anisotropic Yield Behaviour of a Cohesive Bulk Solid", Powder Technology, 64 pp. 183 – 190.

Jenike's Legacy on the Australian Bulk Solids Community

Alan W. Roberts
Emeritus Professor,
Centre for Bulk Solids and Particulate Technologies,
The University of Newcastle, NSW Australia
Email: Alan.Roberts@newcastle.edu.au

Introduction

It is an honour to be invited to present this paper at this special memorial session to commemorate the work of Andrew Jenike which commenced here at the University of Utah some 50 years ago.

Andrew Jenike was, indisputably, the founder of the modern theories and industrial applications of Bulk Solids Handling. His name is synonymous with this subject. Apart from stimulating, on an international scale, much needed research in the many facets of bulk solids handling, his work has had a profound beneficial impact on the vast array of industries world wide that handle and process bulk materials. There is no name which is more widely known and referenced in the research literature on the subject of Bulk Solids than that of Andrew Jenike. His work has had a significant influence on the Bulk Solids research and the industrial applications in Australia, more so than he would have ever realised. It is about the legacy of his work on the bulk solids handling community in Australia about which I wish to speak

Australian Link to Andrew Jenike

There is no doubt that our lives are influenced greatly by the courses of action we take as a result of advice received either directly, or inspiration received indirectly, from others. In my case, and I know that also in the case of others who are present today, the work of Andrew Jenike has had a significant influence. Without a doubt, his research and design procedures have had a major impact on the industrial developments of bulk solids handling operations in Australia. Yet, I am sure Andrew Jenike would never have realised the far reaching influence of his work.

How did the links with Australia commence? To answer this I would like to relate briefly how my association with Andrew Jenike began and how this has led to the professional and industrial development of Bulk Solids Handling in Australia.

It was the period of my PhD research during 1958-61 at the Wollongong University College (Later to become the University of Wollongong) and the years that immediately followed that I first became aware of Andrew Jenike's research. In 1958, I commenced an academic appointment as a lecturer in Mechanical Engineering and was studying part time for a PhD on the subject of the screw conveyor performance when handling grain. During this research I became interested in the influence of the vortex motion of grain as it is conveyed by the rotating screw. This necessitated a knowledge of dynamic stress fields in granular masses during motion. Having no training in the subject of particle technology and bulk solids, I had to teach myself which meant a heavy use of the library facilities. This required making copious hand written notes, since there were no photocopying services available at the time. Obviously the field of Soil Mechanics provided important background material, but of significance was that during my library researches, the name of Andrew Jenike and his research at the University of Utah captured my attention.

Having completed my PhD in 1962 and become firmly committed to the subject of Bulk Solids Handling, I embarked on a period of sabbatical leave in 1963-64 as a Visiting Associate Professor in the Department of Theoretical and Applied Mechanics at Cornell University in Ithaca, New York. The Department ran a weekly seminar session with invited academics and researchers from other American institutions. On one occasion Professor R.T. (Dick) Shield from Brown University came to present a seminar and I was able to speak with him about his collaborative work with Andrew Jenike.

I then had the opportunity of inviting Andrew Jenike to Cornell to present a seminar. Both Andrew and his wife Una drove to Ithaca from Boston and stayed overnight at the University's Statler Hotel on the campus. His seminar was well received and especially for me it was most interesting. I still remember Andrew's words that "it took an Australian to invite him to Cornell". A few months later, my family and I travelled to Boston and visited the Jenikes at 27 Pilgrim Drive, Winchester, Massachusetts. This was a very happy occasion and cemented my resolve to make bulk solids handling my professional field of interest and activity.

After returning to Australia and to Wollongong in July 1964, I continued my research in Bulk Solids together with Peter Arnold, my first PhD student and my now long standing friend and colleague. It was in the latter part of 1964 that we purchased from Andrew a direct shear tester and consolidation bench. Thus began our flow property testing and associated industrial work.

I moved to the University of Newcastle in 1974 to become Head of the Department of Mechanical Engineering and Dean of the Faculty of Engineering. The bulk solids research and associated industrial consulting group TUNRA Bulk Solids Handling Research Associates, often abbreviated as TBS, was established in 1975 as a Division of the University's research company TUNRA Ltd (TUNRA is the acronym for The University of Newcastle Research Associates). The bulk solids research at the University of Wollongong continued on under the leadership of Professor Peter Arnold and the strong, ongoing and successful collaboration with the two Universities, Newcastle and Wollongong has been maintained until this day.

In the early years the industrial interest in the bulk solids work was very slow to be accepted, but gained considerable momentum in the years that followed. The teaching of Bulk Solids Handling has been, for many years, an integral part of the bachelor degree programs in Mechanical Engineering of the Universities of Newcastle and Wollongong. The two Universities also offer a joint masters' degree in this subject which has been run in South Africa as well as Australia. On average 6 industrial awareness seminars are run each year. These are typically 3 to 4 days duration

Bulk Solids Handling was recognized as a professional discipline area in its own right by the Institution of Engineers, Australia, through the establishment of a National Committee for this discipline in 1983. Recently, the National Committee has been replaced by the Australian

Society for Bulk Solids Handling, becoming, amongst other things, the organizing body for the international conference series which began in 1983 with conferences being held, in turn, at Newcastle and Wollongong, every three years.

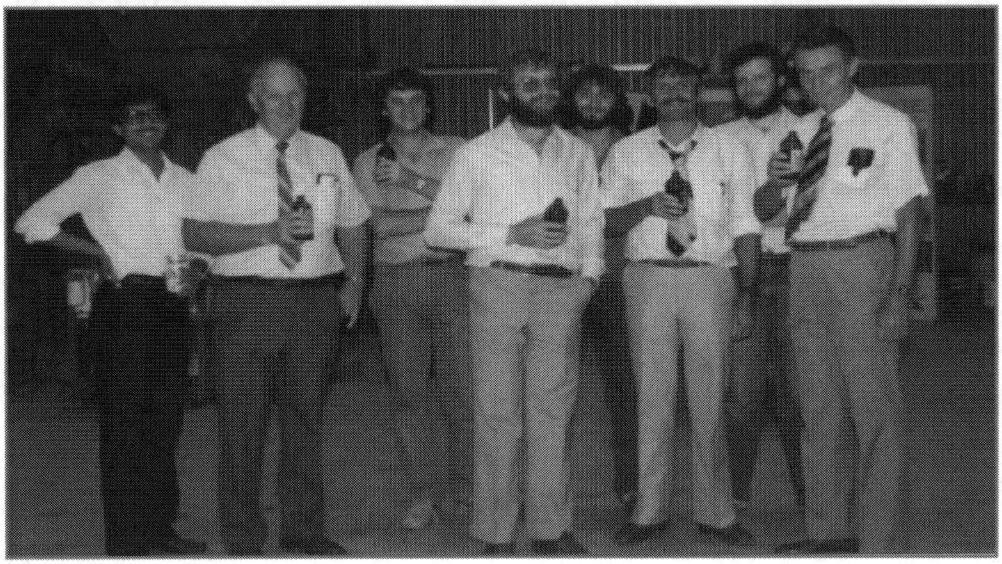

Figure 1. Andrew Jenike during his Visit to Newcastle, Australia, 1984

Andrew and Una Jenike made two visits to Australia, the first for a short stay on a trip that also took in New Zealand. In 1984 they spent 4 months in Newcastle when Andrew accepted our invitation to be a Visiting Research Fellow with the Bulk Solids Handling Group of the Department of Mechanical Engineering. It was a great honor to have both Andrew and Una with us. I can still recall how Andrew was impressed with the impact of his work on level of teaching, research and industrial awareness in bulk solids handling in Australia. Figure 1 shows Andrew Jenike wth the bulk solids group at Newcastle during his 1984 visit.

Bulk Solids Development of Bulk Solids at The University of Newcastle

I would now like to highlight some of the salient aspects of the bulk solids handling work at the University of Newcastle with particular emphasis of the contributions of the TUNRA Bulk Solids group (TBS). As indicated above, TBS was formed in 1975 to become the vehicle for disseminating the findings of this research through professional development courses and consulting. TBS is a totally self supporting organisation, its principal areas of activity being summarized in Figure 2.

The strength of TBS lies in its ongoing research links with the Discipline of Mechanical Engineering and School of Engineering of The University of Newcastle, the research being funded by a range of external grants from a diverse range of granting bodies. TBS has an active involvement in the honours and post graduate research project work of the engineering students. The bulk solids research of the University of Newcastle jointly with the University of Wollongong was recognised by the Australian Research Council (ARC) through the awarding of a Key Centre for Bulk Solids and Particulate Technologies (CBSPT), being one of seven Key Centres established in 1995 from a total of 94 applications. Funding covered a six year period (1995-2001) following which the Centre has been able to continue on a self-funded basis through a range of research grants won competitively from the ARC and other external funding. In effect, CBSPT and is the research arm and TBS is the commercial consulting arm of the combined Bulk Solids activities.

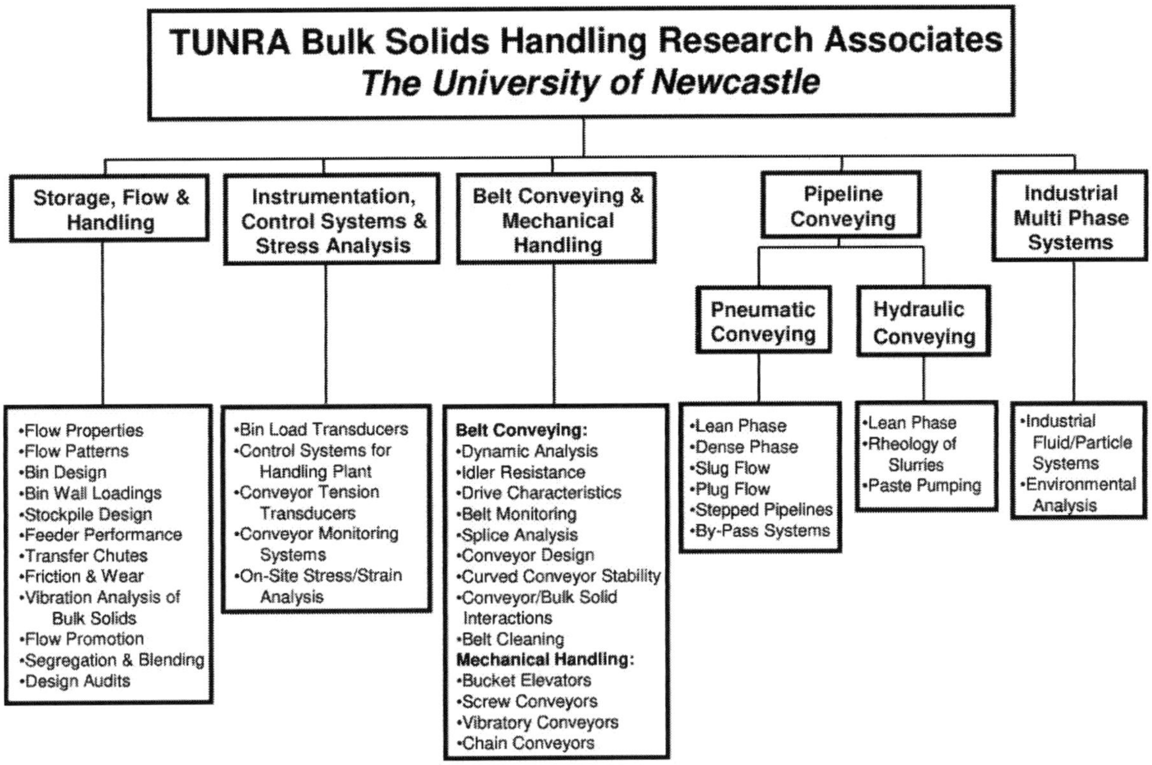

Figure 2. **TUNRA Bulk Solids - Principal Areas of Activity**

With the current resources boom, not only in Australia but world wide, TBS is currently enjoying record years with its consultancy work which is now averaging around 170 projects per year. By way of example, 433 projects were completed over the period 2004 to June 2007. This includes projects for 22 countries other than Australia. The work has been commissioned by 205 companies of which 96 projects were via repeat business. Over its 32 years of operation, TBS has successfully completed some 2500 projects for more than 1000 companies in Australia as well as in 40 countries world wide. The work of TBS focuses on the basic design concepts with recommendations being made for the follow up detailed design for both new and ongoing projects.

While research and higher consulting work is carried out in the five principal subject areas shown in Figure 2, it is the area of Bulk Solids Storage, Flow and Handling which has had the greatest level of activity. It is this area in which the work of Andrew Jenike and his co-workers has had the greatest impact. Some of the main features of the TBS work are summarized below:

Flow Property Testing

The Jenike direct shear test is the principal test adopted by TBS. The equipment is manufactured by TBS and incorporates a load cell with a stiffness of 1 kN/mm. The original direct shear tester developed by Jenike employed a pneumatic load cell for the shear force measurement and a circular chart recorder. Due to the low stiffness of the load cell, the slip-stick effects frequently observable in the shear force results were more a function of the load cell than of the bulk material being tested. In association with the direct shear test, the TBS consolidation benches for time consolidation tests employ pneumatic loading cylinders which control the load on the test samples.

Other equipment developed by TBS include:

- 300 mm diameter Jenike type direct shear tester with hydraulic load cell for testing of bulk solids under high pressures as required in the case of stockpiles
- 300 mm diameter inverted wall friction tester
- Submerged shear test for wet solids
- Vibration shear tester
- Flowability tester based on the unconfined yield strength
- Linear and rotary abrasive wear testers
- Dustiness tester
- Fluidisation testers

Some of the equipment designed and manufactured by TBS is shown in Figure 3.

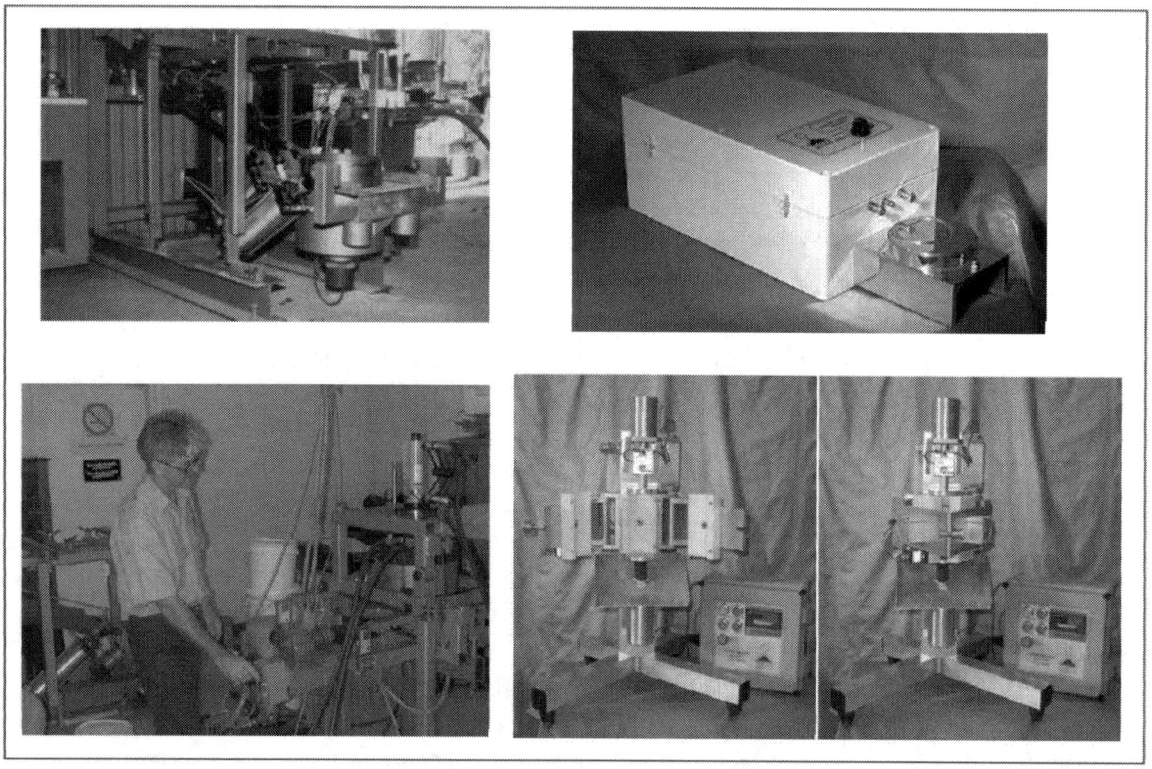

Figure. 3 Flow Property Test Equipment – (Designed and Manufactured by TBS)

Bin, Stockpile and Feeder Design

The principles and theories of mass-flow, funnel-flow and expanded-flow have been applied successfully to a number of industrial bin and stockpile applications. Some examples are illustrated in Figures 4 and 5.

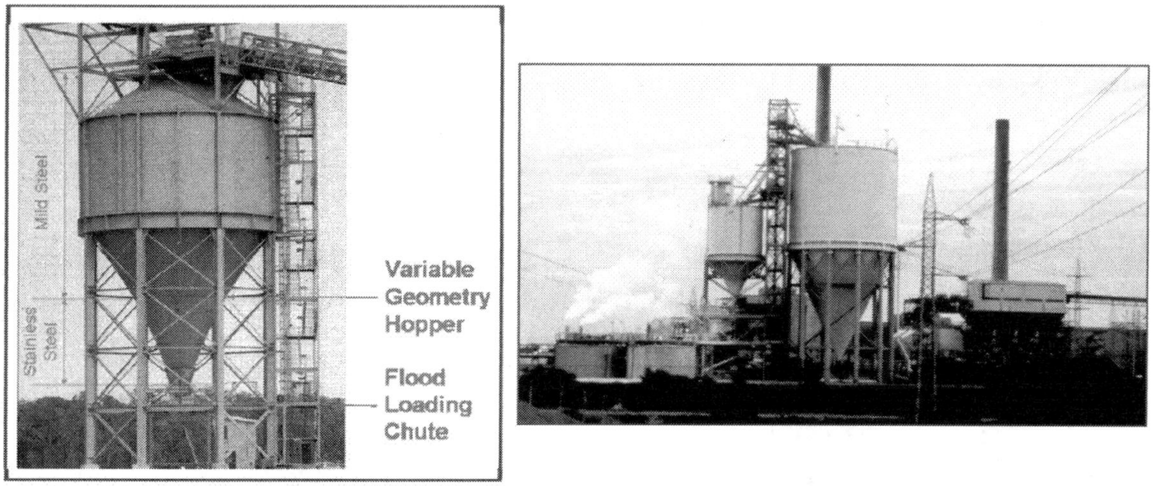

(a) Train Loading Coal Bin - Blair Athol, Qld

(b) Lime Storage Silo and Day Bin – ALCOA, Kwinana, WA

Figure 4. Examples of Bins Designed for Mass-Flow

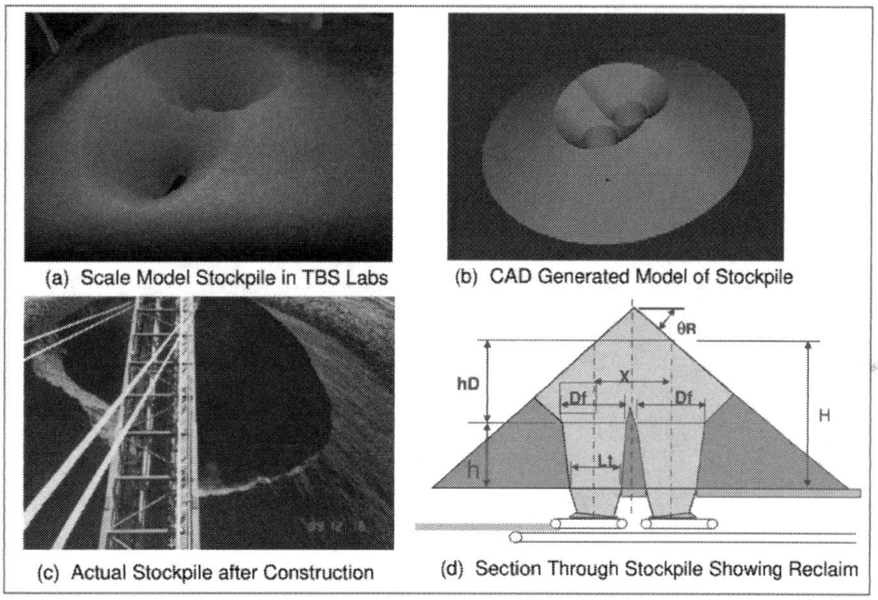

Figure 5. Design of Gravity Reclaim Stockpile to Meet Live Capacity Requirements

Silo Wall Loads

The research on silo wall loads has had an important role in the work of TBS. It is worth noting that the Standards Association of Australia produced a quite definitive Standard, AS3774, dealing with the load on Bulk Solids Containers. This Standard was first published in 1990 with a revised version in 1996. A subject of particular importance concerns the loads due to silo quaking. TBS has had an active involvement in the research and industrial analysis of several silo quaking problems. The test silo used by TBS for the experimental work into silo quaking is shown in Figure 6.

The dynamic loads in bins due to rapid flow rates as occurs in train loading has been an area for study. Figure 7 shows an iron ore train loading bin in which the load out peaks in excess of 60,000 t/h during the first 5 seconds before the steady flow rate of 7.5 t/h due to the rilling action is obtained as each rail wagon is reached. The high impact loads cause significant dynamic loading conditions on the bin structure.

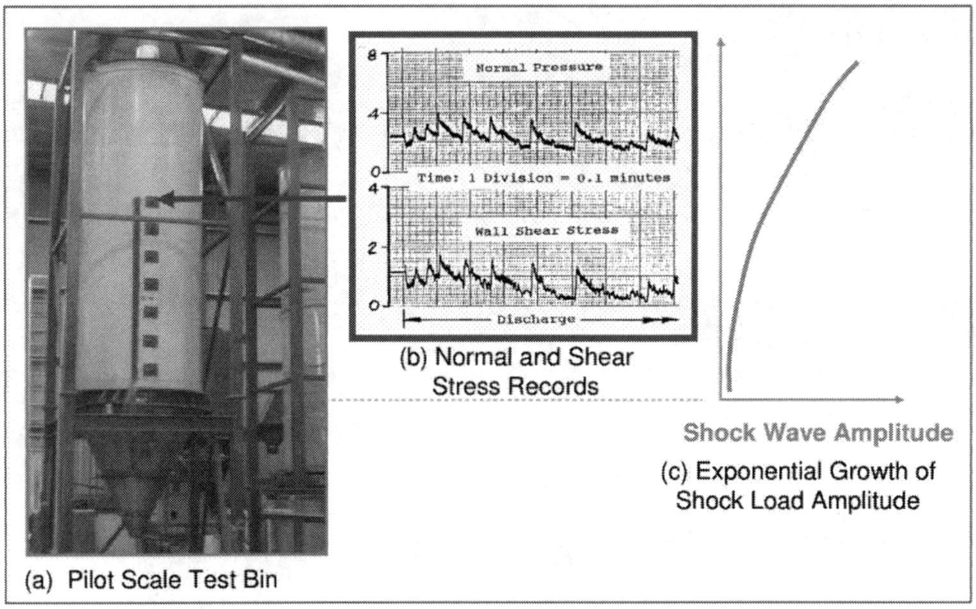

Figure 6. Experimental Studies into Silo Qualing

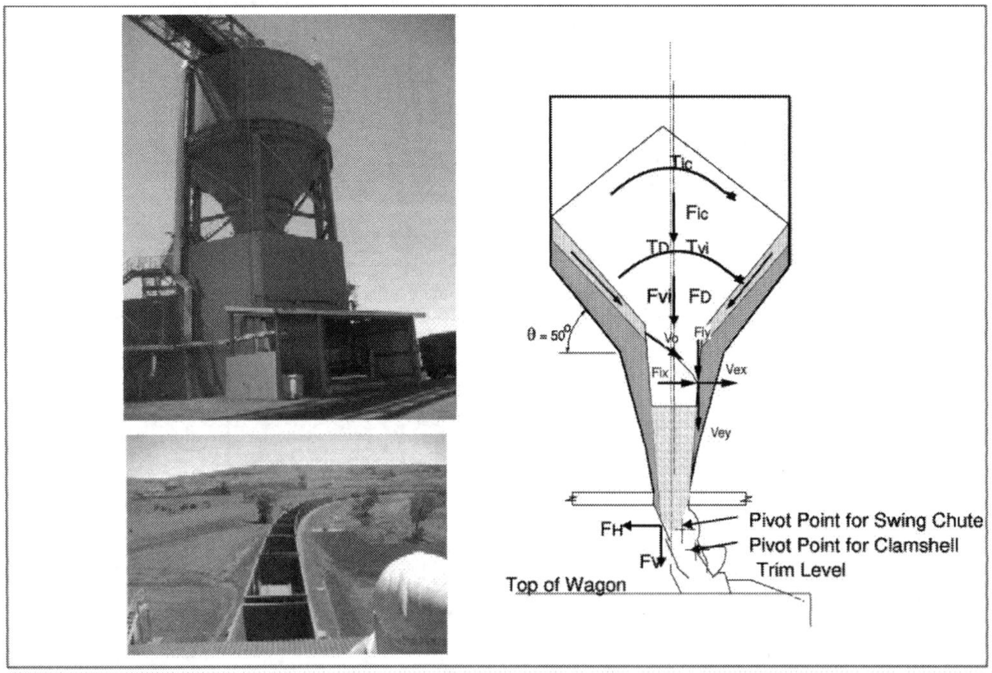

Figure 7. Dynamic Loads on Iron Ore Bin During Flood Loading of Trains

The determination of feeder loads and the performance of feeding equipment has been an area of study. In some cases, scale model testing is found to be useful as illustrated in the performance of large, inclined, open front apron feeders used for feeding ROM ore into primary crushers. An illustration of this area of study is illustrated in Figure 8.

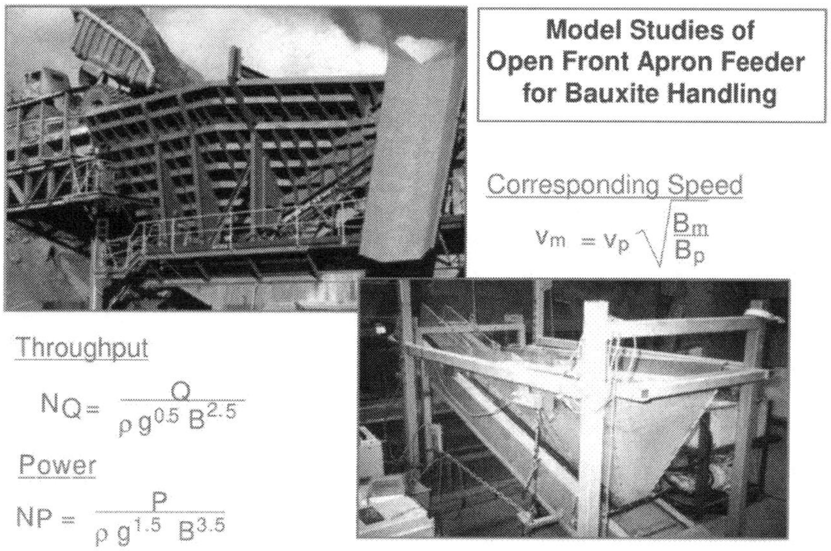

Figure 8. Scale Model Testing of Inclined, Open-Front Apron Feeder

The design of transfer chutes is also an area of ongoing research and industrial design. The approach is to employ dynamic flow analysis which employ the measured flow properties of the bulk solid.

The foregoing examples highlight some of the developments of TBS at the University of Newcastle that have occurred following the initial stimulation of Andrew Jenike's work. As an overriding comment, it is my firm contention that the problems in industry are orders of magnitude more difficult than the level of fundamental research available to solve them. The solution of industrial problems requires the application of research knowledge that is available combined with experience and a thorough understanding of basic mechanics. Since research plays a 'catch up' role, the importance of interaction between research groups and industry in identifying the areas for future research cannot be too strongly emphasised.

Concluding Remarks

There is no doubt that the work of Andrew Jenike laid the foundations of the modern theories and the industrial development of Bulk Solids Handling. The international influence of his work in stimulating research and industrial awareness is very great indeed. When one reviews the original work and publications of Andrew Jenike, notably the University of Utah Bulletins of some 40 years ago, one cannot cease to be amazed at the depth, insight and industrial relevance of his contribution. This contribution provided the foundations of the theoretical and practical development of this subject. His work remains unsurpassed and, I believe, will continue to withstand the passage of time.

As we view the situation today, there is no doubt that the professional discipline of Bulk Solids Handling has come of age.

It has been an honour to have known Andrew Jenike. I thank you for the opportunity to present this tribute to him.

181d the Influence of a. Jenike's Work on the Industrial Design of Silos

Karl Jacob, The Dow Chemical Company, 845 Bldg, Midland, MI 48640

To be added in a future date

Beyond Storage – Design for Mass Flow

Harald Wilms
Zeppelin Silos & Systems GmbH
Leutholdstr. 108
D-880459 Friedrichshafen, Germany

ABSTRACT

The original purpose of silos is to store bulk solids. Design methods as the Jenike design procedure to ensure functionality by mass flow and the Janssen method to calculate pressures in silos are well established. In recent years the advantage of the mass flow concept has been transferred to other applications such as gravity flow blenders and degassing vessels.

1. Introduction

Silos have to be designed under two completely different considerations: Functionality and structural integrity, sometimes also called design for flow and strength [1,2]. While chemical and process engineers are responsible for functionality, civil engineers are responsible for the structural aspects. However, even though these are two completely different fields of engineering, there is a strong interrelation of these two aspects, since the flow in silos influences the pressure distribution and thus the loads on the silo walls. Therefore, a close cooperation between both disciplines is mandatory for building safe and reliable silos [3].

2. The Work of Janssen & Jenike

Pioneering work with continuing impact until today has been provided by Janssen [4] on loads in silos and by Jenike [5,6] on the flow behavior of solids in silos. Both engineers improved the state-of-the-art of their time tremendously and have made silo design much safer and more reliable. Today, silo design is an engineering discipline and not any longer any magic.

2.1 Janssen: Pressure Distributions in Silos

The calculation of pressure distributions in silos dates back to 1895, when Janssen [4] published his method to calculate pressures in silos using the force equilibrium at a slice element. Knowledge of the pressure distribution in silos is mandatory for structural design of silos. Structural design of silos is based on the loads (pressures) exerted by the bulk solids to the silo walls. These loads may be different during filling, storage at rest, and discharge. The fundamental difference to pressure in tanks for liquids with the hydrostatic pressure distribution results from the ability of bulk solids to transmit shear stresses at rest, whereas liquids need a relative velocity to generate shear. This basic difference leads to the completely different pressure distribution in a silo compared to a tank.

A very important effect of this ability to transfer shear stresses at rest is that silo walls are not only subject to internal pressure, but also to axial compression. In metal structures, shear stresses accumulate over height and result in buckling stresses. Also the state of fluidization from rapid filling of powders or from aeration has to be considered.

The filling pressures in a silo can be calculated from integration of the force equilibrium at a slice element spanning the whole cross-section of the cylinder. The force equilibrium results in a differential equation, which can be solved analytically, provided that there is a certain ratio between the horizontal to vertical pressure. This calculation procedure was originally developed by Janssen [4] in 1895 and still is the basis for most international codes.

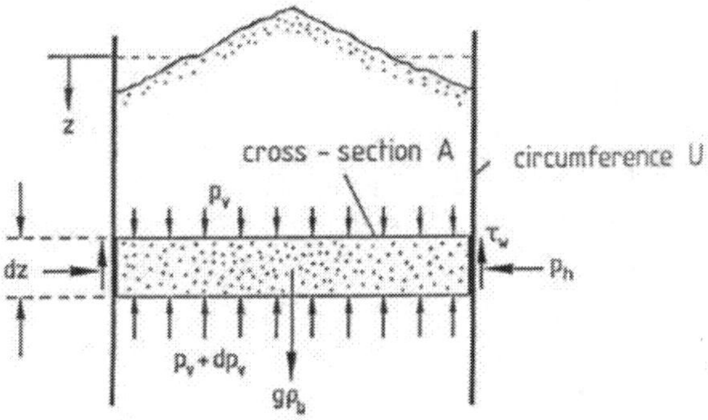

Fig. 1: Force equilibrium at a slice element

The bulk properties needed for calculations are the bulk density ρ_b, the wall friction coefficient μ_w

$$\mu_w = \tan \varphi_w, \quad (1)$$

and the horizontal pressure ratio λ

$$\lambda = p_h / p_v \cong 1 - \sin \varphi_e \, . \quad (2)$$

The force equilibrium at a slice element spanning the full cross-section of the cylindrical part of a silo then yields a differential equation that can b solved to derive the Janssen equation:

$$p_v = g\rho_b D / 4\lambda \tan \varphi_w \{ 1 - \exp(4\lambda \tan \varphi_w z/D)\} \quad (3)$$

The maximum vertical pressure in a silo can be estimated as:

$$p_{v,max} = g\rho_b D / 4\lambda \tan \varphi_w \quad (4)$$

The ultimate pressure in a silo, thus, depends mainly on the properties of the bulk solid and the diameter D of the silo, and only to a lesser extent on the height of the bulk solid level. The total load is considerably lower than in a tank for liquids resulting in a hydrostatic pressure distribution as shown in Fig. 2. Basis for this calculation has been a vessel with a height of 40 m and a bulk solid with a bulk density of 1000 kg/m³. Fig. 2 demonstrates both the effect of the wall friction and of the silo diameter.

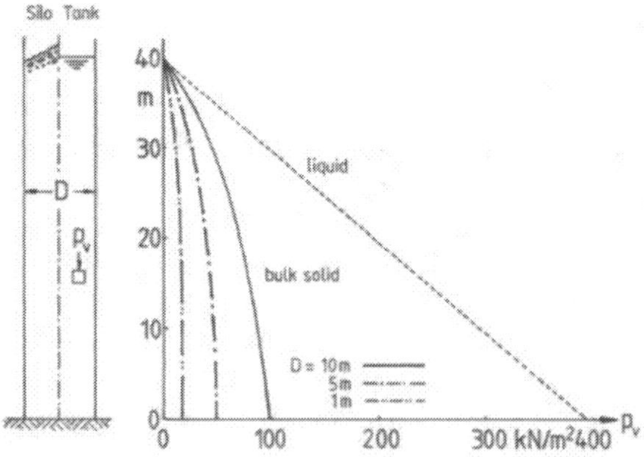

Fig. 2: Pressures in silos and tanks

The pressures calculated with above equations are assumed to be uniformly distributed around the circumference. However, asymmetric flow channels will result in non-uniform pressure distributions as shown in Fig. 3. As can be seen from equations (3) to (5), the pressure in silos depends on the diameter of the respective bulk solids container or flow channel. While filling a silo, the complete silo diameter thus determines the pressures. If an asymmetric flow channel forms, then the pressure in this flow channel of smaller diameter also is smaller. Consequently, the pressure varies around the circumference. In silos made of thin metal plates, this pressure distribution causes local deformation, reducing the buckling strength of the vessel. In thick-walled concrete silos, a bending moment in the shell results, causing cracks.

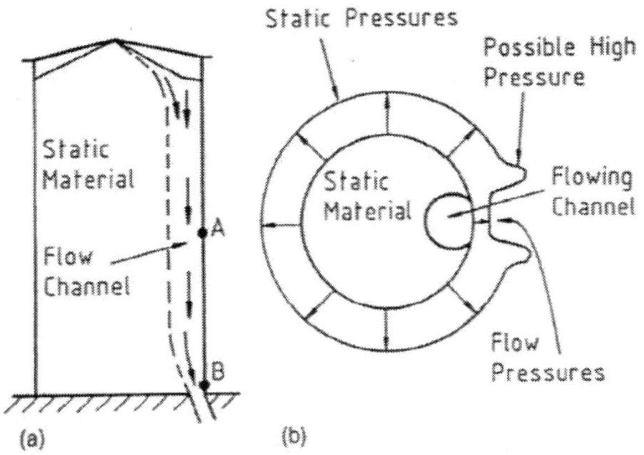

Fig. 3: Non-uniform loads caused by asymmetric flow channels

This design method to calculate pressures in silos based on Janssen's equation is still today the basis of most international codes and standards [7].

2.2 Jenike: Design for Flow

Design for flow includes the requirement

- to design for the correct flow profile, i.e. mass flow or funnel flow,
- to size the silo outlet size for reliable flow, i.e. without arching or ratholing,
- to select appropriate feeders and discharge aids.

Basis for this pioneering work of Jenike is the combination of a theory on flow as well as the measurement technique to determine flow properties with shear testers.

One of the most important results of Jenike's work is the classification of silos according to the flow profile developing during discharge. The two flow profiles, i.e. mass flow and funnel flow, are shown in Fig. 4. In mass flow, all solids are in motion during discharge. Mass flow silos have clear advantages for functionality because they do not exhibit any of the typical funnel flow silo problems such as piping (ratholing), segregation,

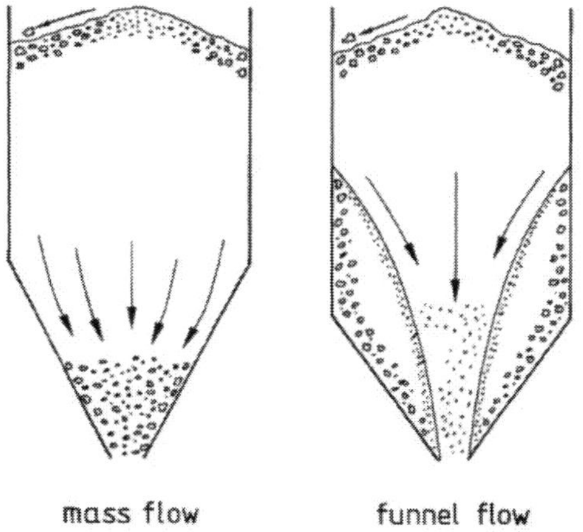

Fig. 4: Flow profiles in silos

flushing, erratic flow, wide residence time distribution, and difficult level control. Arching, as the only problem remaining in mass flow, can be overcome reliably by using the Jenike design procedure on the basis of measured flow properties [1,4,5]. This flow profile normally is preferred for more uniform flow and pressures. In case the hopper walls are to rough or not steep enough, funnel flow will develop during discharge, yielding stagnant zones in the periphery.

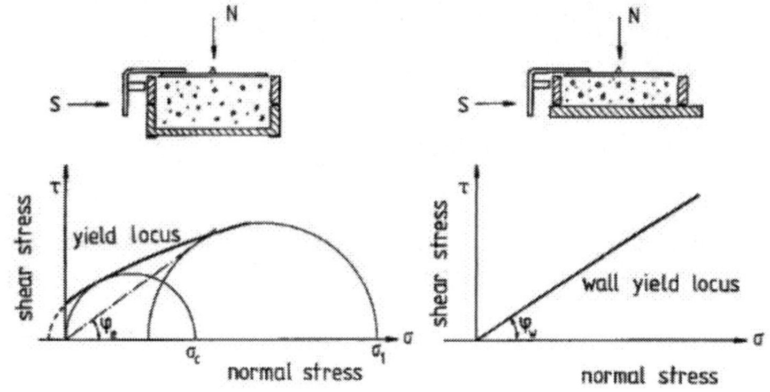

Fig. 5: Jenike shear cell

Key to both design for flow and strength is the knowledge of the relevant flow properties. According to Jenike, these parameters can be measured with shear testers. The most versatile shear tester is the translational Jenike Shear Cell shown in Fig. 5. The upper half of the cell is sheared over the lower half under different normal loads. The lower half of the shear cell is replaced by a sample of the wall material in question for measurement of the wall friction. Respective test procedures have been developed for translational shear testers by the EFCE in Europe [8] and have been modified to become an ASTM standard [9,10].

With this Jenike method silos have been reliably built for mass flow and non-arching in many applications. This is still the most reliable design method for flow. This good experience has been the reason for adopting the mass flow design method to other applications, namely to the design of blenders and their internals as well as for purge silos.

3. Application of Jenike's Theory to Blenders

Jenike has developed his design procedure for gravity flow silos. With respect to blending by recirculation, he did not consider internals to spread the residence time distribution, but only the velocity gradient developing in a mass flow hopper. He had clearly requested that all blending silos need to be designed for mass flow [5].

3.1 Blender Design Principles

Today, there is a variety of blending silos featuring basically two different principles: Either the difference in flow velocity is enhanced by internals and used to spread the residence time distribution, or internal pipes are used for sampling of solids from different levels and thus also spreading the residence time distribution [16].

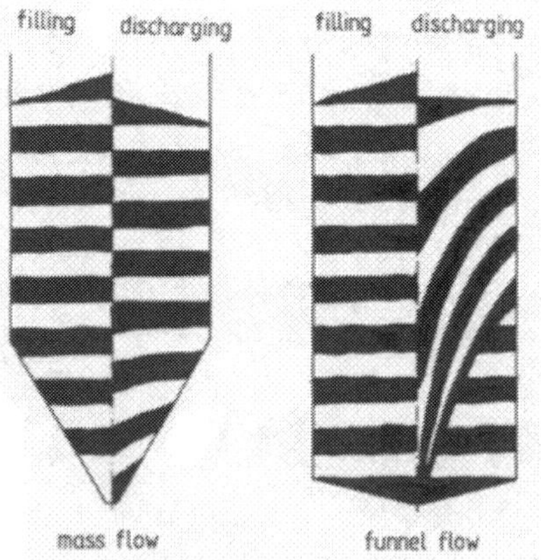

Fig. 6: Blending process in mass flow and funnel flow without internals

If any improvement of blending performance is to be achieved by multiple passes through a blender by recirculation, then any respective gravity blender has to be designed for mass flow. The stagnant zones in funnel flow would prevent that all solids in the blender participate at the blending process. However, since mass flow shows a narrow residence time distribution, a mass flow silo alone does not provide any good blending. Improvement can only result from internals spreading the residence time distribution, typically in form of internal cones or pipes. A funnel flow silo would provide a certain blending effect due to the spreading the residence time distribution. However, this effect is not reproducible because the extent of the stagnant zone cannot be calculated and changes. Fig. 6 shows the effect of mixing by funnel flow compared to mass flow, but also visualizes the effect of a stagnant zone.

Mass flow is also important to generate a more uniform pressure distribution acting on the internals. The limited blending effect of mass flow alone, i.e. without any internals spreading the residence time distribution, also becomes visible form Fig. 7. In this experiment a mass flow silo with a layer of tracers is discharged into a receiving silo by gravity. The layer with a concentration of tracers still is visible in lower silo. This figure demonstrates clearly that mass flow alone does not provide sufficient blending, but that mass flow is needed but also needs additional internals to spread the residence time distribution. Following this approach, a variety of different blenders have been developed and introduced to industry.

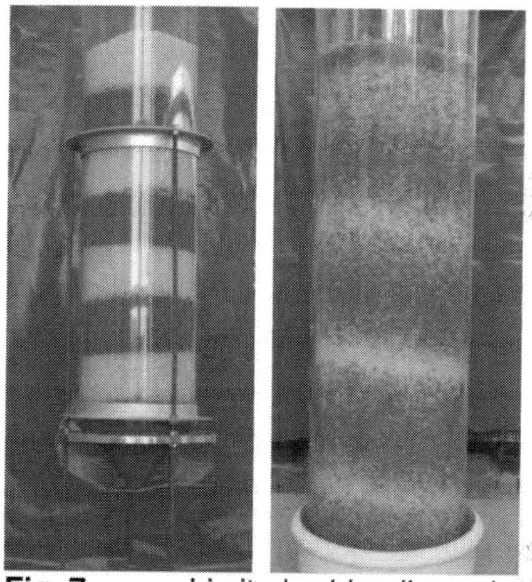

Fig. 7: Limited blending in mass flow

Some if the recent developments of blending silos, all based on mass flow design principles, is shown in Fig. 8. The four blenders to the left all sample from different levels via blending pipes, whereas the design at the right features an internal hopper to generate different flow velocities in the lower section of the blender. It is important to note that the mass flow concept is not only applied to the major hopper of these vessels, but also

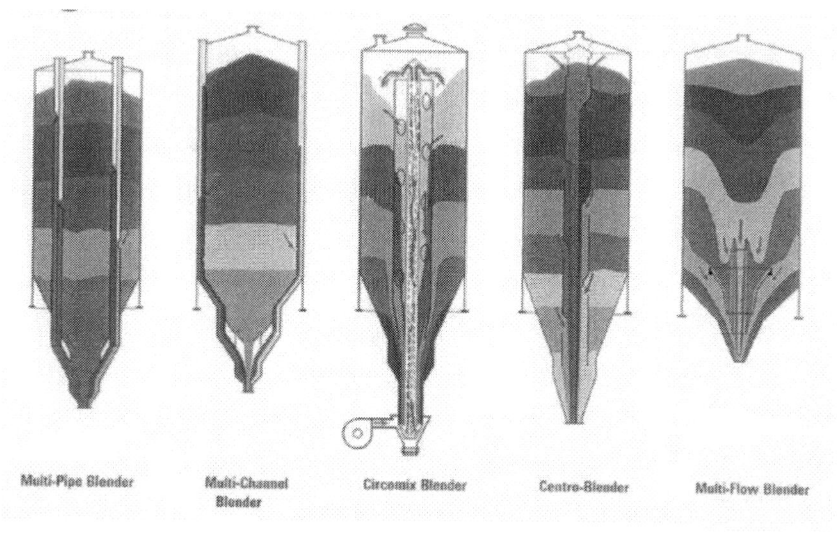

Fig. 8: Different designs of mass flow blending silos

to all blending chambers close to the outlet and all transition sections. The internal baffles of the central blending pipe in the Centro-Blender also fulfill the requirements for mass flow and feature the adventure of outflowing valleys [5], i.e. no bulk solids build-up at the interface between baffle and blending pipe.

One of the most wide-spread gravity blenders with pipe internals is the Multi-Pipe Gravity Blender [12]. The blending concept is based on numerous individual flow channels in a multi-compartment blending pipe ('downcomers') with intake openings at different levels. Free-flowing solids such as polymer pellets always flow into the uppermost intake opening below the surface. Once the blending pipe is full, no bulk solids can flow any longer into this downcomer. This concept is often referred to as the top-hole concept and is shown in Fig. 9: Once the pipe has been filled, no additional pellets from the blue and yellow layers flow any more into the intake openings.

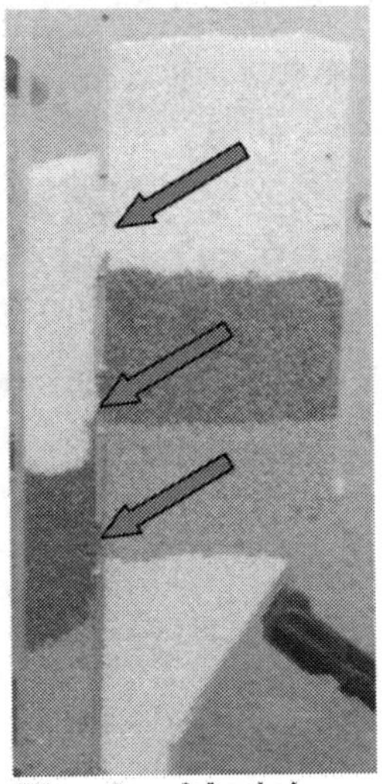

Fig. 9: Model showing the top-hole concept

Typically, each downcomner has several intake openings which only become active if the level of fill drops, e.g. during discharge. The improvement made to the earlier designs was to use a steep mass flow hopper and to adapt the hopper wall inclination individually with respect to the solids wall friction angle. Most of the solids blended in a Multi-Pipe Gravity Blender are free flowing polymer pellets with wall friction angles below 20°. A standard wall inclination of 25° with respect to the vertical then will be sufficient. Other designs featuring a standard 60° opening angle (30° to the vertical) or even flatter hopper walls, will not operate in mass flow. But also free flowing powders, agricultural pellets and other granular media have been homogenized successfully, if the hopper wall inclination is selected for mass flow according to the flow properties of the bulk solids in question.

One significant advantage of the mass flow concept is the more uniform pressure distribution. Funnel flow with partially stagnant zones will result in severe pressure differences acting on the pipes, possibly causing failures such as collapse of the internal structure. However, even in mass flow, the pipes may be subject to different loads resulting in a bending moments. Zeppelin has developed a load model which has been successfully applied to pipe internals in mass flow blenders. The difference in pressure acting on the internal pipe is determined from the difference in filling and discharge pressures according to Fig. 10. This load pattern is based on the fact that in some areas flow into the blending pipes may already start while the surrounding material still is stationary. The resulting

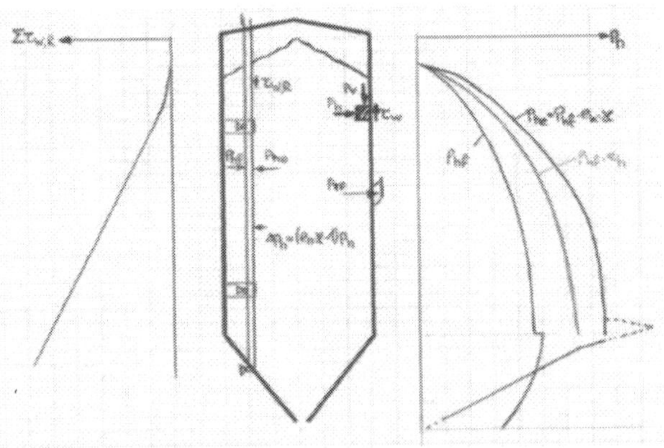

Fig. 10: Load model for internal pipes

forces, laterally and vertically, have to be transferred to the silo walls by means of support brackets as shown in Fig. 11, again being subject to loads. Since also the pressures depend on the flow properties, any changes in the flow properties may not only result in a different flow profile with different blending performance, but may also resulting structure failures due to changes in the load pattern.

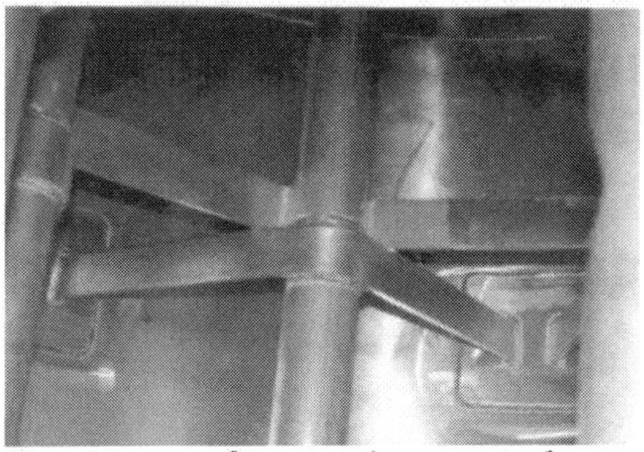

Fig. 11: Structural support of blending pipes

In order to reduce the cost of blending silos, Zeppelin has developed the concept to install the blending pipes directly at the silo wall, i.e. at the periphery. Thus the internal support structures have become obsolete. The blending principle of sampling from multiple layers is maintained for non-segregating bulk solids such as polymer pellets having a very uniform particle size and weight. Even though the blending pipes now cause flow zones in the periphery of the wall, the vertical pipes, solidly welded to the silo wall, serve as vertical stiffeners. The structural integrity and the reliable performance thus both are facilitated by the mass flow concept.

The Circomix Blender features a central pipe for internal recirculation of polymer pellets by means of an airlift. The original design of a previous model called PGM was changed into a mass flow approach after severe structural failures occurred due to the non-uniform pressure distribution associated with the funnel flow concept.

Non free-flowing pellets or cohesive powders such as dryblends cannot be blended in a blender with a relatively large number of small-diameter blending pipes. The Centro Blender is an answer to this problem, also featuring a mass flow design. In this case the bulk solids are sampled into the single central pipe via multiple intake openings in the central pipe. All intake openings are internally covered by a baffle plate spanning the central pipe at decreasing portions, metering the same amount of solids into the blending pipe at each level, as is demonstrated in the test model in Fig. 12. In this case the mass flow profile has some additional advantages: In a circular silo with a round blending pipe, the baffles form an intersection with the pipe called outflowing valleys. According to Jenike [4], this is one of the advantageous designs to prevent build-up in the corners. The flat baffle plates can be designed according to the conditions for mass flow in a plane-strain flow conditions and thus can be less steep than the hopper.

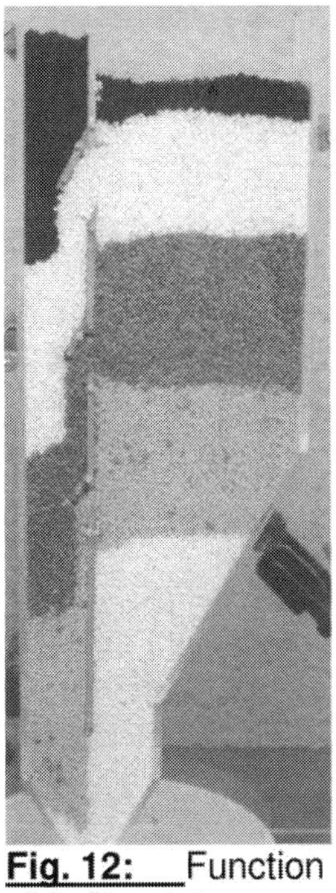

Fig. 12: Function of baffle plates

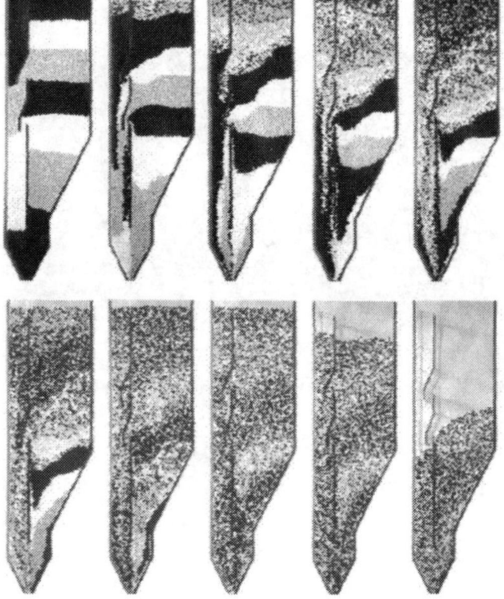

Fig. 13: Blending progress in a Centro-Blender during recirculation and discharge

In Fig. 13 the progress of the blending process with recirculation is demonstrated with black and white pellets in a small model representing half of the Centro-Blender. This means that the central blending pipe with multiple baffled intake openings is on the left side of the model and flow occurs from the silo to the right. The first eight photographs show the blending progress during two full recirculations, whereas the last two photos show the additional homogenization during discharge. A good homogeneity, starting from totally segregated layers, is achieved with discharge after two recirculations, as is typical for gravity blenders.

Powders typically have a rather wide particle size distribution and thus typically tend to segregate during filling. The test results in Fig. 14 reveal that samples of PVC-dryblend taken during the discharge from this blender show almost no change in fines content measured during discharge from a Centro-Blender. Thus, the discharge from a mass flow blender is virtually free from segregation.

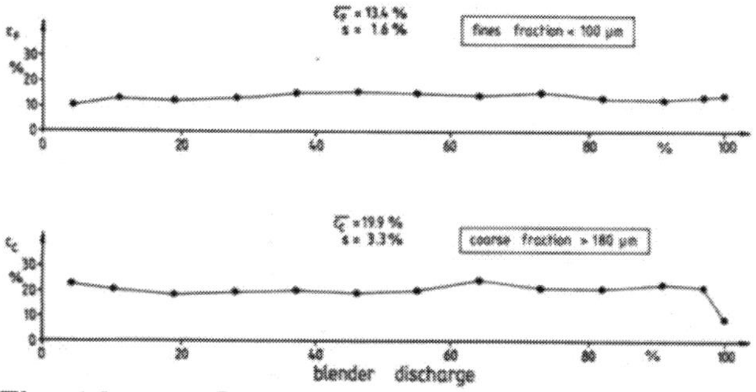

Fig. 14: Segregation-free discharge from Centro-Blender

The cone-in-cone blenders use a different idea for blending: The blending effect is not achieved by withdrawing solids from different zones in the blender, but by generating different velocities resulting in a spread of the residence time distribution. This effect can be achieved by numerous designs of the internals: They could have the shape of hoppers, pipes, baffles or any combination of it. Fig. 15 shows a variety of cone-in-cone blenders, including the Binsert® and the Multi-Flow which was used to generate the profile shown in Fig. 16. The father of all these blenders is the Binsert® developed by Jenike&Johanson [13]. Even though the internal hopper and the outer annulus have the same rate of convergence, i.e. the same relative

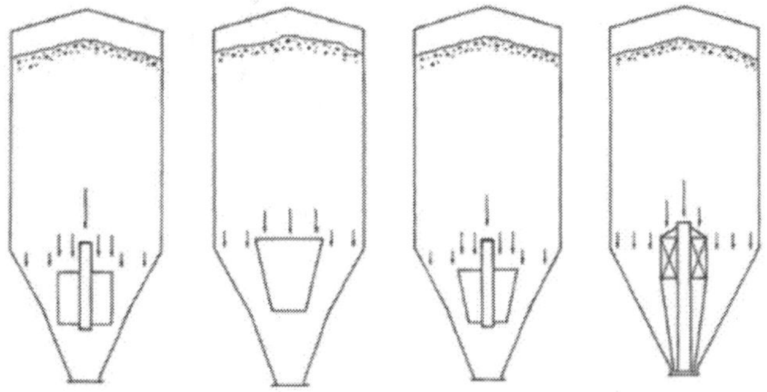

Fig. 15: Various Cone-in-Cone blender designs

reduction in cross-section with height, a velocity distribution and thus a residence time distribution is generated by the lower conical section with its radial velocity field. Both the internal hopper and the out annulus are designed for mass flow. The mass flow hopper beneath the internal blending hopper increases the spread of the flow velocities, since it again allows the solids from the center area to flow faster than the solids from the periphery.

Fig. 16 shows the development of the different flow zones in a Multi-Flow Blender. It becomes obvious that the initially horizontal layers become vertical in the vicinity of the outlet, thus allowing solids from multiple horizontal layers, i.e. different times of filling, to exit the blender simultaneously. From the second picture in Fig. 16 it also is confirmed that the reach of any internal is limited to approximately one silo diameter in height above the internal.

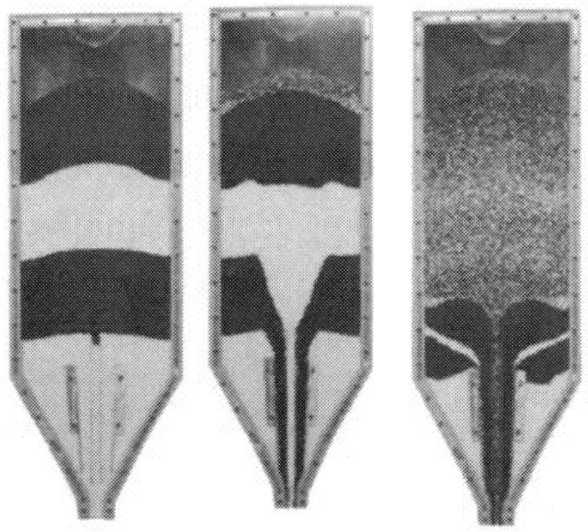

Fig. 16: Development of blending progress in a Multi-Flow Blender

The cone-in-cone concept was also believed to be a simple solution, also for retro-fitting. However, most silos are not operating in mass flow and thus retro-fitting also required modification of the hopper. A standard hopper may be kept in the upper section where the internal hopper improves the development of mass flow, but it is important to consider the steeper hopper required beneath the internals to achieve mass flow in this section.

3.2 Blending Silo Operation

Gravity blenders can easily be incorporated into any process, either batchwise or continuous. Today, still most blenders are operated discontinuously, even though continuous operation results in improved blending efficiency. A very advantageous operation is to use 2 blenders in a row as single pass blenders without recirculation in continuous processes. This leads to better results than batchwise recirculation in 2 blenders working in parallel. With continuous operation, the final fluctuations are much smaller than with a batchwise operation. One very efficient approach is to build 2 blenders on top of each other and thus requiring only one vertical lift of the bulk solids. The options for continuous single-pass blending with two blenders are shown in Fig. 17.

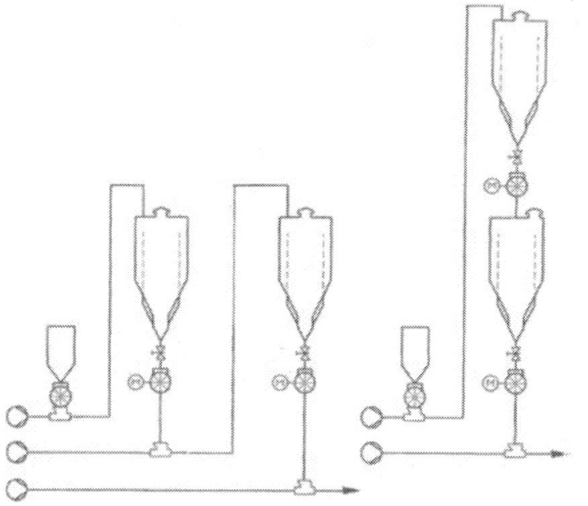

Fig 17: Concepts for continuous single-pass blending

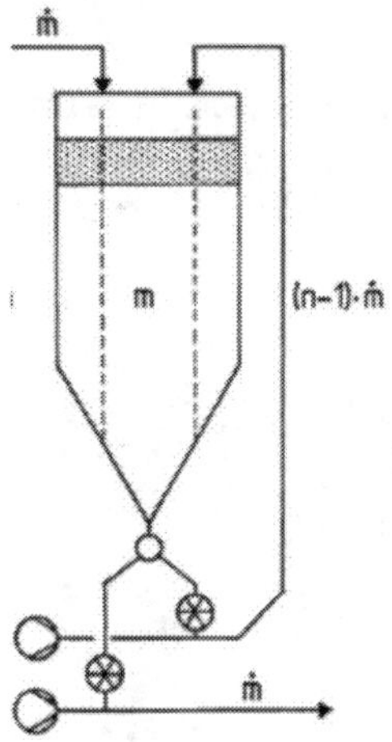

Fig 18: Continuous blending

Most gravity blenders are used in discontinuous operation, i.e. in batchwise mode. However, many applications allow for continuous operation. A respective operation mode is given in Fig. 18. The blender is constantly filled with a fill rate $\dot{m}$ and simultaneously the same flow rate $\dot{m}$ is discharged for single-pass operation. In multiple pass operation, a solids' flow rate of (n-1) $\dot{m}$ is recirculated to achieve n passes. With this procedure, the residence time distribution of any concentration profile is spread much wider, yielding a better blender performance.

A comparison of a single-pass discontinuous (batchwise) and continuous operation is shown in Fig. 19. The solid line and the dotted area gives the inlet concentration profile vs. time for a filling with 90% 'white' plus 10% of 'black' pellets in a Centro-Blender. The fluctuation of the tracer concentration thus is 100 % and the residence time (width) of the fluctuation is 10% of the blender content. The response of the blending system to this fluctuation is shown by the dotted (red) line for discontinuous blending and by the dashed (green) line for continuous blending. The maximum tracer concentration fluctuation in discontinuous mode is reduced to 36% whereas only 18% remains in continuous operation. The residence time of the tracer level is spread from 10% at filling to approximately 75% in discontinuous mode and to 195% in continuous mode.

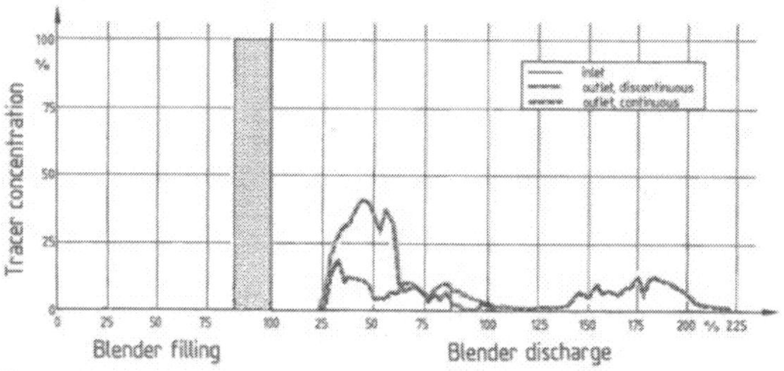

Fig 19: Results from discontinuous and continuous single-pass blending test

4. Application of Jenike's Theory to Fixed Bed Reactors

Many continuous processes require a conditioning stage for drying, heating, cooling, or degassing. The requirement for degassing, deoderizing or devolatizing is growing constantly in the polymer industry to reduce hydrocarbon content in consumer products. These processes require a certain hold-up time at elevated temperatures in a purge bin or at specific ambient conditions in a fixed bed reactor. In order to ensure a uniform residence time for the solids at the specified conditions, the purge bin needs to be designed for mass flow at that respective temperature. Even in mass flow a velocity distribution develops in the converging section. The closer the hopper wall inclination is to the mass flow limit, the wider the residence time distribution becomes. Purge bins thus need a hopper opening angle which is steeper than required only to achieve mass flow. Fig. 20 shows the relative width of the residence time distribution, i.e. that portion of a silo volume into which any tracer would be dispersed.

No internals for injection of purge gas into a degassing silo may destroy the mass flow profile and thus may not protrude into the flow channel. Various concepts are known on how to design the injection of the purge gas [14]. All ideas for these aeration systems focus on not to create any additional convergence of the flow channel in order not to destroy the mass flow profile.

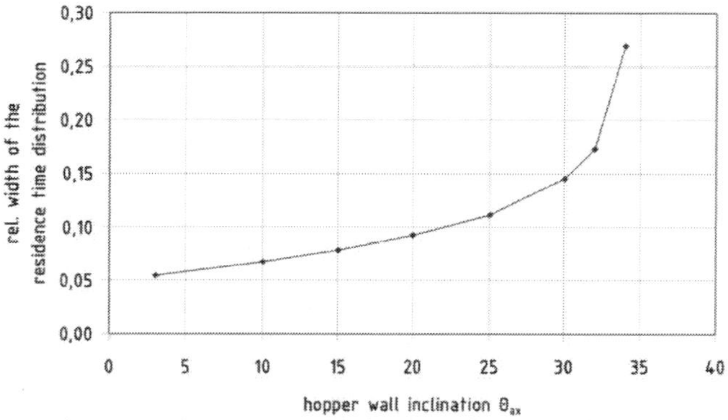

Fig 20: Relative width of residence time distribution (φ_e=44°, φ_w=14°)

In the past the displacement bodies, especially invert cones, have been used to inject the purge gas. The gas then was distributed around the circumference of the invert cone and into the bulk solids (Fig. 21a).The invert cone had the additional advantage of widening the flow channel in case of funnel flow. Newer concepts for gas injection into degassing silos without internals are based on creating an annulus around the hopper by fabricating a stepped hopper, or inserting a conical skirt. An annulus is then formed between the upper and lower hopper and the slope of the pellets. Purge gas then is introduced uniformly into the bulk solids via the large area of the free surface, thus generating a pressure drop as small as possible. The purge gas flows upwards, counter-current to the downwards flow of solids.

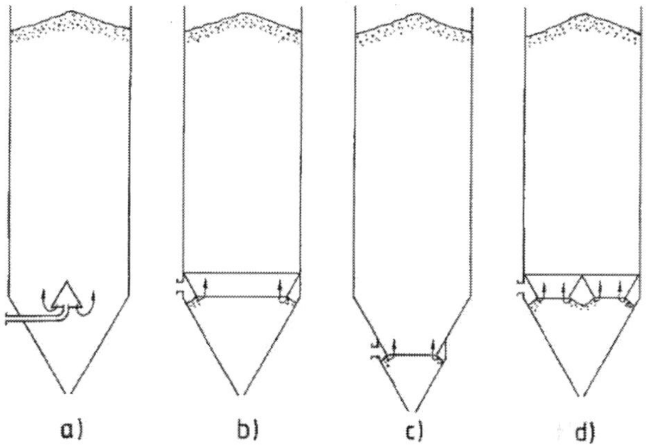

Fig 21: Aeration systems for purge bins

With such a secondary hopper located at the transition between cylinder and hopper a minimum gas velocity and a minimum gas pressure drop are achieved due to the large diameter (Fig. 21b). However, if it is critical to minimize any volume not degassed properly during the start-up of the system, then the aeration ring needs to be located lower in the hopper, as shown in Fig. 21c. If the amount of gas injected via the aeration ring at the transition is not enough, then more purge gas can be injected via a saddle (Fig. 21d) called J-Purge [15].

In some applications it is not sufficient to inject purge gas only at one location in the hopper or at the transition. If, for example for safety issues the aeration has to occur in the lower section of the hopper, then not a sufficient amount of purge gas can be introduced there due to the small cross-section. Exceeding locally the minimum fluidization velocity would increase the pressure drop significantly. Multiple aeration rings

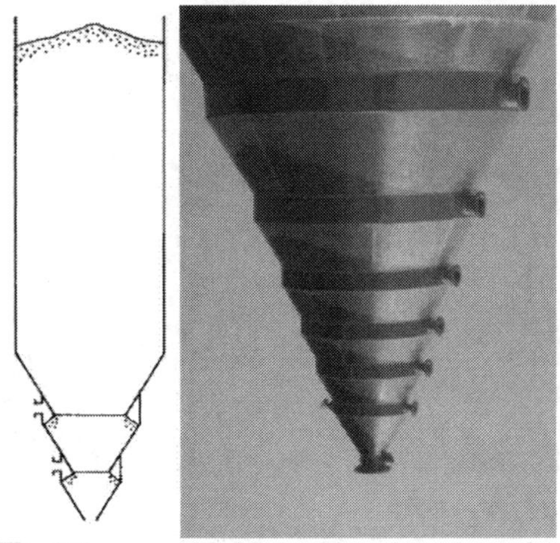

Fig 22: Multiple aeration rings in the hopper

at different levels in the hopper as depicted in Fig. 22 are the best option to solve this problem.

There are two important issues in purge bins: One is the uniform residence time distribution to be achieved through mass flow. The other is the uniform gas distribution to achieve uniform heating, cooling or drying. The introduction of the purge gas via an annulus, as show in Figs. 21 b-c, allows first the uniform distribution around the circumference – much more than any individual gas lances. This also results in the lowest possible injection velocity and thus a minimum in pressure drop. The pressure drop of the purge gas becomes a minimum with the lowest gas percolation velocity. The lowest gas velocity results from a uniform gas distribution across the cross-section of the purge bin. A second effect contributing to the pressure drop and thus the gas distribution is the length of the path for the gas to percolate through the contact bed. Any difference in the length of this path from the aeration point to any point at the top surface is minimized in slender purge vessels, especially with an aeration ring being somewhere between the center and the periphery of the hopper, or with multiple aeration rings. This then will result in a rather uniform propagation front of the purge process.

This has been demonstrated by dry air being injected into a model of a purge bin. The hopper section of this purge bin is filled with dry silica pellet, which are dark orange. The dry air percolating through the dry pellets and then passing into wet silica pellets dries these wet pellets, resulting in a change in color from bright yellow to orange. Fig. 23 shows the propagation of this dry front through the purge bin. The result is a nearly perfect horizontal propagation of the drying process – even in a model with rather low level of fill and exactly as required in purge bins. This experiment has confirmed that a uniform percolation of the purge gas can be achieved – at least in steep mass flow hoppers – and without any additional purge gas injection internals closer to the center line of the purge vessel.

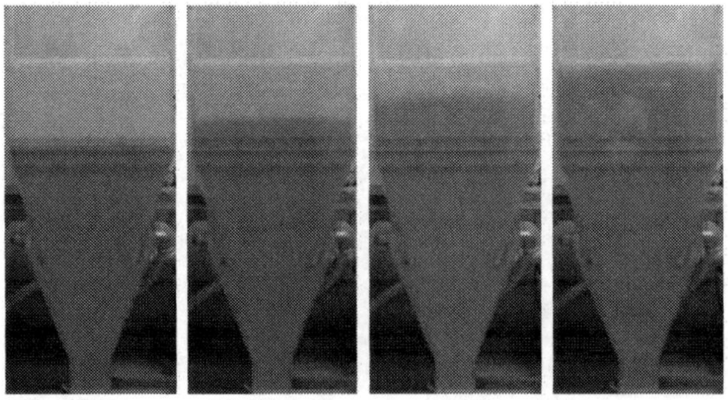

Fig. 23: Uniform propagation of purge zone

5. Summary

The flow properties of bulk solids and their use for designing equipment for reliable and reproducible flow conditions is not limited to storage silos, but is also applicable to other process vessels such as blending silos and purge vessels. Whereas there are still many storage silos being designed and built – on purpose or by chance – for funnel flow, it is becoming more acceptable for blending and purge vessels to use only mass flow and thus steeper hoppers.

6. Literature

[1] Wilms, H., Verfahrenstechnische und statische Aspekte der Siloauslegung, Chemie-Ingenieur-Technik 65(1993),3,284-292

[2] Wilms, H. Silo Design for Flow and Strength – State-of-the-Art and Research Requirements Powder Handling & Processing 6(1994),2,193-197

[3] Schwedes, J. Evolution of Bulk Solids Technology Since 1974 Bulk Solids Handling 3 (1983),1, 143-147

[4] Jenike, A.W. Gravity Flow of Bulk Solids, Bull. 108, Utah Engng. Exp. Station, Univ. of Utah, Salt Lake City, 1961

[5] Jenike, A.W. Storage and Flow of Solids, Bull. 123, Utah Engng. Exp. Station, Univ. of Utah, Salt Lake City, 1964

[6] Janssen, H.A. Getreidedruck in Silozellen Z.-VDI 39, (1895),1045-1049

[7] Wilms, H. Criteria for Evaluation of Silo Design Codes Bulk Solids Handling 11 (1991), 1, 35-39

[8] - Standard Shear Testing Technique for Particulate Solids Using the Jenike Shear Cell, Institution of Chemical Engineers, 1989

[9] - Standard Test Method for Shear Testing of Bulk Solids Using the Jenike Shear Cell, ASTM D6128-97, 1998

[10] Carson, J.W., Wilms H. Development of an International Standard for Shear Testing Powder Technology 167(2006), 1-9

[11] Wilms, H., Brar, B. Gravity Flow Blenders for Polymer Quality Improvement 2nd Intern. Plant Operations & Design Conf. Preprints pp 415-427,AIChE, New York, 1997

[12] Wilms, H. Blending and Homogenizing of Bulk Solids with the Zeppelin Multi-Pipe Gravity Blender Bulk Solids Handling 8(1988),6,733-736

[13] Johanson, J.R. Controlling Flow Patterns in Bins by Use of an Insert Bulk Solids Handling 2 (1983) 3, 495-498

[14] Wilms, H., Schneider, H. Testing and Process Design for Degassing Silos Polypropylene 2006 , MBS, Zurich, 2006

[15] Pittinger, B.H., Carson, J.W., Pescott, J.K., Purutyan, H. Uniform Purging of Resins in Contact Bed Purge Vessels Polymer Engineering and Science 39 (1999),9,1802-1811

181f a Personal Perspective on a. Jenike: the Person and His Work

John Carson, Jenike and Johanson Inc, 400 Business Park Dr., Tyngsboro, MA 01879

To be added at a later date

232a High Throughput Synthesis Of Visible-Light-Active Nanostructured Tiox Photocatalyst In A Flame Aerosol Reactor

Swapnil Dhumal[1], Tyrone Daulton[2], Jingkun Jiang[1], Pratim Biswas[1], and Bamin Khomami[3]. (1) Department of Energy, Environmental, and Chemical Engineering, Washington University in St Louis, St Louis, MO 63130, (2) Department of Physics and Center for Materials Innovation, Washington University in St Louis, St Louis, MO 63130, (3) Department of Chemical and Biomolecular Engineering, University of Tennessee, Knoxville, TN 37996

Titanium dioxide (TiO2) due to its low cost and high chemical stability is widely used as a photocatalyst in a variety of applications ranging from air purification and deodorization to self cleaning coatings on buildings (1,2). However, TiO2 shows photocatalytic activity only under UV light (3.2 eV band gap) which constitutes approximately 5% of the total incident solar energy. As a result, many strategies have been explored to reduce the energy band gap of TiO2 in order to enhance its photoactivity in visible light (approximately 45% of the total incident solar energy). In particular, transition metal (V, Cr, Mn, Fe) doping has been shown to effectively reduce the energy band gap. However, this approach has only had limited success in extending the photocatalytic activity of TiO2 in the visible region due to the thermal instability of the doped materials and the fact that the dopant sites can act as electron-hole recombination centers (3). In the case of oxygen deficient titanium dioxide, represented as TiOx with $x<2$, oxygen vacancy states are formed at 0.75 to 1.18 eV below the conduction band minimum (4). Therefore, introduction of the oxygen vacancy states reduce the band gap (2.02-2.45 eV) with respect to the stoichiometric TiO2 so that photoexcitation of electrons from the valance band can take place with visible light. In this study, we have developed a high throughput flame synthesis technique for production of visible-light-active oxygen deficient titanium dioxide (TiOx with $x<2$) photocatalysts. Specifically, TiOx nanoparticles (size range of 10-50 nm, $1.85< x<1.95$ determined with Energy Loss Spectroscopy) are synthesized by controlling the oxygen concentration in the particle formation zone in a diffusion flame aerosol reactor equipped with a flame quenching system utilizing titanium tetraisopropoxide (TTIP) as a precursor. The absorption spectra of the particles, measured via diffuse reflectance spectrometer, clearly demonstrated light absorption in the visible region. Furthermore, the visible light photocatlytic activity of TiOx particles is demonstrated by degradation of Methyl Orange solutions under visible light illumination. Flame aerosol reactors provide a single step process for synthesis of nanostructured titanium dioxide for visible light activated applications.

References: 1) Thimsen E. and Biswas P. to appear, AIChE J (2007) 2) Almquist, C. and Biswas, P. J. Catalysis 212, 145 (2002) 3) W. Choi, A. Termin, and M. R. Hoffmann, J. Phys. Chem. 98, 13669 (1994) 4) Cronemeyer, D. C., Physical Review 113, 1222 (1959)

232b Synthesis And Characterization Of Cu-SiO2 Core-Shell Nanoparticles

Shampa Aich, Keisha B. Walters, and **Adrienne R. Minerick**. Dave C. Swalm School of Chemical Engineering, Mississippi State University, Box 9595, Mississippi State, MS 39762

Coatings on colloidal nanoparticles can help stabilize particles to enhance their unique optical, catalytic and magnetic properties [1]. Colloidal stabilization is commonly attained via deposition of a homogeneous silica coating by core-shell nanotechnology [2]. Silica has been preferred as the coating material as it has several advantages over other organic or inorganic coating materials, such as chemical inertness, optical transparency and the ability to produce concentrated dispersions of nanosized materials by preventing coagulation during chemical reactions. Depending on chemical affinity, certain metal particles can be directly coated with silica. However, many materials lack this affinity, and in these cases, the materials can be coated with surface functionalizers, which simultaneously act as silane coupling agents. In the literature, nanocoating of colloidal nanoparticles has included: metals (Au, Ag, Fe/Fe2O3), semiconductors (CdS, ZnS, CdSe/ZnSe), polymers (polysterene, PDMS), and inorganic oxides (boehmite rods, gibbsite platelets). However, silica coated copper nanoparticles are absent. In this study, Cu colloidal nanoparticles, ranging in size from 25-125nm (Mesocopper: Purestcolloid Inc.) were coated with a SiO2 nanolayer. Polyvinylpyrrolidone (PVP) and sodium dodecyl benzene sulphonate (SDBS) were used as the surfactants (and simultaneously as surface functionalization linkers) to stabilize the core nanoparticles as well as to facilitate adherence of SiO2 on the particle surface. Tetraethoxysilane (TEOS) was used as the shell building material to synthesize the silica nanolayer. Aqueous ammonia and ethanol were used as the catalyst and the solvent respectively. The resulting core-shell nanostructures were characterized by TEM, SEM, and FTIR. Fourier Transform Infrared Spectroscopy was used to identify chemical groups at each stage of the coating process. UV-visible spectroscopy was used to observe changes in optical properties of the coated nanoparticles, and transmission electron microscope (TEM) as well as scanning electron microscope (SEM) were used to assess average particle diameter, silica shell thickness, and overall morphology of the core-shell nanostructures. TEM was more advantageous than SEM because it allowed the metal core particles to be observed along with the silica layer. The effects of different parameters, such as, concentration of silane (TEOS), concentration of stabilizer (PVP/SDBS) and time of aging, on the growth morphology of the silica nanoshells were compiled.

Supported by the grants from Department of Energy (DOE) Sustainable Energy Research Center (SERC) at MSU.

1. Z. Zhang, R.C. Patel, R. Kothari, C.P. Johnson, S.E. Friberg, P.A. Aikens (2000) J. Phys.Chem. B 104: 1176. 2. L.M. Liz-Marzán, M. Giersig, P.Mulvaney (1996) Langmuir 12: 4329.

232c Mechanistic Study Of Asymmetric Silver Electroless Deposition On Templated Polystyrene Spheres

Jingqin Cui, Department of Chemical Engineering, City College, City University of New York, Steinman Hall, 140th ST. & Convent Ave., New York, NY 10031 and Ilona Kretzschmar, Chemical Engineering, City College, City University of New York, Steinman Hall, 140th ST. & Convent Ave., New York, NY 10031.

The control of the surface morphology of a metal deposit plays an important role in (i) tuning the signal in surface-enhanced Raman scattering studies,[1] (ii) assembly of patchy particles, and (iii) catalysis. Recently, we developed a novel templating procedure to manufacture surface-anisotropic, silver-modified polystyrene spheres via electroless silver deposition.[2] Our data indicated that the silver deposition process might be diffusion-controlled due to the observed dependence of the amount of silver deposited on the stirring rate. Here we report on the effect of reaction conditions on the morphology of the silver deposit. In our experiments, we use 2.4, 0.99, and 0.51 μm polystyrene (PS) spheres partially embedded in polydimethylsiloxane. The templated PS spheres are modified using electroless deposition of silver. The deposition is initiated by addition of a reducing agent, i.e., formaldehyde, to a silver acetate solution in the presence of a complexing agent, i.e., ammonia. In order to determine the function of each component of the electroless bath, we expose the templated PS spheres to changing reaction conditions, namely stirring rate, exposure time, and reactant concentrations/ratios. Variation of the stirring rate and exposure time are efficient and easy to perform experiments to determine the influence of diffusion on the deposition process and the deposit morphology. Stirring rates between 0 and 270 rpm are explored as well as exposure times between 5 and 30 min. In addition, silver salt, formaldehyde, and ammonia concentrations are varied to study the effect on the morphology of the deposit. Silver nanoparticle size and quantity are counted from SEM images. High magnification images are recorded and used to interpret the particle growth mechanism. We will present evidence for a diffusion controlled silver deposition process as well as show that the silver nanoparticle size can be controlled by the reactant/complexing agent ratio. In addition, we will present preliminary data on the effect of the curvature of the spherical substrate on the deposition process.

1. P. Barnickel and A. Wokanu, Mol. Phys. 1989, 67, 1355 2. J.-Q. Cui and I. Kretzschmar, Langmuir, 2006, 22, 8281-8284

232d "Green" Synthesis And Seed-Mediated Growth Of Sodium Carboxymethyl Cellulose Stabilized Metal Nanoparticles

Juncheng Liu and Christopher B. Roberts. Department of Chemical Engineering, Auburn University, 233 Ross Hall, Department of Chemical Engineering, Auburn, AL 36849

Central to the concept of seed-mediated growth of nanoparticles is that small nanoparticle seeds serve as nucleation centers to precisely grow nanoparticles to a desired size. Herein we present an environmentally benign and size-controlled seed-mediated growth method coupled with an organic phase extraction to yield monodisperse dodecanethiol capped Pd nanoparticles. Monodisperse Pd nanoparticles (ca. 3.4 nm, standard deviation: 0.6 nm) were synthesized and stabilized by a "green" and inexpensive polysugar, sodium carboxymethyl cellulose (CMC) in aqueous solution. The monodisperse Pd nanoparticles were employed as seeds on the surface of which fresh Pd^{2+} ions were reduced by weak reducing agent ascorbic acid. This seed-mediated growth procedure results in a complete separation of nucleation and growth, and hence, the atomic-level size controlled synthesis of the Pd nanoparticles. Using different salt concentration in this seed-mediated growth process, different sized Pd nanoparticles were controllably synthesized. The nanoparticles could then be extracted from the aqueous solution to an organic phase by using dodecanethiol as extractant with the assistance of concentrated HCl. The advantage of the seed-mediated growth method used in this study is that the Pd nanoparticles size could be precisely controlled (from 3.4 nm to 9 nm) by simply varying the ratio of seed to Pd salt. Furthermore, the use of solvent (water), capping agent (CMC) and reducing agent (ascorbic acid) in the system provides the "green" and cost-effective merits for this precise size-controlled particle synthesis method.

Optimization of industrial aerosol nanoparticle generation process for creating photocatalytic and anti-microbial surfaces.

Mikko Aromaa[1], Jyrki M. Mäkelä[1], Helmi Keskinen[1], Minna Piispanen[2], Leena Hupa[2], Knut Deppert[3], Jakob B. Wagner[4], L. Reine Wallenberg[4], Stellan Persson[5], Maria Lang[5], Peter Sundberg[5], Joe Pimenoff[6], Thomas Kronberg[7], Viljami Pore[8], Mikko Heikkilä[8], Mikko Ritala[8], Markku Leskelä[8], Mari Raulio[9], Mirja S. Salkinoja-Salonen[9] and Veli-Matti Airaksinen[10].

[1]Institute of Physics, Aerosol Physics Laboratory, Tampere University of Technology, P.O. Box 692, FIN-33101 Tampere, Finland.
[2] Process Chemistry Centre, Åbo Akade Solid State Physics, Box 118, S-221 00 Lund, Sweden.
[4] Polymer and Materials Chemistry, nCHREM, Lund University, Box 124, S-221 00 Lund, Sweden.
[5]Glasforskningsinstitutet, P.O. Box 5157, 350 05 Växjö. Sweden
[6]Beneq Oy, Ensimmäinen savu, FIN-01510 Vantaa, Finland
[7]IDO Bathroom Ltd, Koivuniemi, FIN 10600 Tammisaaari, Finland
[8]Laboratory of Inorganic Chemistry, Department of Chemistry University of Helsinki, P.O. Box 55, FIN-00014 Helsinki, Finland.
[9]Department of Applied Chemistry and Microbiology, University of Helsinki, P. O. Box 56, FIN-00014 Helsinki, Finland.
[10] Micronova, Helsinki University of Technology PO BOX 3500, FI-02015 TKK, Finland

Abstract

Flame methods are a promising way to manufacture nanoparticles in commercial scale. They are widely utilized already. In this study, a process called Liquid Flame Spray, LFS, is used. It is commercially known as nHALO (Hot Aerosol Layering Operation, Beneq Oy). In LFS, liquid precursor is fed into a turbulent hydrogen-oxygen flame. LFS can be used in production of nanoparticles which are deposited on a substrate. Even multicomponent nanoparticles can be manufactured using LFS process.

LFS is a versatile method for nanoparticle generation. In this study, we use LFS to produce an anti-microbial titania-silver nanoparticle coating on ceramic tiles and float glass. The coating is produced by collecting TiO_2-Ag nanoparticles directly from the flame. The particles are deposited on the surface. Some of the excess particles are not properly attached. The coating is then characterized using electron microscopy, contact angle measurements, organic and biofilm removal tests and bacterial tests. Scanning Electron Microscopy (SEM) images indicate that the titania-silver coating covers the tiles even after scrubbing away the loosely adhered particles. Tunnelling Electron Microscopy can be used to obtain information about the particle morphology. TEM micrographs suggest that the coating is formed of 10 to 20 nm TiO_2 particles and small Ag nanoparticles on the surface of TiO_2. The size of the Ag particles is few nanometres.

Keywords: nanoparticle, Liquid Flame Spray, nHALO, anti-microbial coating, titania, silver

Introduction

Flame methods have been used in nanoparticle production for several decades [1]. Multicomponent nanoparticles and deposits have recently been a topic of most of the studies [2-12]. TiO_2 is well known for its photocatalytic properties [13-15]. Addition of silver to titania increases the bacteria killing and biofilm removal efficiency [16-18]. Flame methods are cost efficient and the scale up is rather easy. Therefore, flame methods can also be utilized in an industrial scale. [19] In Liquid Flame Spray (LFS)/ Hot Aerosol Layering Operation (nHALO) [20], the liquid precursor is injected into a high temperature flame (T_{max}~3000 K), where it will evaporate, react and nucleate to form nanoparticles. Titania and silver particles have been studied in recently as well as particle deposition. [12, 21, 22] Also the scale up of the particles to optimize the particle production has been studied [23]. In this study a commercial coating method, nHALO, was used to create photocatalytically active titania-silver deposits for different tile-substrates. nHALO is also known as Liquid Flame Spray. Silver addition in the samples was 0.5-2-wt % of TiO_2. The deposition process parameters were varied in order to find the best deposition conditions. Coatings' functionality will be studied using biofilm removal tests in the next phase of the project, but for now the coatings are only characterized using electron microscopy.

Experimental

In the LFS method liquid precursor is introduced in a high velocity, turbulent, high temperature H_2-O_2 flame. The gas flow rates used in this study are 30 l/min for H_2, 15 l/min for O_2 and 5 l/min for N_2. The maximum temperature is ~3000 K [20]. The luminous part of the flame was about 20 cm in length. One of the combustion gases atomizes the liquid precursor into small droplets. In this study the atomizing gas is H_2. In the flame the droplets evaporate, decompose, nucleate and condense into solid nanoparticles. The nanoparticles are then deposited on a glass and glazed ceramic tile surface. The particles are formed from a precursor solution containing titanium(IV) isopropoxide and silver nitrate in isopropanol. The amount of silver nitrate is 1 wt-% of the titania produced. In this study liquid feed rates of 5 and 10 ml/min are used and the material production rate varies from 50 to 200 mg/min.

The coatings were deposited on five different substrates. Four of them were glazed ceramic tiles and one was float glass. There were two glossy tiles by IDO Bathroom Ltd and another commercial tile. Other two were a commercial tile with a matt glazing and an Åbo Akademi University test tile which lacked zirconia crystals that make the tile white, but make the SEM analysis harder. The sample size was 25 x 25 mm. The deposition was performed in a special nHALO-production line (Figure 1). nHALO is a registered trademark of an industrial technique utilizing liquid flame spray by a company Beneq Oy. The coatings were deposited from two different distances from the torch head 130 and 170 mm, corresponding to flame temperatures of approximately 800 °C and 600 °C. The substrates were pre-heated before deposition, to avoid temperature shock and cracking. Pre-heating temperatures of 500, 550 and 600 °C were used. The pre-heating temperature was measured by a pyrometer from the surface of the float glass sample that was also being coated. After the coating process, the substrates went into the cooling section of the coating line to bring down the temperature in a controlled way in order to avoid cracks and stresses in the samples. The coating speed of the samples was 3 m/min. The coatings were characterized using TEM samples taken from the coating line during the coating process and by SEM analysis performed on the coated samples.

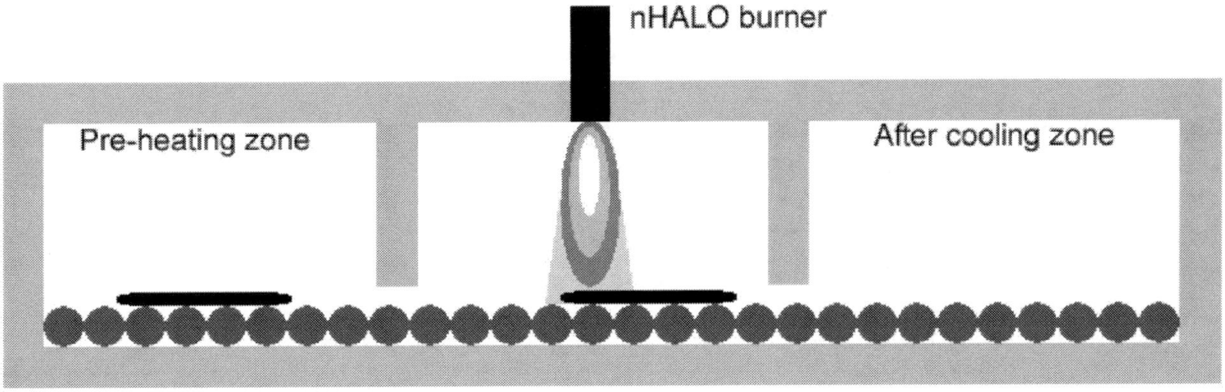

Figure 1. nHALO coating line. The conveyer is moving from left to right (Beneq Oy)

Results and Discussion

In figure 2 there is a SEM analysis from the coated tiles. Figure 2 shows that the coating is uniformly distributed over the ceramic tile. The coating consists of agglomerated titania particles with the primary particle size of less than 20 nm. There are also some larger residual particles but they are covered with smaller nanoparticles and the amount of the residual particles is so small that it does not have a contribution to the antimicrobial performance of the coating.

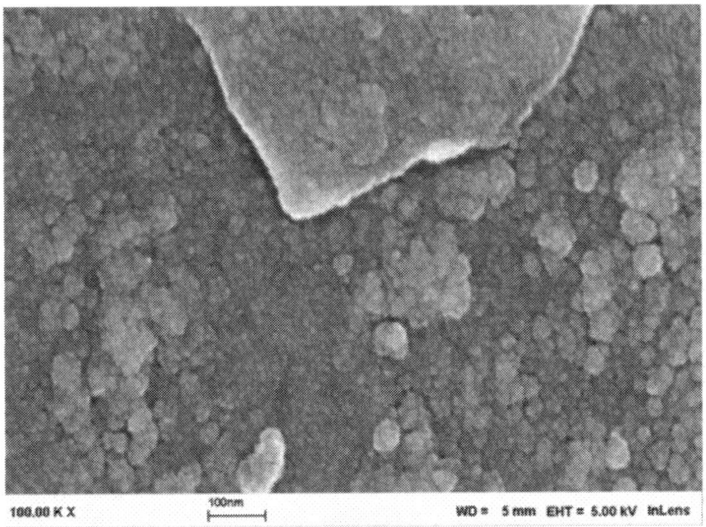

Figure 2. A SEM image of a titania-silver coating on a ceramic tile

TEM analysis of the samples is shown in figures 3 and 4. Figure 3 shows small 1-3 nm silver particles over the larger titania particles. Silver particles also appear on the surface not covered by titania particles. TEM analysis reveal that there are both anatase and rutile particles in the coatings. Figure 4 showing a HRTEM picture of an anatase particle.

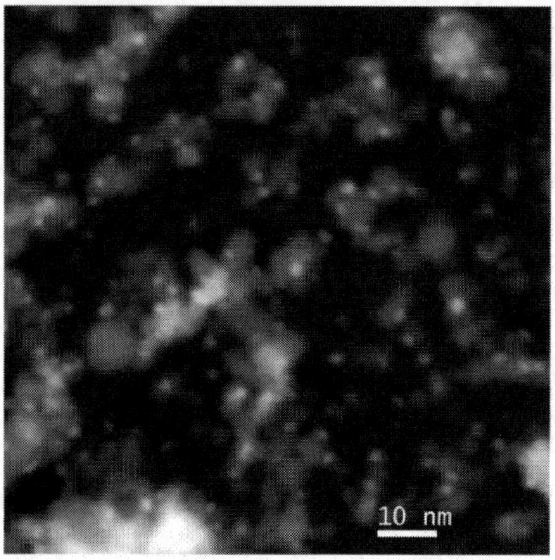

Figure 3. A high angle annular dark field TEM image of the coating. The small bright spots are silver particles.

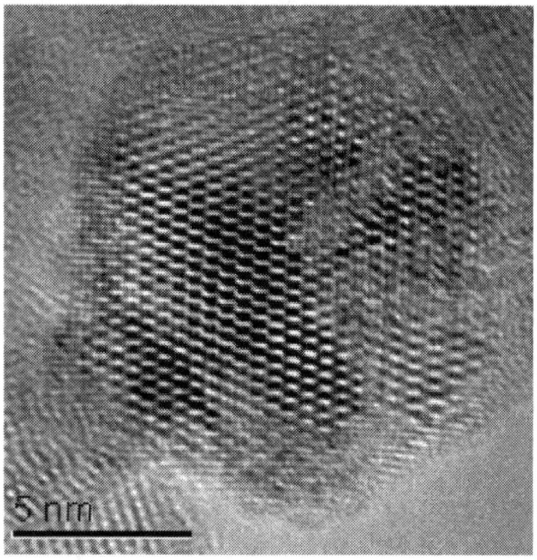

Figure 4. An anatase particle

The float glass samples were coated to see the colour effect of the coating. Using some parameters the glass samples had some white colour due to titania particles. Most of the samples were transparent. Also the effects of silver dissolving in the glass were investigated by human eye. Almost all the samples had some yellow shades. This is an indicator of colloidal silver inside the glass matrix. However, in the ceramic tiles the yellowish colour was not observed.

At the moment a systematic series of the coatings has been manufactured. The photocatalytic performance of the samples is being analyzed in bio- and organic film reduction tests. This set of coatings also shows the reproducibility of the samples as there are several parallel samples.

Acknowledgements: This work has been funded by Nordisk InnovationsCenter (NICe) project FUNCOAT.

References

[1] S.E. Pratsinis, Prog. Energy Combust. Sci., 24 (1998) 197.
[2] S.H. Ehrman, S.K Friedlander and M.R. Zachariah J. Aerosol Sci. 29 (1998) 687.
[3] T. Johannessen and S. Koutsopoulos. J. Catal. 205 (2002) 404.
[4] H. Keskinen, J. M. Mäkelä, M. Vippola, M. Nurminen, J.K. Liimatainen, T. Lepistö , and J. Keskinen, J. Mater. Res., 19 (2004) 1544.
[5] W. Y. Teoh, L. Mädler, D. Beydoun, S. E. Pratsinis, R. Amal, Chem. Eng. Sci., 60 (2005) 5852.
[6] H. Schulz, L. Mädler, R. Strobel, R. Jossen, S. E. Pratsinis, T. Johannessen, J. Mater. Res., 20 (2005) 2568.
[7] W. J. Stark, J. D. Grunwaldt, M. Maciejewski, S. E. Pratsinis, A. Baiker, Chem. Mater., 17 (2005) 3352.
[8] R. Jossen, R. Mueller, S. E. Pratsinis, M. Watson, M. K. Akhtar, Nanotechn., 16 (2005) 609.
[9] D. Perednis, O. Wilhelm, S. E. Pratsinis, L. J. Gaukler, Thin Solid Films, 474 (2005) 84.
[10] A. Teleki, S. E. Pratsinis, K. Wegner, R. Jossen, F. Krumeich, J. Mater. Res., 20 (2005) 1336-1347.
[11] R. Strobel , S. E. Pratsinis, A. Baiker, J. Mater. Chem., 15 (2005) 605.
[12] H. Keskinen, J.M. Mäkelä, M. Aromaa, J. Ristimäki, T. Kanerva, E. Levänen, T. Mäntylä and J. Keskinen, In press (2006), to be published in J. Nanoparticle Res.
[13] A. Fujishima, K. Honda, Nature 238 (1972) 37.
[14] A. Fujishima,, T. N. Rao, D. A. Tryk, J. Photochem. Photobiol. C 1 (2000) 1.
[15] A. Mills, S. Le Hunte, J. Photochem. Photobiol. A: Chem. 108 (1997) 1.
[16] C. He, Y. Yu, X. Hu and A. Larbot, Appl. Surf. Sci., 200 (2002) 239.
[17] V. Vamathevan, R. Amal,, D. Beydoun, G. Low and S. McEvoy, J. Photochem. and Photobiol.A: Chemistry 148 (2002), 233.
[18] A. M. Mulligan, M. Wilson and J.C. Knowles, J. Biomed. Mat. Res.. Part A. 67(2) (2003) 401.
[19] K. Wegner, S. E. Pratsinis, Powder Technol., 150 (2005), 117.
[20] J. Tikkanen, K.A. Gross, C.C. Berndt, V. Pitkänen, J. Keskinen, S. Raghu, M. Rajala and J. Karthikeyan, Surf. Coatings Technol., 90 (1997) 210.
[21] J.M. Mäkelä, S. Hellstén, J. Silvonen, M. Vippola, E. Levänen, and T. Mäntylä, Mater. Lett., 60 (2006), 530.
[22] H. Keskinen , J.M. Mäkelä S. Hellsten, M. Aromaa, E. Levänen and T. Mäntylä Electrochem. Soc. Proc. Vol. 2005-09, 491-498.
[23] M. Aromaa, H. Keskinen, J.M. Mäkelä, Biomolecular Engineering (2007) doi:10.1016/j.bioeng.2007.08.004

232f Growth Mechanisms and Properties of Metal/semiconductor Hybrid Nanocrystals

Aaron E. Saunders and Uri Banin. Institute of Chemistry, The Hebrew University of Jerusalem, Jerusalem, 91904, Israel

The combination of different material domains into a single "hybrid" colloidal nanocrystal offers an interesting material platform, both for studying the fundamental behavior of coupled nanocrystal systems, as well as for creating particles with multiple functionalities. Inorganic colloidal nanocrystals with both metallic and semiconducting domains make up an interesting subset of the general class of hybrid nanocrystals; when metallic and semiconducting domains of similar size are joined on the nanometer scale, the resulting hybrid nanoparticles exhibit overall optical properties which are not simply linear combinations of the optical properties of the individual components. Other properties of the resulting hybrid nanocrystals make them intriguing possibilities for use in directed assembly techniques, or as components in catalytic and photovoltaic devices.

In our work on binary metal/semiconductor nanocrystal systems, we have identified three dominant growth mechanisms: (1) surface growth of material B onto preformed nanocrystals of material A (e.g. Au onto CdS), (2) surface growth of B followed by diffusion into material A (e.g. Au onto InAs), and (3) simultaneous nucleation and growth of both components A and B (e.g. Ag and AgBr). Different material combinations which highlight each of these mechanisms will be discussed, as well as the results of optical and structural characterization. Understanding the underlying mechanisms of nanocrystal growth may allow for further tuning of the morphology and functionality of hybrid nanocrystals.

241a Panel Discussion - Various Panelists

Shrikant Dhodapkar, The Dow Chemical Company, B-1402, Dow Chemical Company, Freeport, TX 77541 and Harald Wilms, Zeppelin Silos & Systems GmbH, Leutholdstr. 108, Friedrichshafen, 88069, Germany.

Various panelists will present their view points which will be followed with discussion.

285a Unsteady Shear Flow Of Assemblies Of Cohesive Granular Materials

Lee R. Aarons and S. Sundaresan. Chemical Engineering, Princeton University, Princeton, NJ 08544

In the present study we have examined through Discrete Element Method (DEM) simulations the dynamic response of dense assemblies of cohesive particles to time-varying shear flow. Such simulations reveal the essential characteristics that a dynamic continuum model for granular rheology should possess. For example, it is well known that in the quasistatic regime of flow the stresses in a dense granular assembly are independent of the steady shear rate. Thus, if a step change is made in the shear rate, the asymptotic values of the stresses before and after the change will not differ. However, there can be significant dynamic excursions and the nature and extent of such excursions are not well understood. The present study examines the dynamic response of the stresses and the microstructure to imposed changes in the shear rate.

We consider DEM simulations [1] of the plane shear flow of particle assemblies in periodic domains, employing Lees – Edwards boundary conditions [2] across the faces. Simulations were performed for different strengths of cohesion, shear rates, and particle volume fractions. In these simulations, assemblies were subjected to oscillatory shear or step changes in shear rate, and we have tracked the response of the normal and shear stresses. All the simulations were done at small shear rates, representative of the "quasistatic regime".

Under constant shear [3], it was found that steady state was achieved for a given simulated system after shearing for a time equal to about the inverse of the shear rate. Similarly under oscillatory shear, we found that if the oscillation frequency was chosen to be larger than the maximum shear rate, the steady state stress achieved in the steady shear simulations was never achieved, regardless of the number of oscillations performed. At lower frequencies, the steady state stress value was achieved, and the stress fluctuated about that value until shear was reversed. Upon shear reversal, the normal stress rapidly dropped to a negative value, such that the assembly was in tension. Afterwards, the normal stress increased monotonically towards the steady state value until the shear reversal occurred again. As cohesion was increased, the normal stress started to drop off just before shear reversal (i.e. at very small shear rates), even if the steady state stress was reached and even though the shearing occurred in the quasistatic regime.

The behavior of cohesive materials subject to a step change in shear rate depended on the level of cohesion. At low levels of cohesion, the change in shear rate was followed by a large spike in the normal stress. Investigations into the microstructure of these systems have shown a distinct correlation between the behaviors of the force chains and the normal stress. After the change in shear rate but before the spike in stress, the majority of the stress is carried by the particles at the top and bottom of the simulated system. At the top of the spike, the stress became more evenly spread out across the assembly. After the stress spike, the majority of the stress was once again confined to force chains. At higher levels of cohesion, none of this happened. There was no large spike in stress and the force chains were virtually unaffected by the change in shear rate. Thus, cohesion serves to mitigate very large fluctuations that can occur under dynamic conditions, suggesting that one can beneficially exploit modest levels of cohesion to suppress stress fluctuations.

In this presentation, we will also compare these results with the hypoplastic model for slow flow of granular materials proposed by Wu et al. [4]. It will be shown that such models are miss important qualitative features.

[1] P. A. Cundall, O. D. L. Strack, A discrete numerical model for granular assemblies, Geotechníque 29 (1) (1979) 47-65.

[2] A. W. Lees, S. F. Edwards, The computer study of transport processes under extreme conditions, Physics C: Solid State Phys. 5 (1972) 1921-1929.

[3] L. Aarons, S. Sundaresan, Shear flow of assemblies of cohesive and non-cohesive granular materials, Powder Technol. 169 (1) (2006) 10-21.

[4] W. Wu, et al., Hypoplastic constitutive model with critical state for granular materials, Mechanics of Materials 23 (1) (1996) 45-59.

285b Dem Simulation On Pattern Formation Of Radial Segregation Of Granular In Rotating Drums

Yongzhi Zhao and **Yi Cheng**. Department of Chemical Engineering, Tsinghua University, Beijing, 100084, China

Segregation and de-mixing often occur when a granular mixture is vibrated or made to flow. Two types of particle systems, i.e., D-systems and S-systems, have received more attention in recent years, where particles have the same size and different density in D-systems, and have the same density and different size in S-systems. Both systems present theoretical challenges in understanding the physical mechanisms behind the segregation and de-mixing, of which the S-systems often exhibit interesting segregation patterns. For example, a binary-solid mixture in rotating drums generally shows two cases: one is the axial segregation when the axis of the drum is much longer than the diameter and the other is the radial segregation when the axis of the drum is much shorter. In the latter case, rotation of a mixture of particles in different sizes in a thin drum often results in radial segregation with the smaller particles concentrating in a central core and the larger particles in the periphery. However, some different flow patterns may appear under certain conditions. For instance, the pattern of radial segregation could reverse at high rotational speed, i.e., smaller particles at the periphery and larger particles in the core. When the drums are slightly more than half-filled and the rotational speeds are low that just in the avalanching regime, radial streaks, called petals patterns or sun patterns, would be found in thin drums.

This work focused attention on the dynamics of pattern formation during the radial segregation of a binary-solid system in thin rotating drums. A discrete element model (DEM) considering the normal contact force, the tangential contact force, and the rolling friction torque for granular dynamics was employed to simulate the segregation of the S-type binary granular mixtures in a horizontal drum. The simulation results successfully predicted the dynamic evolution during the pattern formation of radial granular segregation started from a uniformly mixed binary-solid system as the initial state. In particular, the dependence of different patterns on the rotary speed and the load of the drum was well predicted, which agreed well with the experimental observations in the literature. As a higher rotary speed is applied, the moon pattern that the smaller particles concentrating in a central core and the larger particles in the periphery appears. As the rotary speed is lower, the petals pattern forms in the thin drum, and the number of petals increases as the rotary speed increases. If the particles loaded in the drum are few, the petals pattern cannot form in the drum. As all these results were obtained from DEM simulations, micro-scale analysis can be made to understand the evolution and formation of the segregation patterns.

A Novel Method for Prediction of Particle Dispersion in a Planar Jet Using ODT Model

Yuxin Wu[1], P. J. Smith[2], J. Thornock[2], Guangxi Yue[1], Jiansheng Zhang[1]

1. Key Laboratory for thermal science and power engineering of ministry education
Department of Thermal Engineering, Tsinghua University, Beijing, 100084, P.R.CHINA
2. Department of Chemical Engineering, University of Utah, SLC, USA, 84108

Abstract The One Dimensional Turbulence (ODT) model was used to study a particle laden flow in a planar jet. As an outgrowth of the LEM (Linear Eddy Model) model, the ODT model maintains a distinction among the turbulence, molecular diffusion and chemical reaction scales. Additionally, the turbulent mixing process is only considered in one-dimension, where eddy events are stochastically represented, which allows for representation of a large range of time and length scales with relatively small computing requirements. In this study, the particle and fluid time scales are split by advancing the fluid phase independently of the particle phase. When an eddy was judged to occur, it is assumed that the eddy will always exist at this area. Therefore locations of the eddy occurrences in the fluid phase are recorded after each fluid advancement step. After the fluid phase has advanced, particles are tracked through the established flow field, interacting with the eddy occurrences. When a particle encounters an eddy, its motion is affected by the eddy velocity, which is a combination of the local gas flow field and turbulent mixing. With this method, eddy effects on the particle motion are considered.

Different intensities of eddy effects on particle motion were compared. Particle dispersion rate is proportional to eddy effects. This suggests that turbulent eddies always have a positive effect on particle dispersion. Besides that, particle diameter has an important influence on particle-eddy interactions. Small particles are not sensitive to eddy effects, whereas medium particles are very sensitive to eddy effects. According to Budilarto's observation, a medium eddy shape factor is selected. Based on this eddy shape factor, the final results shows that particle diameter always has a negative effect on particle dispersion, causing large particles to concentrate at center of the jet while small particles disperse to the edge. A 2-way The ability of this ODT model to capture these dispersion effects provides motivation for applying it to particle-gas chemical processes, such as coal combustion and gasification where particle clustering and dispersion has been observed.

1. Introduction

Gas solid flows, defined as multiphase flows consisting of solid particles dispersed throughout a gas medium, have been used extensively in the chemical process industries. For most gas-solid flows, particles move in a strong turbulent gas stream. When particles are introduced into a turbulence flow, they will interact with turbulent eddies, and their motion will be altered by the turbulent motion of the gas. As an outgrowth of the LEM (Linear Eddy Model) model, the ODT (One Dimensional Turbulence) model[1] maintains a distinction among the turbulence, molecular diffusion and chemical reaction scales. Due to its low dimensional characteristic, ODT model costs low computation time while maintaining complete turbulent information throughout a big range of time and space scales. This is a big advantage to consider particle-gas phase turbulent interactions.

Some attempts have been carried out in the past yeas. Schmidt[2] proposed a new method to consider particle motion in a temporally developed turbulent channel flow based on Lagrangian frame. One way coupling is achieved by following motions of the particle as dictated by the particle dragging law. The fluid motion equations are solved with ODT model without considering particle momentum source and the gas velocities are used as an input into solving the particle dragging law. Kerstein[3] studied bi-dispersion and monodispersion for low initial particles and indicated that the ODT model is useful for clarifying the origin of clustering. In this paper, a new method was proposed to simulate particle dispersion in a spatially developed planar jet. The effect of particle size on particle dispersion was studied and a Particle Size Distribution (PSD) was acquired.

2. ODT Model

ODT is a Monte Carlo method for turbulent mixing of velocity and scalar fields[4]. It makes a representation of turbulent advection by a postulated stochastic process rather than an evolution equation. Viscosity is incorporated as a concurrent deterministic process, governed by an evolution equation of conventional form. It had been used to compute statistical properties of velocity and scalar fields in stationary and decaying homogeneous turbulence, shear flow, and various buoyant stratified flows [1][5].

2.1 Model implement

For a steady-state spatially developed turbulent planar jet, ODT generates realizations parameterized by (y, x) instead of (y, t), where x is the streamwise coordinate, y is the transverse direction. Turbulence was only considered in y direction. The field defined on the one-dimensional domain is evolved by two separate mechanisms: molecular evolution and a stochastic process representing advection. Assuming a constant density and viscosity case, a boundary layer equation which keeps a conservation form, was used to interpret molecular evolution process, shown as Eq(1)

$$\frac{\partial(uu)}{\partial x}=\upsilon\frac{\partial^2 u}{\partial y^2} \tag{1}$$

where υ is the kinematic viscosity.

Followed by this diffusion process, the statistic events representing advection may be interpreted as the model analogue of individual turbulent eddies. Each event is characterized by three properties: an eddy length scale, an eddy time scale and a measure of kinetic energy. When an eddy event was judged to occur, the effects of the eddy is represented by a triplet map process, which consists of two mathematical operations. One operation is a measure preserving map(so that all y-integrated quantities obey the applicable conservation laws) representing the fluid motions associated with a notional turbulent eddy. This operation will be implemented on all scalars. The other is a modification of the velocity profiles in order to implement energy transfers prescribed by the dynamical rules, which is only operated on velocities. These operations are represented symbolically as[6,7]:

$$\begin{aligned} &v_i(y)->v_i(f(y))+c_iK(y)\\ &\theta(y)->\theta(f(y)) \end{aligned} \tag{2}$$

Where

$$f(y)=y_0+\begin{cases} 3(y-y_0) & y_0\le y\le y_0+1/3l\\ 2l-3(y-y_0) & y_0+1/3l\le y\le y_0+2/3l\\ 3(y-y_0)-2l & y_0+2/3l\le y\le y_0+l\\ (y-y_0) & otherwise \end{cases} \tag{3}$$

$$K(y)=y-f(y) \tag{4}$$

$$c_i=\frac{27}{4l}\left(-v_{i,k}+\mathrm{sgn}(v_{i,k})\sqrt{(1-\alpha)v_{i,k}{}^2+\alpha/2(v_{j,k}{}^2+v_{k,k}{}^2)}\right) \tag{5}$$

The sign ambiguity in Eq.(5) is resolved by requiring that $c_i\to0$ as $\alpha\to0$.

In numerical implication, for each ODT step, or Δx, a molecular evolvement and triplet map are completed and velocities are renewed. During molecular evolution and advection process, in order to keep the principle of mass conservation and momentum conservation, cell size should be changed on 1D domain to keep the mass flux in each cell a constant value. This makes the calculation domain change after each step, thus cells should be added or cut to assure the domain width is invariable. Actually, this process has an obvious physical meaning. In a planar jet, as jet flow develops, the center velocity decreases and axial momentum diffuses to the co-flow. As a result, the co-flow gas will be engulfed to the jet flow and jet flux will increase continuously. The change of domain cell size and its number embodies this kind of engulfment. Correspondingly, this process arouses an additional transverse velocity. In the model, this transverse velocity, v_{diff}, is calculated in an indirect way. After each ODT step, the displacement of the corresponding cells is recorded and v_{diff} is calculated as the division of this displacement by ODT time.

2.2 Eddy Selection

Up to now, the only remaining question is to judge if an eddy will occur or not. This problem is solved by a stochastic Poisson process. Firstly, a random set of eddy start position and eddy length is sampled based on Possibility Density Function (PDF) of $f(l)$ and $g(y_0)$ at each stir time step Δt_{stir}. Although the form of $f(l)$ and $g(y_0)$ has no effects on the final results, its selection determines the calculation efficiency. In this simulation work, $f(l)$ and $g(y_0)$ are as follows[4]:

$$f(l)=cl^{-2}\exp(-2L_p/l) \tag{6}$$

$$g(y)=\frac{1}{domain\_y-l} \tag{7}$$

Where c is a constant number, which ensure the integration of $f(l)l$ through eddy length is unit, L_p is the eddy length that $f(L_p)$ is largest.

Based on the given eddy length l and eddy position y_0, eddy rate distributionλ,which means the occurrence frequency for eddies whose length is l at y_0 , can be calculated by Eq.(8)

$$\begin{aligned} \lambda(y_o,l;t)&\equiv\frac{C}{l^2\tau(y_o,l;t)}\\ &=\frac{C\upsilon}{l^4}\sqrt{\left[(1-\alpha)v_{2,K}^2+\frac{\alpha}{2}(v_{1,K}^2+v_{3,K}^2)\right]\left(\frac{l}{\upsilon}\right)^2-Z} \end{aligned} \tag{8}$$

Where C, α and Z are constant number, C=8.9, α=1/7, Z=0.02, $v_{i,k}$ is calculated as follows:

$$v_{i,K}\equiv\int u(y)K(y)\mathrm{d}y=\frac{4}{9l^2}\int_{y_0}^{y_0+l}v_i(y)\left[l-2(y-y_0)\right]\mathrm{d}y \tag{9}$$

When eddy stir time step Δt_{stir} is much less than ODT time step, the sampled eddy occurrence possibility can be calculated by Eq(10).

$$P(y_o,l;t,\Delta t_{\mathrm{stir}})=\frac{\lambda(y_o,l;t)\Delta t_{\mathrm{stir}}}{f(l)g(y_o)} \tag{10}$$

A random number P_{ram} is generated to represent the possibility of the occurrence of this eddy. If $P(y_0,l;t,\Delta t_{\mathrm{stir}})> P_{\mathrm{ram}}$, the eddy occurs and its effects on flow field should be considered.

3. Particle Tracking

3.1 Transformation from t-x and t-y coordinates to y-x coordinates

When ODT method was implemented on gas phase, the complete turbulent information was recorded as a series of eddy events, which are assumed to occur instantly at a certain point. However, it will take a period of time for each particle tracking step. The arising question is then how to consider particle-eddy interaction during particle integration process. For a temporally developed turbulent channel flow, J. Schmidt [2] made an assumption that each eddy occupies a time-space region. Based on this assumption, eddy effects were considered when a particle and an eddy occupy the same time-space point during particle tracking process. However, for a spatially developed turbulent planar jet flow, ODT line develops in x direction. During stochastic Poisson process, the eddy stir time step Δt is affected by sampled eddy position and local velocity. The *time-x* (t-x) coordinate for the gas phase is determined by eddy judging process and is fixed in a certain realization. While for a particle, its *t-x* coordinate is determined by its position and integral time step, which is unique for each particle and is different from *t-x* coordinate of the gas phase. As is shown in Fig.1, the *t-x* coordinate develops along the diagonal line, during a same *x* step, the time duration of the gas phase (Fig.1(a)) and of each particle (Fig.1(b)) are different. This differences makes eddies and particles can't meet together at the same space-time point.

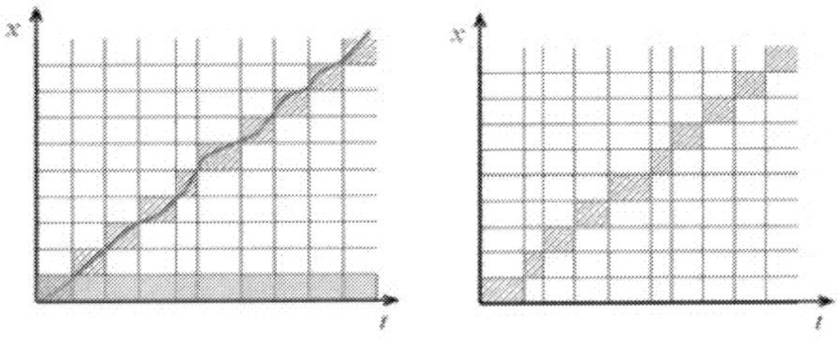

(a) gas phase (b) a certain particle

Fig.1 t-x coordinates for gas phase and one particle

Alternatively, there is another method based on the concept of eddy distribution rate λ to resolve this problem. In numerical implementation, a Poisson process is adopted to avoid time consuming process of double integration ofλ. For a certain *x* step (height) of a steady state planar jet, eddies occur and vanish without interruption through the whole time, which is shown as the bottom line in Fig.1(a). In ODT method, however, eddy judging process was implemented only through a part of time. The whole time line is represented by a diagonal hatched block, thus eddy rate distribution through the whole time was also determined by this time step. If the whole time line is assumed to be composed of the same diagonal hatched blocks, or the same eddy rate distribution repeats through the whole time, then any eddy judged to occur at a certain position and a time can be regarded to appear at this place through the whole time. Additionally, although an eddy occurs on a certain line in y direction, its region is supposed to be a rectangular in *x-y* coordinate. The ODT line which eddy occurs is regarded as the eddy beginning position in x direction. There is an eddy shape factor to determine the ratio of eddy height to eddy length. In this simulation, the eddy shape factor β is set to 0.6. Once a particle moves into this region, the eddy effects should be considered. Based on these two assumptions, the *x-t-y* coordinates is transformed to x-y coordinates. This transformation is shown in Fig.2.

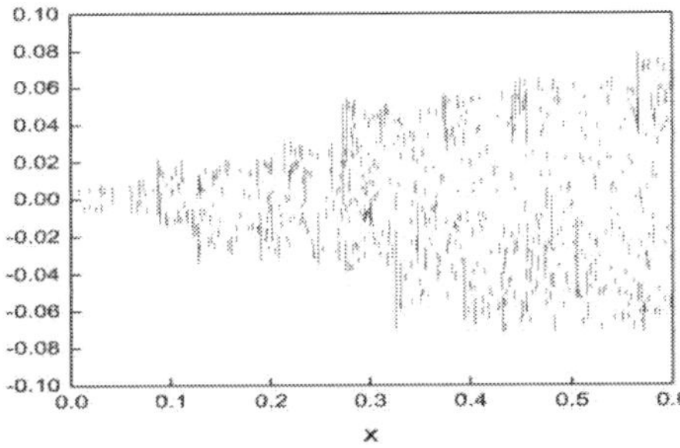

(a) Eddy distribution through a realization

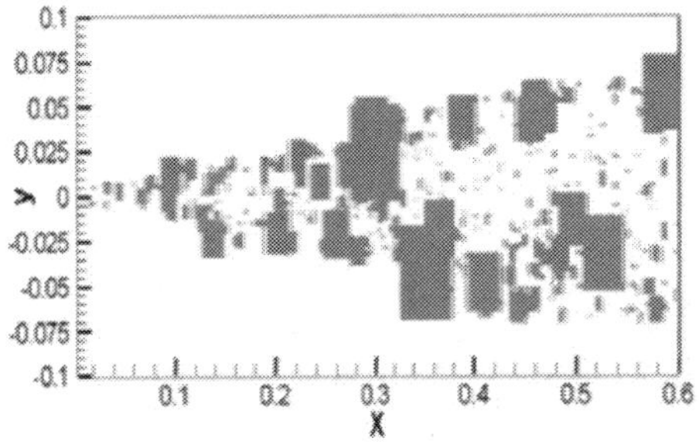

(b) transformed turbulent information

Fig.2 Transformation of turbulence information through x-y coordinates

Fig.2(a) is the eddy distribution acquired by Poisson process in a certain realization, after transformation, the same eddy distribution is regarded as in steady-state in *x-y* coordinates represented by a series of rectangular regions, shown as Fig.2(b). In the following particle integration process, when a particle moves to a position where is occupied by an eddy, the eddy-particle interactions is involved and turbulent influence on particle motion can be captured. .

3.2 Particle-eddy interaction

Particle motions can be predicted by solving its force balance equations in *x* and *y* directions, shown as Eq.(11)

$$\frac{\mathrm{d}u_{\mathrm{p},i}}{\mathrm{d}t}=\frac{f}{\tau}\left(u_{\mathrm{g},i}-u_{\mathrm{p},i}\right)+\frac{g_i(\rho_\mathrm{p}-\rho_\mathrm{g})}{\rho_\mathrm{p}}+\frac{F_i}{m_\mathrm{p}} \tag{11}$$

The three terms in right part of the equation are drag force, gravity force and body force separately. The suffix "p", "g" and "*i*" denotes particle, gas and *i*th direction.

When a particle moves to the position where an eddy occupies, its axial motion is still determined by local gas axial velocity, but its transverse motion is merely affected by the eddy, because it is in this direction that turbulence is involved. The eddy effects on particle motion is considered by substituting $u_{g,y}$ with v_{eddy} in the drag force term of Eq.(11) in *y* direction. v_{eddy} is the instant eddy *y* velocity at that position when particle get there. It can be calculated by Eq.(12).

$$v_\mathrm{eddy}=\left\langle u_{\mathrm{g},y}\right\rangle+\left\langle v_\mathrm{diff}\right\rangle+\frac{\Delta y}{t_\mathrm{e}}=\frac{\int_{y_0}^{y_0+l}\left((u_\mathrm{g}(y,t)+v_\mathrm{diff}\right)\mathrm{d}y}{l}+\frac{\Delta y}{t_\mathrm{e}} \tag{12}$$

Formally, the particle motion in an eddy is determined by 3 factors. The first term is the average gas velocity of the eddy in *y* direction, the second term is the average diffusion velocity of the eddy due to the engulfment of the jet flow, the third term is gas phase parcel displacement velocity due to triplet map in advection process, where Δy is gas parcel displacement, t_e is the time period that the parcel spends to finish the movement. This movement also drives the particle motion in the eddy. Generally, the first two terms are smaller compare to the third term.

For a very small particle, it will follow the gas flow closely. This physical phenomena can be used to determine the particle-eddy interaction time t_e. According to Eq.(3), for any gas parcels in an eddy, there are three possibilities for its motion, shown as Fig.3. Although there is only one possibility for gas parcels movement due to its serial number in the eddy during numerical implementation, a particle motion at the same place should be considered as a random process. Thus the particle displacement Δy in the eddy due to triplet map is determined by a random selection process. If a low inertia particle comes into the eddy from the bottom line, and the random selection process shows it will be mapped to the second position in Fig.3, this particle motion will follow the solid line when it is this eddy. The particle-eddy interaction time is then the duration that the gas parcel or the particle goes through the eddy height. Since the particle axial velocity is the same with local gas velocity, and the eddy height is determined by its length and its shape factor β. The particle-eddy interaction time t_e can be calculated as:

$$t_e = \frac{l\beta}{u} \qquad (13)$$

Where u is local axial gas velocity, l is eddy length, β is the eddy shape factor.

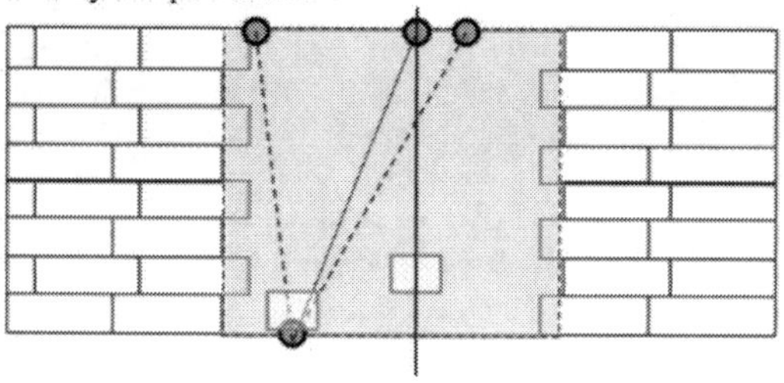

Fig.3 three possibilities for particle motion due to triplet map

Acutally, not all of the particles enter an eddy from bottom. A lot of particles hit an eddy at the eddy edge. Then the gas parcel displacement should be modified by multiplying a ratio of the distance from particle x position to eddy end position in x direction to the eddy height.

3.3 2-way coupling between particle gas phase

There are three methodologies in particle tracking method: one way coupling method, two way coupling method and four way coupling method. In one way coupling method, only effects of gas phase to particles are considered. By two-way coupling, the particles experience drag from the fluid as well as the fluid experiencing turbulence modulation from the presence particles. Also, the particle-particle interactions should be considered in a four way coupling method. In this model, we only discuss a 2-way coupling method, although a 4-way coupling method can be extended.

During particle tracking process, particle motion will affected by drag force and body force. Its momentum change can be calculated in each integration step as:

$$Mom_p = m_p(u_i^{j+1} - u_i^{j}) \qquad (14)$$

Where Mom_s is momentum change of the particle, kgm/s^2, m_p is particle mass flow rate represented by this particle, kg/s, u_i is particle velocity in i direction, m/s, j is particle integration step. This change of particle momentum is regarded as the gas phase momentum source given by this particle to the cell which the particle is in. If during a certain particle integration step, the particle goes into next cell, the momentum source is split by two cells according to linear assumption. Based on this method, the particle momentum source of each cell can be calculated as a summation of all particle momentum changes in this cell. The renewed gas velocity can then be calculated by Eq.(15)

$$\frac{\partial(uu)}{\partial x} = \frac{\mathrm{sum}(Mom_p)}{\rho_g \Delta y \Delta x \Box} \qquad (15)$$

Where Δy is the cell size, Δx is ODT length step, 1 means the unit length in z direction. During this procedure, the turbulent flow property is not changed by Eq.(15). However, it does be changed in an indirect way: the renewing of gas velocity by particle source term changes the gas velocity distribution thus changes the eddy selection results in the next ODT step. This made the eddy distribution rate change and the gas turbulent flow properties is changed.

4. Results and discussion

Simulations were carried out for a co-flow planar jet, shown as Fig.4. The calculation domain is a $40b_0 \times 60b_0$ rectangular. b_0 is jet width, b_0=1cm. The average jet velocity is 20m/s, its landscape profile across the jet is acquired by FLUENT simulation of a channel flow. The coflow velocity is 2m/s with a small disturbance. Both the initial ODT cell size and ODT developed step are 0.1mm, thus there are 4000 cells on the initial ODT line and 6000 ODT steps through the calculation.

Particles are injected from the jet nozzle with a uniform mass flow rate distribution in each cell. The particles in each cell are divided into 18 streams, each of which has the same mass flow rate. The mass weighted average particle diameter is 70 μm. The initial particle diameter and mass flow rate distribution is listed in Tab.1. Based on these initial conditions, the particle stocks number can be estimated: to regard b_0 as the large eddy length scale and assume the fluctuation velocity is about 3% of initial axial velocity. The stocks number for small, medium and large particles are about 0.05, 2.0 and 7.0 respectively.

The calibration of realizations was carried out. There is a good agreement of the average values based on 500/1000/2000 realizations. Since a more smooth result can be acquired by larger realizations, the final calculation was based on 4800 realizations. It will take 5 hours for 100 parallel processors to acquire the final results.

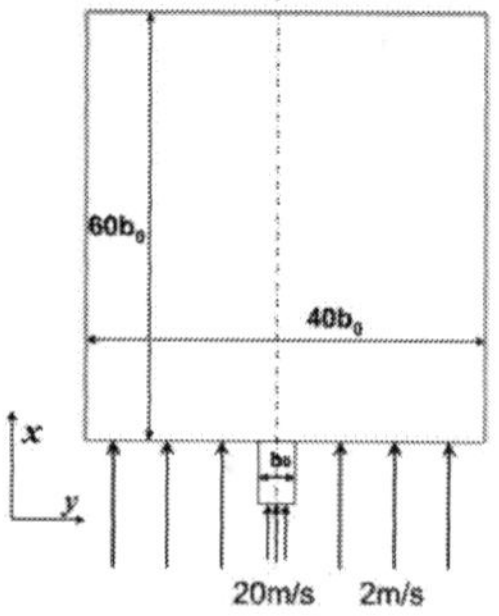

Fig.4 Calculation domain

Tab.1 particle initial conditions in each cell of jet nozzle

Diameter μm	Streams in each cell	Mass flux Kg/(m²•s)	Load rate kg/kg
10	6	8.6	0.333
70	6	8.6	0.333
130	6	8.6	0.333

4.1 characteristics of gas flow in jets

For planar jets, flow field is predominated by its momentum flux. The non-dimensional scalars such as axial velocity, Reynolds stress, et al, show a good self similarity at the developed region which is far from the nozzle. This similarity of axial velocity is repeated by ODT simulation. For co-flow jets, a lot of experimental work had been done in 1960s. The flow field can be predicted by semi-empirical theories based on the principles of self similarity of the jet flow and momentum flux conservation at each height. For example, Bradbury[8] proposed Eq.(16) and Eq.(17) to predict the center line velocity decay and half width of the jet flow based on a large amount of experimental data. The half width of a co-flow jet is defined as the location where the excess velocity is half of the excess velocity on the center line. In Eq.(16) and Eq.(17), u_c and u_∞ are center line velocity and co-flow velocity respectively, $u^* = \sqrt{u_{jet}(u_{jet} - u_{co})}$, b_e is the jet width.

$$u_c - u_{co} = u^*\left(0.188(x - x_0)/b_e\right)^{-1/2} \tag{16}$$

$$b_{half} = \frac{0.103(x - x_o)}{1 + 0.532 u_{co}\sqrt{(x - x_o)/b_e}\,/u^*} \tag{17}$$

The comparison of center line axial velocity profile and half width of the jet predicted by ODT Model, FLUENT simulation using standard *k-ε* model and Semi-empirical theory (Eq.16 and Eq.17) is shown in Fig.5. ODT model underestimates the spreading rate of half width. But there is a good agreement of center line axial velocity profile between three models.

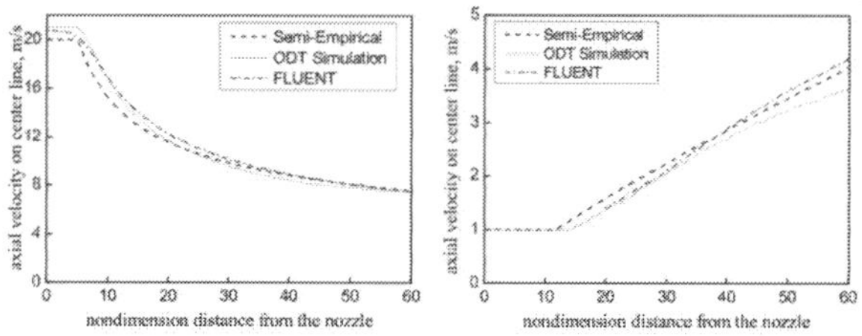

(a) centerline line velocity decay (b) half width

Fig.5 Comparison of center line velocity and half width acquired by ODT Model, FLUENT Simulation and Semi-empirical theory based on experimental data.

In ODT model, turbulent mixing process is represented by a series of eddy events. Each eddy event leads to a local mixing of each scalar and kinetic energy redistribution. Thus eddy density is introduced to denote the intensity of local mixing process. At each realization, if a point is occupied by an eddy, the summation of local eddy density at this point will increase by 1. The eddy density can then be calculated by using the final summation of eddy density divided by total realizations. The contours of eddy density give a good representation of mixing intensity of the jet flow, shown as Fig.6(a). Fig.6(b) shows the location of half width and the maximum eddy density at each height. The solid line falls into the dot region, which shows an immediate conclusion that the most intensively turbulent mixing region happens around the shear layer but not at center line.

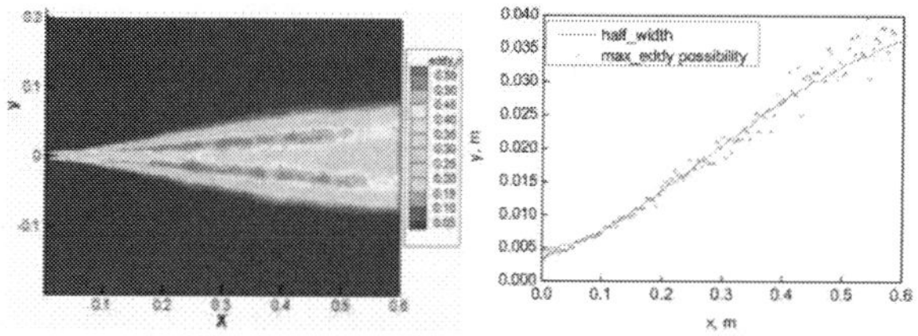

(a) contours of eddy density

(b) half width and maximum eddy density location

Fig.6 Eddy distribution and relationship between edds and sheer layer

4.2 Particle dispersion in planar jets

In a turbulent jet flow, particle motion is affected by eddies continuously, which makes particles disperse in the jet flow. The particle-eddy interaction is considered through a modified drag law, as is described in section 3.2. The intensity of eddy effects on particles is represented by eddy shape factor *β*. A small *β* means a high eddy effect on particles, because it will take less time for this particle to across the eddy while keep the same displacement in ODT line direction. This creates a larger relative velocity between particles and eddies. The eddy effects on particle dispersion of different particle size are shown in Fig.7. Each line in a picture represents particle mass flux of a certain particle size at the distance of 50 times of jet width from the nozzle based on an eddy shape factor. Particle dispersion always increases with a decreasing *β*, as is expected. This means that eddies has a positive effect on particle dispersion.

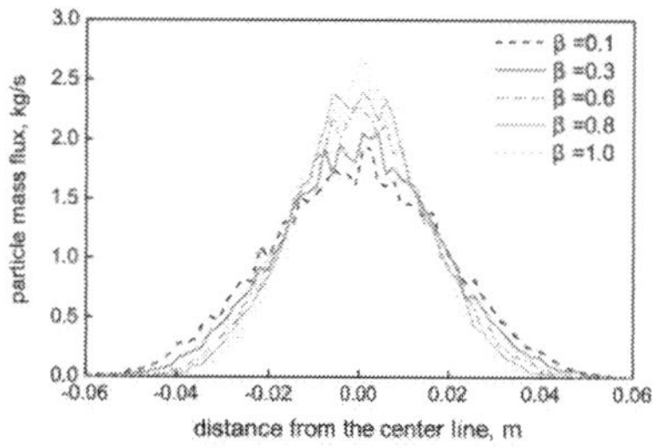

(a) small particle, d_p=10μm

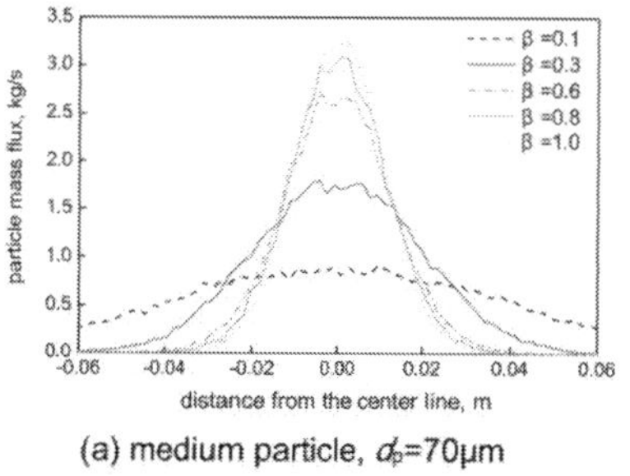

(a) medium particle, d_p=70μm

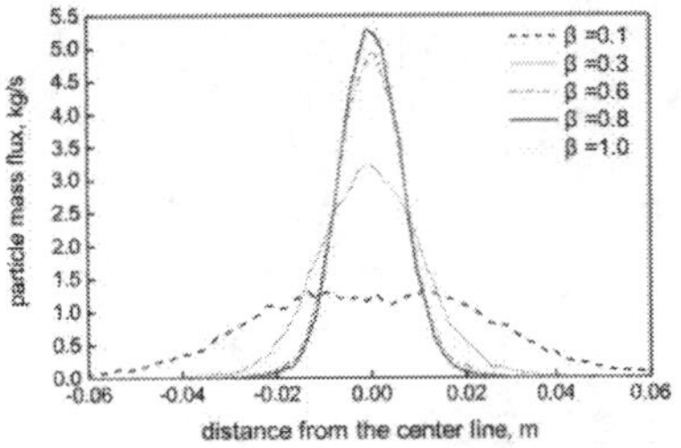

(a) large particle, d_p=130μm

Fig.7 eddy effects on different particles

Comparing Fig.7(a), (b) and (c), the dispersion of small particles is not sensitive to β, while medium particle is very sensitive to the eddy shape factor. This difference is caused by different particle inertias. When particles enter an eddy, a large and heavy particle has so large inertia that its motion is hard to be affected by the eddy, which has a negative effect on particle dispersion. However, when the large particle leaves the eddy, it also takes a long time for the large particle to follow the local velocity. At this time, large inertia will have a positive effect on particle dispersion since the holistic particle transverse velocity at the moment when it leaves an eddy is always departure from the center line. These two effects caused by particle inertia are opposite and dominate particle motion at the same time. For small particles, although eddies have strongest effects on them, they will follow the local gas velocity quickly after they leave the eddy. Little inertia has both large positive and negative effects on particle dispersion. For medium particles, its motion is both affected by eddies and local gas velocity when it leaves eddy moderately, this makes the total motion of the medium particle sensitive to eddies.

Fig.8 shows the average particle diameter profile at the distance of 50 times of jet width from the nozzle. Both mass weighted average particle diameter (d10) and Saunter diameter (d32) are compared based on a small β and large β. Since medium and large particles are sensitive to eddy shape factor, their dispersion will increase sharply at a small β. This makes average particle diameter profile a "U" shape when β is 0.1, as is shown by round points in Fig.8. However, when eddy effect is not that intensive, large particles will concentrate in the center area, and the average particle diameter appears a Gaussian shape, which is observed by Budilarto[9] for a round jet.

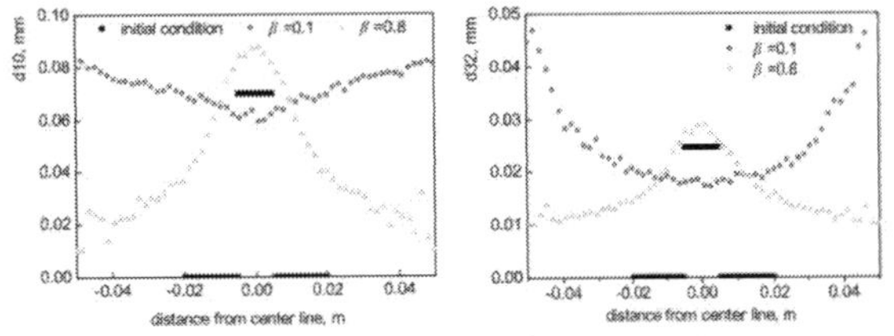

(a) d10 (b) d32

Fig.8 Effects of particle size on particle dispersion

When particles are injected from nozzle, its dispersion develops with jet flow. This development is shown in Fig.9. In Fig.9(a), each line represents the mass flux of a certain particle size at a certain height. Dash dot lines are at the distance of 10 times of jet width from the nozzle, while continuous lines represent distance of 50 times of jet width from the nozzle. At regions near the nozzle, there is a double-peak shape for particle mass flux profile. Then the Gaussian distribution appears for all particles at the distance far from the nozzle. This means that particles will first gather at the edge of the shear layer, then disperse both to and away from the center. Fig.9(b) shows the comparison between the positions at each height where maximum eddy density and maximum particle number density for each particles appear. Particle dispersion is slower than momentum dispersion. Besides that, particle dispersion is not the exact case in the near nozzle region.

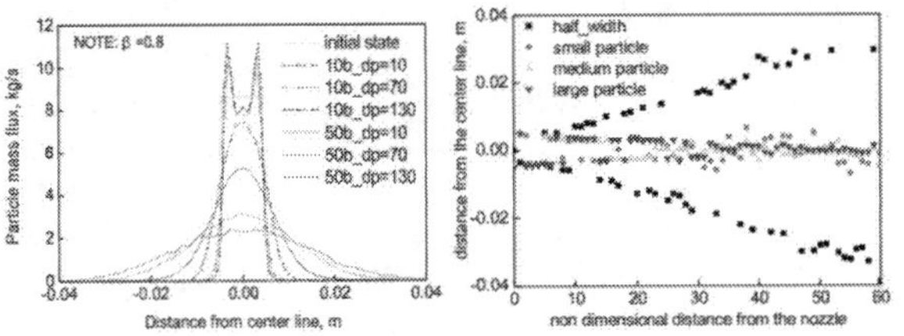

Fig.9 dispersions of small, medium and large particles at different height when β=0.8

4.3 Effects of particle source term on jet flow field

In this model, a 2-way coupling method was proposed to consider interactions between particles and gas phase. The particle load rate is set to 2. Particle source terms are recorded during particle tracking process and are used to modify gas flow field. Actually, particle source term is determined both by gas phase and particle initial conditions. In this simulation, the initial particle velocity is set to equal to jet velocity, this makes particle promotion be held back by gas phase. Thus there is always a positive particle source term for gas phase. Both the spreading rate and axial velocities increase for the gas phase, as is shown in Fig.10. Besides that, the particle source term strength the turbulent mixing process since the selected eddy number increases 1/3 of the original value. The comparison of the normalized velocity profile at the same height with and without considering particle source term is almost the same, this means although the particle momentum source term concentrate at the center region, turbulent mixing process spread the momentum source term out and keep the velocity profile always self similar.

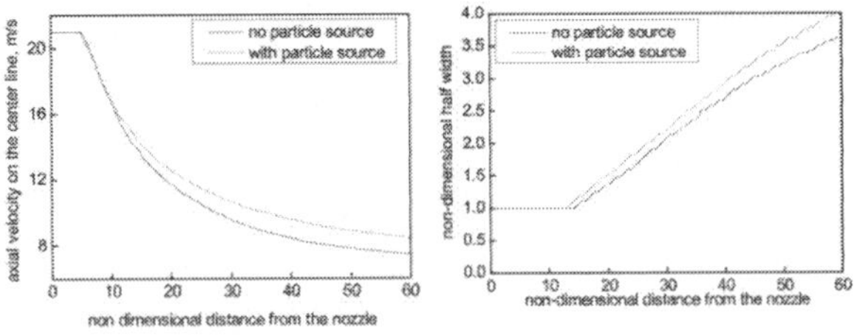

(a) center line velocity (b) half width

Fig.10 effects of particle moment sour term on jet flow field

5. Results and discussion

The One Dimensional Turbulence (ODT) model was used to study a particle laden flow in a planar jet. A new method was proposed to record the total turbulence information of the flow field, which is used to study the particle eddy interaction based on a modified drag law. With this method, eddy effects on particle motions are considered.

Different intensities of eddy effects on particle motion were compared. Particle dispersion rate is proportional to eddy effects. It suggests that turbulent eddies always have a positive effect on particle dispersion. Besides that, particle diameter has an important influence on particle-eddy interactions. Small particles are not sensitive to eddy effects, whereas medium particles are very sensitive to eddy effects. According to Budilarto's observation, a medium eddy shape factor is selected. Based on this eddy shape factor, the final results shows that particle diameter always has a negative effect on particle dispersion, causing large particles to concentrate at center of the jet while small particles disperse to the edge. A 2-way The ability of this ODT model to capture these dispersion effects provides motivation for applying it to particle-gas chemical processes, such as coal combustion and gasification where particle clustering and dispersion has been observed.

References

[1] Alan R. Kerstein. One-Dimensional Turbulence: Model Formulation and Application to Homogeneous Turbulence, Shear Flows, and Buoyant Stratified Flows. J. Fluid Mech, 1999(392):277-334

[2] J.R. Schmidt, O. L. Wendt, A. R. Kerstein. A Novel Method for the prediction of Particle-Laden Turbulent Channel Flow Using the One Dimensional Turbulence Model. 4th International Conference on Heat Transfer, Fluid Mechanics and Thermodynamics, 2005:

[3] A.R. Kerstein, Steven Krueger. Clustering of randomly advected low-initial particles: A solvable model. Physical Review, 2006(73):025302.

[4] Alan R. Kerstein, W.T. Ashurst, Scott Wunsch et al. One-Dimensional Turbulence: Vector Formulation and Application to Free Shear Flows. J.Fluid Mech. 2001(447):85-109.

[5] Wm.T.Ashurst, Alan R. Kerstein. One-Dimensional Turbulence: Variable-Density Formulation and Application to Mixing Layers. Physics of Fluids. 2005(17):025107

[6] Alan R. Kerstein, Thomas D. Dreenben. Prediction of Turbulence Free Shear Flow Statistics Using a Simple Stochastic Model. Physics of Fluids[J], 2000(12):418-424

[7] Scott Wunsch, Alan Kerstein. A model for layer formation in stably stratified turbulence. Physics of Fluids[J], 2001(13):702-712.

[8] Bradbury, L. J. S., and Riley, J., The spread of a turbulent plane jet issuing into a parallel moving airstream, *J. Fluid Mech.* 27(2):381-394 (1967).

[9] S.G. Budilarto. An experimental study on effects of fluid aerodynamics and particle size distribution in particle laden jet flows. PHD Thesis, Purdue University, 2003

285d Computer Aided Analysis in Mechanical Operations – Particle Technology and Particle Dynamics

Asha Immanuel Raju Chaduvula[1], Jagannadha Rao Patruni[2], Ravi Prasad Andra[1], Murali Yugandhar Nikku[3], Kiran Kumar Nalla[4], Srinivasa Babji Josyula[2], and Israel Yedluri[5]. (1) Chemical Engineering, Andhra University, Andhra University, Visakhapatnam - 530 003, AP, INDIA, Visakhapatnam, 530003, India, (2) Chemical Engg, Andhra University, Andhra University, Visakhapatnam - 530 003, AP, INDIA, Visakhapatnam, 530003, India, (3) Chemical Engineering Department, Andhra University, Andhra University, Visakhapatnam - 530 003, AP, INDIA, Visakhapatnam, 530003, India, (4) Chemical Engg, Center for Energy Engg, Room No: 109, Andhra University, Visakhapatnam, 530003, India, (5) Department of Geophysics, Andhra University, Visakahpatnam, 530003, India

ABSTRACT

Today's Era is commonly known as computer Era. The computations involving specific surface of particle mixtures, mean particle diameter, total number of particles, Screen effectiveness and effectiveness of screener in particle technology and terminal and settling velocities, specific gravity of the medium and displacement of solid particle in particle dynamics are estimated at a more speed with the help of the computer aided analysis in which a computer flow diagram will be prepared. A computer flow diagram is simply nothing but the flow sheet illustrating all the steps of the computation process and the detailed procedure followed in each step. In the present paper numerous problems are taken, coded in 'C' language, tested with the manual and computer calculations and the results are compared and discussed provided with the application of numerical methods like Interpolation, extrapolation, Simpson's, Euler methods etc. The application of numerical methods is also incorporated for example calculating the area under curve. The main reason for the computer aided analysis in mechanical operations especially Particle Technology and Particle Dynamics is the repetition of the calculations.

Modeling Precipitation During Flow in Porous Medium

Terry A. Ring and Milind Deo
Department of Chemical Engineering
University of Utah
50 S. Central Campus Drive
Salt Lake City, UT 84112

Abstract

The generation of particulates during flow alters the rheological properties of the fluid as precipitate volume fraction increases. In a multi-physics modeling environment, precipitation (of asphaltene or wax or hydrates) is induced in a porous oil reservoir due to the pressure drop associated with the flow. The solids produced alter the fluid rheology (Newtonian to Careau), increase both the low and higher shear rate viscosity and slow the fluid flow. In some cases where a large volume fraction of particles is precipitated the flow is completely chocked off by the increase in viscosity caused by the precipitation.

Introduction

During the pumping of an oil reservoir in secondary and tertiary oil recovery, the pressure is large at the well where the forcing fluid (e.g. water, surfactants in water, brine or CO_2) is sent down and much smaller at the producing well where fluid is being pumped out. This large pressure drop in the reservoir can cause certain constituents of the crude oil (asphaltene or wax or hydrates) to be soluble under high pressure conditions and insoluble under low pressure conditions. As a result, the flow field induces precipitation of a constituent of the oil. The impact of precipitation is usually to plug the formation and prevent further flow.

This work uses a multi-physics modeling environment to simulate the flow of an oil suspension of particulates in a silicate rock formation. In the flow at high pressures there are no particulates in the flow at a given pressure a second (solid) phase is precipitated out suddenly in the form of very fine particulates which forms a suspension in the remaining oil.

Model

The model starts with flow in a porous medium modeled by the Navier-Stokes equation for the momentum balance. The porous medium was created in the computer simulation by taking a cut surface of a silicate rock and pressing it against a silicon wafer and adding silicon etch solution to the pores in the rock to remove the silicon where the silicate rock grains are not in contact with the silicon surface. The electron micrograph of etched silicon surface which is a template of the original rock was digitized[1] and image processed to identify the surfaces of

[1] Keller, A. A., Blunt, M.J. and Roberts, P.V., "Transport in Porous Media, 26,277-97,1997.

the templated silica grains and the pores. Figure 1 shows the electron micrograph of the etched silicon wafer surface used for the simulation. The pore throats vary from 1 to 100 microns in diameter. A 640 microns by 320 microns section of the pore geometry, shown in Figure 1 was divided into 10,033 mesh elements for subsequent multi physics calculations.

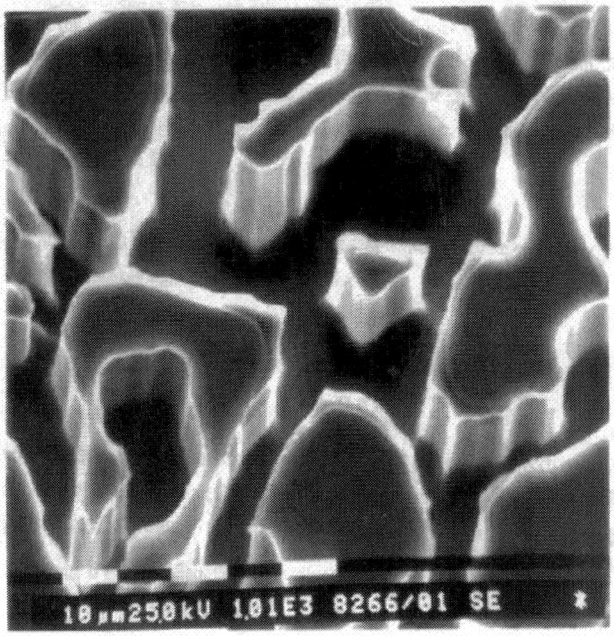

Figure 0-1: Scanning Electron Microscope image of the repeat pattern in the silicon wafer. Scale shown at bottom indicates that pore throat and body dimensions are on the order of 1-100 m. From Keller et al. (1997).

In the flow simulation the walls of the pores have a no slip boundary condition. Initially the pores are filled with a fluid, oil, which does not contain particles at a given pressure. At one edge of the porous field, an increased pressure is applied and new oil containing a component that will precipitate at a volume fraction of 50% is allowed to flow into the porous medium. At a specific pressure 20% of the pressure difference in the porous medium, precipitation is allowed to take place at a rate given by:

$$\text{Rate}=d\Phi_s/dt = k_p{}^* \varphi(\Phi_f - 0.1)^*(1 - \varphi(\Phi_f - 0.5))^*(1 - \varphi(P-P_o - 0.5^*\Delta P)) \qquad [1]$$

where

Φ_s = volume fraction solids

K_p = precipitation rate constant, k=0.1 1/s

Φ_f = volume fraction of precipitating component in the fluid (oil)

$(\Phi_f - 0.1)^*(1 - \varphi(\Phi_f - 0.5))$ = product of Heavyside functions that turns on the precipitation when the impurity volume fraction is in the range of 0.1 to 0.5.

$\varphi(P-P_o - 0.5^*\Delta P)$ = Heavyside function that turns on precipitation when pressure is greater than a threshold pressure given by $0.5^*\Delta P$

P_o = reservoir pressure (Pa)

ΔP = applied pressure minus reservoir pressure (Pa)

This rate operates between a volume fraction of 0.1 (the equilibrium value) and 0.5 (the maximum inlet concentration of insolubles). The amount precipitated is modeled with a simultaneous mass balance assuming a zero diffusion coefficient.

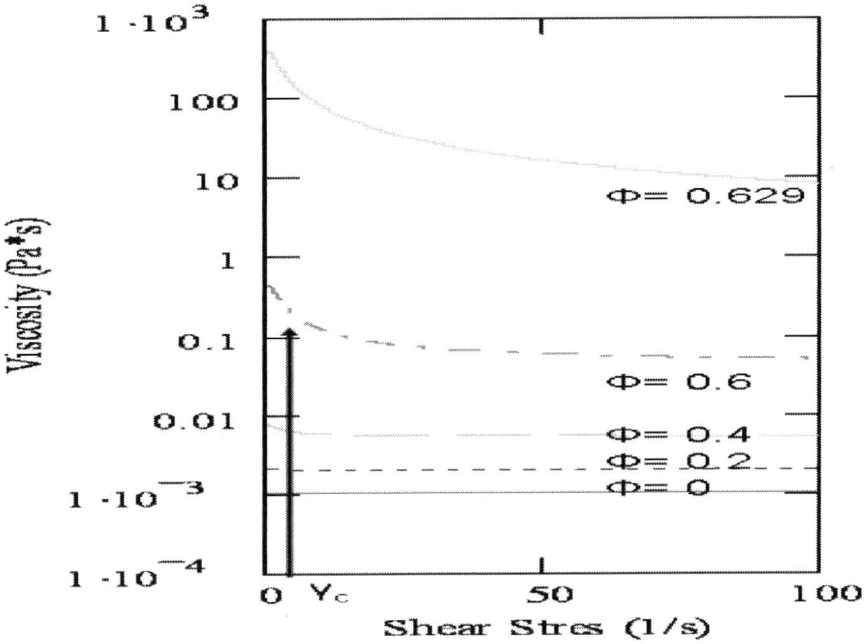

Figure 2 Carreau Rhelogical Formulation for Precipitation in Porous Media

The precipitation of fine particulates alters the viscosity of the fluid moving it from a Newtonian Fluid to a non-Newtonian fluid. The nature of the fluid is modeled using the Carreau rhelogical formulation with the viscosity given by:

$$\eta(\gamma,\phi_s) = \eta_\infty(\phi_s) + \frac{(\eta_o(\phi_s) - \eta_\infty(\phi_s))}{\left(1+\left(\frac{\gamma}{\gamma_c}\right)^2\right)^{1/2}} \qquad [2]$$

where

γ= shear rate (1/s)

γ_c= critical shear rate, γ_c=0.5 1/s depends upon the particle size

Φ_s = volume fraction fine particulates

$\eta_\infty(\phi_s) = \frac{\eta_s}{\left(1-\frac{\phi_s}{0.71}\right)^2}$ the viscosity at large shear rate

$\eta_o(\phi_s) = \frac{\eta_s}{\left(1-\frac{\phi_s}{0.63}\right)^2}$ the viscosity at small shear rate

With the Carreau rheological formulation, we see that the fluid is shear thinning and the low and high shear rate viscosities increase with the volume fraction solids, see Figure 2. As more and more material precipitates out, the viscosity increases substantially. If all solids are precipitated an increase in viscosity of approximately three orders of magnitude at low shear rates and increases by two orders of magnitude at high shear rates.

Numerical Methods

The solution to the combined transient mass and momentum balance equations was performed in Comsol 3.3, a finite element method based on Hermit polynomial collocation using a variable time step algorithm. The transient momentum balance equation was solved using 10,033 mesh elements using a calculation tolerance of 10^{-8}. The applied pressure gradient was 715 Pa over a distance of 640 microns in the porous structure from the right boundary to the left boundary so that flow was from right to left. The fluid (oil) viscosity was 1 m Pa*s without particles present and with particles, the viscosity was determined from the Carreau rheological formulation which is given in equation 2 and plotted in Figure 2. The transient mass balance equation was solved over the same set of mesh elements using an initial concentration of particles that was zero. At the right boundary, a high (50% volume fraction) insoluble concentration with some 10% by volume particles that are already precipitated out from solution was imposed at time zero. This material flowed into the porous media due to convection displacing the insoluble free oil. The Brownian diffusion coefficient of the particles was assumed to be 10^{-8} cm^2/s.

Numerical Results

The solution to combined mass and momentum balance for this situation is shown in Figure 3. In addition to the insoluble concentration plotted with the red to blue color field, this figure shows the velocity vectors as arrows overlaid on this plot. In this figure, we see that the flow moves from the right side to the left side with increasing time. Figure 3a is for 30 s and Figure 3b is for 60 s. With increasing time the flow has zones where precipitation occurs and this slows the flow in these regions. The plugging occurs first in the lower left corner and mores upward at the right boundary plugging off all the flow except the very top left pore in 60 s. Contrast Figure 3 with Figure 4 where the rate constant in equation 1 is set to zero. In Figure 4 the flow of insoluble material passes from the left boundary to the right boundary with ease in 60 s. Comparing the velocity profiles in the two cases we can see that the particles plug the porous structure to flow at the pore necks. Precipitation increases the viscosity at all shear rates but most at low shear rates. It is surprising to note that the biggest effects are at the pore necks where the velocity gradient or shear rate is higher. The precipitation initiates at the center of the pore where the shear rate is lowest and the viscosity is the highest. High velocities also develop in wide channels where the pressure gradient is relatively small due to multiple tributaries feeding the wide channel. This has the effect of lowering the flow at the bottlenecks more than in the large pores where the velocity gradient is lower. There also some interesting observations to be noted in addition to the effects of precipitation on the flow.

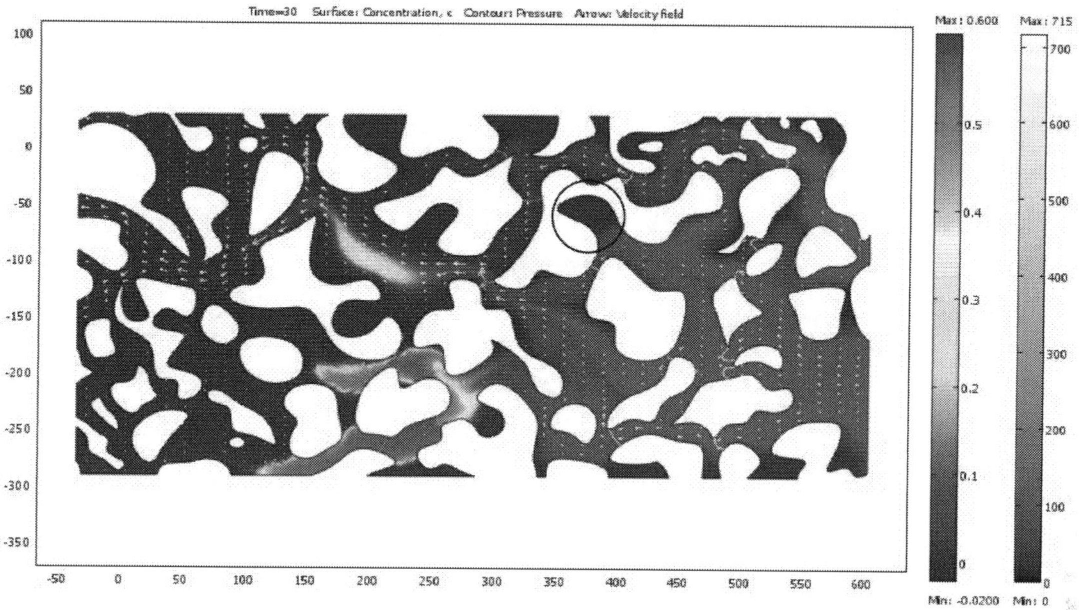

3a

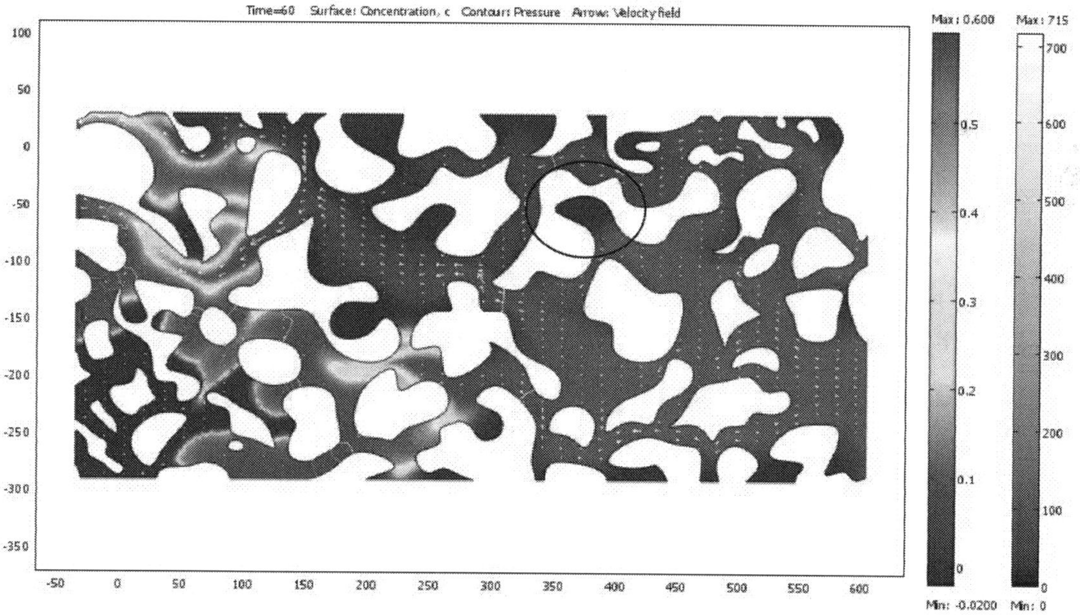

3b

Figure 3 a & b Plot of the concentration profile superimposed on a velocity vector plot in the porous media at 30 s (a) and 100 s (b) with precipitation taking place.

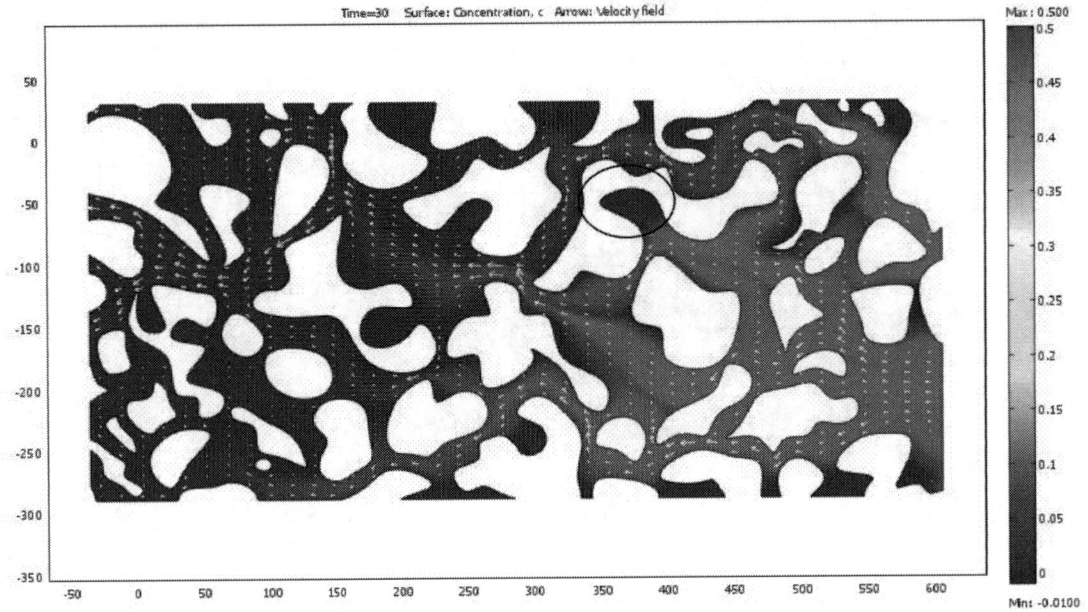

4a

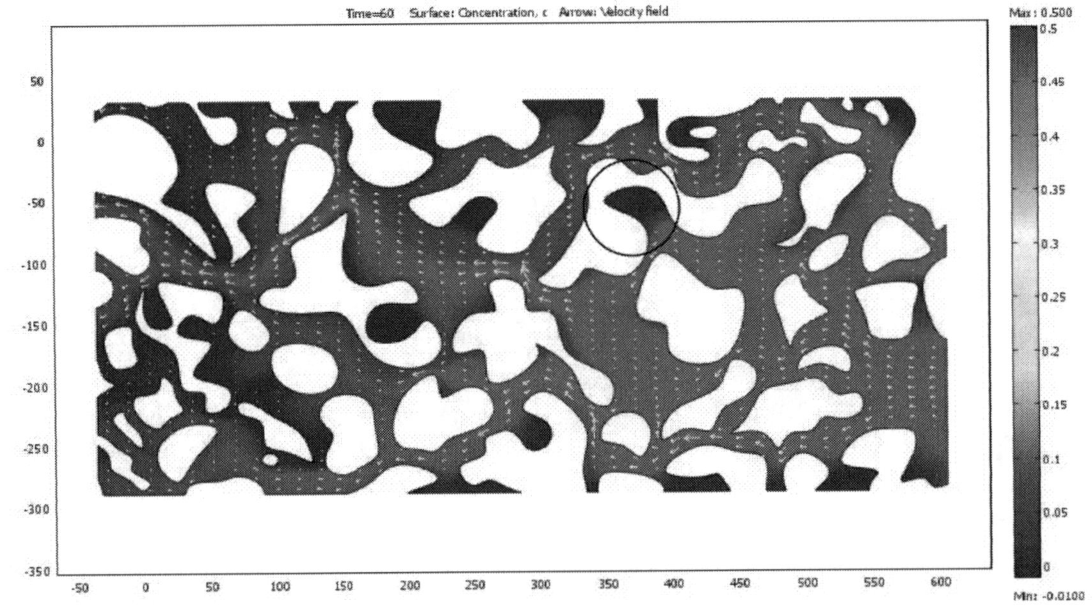

4b

Figure 4 a & b Plot of the concentration profile superimposed on a velocity vector plot in the porous media at 30 s (a) and 100 s (b) without precipitation taking place.

There are several dead end pores that do not empty when the flow sweeps by them see circles in Figures 3 and 4. This is an indication of how difficult it is to remove all the oil from a reservoir. The removal from these dead end pores is more difficult when the pores plug due to precipitation.

The integral of the x velocity at the outlet of the pore structure for these two situations was determined for the left hand boundary in Figures 3a&b and 4a&b. The integral of the outlet flux as a function of time is plotted in Figure 5 for the cases with and without precipitation. In the case without precipitation the outlet velocity decreases from the initial flow state where there are no precipitates to a lower asymptotic value due to precipitates at 10% by volume (the initial condition) while the case with precipitation shows a steep decrease as the structure plugs up at 20 s. After 40 s the flow is variable as periodically the pores plug and then the pressure gradient unplugs them.

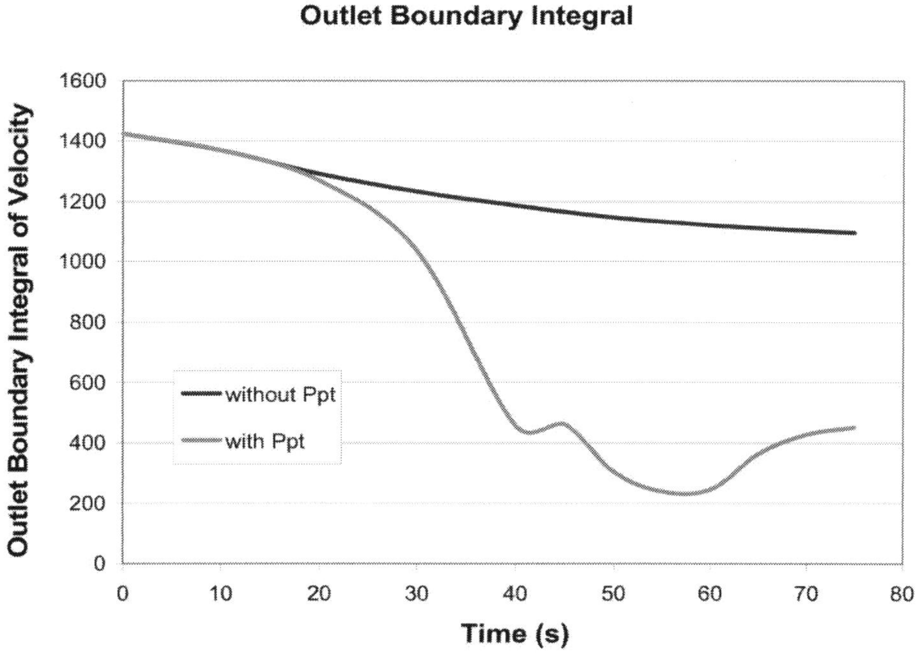

Figure 5 Outlet velocity determined from a boundary integration of the x velocity as a function of time for conditions with precipitation (magenta line) and without precipitation (blue line).

Conclusions

A multi physics model of the plugging of porous media by precipitation has been demonstrated using a pressure sensitive constituent in the crude oil. The results show that the biggest effects are at the center of pore necks where the velocity gradient or shear rate is low and the viscosity is high thus lowering the flow at the bottlenecks which leads to flow that is effectively plugged by the precipitates.

285f Validation Of Barracuda Cpfd Dense-Phase Particle Physics Models

Ken Williams, CPFD Software, LLC, 10899 Montgomery Blvd. NE, Suite B, Albuquerque, NM 87111 and Ray A. Cocco, Particulate Solid Research, Inc., 4201 West 36th Street, Suite 200, Chicago, IL 60632.

A unique experimental data has been obtained from a 5 ft diameter deep-bed reactor operated at PSRI. The data include particle entrainment rates and local pressure gradients within the bed for gas superfacial velocities ranging from 1.5 to 3.0 ft/s. The commercial engineering software package Barracuda-CPFD [1] was used to perform fully three-dimensional simulations of the bed behavior, and direct comparisons are made to entrainment and pressure gradients. The bed's turbulent fluidization behavior is complex with local gas streamers and bypassing, and Barracuda reveals critical details of these phenomena.

Web Page: www.barracuda-cpfd.com

328a Polymer/titania Composites Produced By Atomic Layer Deposition

Xinhua Liang, Chemical and Biological Engineering, University of Colorado, 1111 Engineering Drive, Campus Box 424, Boulder, CO 80309-0424, David M. King, Department of Chemical Engineering, University of Colorado, 1111 Engineering Dr., Campus Box 424, Boulder, CO 80309, Steven M. George, Department of Chemistry and Biochemistry, University of Colorado, Campus Box 215, Boulder, CO 80309, and Alan W. Weimer, Chemical and Biologicial Engineering, University of Colorado, 1111 Engineering Drive, Boulder, CO 80309-0424.

Micron-sized HDPE particles are coated with TiO2 films by atomic layer deposition (ALD) in a scalable fluidized bed reactor. The coating experiments are carried out by alternatively exposure to titanium tetraisopropoxide (TTIP) and hydrogen peroxide at the temperature of 77 °C, which is lower than the softening/melting point of the HDPE particles. The coated particles are characterized using Fourier Transform Infrared Spectroscopy, Scanning and Transmission Electron Microscopy, Surface Area, X-Ray Photoelectron Spectroscopy, X-Ray Diffraction and Induced Coupled Plasma-Atomic Emission Spectroscopy. The coated polymer particles are extruded into films by a heated extruder. During the high shear/high stress extrusion process, the shell on the polymer particle surface will crack and the shell remnants will be dispersed homogeneously throughout the polymer matrix. By means of this novel technique, well-dispersed polymer/TiO2 nanocomposites are formed. The photoactivities of TiO2 make this nanocomposites have advanced features, such as self-cleaning and antibacteria. The photoactivities of TiO2 coated HDPE particles and HDPE/TiO2 nanocomposite are tested by measuring the decomposition of methylene blue in its aqueous solution under UV irradiation.

328b Passivated Pigment-Grade TiO_2 Nanoparticles By Atomic Layer Deposition Barrier Films Using A Scalable Fluidized Bed Reactor

David M. King, Xinhua Liang, and Alan W. Weimer. Department of Chemical and Biological Engineering, University of Colorado, 1111 Engineering Dr., Campus Box 424, Boulder, CO 80309

Atomic layer deposition (ALD) of nanothick SiO_2 barrier films is utilized here for the passivation of ultrafine and nanosized TiO2 pigment particles. Conformal, pinhole-free films have been successfully deposited in a scalable fluidized bed reactor. Both anatase and rutile substrates were coated for this work. Films as thin as 2 nm, as verified by HRTEM, eliminated 98% of the photoactivity as compared to the uncoated substrate. Sulfuric acid digestion studies were performed using ICP-AES to monitor the dissolution rate of the TiO_2 core. The 2 nm SiO_2 films prevented > 99% of the TiO_2 cores from dissolving, which also verified the conformal, pinhole-free nature of these films. In situ mass spectrometry was used to monitor reaction progress and to prevent excess precursor usage and process time. This residual gas analysis technique is advantageous to minimize process costs and is applicable to fluidized bed or similar flow-type reactor configurations at any scale. The fabrication of colored, passivated pigments is also possible by first depositing an appropriate number of atomic layers of chromophore-based transition metal oxides (i.e. Al, Cr, Mn, Ni, V, etc.); then the passivating film can be applied in situ. SiO_2 ALD films are very efficient and can be further optimized to meet the needs of all aspects of the pigment industry. Surface passivation can be extremely economical in large-scale batch processing using the process control measures presented in this work.

328c Size Classification of Nano and Submicron Particles in An Ultracentrifuge

Rose Tippmann[1], Zac Doorenbos[1], Jan Puszynski[1], and Deepak Kapoor[2]. (1) South Dakota School of Mines & Technology, Rapid City, SD 57701, (2) Armament Research, Development and Engineering Center, Picatinny, NJ 07806

Many nano-powders produced by wet or high temperature processes result in a relatively wide particle size distribution. The newly proposed classification method uses ultracentrifugal force for interfacial particle transport between two immiscible liquids. This process allows both batch and continuous multiple size fractionation of a particulate slurry. The effect of key parameters such as, type of liquids, use of dispersants, initial deagglomeration of particles by attrition milling, ultracentrifuge speed and run time, and the initial slurry concentration in the top liquid on separation efficiency will be presented for various powder systems. Current results show a separation at 40 nm for an initial slurry with a particle size distribution from ~25-100 nm. This method for the classification of nano-materials may result in a narrow particle size distribution that is frequently desired for many energetic and structural applications.

328d Evaluation Of Polydisperse Kinetic Theories

Sofiane Benyahia, National Energy Technology Laboratory, Department of Energy, 3610 Collins Ferry rd., Morgantown, WV 26505

Powders found in nature and used in industry have wide particle size distributions (PSD). Some powders are not even homogenous: different cuts in a PSD may have different densities as is the case of coal commonly used in industrial transport gasifiers. In this study, several kinetic theories used to model polydisperse systems are evaluated based on comparison with published molecular dynamic data obtained in a simple shear flow. The most accurate theory is then used to model a dilute flow of bidisperse powders in a circulating fluidized bed. Computational results show large particles in the mixture segregate more at the walls of the riser similar to experimental measurements. The mechanism of the preferential lateral segregation of large particles at walls is due to two terms in the solids-solids drag that contain solids number density and granular temperature gradients. Without these terms, the theory predicts segregation of smaller particles at the walls due to solids pressure. This explains why previously published computational results based on other kinetic theories failed to predict the experimentally observed lateral segregation trends.

328e Characterizing Steady-State Size and Structure of Aggregates Outside the Limits of the Rayleigh-Debye-Gans Theory with Small-Angle Static Light Scattering and Image Analysis

Lyonel Ehrl, Miroslav Soos, Amgad Salah Moussa, and Massimo Morbidelli. Institute for Chemical and Bioengineering, Department of Chemistry and Applied Biosciences, ETH Zurich, Wolfgang-Pauli-Str. 10, HCI F 138, Zurich, 8093, Switzerland

The steady-state size and structure of aggregates outside the limits of the Rayleigh-Debye-Gans theory produced under turbulent conditions in stirred tank (using polystyrene latexes with primary particle sizes equal to 810nm, 420nm, and 120nm) was studied experimentally for various values of the volume average shear rate and solid volume fraction. The steady-state aggregate structure was characterized by image analysis of two-dimensional pictures of the aggregates obtained by confocal microscopy and compared to the value of the scaling exponent evaluated from light scattering data. The mass fractal dimension showed to be invariant to primary particle size and applied shear rate with a value equal to 2.64±0.18 for all primary particle sizes investigated. The cluster mass distributions were characterized by two characteristic moment ratios, i.e., the root-mean-square radius of gyration and the zero angle intensity, obtained from small-angle static light scattering. The root-mean-square radius of gyration for all studied conditions follows a power-law scaling with the volume average shear rate with an exponent equal to -1/2, which is in agreement with data published in the literature. Relating the two measured moments of the cluster mass distribution with the independently measured aggregates structure can be used to evaluate the effect of multiple scattering within the aggregates.

328f Comparison and Experimental Validation of Aggregation Efficiency Models

Amgad Salah Moussa, Matthäus U. Bäbler, Miroslav Soos, and Massimo Morbidelli. Institute for Chemical and Bioengineering, Department of Chemistry and Applied Biosciences, ETH Zurich, Wolfgang-Pauli-Str. 10, HCI F 133, Zurich, 8093, Switzerland

Abstract:

Aggregation kinetics of fully destabilized polystyrene latex particles in a turbulent flow was studied experimentally in a batch stirred tank. Small angle static light scattering (SASLS) was used to monitor the time evolution of two independent moments of the cluster mass distribution (CMD), namely, the mean radius of gyration and the zero angle scattered light intensity. The obtained data and data from the literature[1], five sets in total, were used to evaluate different models to describe aggregation efficiency: Adler's solid sphere model[2], Kusters's core-shell model[3], and Bäbler's permeable sphere model[4]. Aggregation was simulated using population balance equation. Results of the simulation, using different aggregation efficiency models, were compared to the experimental data. It was found that only the models that incorporate permeability were able to predict the experimental data, i. e., Kusters's and Bäbler's models. Among these, Bäbler's model was able to better represent the aggregation efficiency.

1. Moussa, A. S.; Soos, M.; Sefcik, J.; Morbidelli, M., Effect of Solid Volume Fraction on Aggregation and Breakage in Colloidal Suspensions in Batch and Continuous Stirred Tanks. Langmuir 2007, 23, (4), 1664-1673.

2. Adler, P. M., Interaction of Unequal Spheres .1. Hydrodynamic Interaction - Colloidal Forces. Journal of Colloid and Interface Science 1981, 84, (2), 461-474.

3. Kusters, K. A.; Wijers, J. G.; Thoenes, D., Aggregation kinetics of small particles in agitated vessels. Chemical Engineering Science 1997, 52, (1), 107-121.

4. Bäbler, M. U.; Sefcik, J.; Morbidelli, M.; Baldyga, J., Hydrodynamic interactions and orthokinetic collisions of porous aggregates in the Stokes regime. Physics of Fluids 2006, 18, (1).

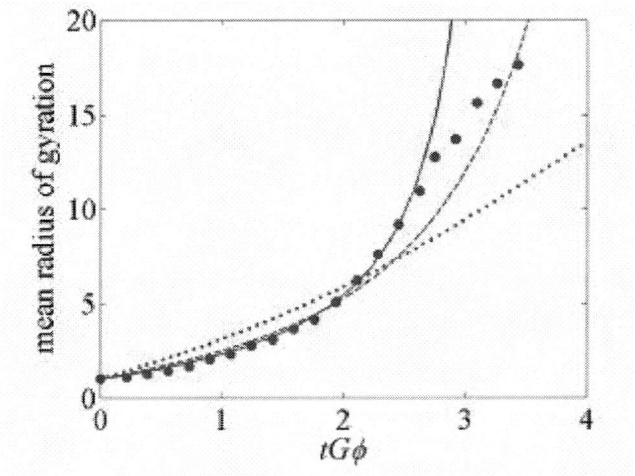

Figure 1 Mean radius of gyration as a function of dimensionless time. Solid circles: experiments; solid line: Bäbler's model; dashed line: Kusters's model; dotted line: Adler's model.

328g An Investigation of Interfacial Particle Phenomena Cohesiveness on Segregation

David Ely[1], Edwin Garcia[2], and M. Teresa Carvajal[1]. (1) Industrial and Physical Pharmacy, Purdue University, 575 Stadium Mall Drive, West Lafayette, IN 47907, (2) Materials Engineering, Purdue University, 575 Stadium Mall Drive, West Lafayette, IN 47907

A significant portion of blending research has focused on the drum mixer because of its very simple geometry and fascinating segregation dynamics. Research results that elucidate our understanding of flow dynamics, mixing, and segregation in the drum blender can then be extended to more complex geometries of tumbling blenders such as bin blenders and twin shell blenders where similar segregation patterns arise. Two main segregation patterns have been well established in drum mixers. And are the focus of this study: radial segregation and axial segregation. Since the 1950's and probably before, differences in particle sizes and densities between constituent particles comprising mixtures have been considered to be key parameters that frequently accompany segregation. While there have been numerous correlations made between these differences and segregation, there are numerous exceptions, and many other correlations exist as well such as friction and angles of repose. More recently, some nice correlations have been made between the dynamic angles of repose and axial segregation. Changing the speed of rotation until the dynamic angle of repose is matched however, will reverse the axially segregated state.1 Undoubtedly, dynamic angles of repose are, in part, related to interparticle friction, but this has not yet been fully investigated. Radial segregation may also be related to friction. To date, most experiments have correlated size and density differences to radial segregation. However, Monte Carlo simulations can model radial segregation using different coefficients of friction. Surface velocities also differ between powders and are likely responsible for radial segregation. Other interparticle forces that are considered for powders include electrostatic forces and Van der Waal's forces. In fact, they directly affect the frictional forces by the Coulomb/Amonton friction equation. For fine powders, electrostatic and Van der Waal's forces will contribute appreciably to the cohesiveness. The coefficient of friction is dependent on the materials at the interface. As an example, the coefficient of friction for an ice/copper interface is different than the coefficients of friction for ice/ice or copper/copper interfaces. In addition, it is dependent on the roughness of the mating surfaces. That is, two copper/copper interfaces can have two different coefficients of friction if they have different scales of roughness even though the materials at the interfaces are the same. Therefore, surface roughness is also considered to correlate friction differences to segregation. Particle shape and size also affect particle mobility. Irregularly shaped particles will have far less mobility than spherical particles made of the same material. Furthermore, large diameter particles have higher mobility than small diameter particles. Various techniques are used to assess particle size, shape, texture and friction on the various particulate systems followed by the use of a quantitatively model of the dry powder blending and segregation from first principles by integrating a model that benefits from resolving the local particle-particle interactions through Discrete Element Methods (DEM), while resolving the local thermodynamic state (mechanical, chemical, etc.) through the use of the Finite Element Method (FEM). Bulk powder properties such as shear stresses under different states of compaction and internal friction parameters are probed.

328h Physico-Chemical Properties As Early Risk Indicators And A Basis For Classification Of Nanomaterials

Ludwig K. Limbach and Wendelin J. Stark. Chemistry and Applied Biosciences, Swiss Federal Institute of Technology (ETH Zurich), Wolfgang-Pauli-Str. 10, ETH Hönggerberg, Zurich, 8093, Switzerland

The board and interdisciplinary area of nanotechnology has witnessed an unprecedented explosion of research and product development within the last few years. High expectation and enthusiasm were fuelled through experimental success stories and received a mainly positive response in society. Increasingly more critical voices and a growing body of scientific evidence for nanomaterial related adverse health effects, however, stay in clear contrast to the ongoing rapid market introduction of nanomaterials in numerous consumer products and demand for a more thorough risk evaluations prior to application. As a result, time- and cost-effective research instruments are required as tools for a sustainable and safety oriented product development both in academia and industry. The present contribution investigates a very rapid method to assess critical materials in an early stage of development or product design. The use of physico-chemical properties as primary decision parameters allow a first step of material risk evaluation at the design stage using predominantly accessible data. Based on the current rapidly increasing number of toxicological studies we propose a mapping of nanomaterials along their persistence and chemical activity within biological systems.

References:

L.K. Limbach, P. Wick, P. Manser, R. Grass, A. Bruinink, W.J. Stark, Exposure of Engineered Nanoparticles to Human Lung Epithelial Cells: Influence of Chemical Composition and Catalytic Activity on Oxidative Stress. *Env. Sci. Technol published online* **(2007)** DOI: 10.1021/es062629t

L.K Limbach, Y. Li, R.N. Grass, T.J. Brunner, M.A. Hintermann, M. Muller, D. Gunther, W.J Stark, Oxide nanoparticle uptake in human lung fibroblasts: Effects of particle size, agglomeration, and diffusion at low concentrations, *Env. Sci. Technol* 39, 9370-9376 **(2005)**.

T.J. Brunner, P. Wick, P. Manser, P. Spohn, R.N. Grass, L.K. Limbach, A. Bruinink, W.J. Stark, In Vitro Cytotoxicity of Oxide Nanoparticles: Comparison to Asbestos, Silica, and the Effect of Particle Solubility, *Env. Sci. Technol.*, 40 (14), 4374-81 **(2006)**.

P. Wick, P. Manser, L.K. Limbach, U. Dettlaff-Weglikowska, F. Krumeich, S. Roth, W.J. Stark, A. Bruinink, The degree and kind of agglomeration affect carbon nanotube cytotoxicity, *Toxicol. Lett.*, 168 (2), 121-31 **(2007)**.

328i Charging And Agglomeration During Granular Flow

Keirnan R. LaMarche, Shejal K. Shah, Troy Shinbrot, and Benjamin J. Glasser. Department of Chemical and Biochemical Engineering, Rutgers University, 98 Brett Rd., Piscataway, NJ 08854

The effects of electrostatics can have a large impact on granular flows. These effects can cause segregation and agglomeration and have an important impact on many industrial processes. The charging of grains also leads to dangerous electric discharges and dust explosions. Many industries spend large sums of money to prevent or control electrostatic charging. The pharmaceutical industry is especially susceptible because it requires that small amounts of drug be mixed with large amounts of excipient and the unpredictable nature of electrostatics can cause segregation and product non-uniformities.

How charge is accumulated in the flow of granular materials through a stainless steel cylinder was analyzed. It has been found that the net charge accumulated by the grains was dependent on the surface area of the cylinder initially covered by particles. In mixtures that were both homogeneous and heterogeneous, only those grains in contact with the walls of the cylinder became charged. The preferred charging of one component in a mixture of granular materials based on composition or size could result in segregation, producing a change in product concentration. In bi-disperse mixtures, the smaller particles were observed to segregate toward the walls; only these smaller particles charged, even in a mixture of 50% larger particles by volume. These experiments show the possibility for the segregation of smaller particles due to electrostatic effects.

We also examined the effects of non-uniform electric fields on uncharged grains. A large non-uniform electric field can polarize particles and cause them to agglomerate or adhere to surfaces. The amount of materials that adhered to the surface of a rod in a strongly non-uniform field were measured. These forces acting on the flow of granular materials, both down an inclined plane and from a hopper, created agglomerates, which had a major impact on the behavior of the flows.

328j Influence of Particle Surface Characteristics on Sintering of Elastomer Clusters

Cornelius Gauer, Hua Wu, Marco Lattuada, and Massimo Morbidelli. Institute for Chemical and Bioengineering, Department of Chemistry and Applied Biosciences, ETH Zurich, Wolfgang-Pauli-Str. 10, HCI F 138, Zurich, 8093, Switzerland

Coagulation is frequently applied to polymer particles e.g. as a separation step after emulsion polymerization. The resulting cluster structure greatly influences the properties of the resulting material. Universal cluster structure has been found for aggregates of rigid particle systems [1]. In contrast, soft particle clusters may undergo densification after aggregation due to polymer chain diffusion. The extent of this sintering or coalescence process is governed by surface properties of the primary particles.

In order to investigate this phenomenon the aggregation kinetics of industrial elastomer latices has been studied within the industrially relevant aggregation regime of diffusion limited cluster aggregation (DLCA) in quiescent fluid. The aggregation extent was followed by particle sizing via dynamic light scattering (DLS). Simulation of the aggregation using the kinetic approach (population balances) allows an interpretation of the light scattering result in terms of cluster structure, which is described by the mass fractal dimension – the key parameter for fitting the average hydrodynamic radius resulting from simulation to experimental DLS data. Additionally the cluster structure has been probed by cryogenic electron microscopy. Our findings indicate that cluster structures evolving during soft particle aggregation strongly depend on the surface properties of the elastomer particles with respect to the latex stability. Bare elastomer particles, stabilized with surfactants (mobile charges) only, allow formation of dense, sphere-like clusters due to very strong sintering of primary particles. On the other hand sintering is practically absent if the particles carry a sufficiently high number of fixed charges. In agreement, latex particles that are stabilized by both fixed charges and surfactants form clusters of intermediate sintering extent.

References: [1] M.Y. Lin, H.M. Lindsay, D.A. Weitz, R.C. Ball, R. Klein, P. Meakin, Nature 339 (1989) 360-362.

328k Application Of The Attainable Region (Ar) To Optimize Ball Milling

Matthew J. Metzger[1], Benjamin Glasser[1], David Glasser[2], Ngangezwe Khumalo[3], Brendon Hausberger[4], and Diane. Hildebrandt[5]. (1) Chemical and Biochemical Engineering, Rutgers University, 98 Brett Rd, Piscataway, NJ 08854, (2) School of Chemical and Metallurgical Engineering, University of the Witwatersrand, Johannesburg, 1 Jan Smut Avenue, Johannesburg, South Africa, (3) University of the Witwatersrand, School of Process and Materials Engineering, Wits University, P. Bag 3, Wits, Johannesburg, 2050, South Africa, (4) Centre of Material and Process Synthesis, School of Chemical and Metallurgical Engineering, University of the Witwatersrand, Johannesburg, 1 Jan Smut Avenue, Johannesburg, South Africa, (5) Centre of Material and Process Synthesis, University of the Witwatersrand, Private Bag 3, WITS 2050, Johannesburg, South Africa

Milling of particles is important to a large number of industries including the mineral processing, pharmaceutical, fertilizer and energy industries. Often it is desirable to produce particles of a certain size range while minimizing process time and/or energy usage. Production of fines can lead to powder handling problems and inefficient processes. In the pharmaceutical industry, excessive production of fines during milling of the drug crystals can increase risk of dust explosions. In mineral processing, the milling process is extremely energy intensive with large amounts of energy lost to hear or noise. There is a need to better understand milling in order to optimize milling processes and reduce energy usage.

Presented here is the application of the Attainable Region (AR) approach to optimizing the milling process. Previously, the AR approach has been used to optimize complex reactor networks and separation systems. Due to the similarities between reaction networks and the comminution process, the AR lends itself well to determining the optimal control strategy for a milling project. Though the optimization done in this work is for a ball mill and a specific material, the concepts can be applied to any system consisting of breakage and classification.

We have experimentally determined the optimal control strategy for achieving a certain size distribution with the use of the least amount of energy. The control strategy recommends a variation of speed throughout the process - fast speed at first, followed by a slightly slower speed, and then the lowest speed - to achieve the maximum amount of material of a certain size. Benefits of this approach can be realized anywhere milling is found, as a size profile over time can be predicted. Therefore, a design methodology can be constructed without complicated mathematical tools.

328l Scale-Up Of A Nanoparticle Size-Selective Fractionation Process Using CO_2-Expanded Liquids

Steven R. Saunders and Christopher B. Roberts. Department of Chemical Engineering, Auburn University, 212 Ross Hall, Department of Chemical Engineering, Auburn, AL 36849

Nanoparticle properties depend heavily on their size and size distribution. While solution based synthesis techniques are popular due to their simplicity, they often result in particles with a wide size range. Our group has developed a technique for the size-selective fractionation of ligand-capped nanoparticles that utilizes the tunable thermophysical properties of gas-expanded liquids (J. Phys. Chem. B, 109, 2005). This nanoparticle size separation process is based on the controlled reduction of the solvent strength of an organic phase nanoparticle dispersion through increases in concentration of the antisolvent CO_2 via pressurization. These changes in solvent strength affect the subtle balance between the osmotic forces of repulsion (originating from the solvation of the nanoparticle ligand tails) and the van der Waals forces of attraction between different sized particles necessary for a stable dispersion. As such, increasingly smaller sizes of particles can be controllably precipitated from the dispersion through modest increases in CO_2 pressure, resulting in the separation of particle dispersions typically ranging from 2nm to 12nm into ±1nm fractions. Recently, we have demonstrated effective separations of metal or semiconductor nanoparticles using a novel spiral tube apparatus within a high pressure vessel. This initial apparatus was capable of handling ~200 μ L (2 mg) quantities of the nanoparticle dispersions. It is desirable to scale up this process to handle larger quantities for applications such as catalysts, sensors, semiconductors, and optics. The scaled up apparatus consists of three vertically mounted high pressure vessels connected with high pressure needle valves that allow for sequential isolation and separation of the fractionated nanoparticle dispersion. This process, operated at room temperature and pressures between 400 PSI and 700 PSI, can result in a size-selective separation of ~20 mL (200 mg) of a nanoparticle dispersion, a scale up factor of 100. This process can be easily modified by increasing the number of the vessels or by increasing the size of the vessels allowing for more fractions or separation of larger samples, respectively. Comparisons of the results for the size-selection separation of a silver nanoparticle dispersion in both the original process and the scaled up process will be discussed.

Powder flow pattern in fluid bed using Near Infrared

Teófilo Donaires, Alvaro Realpe and Carlos Velázquez. University of Puerto Rico, Mayaguez, Puerto Rico

ABSTRACT

Fluid Bed is equipment frequently used as dryer or granulator in the industry of foods, pharmaceutical, agriculture. In a fluid bed a product you fluidization for the action of an upward flow of gas. The movement mechanical of the fluidize reforces this effect. Of this form, the gas and the product they mix intensely and a high thermal transfer and a good speed of physical reaction are gotten this way. Elegized a good combination of speed of the gas and of the movement mechanical they can be processed with success granular products with a great variety of grain sizes although forming a minimum quantity of powder.
Therefore it is necessary to study the whole fluidization pattern with the purpose of to assure conditions of appropriate operation and to reduce the segregation problem, canalization and agglomeration.

The objective was to study fluidization pattern using the near infrared, could observe with a size of particles of 30μm and 30 g the coefficient correlation is 97.74%, with 50μm and 50 g the coefficient correlation is 96.59%, with 50 μm and 150g the coefficient correlation is 65.48% and with 50μm and 200g the coefficient correlation is 62.34%.
We can conclude work presently when the particles are of smaller size and smaller quantity of mass, we can achieve a good fluidization pattern, as it is increased the size and the quantity of material the fluidization pattern it reduce.

Key word: Fluid bed, fluidization, pattern, NIR

INTRODUCTION

Fluidization is the operation by which solid particles are transformed into a fluidlike state through suspension in a gas or liquid [1].That makes it potentially important for the chemical industry, food, and pharmaceutical industry. The high heat- and mass-transfer coefficient is one of the main advantages of fluidization. It is run by air fluid and the air fluid is variable. The reason monohydrated lactose is used is because is very cohesive [1]. Other properties that are important are a fluidized bed is the uniformity of the temperature. An abrupt change slows the speed that is why the system can be controlled much easier. Because a homogenous distribution exists the temperature implies that all fluidized bed is being processed in the same manner, thus making sure of the quality of the product1 [1].

$$F_k = S\left[p_1 - p_2 - L_b(1-\varepsilon)(\rho_p - \rho_f)g\right] \qquad (0.1)$$

Where: F_k force; S area; p_1, p_2 pressures; L_b longitude; ε void fraction; ρ_g gas density; ρ_f fluid density; g gravity.

The fluidization in the fluidized bed is processed is for the coating of air suspension with emphasis in the batches of ideal fluid. The phenomenon occurs when the coating becomes a fluid and coats the coat. The key to success is using sophisticated equipment so it can continue to be observed adequately [1-2]

Minimum Fluidization Velocity

One of the basic parameters to be determined when designing bubbling fluidized-bed systems is the minimum fluidization velocity, U_{mf}. The effect of temperature and pressure on U_{mf}.

This equation can be used to calculate the minimum fluidization velocity u_{mf} if the void fraction u_{mf} at incipient fluidization is known.

$$\left(\rho_p-\rho_f\right)g=\frac{\rho_f u_{mf}^2}{\Phi_s D_p \varepsilon_{mf}^3}\left[\frac{150\left(1-\varepsilon_{mf}\right)\mu}{\Phi_s D_p u_{mf}\rho_f}+1.75\right] \qquad (0.2)$$

where: ρ_g gas density; ρ_f fluid density; g gravity; u_{mf} minimum velocity; D diameter; μ viscosity, and Φ porosity fraction [24-29].

$$U_{mf}=\frac{\mu}{\rho_g d_p}\left[\frac{(33.7)^2+0.0408 d_p^3\rho_g\left(\rho_p-\rho_g\right)g}{\mu^2}\right]^{\frac{1}{2}} \qquad (0.3)$$

Fluid bed dryers

Dryers in which the solid are fluidized by the drying gas find applications in a variety of drying problems [24].

The gas fluidized bed is characterized by having good heat transfer properties between the fluidized layer and heating or cooling surfaces and extensive work has been done in order to develop equations for the estimation of the heat transfer.

$$-D_i\frac{dC_i}{dz}=U_{pi}C_i \qquad (0.4)$$

where D_i is the dispersion coefficient of the particle species, C_i is its concentration, and z is the axial distance.

Homogenization in the Fluidized Bed

The fluidized bed has a hydrodynamic behavior non linear with a bubbly movement. The movement of the particles corresponds to the movement of the phases of the fluid that has been modeled for particles [3, 4]. An extended variety of experiments is reported that studies the influence that has interior forces in the fluidized beds. The behavior of the particles of the fluidization aspect of the fine powders [5].

These have been analyzed using the gas theory of chaos, using dominant frequencies to analyze by infrared spectroscopy [6]. This can appear in fluidized industrial beds [7]. That is for purposes of the active ingredient for respective coating along with the control of the particles [8]. The evaluation of the movement and distribution of the particles in the fluidization is a fundamental aspect. This action depends on the culmination of the elaboration of the products and the quality of the product that is obtained. The complementary evaluation of the dynamic fluid showed maximum pressure, low approximately the same amount in both components of the particles [9]. The weight has to be evaluated in fractions and sizes of powders of through group C in the fluidized bed [10]. This has to do with the nonlinear interactions of two independent mediums with there own movements and tendencies, which are also pertinent in the individual movement of the fluids and particles [11, 12, and 13]

Recent advances have been taken place and a lot of interest shown in the biochemical world. Many people are trying to develop novelties in particles to manufacture medicines for asthma is infectious respiratory illnesses. They need accurate and constant inhalation alongside with reliable inhaling equipment [14, 15, and 16].

METHODOLOGY

Equipment and Instrument

The equipment and instruments that will be used belong to the University of Puerto Rico, Mayagüez Campus, Department of Chemical Engineering, and Pharmaceutical Engineering Research Laboratory.

Fluidized Bed

The minimum fluidization air flow determined by test trials was (0.2 m/s) and maximum fluidization of air flow is 2.13 m/s. The range temperature between 25°C and 80°C. The more must be sufficient to presume a good mixing of particulates while flowing and a minimum effect of segregation and fluidization. The controllers can be operated manually and automatically. The air fluid is situated in the inferior part, and in the superior part you can find the air filter.

Calibration and Adaptation

In this phase calibration and adaptation take place. All equipment used has to be given maintenance, calibrated, and carefully checked before using it. This is of utmost important when fluidization rehearsals are going to be done. The equipment must be in good working order along with the control instruments, the pressure, temperature; the fluid air velocity must be checked including doing the necessary calibration and adaptation.

Near-Infrared Instrumentation and Calibration

Full NIR spectra were measure using a Fourier Transform (FT)-NIR spectrometer (Bruker Optics™ model Vector 22/N). This spectrometer uses He-Ne laser light, a 429 InGaAs diode detector, a quartz beam splitter and a frequency range of 800 to 4200 cm^{-1}.

MATERIAL PREPARATION

Lactose monohydrate

Pharmaceutical grade lactose is highly pure lactose specifically produced to meet the standards of identity and purity set down in the lactose monographs of the various pharmacopoeias. Lactose monohydrate.

Density of Powder

There are two ways to define the volume and/or density of a powder. One is the packing volume, or density. This refers to the volume/density of the powder including the voids.

Distribution of Particles by Sizes

Properties of certain products depend on the dimension or surface of the particles and how they are composed. The first issue, is the interval of measured needed, that means the size of the particle either minimum or maximum. The precision required determines what method of analysis should be used, and also the amount of samples that should be analyze daily.

Experimental Design

Designing an experiment is planning everything meticulously and most important is to have all the information needed that is pertinent to the experiment [24].

Table1. Experimental Design

Factors	Radial position(cm)					Axial position				Powder mass(g)			Particle size (μm)
Levels	0	2.5	5.0	7.5	10	Low	Med1	Med2	Top	10	30	50	<44

Experimental Procedure

The experimental part will be done according to the planned experimental design proposed.

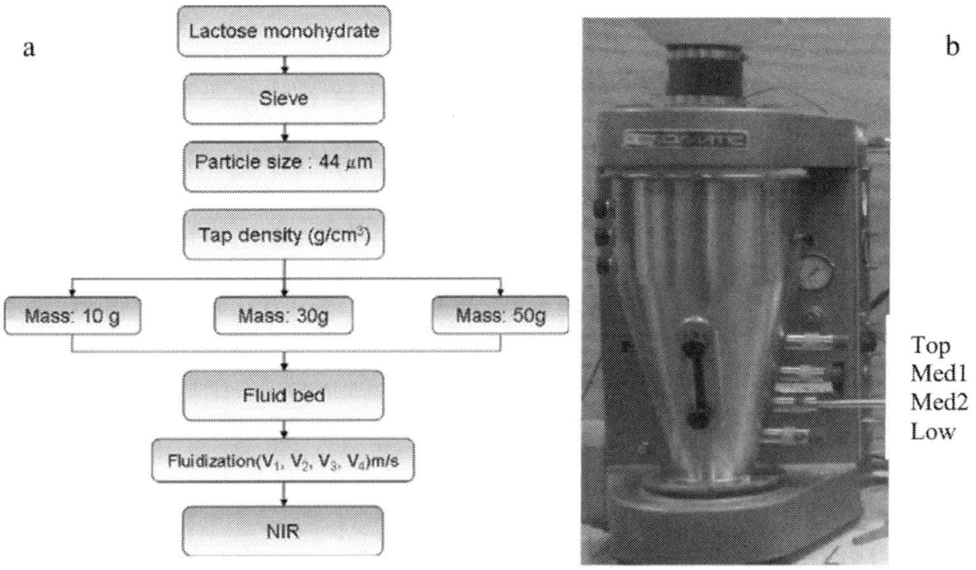

Figure 1 a. Experimental procedure; b. Fluid bed processing

EXPERIMENTAL RESULTS

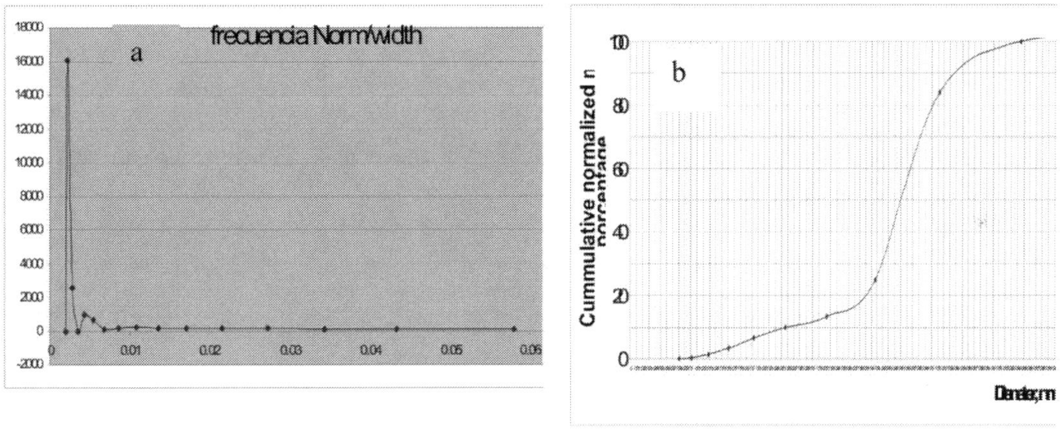

Figure 2 a. Particle size distribution of lactose monohydrate; b. Diameter average

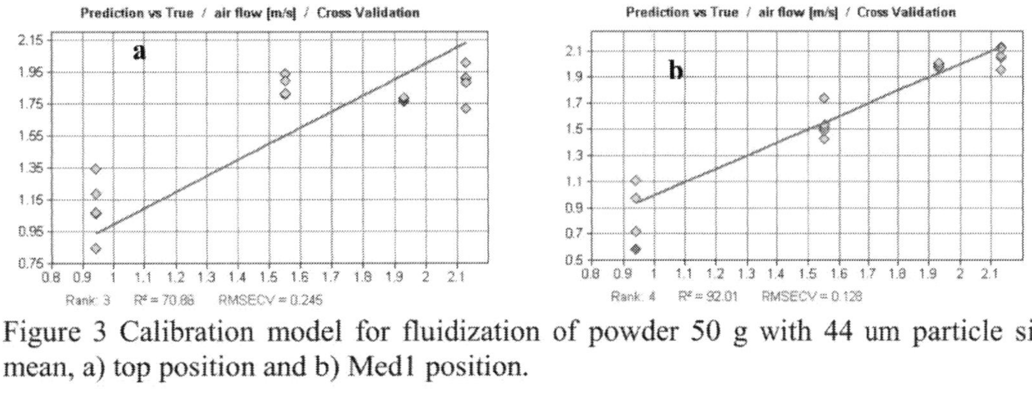

Figure 3 Calibration model for fluidization of powder 50 g with 44 um particle size mean, a) top position and b) Med1 position.

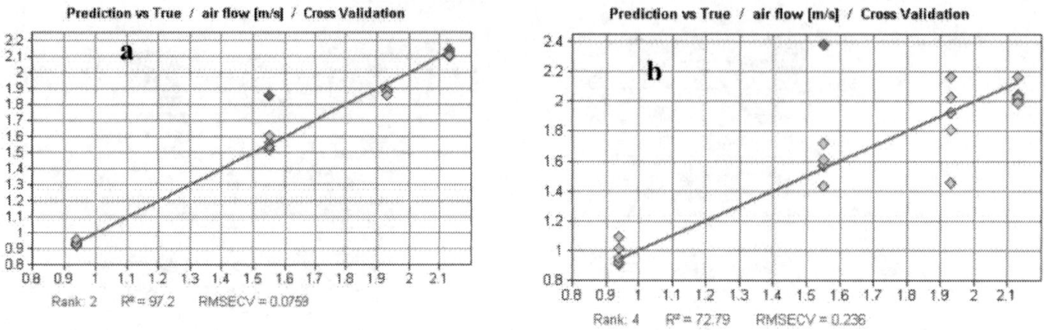

Figure 4 Calibration model for fluidization of powder 50 g with 44 um particle size mean, a) Med2 position and b) low position.

Analysis data

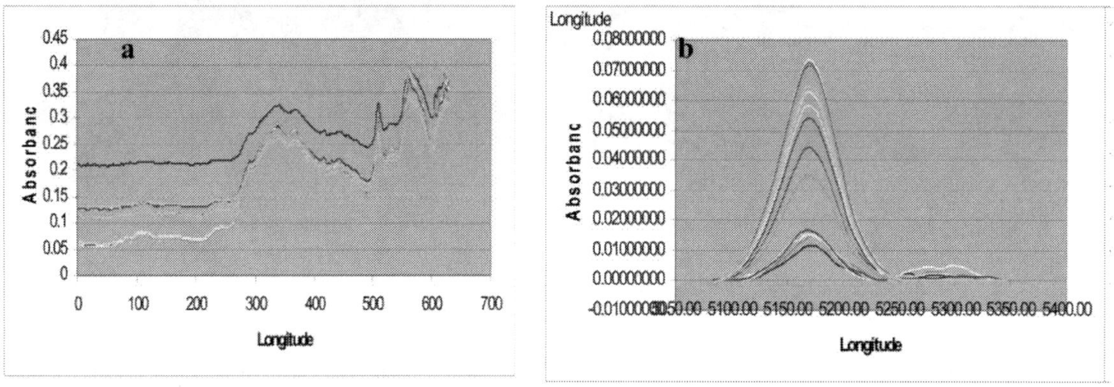

Figure 5 a) lactose spectra in the fluid bed and b) absorbance change in different positions in the fluid bed

CONCLUSION

- The Near Infrared Spectra can used to determine the fluidization pattern of powder particle in a fluid bed.
- The powder particle if are smaller size and smaller quantity of mass is good fluidization pattern.
- The powder particle if are increased the size and the quantity of Material the fluidization pattern reduce.
- The radial position of Near Infrared Spectra not change fluidization pattern

REFERENCES

[1] Kunii,D. and Levesnpiel, O., *Fluidization Engineering*, Butterworth-Heinemann series in Chemical Engineering, 1991.

[2] E. Teunou and D. Poncelet, *Journal of Food Engineering*, Volume 53, Issue 4, August 2002, Pages 325-340.

[3] P. Frake, D. Greenhalgh, S. M. Grierson, J. M. Hempenstall and D. R. Rudd, *International Journal of Pharmaceutics*, Volume 151, Issue 1, 12 May 1997, Pages 75-80.

[4] Stefano Brandani and Kai Zhang, *Powder Technology,* Volume 163, Issues 1-2, Fluidization and Fluid Particle Systems, 25 April 2006, Pages 80-87.

[5] K. Otawara, L. T. Fan, A. Tsutsumi, T. Yano, K. Kuramoto and K. Yoshida, *Solutions, & Fractals*, Volume 13, Issue 2, February 2002, Pages 353-362.

[6] Giovanna Bruni, Paola Lettieri, David Newton and Diego Barletta, *Chemical Engineering Science*, Volume 62, Issues 1-2, Fluidized Bed Applications, January 2007, Pages 387-396.

[7] Lu Huilin, Dimitri Gidaspow and Jacques Bouillard, *Powder Technology*, Volume 123, Issue 1, 18 February 2002, Pages 59-68.

[8] F.X. Llauro and M.F. Llop, *International Journal of Multiphase Flow*, Volume 32, Issue 12, December 2006, Pages 1397-1404.

[9] Ryuji Kikuchi and Atsushi Tsutsumi, *Chemical Engineering Science*, Volume 56, Issue 23, December 2001, Pages 6545-6552.

[10] Marcello Nitz and Osvaldir Pereira Taranto, *Journal of Food Engineering*, Volume 80, Issue 1, May 2007, Pages 249-256.

[11] Jiliang Li, Tsutomu Nakazato and Kunio Kato, *Chemical Engineering Science*, Volume 59, Issue 13, July 2004, Pages 2777-2782.

[12] Miroslav Puncochar and Jiri Drahos, *Chemical Engineering Science*, Volume 58, Issue 12, June 2003, Pages 2515-2518.

[13] Shirzad Azarmi, Xia Tao, Hua Chen, Zhaolin Wang, Warren H. Finlay, Raimar Lobenberg and Wilson H. Roa, *International Journal of Pharmaceutics*, Volume 319, Issues 1-2, 17 August 2006, Pages 155-161.

[14] Igor Sidorenko and Martin J. Rhodes, *Powder Technology*, Volume 141, Issues 1-2, 18 March 2004, Pages 137-154.

[15] Kunn Hadinoto, Ponpan Phanapavudhikul, Zhu Kewu, and Reginald B. H. Tan, *Ind. Eng. Chem. Res.* 2006, *45,* 3697-3706.

[16] Albert H. L. Chow,Henry H. Y. Tong, Pratibhash Chattopadhyay, and Boris Y. Shekunov, *Pharmaceutical Research*, Vol. 24, No. 3, March 2007.

[17] Hickey AJ. Factors Influencing the Dispersion of Dry Powders as Aerosols. *Pharm Technol* 1994;18(9):58-82.

[18] Hindle M, Byron PR. Dose emissions from marketed dry powder inhalers. *International J Pharm*. 1995;116:169-77.

[19] Hickey AJ, Dunbar CA. A New Millennium for Inhaler Technology. *Pharm Technol.* 1997;21(6):116-25.

[20] Trofast EAC, Briggner LE. Astra Aktieholag. *Process for Conditioning Substances*. US Patente 5,709,884. 1998.

[21] Clark AR , Hsu CC, Walsh AJ. Genentech, Inc. Preparation of Sodium Chloride Aerosol Fromulations. US Patente 5,747,002. 1998.

[22] Trofast EAC, Falk EJ, Astra Aktieholag. Agglomeration of Finely Divided Powders. US Patente 5,551,489. 1996.

[23] Satoru Watano, Yasushi Imada, Kenji Hamada, Yoshihiro Wakamatsu, Yoshiaki Tanabe, Rajesh N. Dave and Robert Pfeffer *Powder Technology*, Volume 131, Issues 2-3, 23 April 2003.

[24] Warren L. McCabe, et al. Unit Operations of Chemical Engineering. *McGraw-Hill,* Inc. 1993.

328n Simulation Of Drying Of Pharmaceutical Powders Via Dem

Eva Kalvoda, Institute for Resource Efficient and Sustainable Systems, Graz University of Technology, Inffeldgasse 21/B, Graz, 8010, Austria and Johannes G. Khinast, Institute for Process Engineering, Graz University of Technology, Inffeldgasse 21/B, Graz, A-8010, Austria.

Drying of powders is an important process in many industries, especially in the pharmaceutical industry. In many dryers agitators are used to improve heat and mass transfer. Neither the drying process nor particle mixing mechanisms are fully understood until now. The moisture between the particles, the non-spherical particle shapes and the particle growth in the case of crystallisation make it difficult to predict the particle flow in the dryer. This makes it hard to optimise or to scale-up processes.

To be able to understand the drying process better, computer simulations were used to examine the particle flow in the dryer and the drying process itself. In this work a discrete element method code was adapted to be able to model the powder flow in the dryer and the particle mixing mechanisms. Experimental studies were carried out in order to get more insight into particle flows and to be able to validate the computational results. [1,2]

The DEM code developed during this research project will be able to model agitated powder flows and can be used to examine segregation mechanisms and cohesion in particle flows.

[1] A. Lekhal, S.L. Conway, B.J. Glasser, J.G. Khinast, 2006, Characterization of Granular Flow of Wet Solids in a Bladed Mixer, AIChE J. 52. [2] S.L. Conway, A. Lekhal, J.G. Khinast, B.J. Glasser, 2005, Granular Flow and Segregation in a Four-bladed Mixer, Chemical Engineering Science 60.

328o Study of the Lubrication Effect on the Feedframe

Rafael Mendez-Roman[1], Alvaro Realpe[2], Carlos Velazquez-Figueroa[1], and Fernando J. Muzzio[3]. (1) Chemical Engineering, University of Puerto Rico- Mayaguez Campus, Miradero, Road 108, Km 3.0, Mayaguez, PR 00680, (2) Chemical Engineering, University of Puerto Rico, Miradero, Road 108, Km 3.0, Mayaguez, PR 00680, (3) Chemical & Biochemical Engineering, Rutgers University, 98 Brett Road, Piscataway, NJ 08854

The interest on granular material behavior has been growing in the last years especially for understanding and prediction of powder flow, mixing, and compaction. The advance obtained from these processes let us focus in one problem the pharmaceutical industry is facing. This is the mechanical properties of tablets and their final composition. When processing tablets, lots of them are destroyed because of mechanical failures. Some of the variables that can affect the final properties of a tablet may be present when feeding the die. Feedframe is a common way used to feed the powder to the tablet press dies. There exist different designs of feedframes between manufacturers. These equipment variations can make a difference in the powder flow into a die. The filling of the dies also is an essential part to control the final properties of tablets. Because of the complexity of powder flow through the feedframe there is no actual model or principle to describe it. To understand the causal effect, this project focuses on studying the lubrication effect on the feedframe. Also examine the powder flow through it varying revolutions per minute (rpm). These variables have a strong effect on the compression of powders and thus in the final properties of tablets.

328p Boron-Rich Fluorescent Silica Nanoparticles For Neutron Capture Therapy

Alexis H. A. Mollard, Erik M. Brozek, and Ilya Zharov. Department of Chemistry, University of Utah, 315 So. 1400 E. Rm. 2020, Salt Lake City, UT 84112

The selective delivery of large numbers of boron atoms to cancerous cells is a requirement for the success of Boron Neutron Capture Therapy (BNCT).

To this end, we have synthesized monodisperse boron-rich silica nanoparticles, either by a sol-gel process involving boric acid and alkoxysilanes, or by surface-initiated atom-transfer radical polymerization of boron-containing monomers. Several fluorone-based dyes (fluorescein, rhodamines) were also incorporated within the silica matrix, resulting in a strong and stable fluorescence which allows direct imaging of the particles by confocal microscopy. Further functionalization of such particles with tumor-targeting moieties such as integrins is currently under investigation.

Rheological Properties and Characterization of Alginate/Bentonite Biocomposite Systems

Birgul Benli[1], Mehmet S. Celik[2]

[1]Chemical Engineering Department, Istanbul Technical University, Maslak, Turkey.

[2]Mineral Processing Engineering Department, Istanbul Technical University, Maslak, Turkey.

Abstract

In this study, the effects of alginate, a biopolymer, on the bentonite (MMT) suspensions and the films of alginate/bentonite biocomposites were investigated. Physical properties of alginate/MMT biocomposites with $CaCl_2$ as a crosslinking agent and MMT as a filler have been studied and compared with pure alginates. In this case, the viscosity and zeta potential measurements were performed. The aim for these experiments was to see the dependence of alginate addition on viscosity and electrokinetic properties of MMT suspensions. The results showed that while alginate made the zeta potential of bentonite suspensions more negative, it decreased the viscosity of bentonite suspensions. The characterization of biocomposite was performed with different methods such as X-ray Diffraction (XRD) and Thermogravimetric Analysis (TGA).

1. Introduction

Biocomposites have been used as a short term application materials (e.g., packaging) since they are biodegradable. With this application, biocomposites are an alternative to petroleum-based polymers (traditional plastics). Development of the biodegradable plastics gives one of the approaches to solving the ecological problem (plastic waste disposal). One of the production methods of the commercially successful biodegradable plastics is prepared from chemically modified natural products (e.g., starch, cellulose, chitin- chitosan or soy protein). The addition of nanofillers into a biodegradable polymer matrix leads to the creation of a novel class of materials, called nano-biocomposites with an environmental approach (1). Biocomposites briefly consists of clay and biopolymer mixtures. Biopolymers are generated from renewable natural sources, which are often biodegradable and nontoxic. They can be produced by biological systems (i.e., microorganisms, plants, and animals), or chemically synthesized from biological materials (e.g., sugars, starch, natural fats and oils, etc.) (2). Recent studies have been previously reported for the elaboration and characterization of these nanomaterials, based on epoxy (3), rubber, (4), PVA (5,6), aromatic polyester (1), PLA (7), chitosan (8-13).

Alginates are a family of unbranched binary copolymers consisting of 1-4 glycosidically-linked-β-Dmannuronic acid acid (M) and its C-5 epimer α-L-guluronic acid (G) as shown in Figure 1. Alginates bound with monovalent ions are generally water soluble whereas bivalent cations like Ca^{+2} form hydrogels, which are partially soluble in water (14). Although alginate which is a naturally occurring polysaccharide is widely used in biomedical applications, drug delivery and cell immobilization for decades, there are limited studies on the preparation and characterization of alginate biocomposite systems (14,10,15,16).

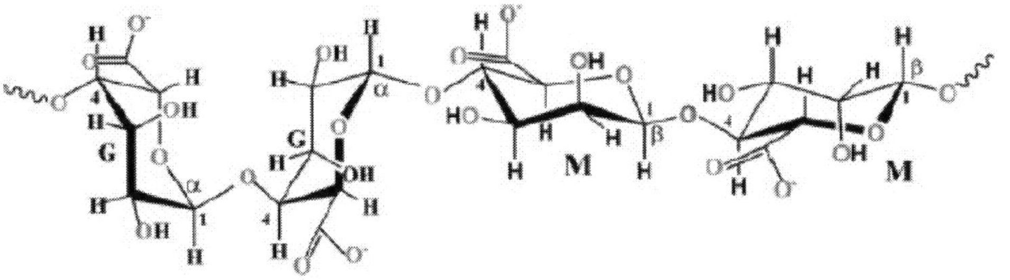

Figure 1. Structure of Alginate.

Bentonite, a trade name of montmorillonite, is best known as a member of the smectite group due to its high swelling capacity, the ability to adsorb significant amounts of polymer on its surfaces, and the formation of a gel-like structure with yield characteristics and viscoelastic properties at relatively low clay concentrations. The bentonite structure classified as dioctahedral, have two thirds of the octahedral sites occupied by trivalent cations as seen in Figure 2. Dioctahedral montmorillonite has its structural charge originating from the substitution of Mg^{+2} for Al^{+3} in the octahedral sheet (17). Bentonite is the most widely used layered silicate in polymer nanocomposites. The use of clays in the preparation of nanocomposites has recently been increased due to better access to relatively well characterized clays (9).

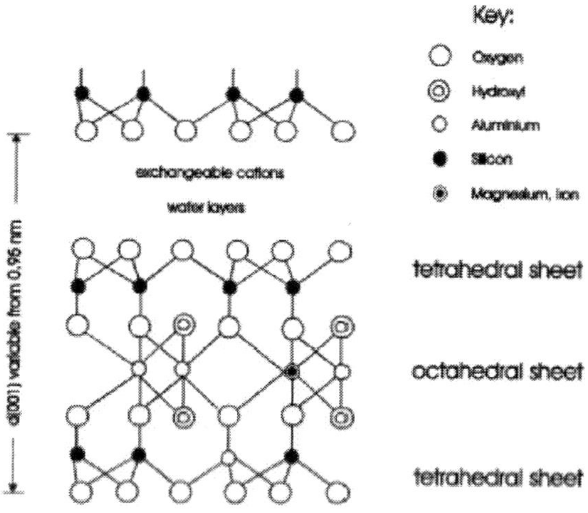

Figure 2. Structure of bentonite.

The aim of this study is to investigate the effect of biopolymer alginate on rheological properties of bentonite suspensions and also develop some methodology on the preparation and characterization of alginate/bentonite biocomposites. Towards this aim, direct nucleation of bentonite on alginate biopolymers was carried out to provide an excellent example of self assembling process. The driving forces behind assembling process are essentially hydrogen-bonding, van der Waals, electrostatic forces and electron-transfer interactions (14). Characterization and thermal behaviors of the prepared biocomposites were investigated using the rheological and electrokinetic properties along with data on X-ray Diffraction (XRD) and Thermogravimetric Analysis (TGA).

2. Experimental

2.1. Materials

Natural sodium bentonite sample obtained from Resadiye Region of Turkey was used after a rigorous purification process through hydrocyclones. Sodium alginate purchased from Sigma-Aldrich as alginic acid sodium salt from brown algae was used in the biocomposite studies. It is specifically used for immobilization and described as Biochemika- a straight-chain, hydrophilic, colloidal, polyuronic acid composed primarily of anhydro-β-D-mannuronic acid residues with 1→4 linkage.

2.2. Preparation of Bentonite Sample

Raw bentonite sample used in this study was dispersed in tap water at room temperature for 48h and then subjected to a multistage purification process by hydrocyclone at different apex and vortex diameters and inlet pressures (18,19). Under the optimum conditions, a concentrate with cation exchange capacity of 98 meq/100 g and average particle size of 2.5 μm was obtained; this indicates that the sample labeled as MMT is rather pure. The sample was dried and mildly ground for further use.

2.3. Preparation of Biocomposite Systems

Two different types of alginate/MMT biocomposite systems involving suspension and films respectively were used. In the first part of the experiments, biocomposite suspensions were prepared for the rheology and zeta potentials measurements. For this purpose, the polymer powder was dissolved in water using a mechanical stirrer at high speeds. The prepared polymer was left overnight before its use. The zeta potential measurements were performed in the presence of 1 wt% of alginate solution and 2 wt% of MMT suspension as a function of alginate concentration, and the final concentration of MMT suspension was adjusted as 1wt%. Then, the suspensions were diluted 50 times before doing the measurements. In the rheology measurements, MMT was added to the alginate solution in different ratios.

In the second part of the study, the biocomposites films were prepared by casting and solvent evaporation technique at room temperature. The desired quantity of sodium alginate was hydrated overnight in demineralized water to facilitate the dissolution of alginate and also to obtain the solutions of 1-3 wt%, respectively. The dissolved polymer solutions were cast onto a clean glass plate at room temperature for XRD studies. Also, biocomposite films were casted on a Teflon mold for other characterization studies. Then, all samples were soaked in $CaCl_2$ solution bath (20 wt%) for half an hour. The films were rinsed twice with ethyl alcohol to strengthen the film, and washed with deionized water to remove the remaining traces of solvent. The film thus formed was dried at room temperature for 3 days. The product was characterized using XRD (Rigaku, Miniflex model X-ray diffractometer) equipped with Cu-Kα radiation at room temperature; this is to identify the smectite groups and other associated impurities.

2.4. Characterization of Biocomposites

The chemical composition (wt %) of the purified bentonite (Na- montmorillonite) is given as: SiO_2 60.11; Al_2O_3 17.90; TiO_2 0.31; Fe_2O_3 3.66; MnO 0.02; MgO 2.33; CaO 0.75; K_2O 0.31; Na_2O 2.54; P_2O5 0.04; ignition loss 12.00. Chemical composition together with XRD was used to identify the composition of bentonite.

The flow behavior of the dispersions was measured using Brookfield type low-shear rheometer. The zeta potential measurements were carried out using a Zeta PALS instruments. Thermal stability of the sample was investigated using the thermo gravimetric analysis (TGA).

3. Results and Discussion

3.1. Rheological and Electrokinetic Studies of Alginate/MMT Biocomposites

The apparent viscosity and zeta potential measurements of alginate/MMT biocomposites was carried out in order to understand the effect of alginate on MMT dispersions (1 wt%). Figure 3 shows the variations in the apparent viscosity and zeta potentials of suspensions of bentonite/alginate mixture as a function of the concentrations of the alginate. The viscosities of pure alginate solution (1 wt%) and pure MMT dispersions (1 wt%) was measured as 1.496 PaS and 1.171 PaS, respectively.

Figure 3 (a) shows that the apparent viscosity of the suspensions decreases as a function of alginate concentration and then remains constant above about 0.5 alginate ratio. Zeta potential profile shown in Figure 3b, on the other hand, indicates that zeta potential values of MMT decreases with increasing the alginate concentration; This clearly indicates the incorporation of oxygen containing negatively charged groups into the structure of bentonite.

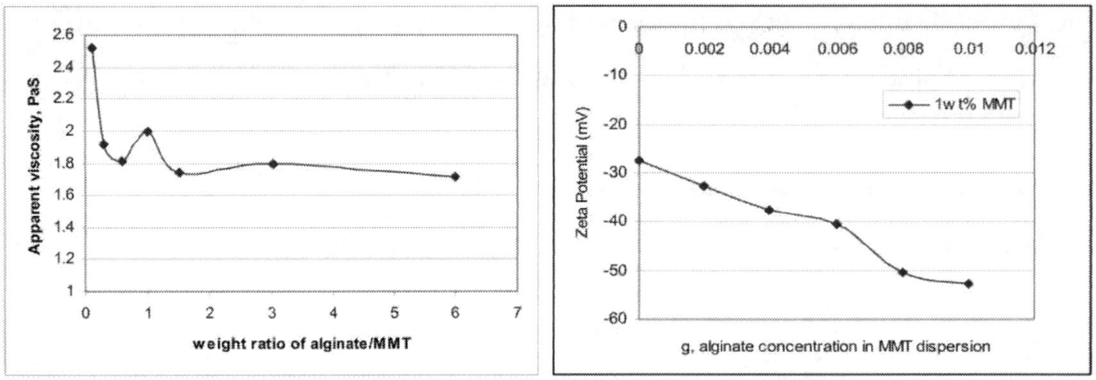

(a) (b)

Figure 3. The changes of alginate/MMT systems, (a) Apparent viscosity, (b) Zeta potential (pH = 7.4 ± 0.2)

3.2. Nanostructure and Morphology

The biocomposite films were prepared using the same method given in Section 2.3. Casted biocomposite films and solutions of alginate/MMT are seen in Figure 4.

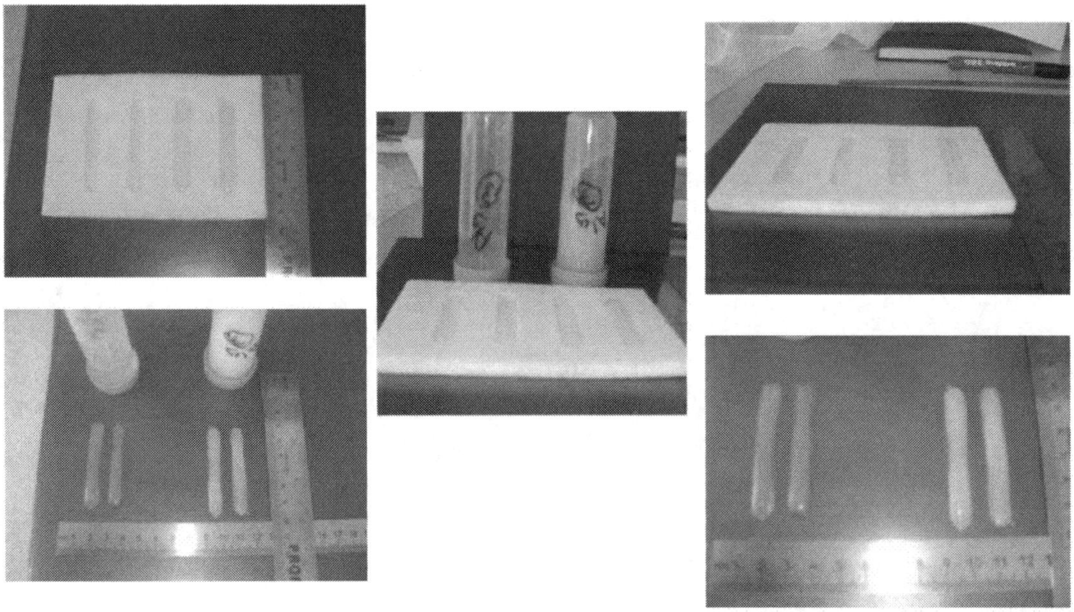

Figure 4. The comparison of biocomposites prepared from pure alginate and 3.33 w/w% MMT/alginate.

Figure 5 reveals that the amount of clay particles in the alginate films affects the physical properties of the films such as density and optical clarity.

The alginate/MMT biocomposites have been characterized with XRD which are the most widely used methods to study the structure of nanocomposites. Figure 5 illustrates the XRD patterns of MMT and alginate/MMT biocomposites at different concentrations.
The XRD patterns of MMT shows a reflection peak at about $2\theta \approx 12°$ which is the characteristics peak for the MMT The XRD patterns of alginate films shows the characteristic crystalline peaks at around $2\theta \approx 32°$, 45°, and 55° as seen in Figure 6. It is important to note that XRD pattern of biocomposites didn't detect the characteristics peak for MMT.

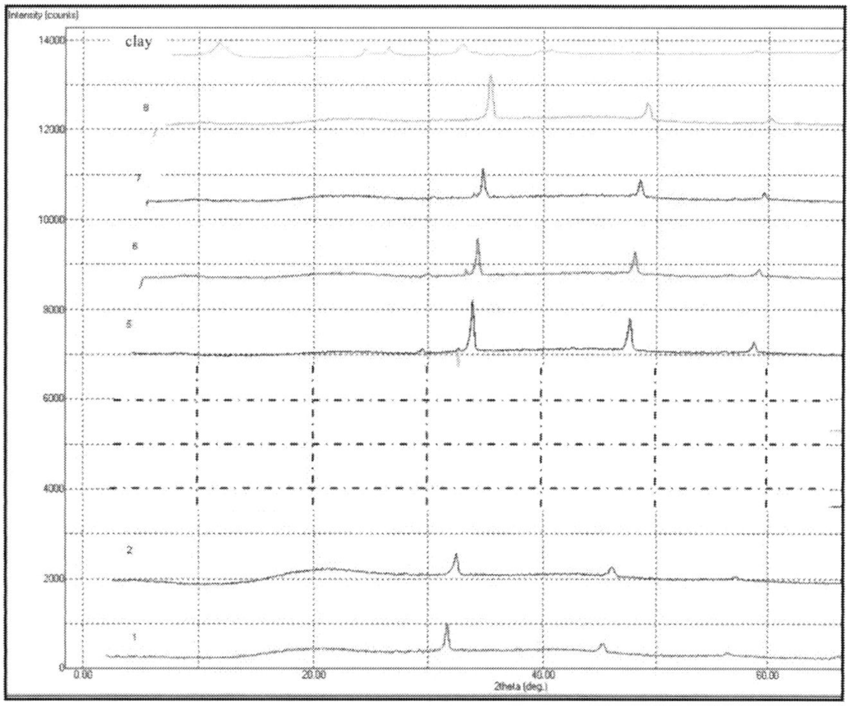

Figure 5. XRD patterns of the bentonite (clay), pure alginate ratios (1,2) and its biocomposites (5-8) with different ratios. In this figure, the patterns are labeled as follows; pure MMT suspension (clay), pure alginate solution with different ratios 1 wt% (1), 3 wt% (2), and biocomposites with different ratios, 0.03g MMT in 1 wt% alginate (5), 0.03g MMT in 3 wt% alginate (6), 0.01g MMT in 1 wt% alginate (7), 0.01g MMT in 3 wt% alginate (8), respectively.

3.4. Thermal Stability of Biocomposites

The thermal stability of alginate and its biocomposites was investigated using TGA under nitrogen. As seen in Figure 6, the biocomposites have more thermal stability than the pure alginate films because inorganic filler such as MMT have good thermal stability. This increase in thermal stability can be attributed to the high thermal stability of clays and the dispersion of clay particles within the alginate alginate.

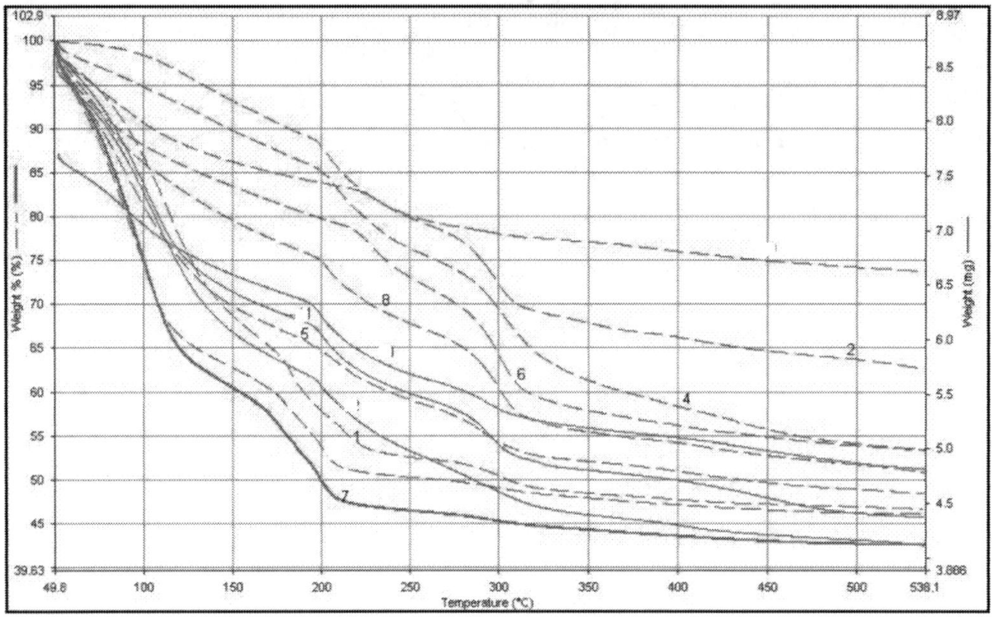

Figure 6. TGA curves of alginate and its biocomposites with MMT. In this figure, the patterns are labeled as follows; pure MMT suspension (clay), pure alginate solution with different ratios 1 wt% (1), 3 wt% (2), and biocomposites with different ratios, 0.03g MMT in 1 wt% alginate (5), 0.03g MMT in 3 wt% alginate (6), 0.01g MMT in 1 wt% alginate (7), 0.01g MMT in 3 wt% alginate (8), respectively.

4. Conclusions

The results of this study can be summarized as follows.

1. Biopolymer alginate increased the zeta potential of bentonite suspensions, but decreased the viscosity of bentonite suspensions.
2. The casting and solvent evaporation technique can be conveniently used in the preparation of the alginate and alginate/MMT biocomposites.
3. The thermal stability increased with increasing the ratio of MMT in the biocomposites.
4. The physical properties of biocomposites can be changed by adjusting the level of MMT concentration in the biocomposite matrix.

Alginate/MMT biocomposites are expected to have potential applications in the area where environmentally friendly materials are preferred.

Acknowledgements

The authors thank Isik Yavuz for XRD, TGA measurements, and Huseyin Sezer for XRD spectrum measurements.

References

1. Chivrac, F., Kadlecova, Z., Pollet, E., Averous, L., 2007. "Aromatic copolyester-based nano-biocomposites: Elaboration, Structural characterization and properties" J. Polym., Environ., Uncorrected proof.
2. Flieger, M., Kantorova, M., Prell, A., Rezanka, T., Votruba, J., 2003 "Biodegradable plastics from renewable sources", Folia Microbiol., 48 (1) pp:27-44.
3. Uyanik, N., Erdem, A. R., Can F., Celik, M.S., 2006. "Epoxy nanocomposites curing by microwaves", Polymer Engineering and Science, 46 (8) pp:1104-1110.
4. Maiti, M., Bhowmick, A.K., 2006. "New insights into rubber-clay nanocomposites by AFM imaging", Polymer, 47, pp:6156-6166.
5. Yurudu, C., Isci, S., Unlu, C., Atici, O., Ece, O.I., Gungor, N., 2006. "Preparation and characterization of PVA/OMMT composites", Journal of Applied Polymer Science, 102, pp:2315-2323.
6. Yu, Y., Lin, C., Yeh, J., Lin, W., 2003. "Preparation and properties of poly (vinyl alcohol)-clay nanocomposite materials", Polymer, 44, pp:3553-3560.

7. Pluta, M., Paul, M.A., Alexandre, M., Dubois, P., 2006. "Plasticized polylactide/clay nanocomposites. 1. The role of filler content and its surface organo-modification on the physico-chemical properties", Journal of Polymer Science Part B:Polymer Physics, 44, pp:299-311.
8. Gunister E., Pestreli, D., Unlu, C.H., Atici, O., Gungor, N., 2007. "Synthesis and characterization of chitosan-MMT biocomposite systems", Carbohydrate Polymers, 67, pp:358-365.
9. Darder, M., Lopez-Blanco, M., Aranda, P., Aznar,A.J., Bravo, J., Ruiz-Hitzky E., 2006. "Microfibrous chitosan-sepiolite nanocomposites", Chem. Mater., 18, pp:1602-1610.
10. Smitha, B., Sridhar, S., Khan, A.A., 2005. "Chitosan-sodium alginate polyion complexes as fuel cell membranes", European Polymer Journal, 41, pp:1859-1866.
11. Wu, T., Wu, C., 2006. "Biodegradable poly (lactic acid)/chitosan-modified montmorillonite nanocomposites: Preparation and characterization", Polymer Degradation and Stability, 91, pp:2198-2204.
12. Wang, S., Chen, L., Tong, Y., 2006. "Structure-property relation in chitosan-based biopolymer/montmorillonite nanocomposites", Journal of Polymer Science: Part A: Polymer Chemistry, 44, pp:686- 696.
13. Wang, F., Shen, L., Tong, Y., Chen, L., Phang, I.Y., Lim, P.Q., Liu, T.X., 2005. "Biocomposite chitosan/montmorillonite nanocomposites: Preparation and characterization", Polymer Degradation and Stability, 90, pp:123-131.
14. Parhi, P., Ramanan, A., Ray, A.R., 2006. "Preparation and characterization of alginate and hydroxyapatite-based biocomposite", Journal of Applied Polymer Science, 102, pp:5162-5165.
15. Caykara, T., Demirci, S., Eroglu, M.S., Guven, O., 2005. "Poly (ethylene oxide) and its blends with sodium alginate", Polymer, 46, pp:10750-10757.
16. Xiao, C., Lu, Y., Liu, H., Zhang, L., 2000. "Preparation and physical properties of blend films from sodium alginate and polyacrylamide solutions", Pure Applied Chemistry A, 37(12), pp:1663-1675.
17. Luckham, F.L., Rossi, S., 1999. "The colloidal and rheological properties of bentonite suspensions", Advances in Colloid and Interface Science, pp:82, 92.
18. Bain, J.A., and Morgan D. J., (1982),"Laboratory separation of Clays by Hydrocycloning", Clay Minerals, 18, pp: 33-47.
19. Bloodworth, A.J., Morgan, D.J., and Briggs, D.A, (1989), "Laboratory Processing Trials on kaolin-Bearing Sandstones from Pugu, Tanzania, Using conventional and New Hydrocyclone Bodies", Clay Minerals, 24, pp: 539-548.

328r A Comparison of Atomic Force Microscopy (Afm) and Dynamic Light Scattering (Dls) Methods to Characterize Nanoparticle Size Distributions

Natasha Starostin[1], Paul West[1], Chris Hoo[2], and Martha Mecartney[3]. (1) R&D, Pacific Nanotechnology Inc., 17981 Sky Park Circkle, Suite J, Irvine, CA 92614, (2) Department of Chemical Engineering and Materials Science, UC of Irvine, 916 Engineering Tower, Irvine, CA 92697, (3) Chemical Engineering and Material Science, UC of Irvine, 916 Engineering Tower, Irvine, CA 92697

This paper compares the capabilities of dynamic light scattering (DLS) analysis and atomic force microscopy (AFM) analysis for characterizing size distributions of various nanoparticle mixture in the size range of 1 – 100 nm. The bimodal system of polymer nanospheres, three-modal mixture of colloidal gold and CdTe quantum dots were characterized by both techniques. Average DLS values for mono-size dispersed particles of polymer and gold are slightly higher than the nominal values where AFM values were slightly lower than the nominal values. Bimodal and three-modal distributions were easily identified with AFM, but DLS results were skewed toward larger particles. AFM average particle size for mono-size dispersed particles of CdTe quantum dot is in a good agreement with the nominal values, while average DLS values were found to be erroneous. AFM characterization of nanoparticles using automated analysis software provides an accurate and rapid analysis for nanoparticle characterization.

328s Flow-Field Flow Fractionation to Separate and Characterize Single Wall Carbon Nanotubes by Length

Jaehun Chun, Jeffrey A. Fagan, Erik K. Hobbie, and Barry J. Bauer. Polymers Division, National Institute of Standards and Technology, 100 Bureau Dr., Gaithersburg, MD 20899-8542

Flow-Field Flow Fractionation (Flow-FFF) has been a useful technique to separate various colloids, particles, and macromolecules such as polymer latexes and proteins by their sizes, taking advantage of a balance between random thermal and uniform flow-driven motions on an object of interest. We exploit Flow-FFF, equipped with on-line multi-angle light scattering (MALS) and UV-Vis photodiode array, to separate single wall carbon nanotubes (SWNTs) dispersed in aqueous medium by the use of DNA. Separations are made of both unfractionated SWNTs and SWNT fractions made by use of size exclusion chromatography (SEC). Rod lengths of the SWNTs are calculated from MALS and those are compared to rod lengths from hydrodynamic models based on latex sphere calibrations. Flow-FFF is capable of probing subtle differences in SWNT size distributions, which suggests that Flow-FFF may be a simple method of SWNT identification and quality assessment.

UNIVERSITY OF CALGARY

In-Situ Removal of $H_2S_{(g)}$ Using Ultradispersed Iron Oxide Nanoparticles

Maen M. Husein, Luis Patruyo and Pedro Pereira
Department of Chemical and Petroleum Engineering, University of Calgary, Calgary, AB, Canada

Introduction

The worldwide demand on oil is increasing, while the sources of light crude oil are depleting. Refineries are more and more required to handle oilsands, bitumens and heavy oils as feasible substitutes for crude oils. Economic and environmental limitations provide motivations for developing new and innovative in-situ technologies for recovery and upgrading, which would provide access to the large percentage, about 93%, of bitumen resources that are too deep to mine. In addition, the demand for conversion of residues of crude oil distillation into clean and more valuable liquid products is continuously increasing as such residues represent significant crude portion, up to 85 vol. %, and its disposal treatment is still not yet satisfactorily developed. An envisioned application of ultradispersed nanoparticles has to do with the upgrading of oil in-situ, whereupon active surface nano-structures, small enough to breakthrough the reservoir formation, interact with the heavy oil molecules promoting upgrading reactions. In-situ upgrading spares the environment the impact of on-surface upgrading effluent stream. Nonetheless, in-situ upgrading may pose different environmental concerns. Many of the upgrading side products may remain contained within the reservoir in a chemically active form posing high risk of leaking into adjacent water aquifers, or changing the structure of the formation. These products could also come to the surface, thus increasing the need for oil production effluents cleaning and affecting the surface environment and/or posing multiple location disposition problems usually more costly to solve than single centralized ones as they appear in conventional upgrading.

This work introduces a novel microemulsion-nanoparticle based approach for in-situ removal of $H_2S_{(g)}$ associated with in-situ heavy oil upgrading. Oil reservoirs have very complicated structures combining oil, water, aqueous species, and sand, in addition to water-in-oil, (w/o), emulsions/microemulsions. The current study employed a heavy hydrocarbon as the continuous oil phase to better model oil reservoirs. The (w/o) microemulsions were used to form and stabilize ultradispersed colloidal iron oxide/hydroxide nanoparticle absorbents. Ultradisperse absorption combines two advantages; it provides a huge surface of contact with the absorbent particles, and decreases the distance traveled between the reacting species and the surface. More importantly, the resultant solid forms of sulfur may be either surface produced or retained in the reservoir, in both cases under a relatively chemically inactive solid form, which is naturally abundant in most surface and underground environments.

Objectives

The study aimed at evaluating the in-situ removal of $H_2S_{(g)}$ by means of ultradispersed colloidal nanoparticle absorbents. The effect of temperature, concentration of colloidal nanoparticles, and origin (synthesized versus commercial) of absorbent on the removal efficiency of $H_2S_{(g)}$ was estimated.

Experimental Methods

(w/o) Microemulsions were used to prepare iron oxide/hydroxide nanoparticles. The $H_2S_{(g)}$ removal efficiency of the microemulsion-prepared nanoparticle absorbents was compared with that of commercially available iron oxide nanoparticles. The size of the commercial nanoparticles ranged between 25-50 nm, and the role of the microemulsions was to disperse the commercial iron oxide nanoparticles. The $H_2S_{(g)}$ absorption experiments were carried out in the experimental set-up shown in Figure 1. The gas from a cylinder containing 200 ppm H_2S and 1000 ppm N_2 in He (balance) was allowed into the reactor. The gas bubbling flow was controlled from the outlet by a flow controller (Brooks Instruments, 5850S). Temperature control was achieved by a heating coil wrapped around the Parr reactor (Parr Instrument Company). The exit stream of the reactor was allowed into a gas chromatograph SRI 8610C. The detection limit of the GC ranged up to 400 ppm $H_2S_{(g)}$ at the selected operating settings. The 100 ml volume of the reactor was occupied by 50 ml of the liquid phase, unless otherwise noted.

Results and discussion

Effect of origin of nanoparticles. Figure 2 shows breakthrough curves obtained when the gaseous mixture of H_2S and N_2 was allowed into reactor containing the oil phase, the microemulsion, the microemulsion containing microemulsion-synthesized FeOOH nanoparticles, and the microemulsion containing commercial iron oxide nanoparticles. The other experimental variables were 25 °C, 200 psi, 40 sccm and 60 rpm stirring.

Figure 2 shows that the breakthrough time was minimum for the oil phase and maximum for the microemulsion with microemulsion-synthesized FeOOH nanoparticles. The breakthrough time obtained for the oil phase is attributed to the retention time of $H_2S_{(g)}$ in the reactor which was magnified by the solubility of $H_2S_{(g)}$ in the oil. The slightly higher breakthrough time of $H_2S_{(g)}$ in the microemulsion system is probably due to the higher solubility of the gas in the system in the presence of water. Dispersion of commercial iron oxide nanoparticles within the microemulsion system did not seem to alter retention time of $H_2S_{(g)}$. The breakthrough curve for

$H_2S_{(g)}$ in the microemulsion and in the microemulsion containing the commercial iron oxide are almost identical. This indicates that the interaction of $H_2S_{(g)}$ with commercial iron oxide nanoparticles was kept minimal, probably due to insufficient exposure of the two species. Comparison of the previous results with the run containing the microemulsion with microemulsion-synthesized iron oxide/hydroxide reveals high interaction between $H_2S_{(g)}$ and the colloidal nanoparticles. Calculations based on reaction R1 below,

$$2FeOOH_{(s)} + 3\,H_2S_{(g)} \rightarrow Fe_2S_{3\,(s)} + 4\,H_2O_{(l)} \qquad \text{(R1)}$$

showed that the breakthrough time of $H_2S_{(g)}$ corresponded to the stoichiometric amount. This calculation clearly indicates that the microemulsion system was able to make all iron oxide/hydroxide available for the reaction without any mass transfer limitations. This attests to the advantages of using the microemulsion itself to synthesize the ultradispersed absorbents. When nanoparticles are synthesized within microemulsions, the microemulsions provide excellent control over particle size and protect the colloidal particles from aggregation by virtue of their protective surfactant layer (Nassar and Husein, a,b).

Effect of temperature. The effect of temperature was studied by comparing $H_2S_{(g)}$ breakthrough curves in microemulsions containing microemulsion-synthesized FeOOH nanoparticle absorbents at three different temperatures: 25 °C, 60 °C and 120 °C. The other experimental variables were 200 psi reactor pressure, 40 sccm gas flow rate and 60 rpm stirring. The three breakthrough curves are plotted in Figure 3.

The figure shows that breakthrough time decreases as the temperature increases. This decrease is more significant between 25 °C and 60 °C. Nonetheless, the two curves seem to level off at a ratio of outlet to inlet concentration of $H_2S_{(g)}$ of 1. The ratio of outlet to inlet concentration of $H_2S_{(g)}$ seem to level off at around 0.75 for the 120 °C curve. The decrease in retention time with temperature can be attributed to less accessible FeOOH nanoparticles. Separate runs showed that the microemulsion collapses and two phases form when temperatures were fixed at 60 or 120 °C. This collapse of the microemulsion system promotes particle aggregation and poses interphase mass transport limitations between the gas and the absorbents. The lower value of outlet concentration of $H_2S_{(g)}$ at 120 °C is attributed to a side reaction between $H_2S_{(g)}$ and the microemulsion constituents. Calculations using the thermodynamic equilibrium package Factsage® indicated that methyl mercaptan, (CH_3SH), and/or dimethyl mercaptan, $(CH_3)_2S$ are possible sulphur compounds. These calculations are shown in Figure 4. It is worth noting that the side product only appeared after 1.45 h of run time indicating that the reaction of iron sulphide formation was predominant, as long as there accessible colloidal nanoparticles.

Conclusions

In-situ heavy oil upgrading may leave the reservoir with many chemically active pollutants. These pollutants may change the formation structure and/or find their way to adjacent water aquifers or the surface. The current work employed (w/o) microemulsions to simulate heavy oil reservoirs, and showed that $H_2S_{(g)}$ could be potentially removed in-situ by means of colloidal ultradispersed nanoparticle absorbents. These absorbents are much more effective when they are formed within the microemulsions, since their colloidal stability is higher. The temperature negatively impacted the absorption capacity, due to its effect on the stability of the microemulsions. The collapse of the microemulsion to two phases at 60 and 120 °C resulted in interphase mass transport limitations, which prevented further reaction of FeOOH.

Acknowledgement

We would like to thank the financial support from the Alberta Ingenuity Centre for In-Situ Energy (AICISE) and the Centre for Environmental Engineering Research and Education (CEERE) at the University of Calgary.

References

Nassar and Husein, Langmuir, in press, a.

Nassar and Husein, Journal of colloid and interface science, in press, b.

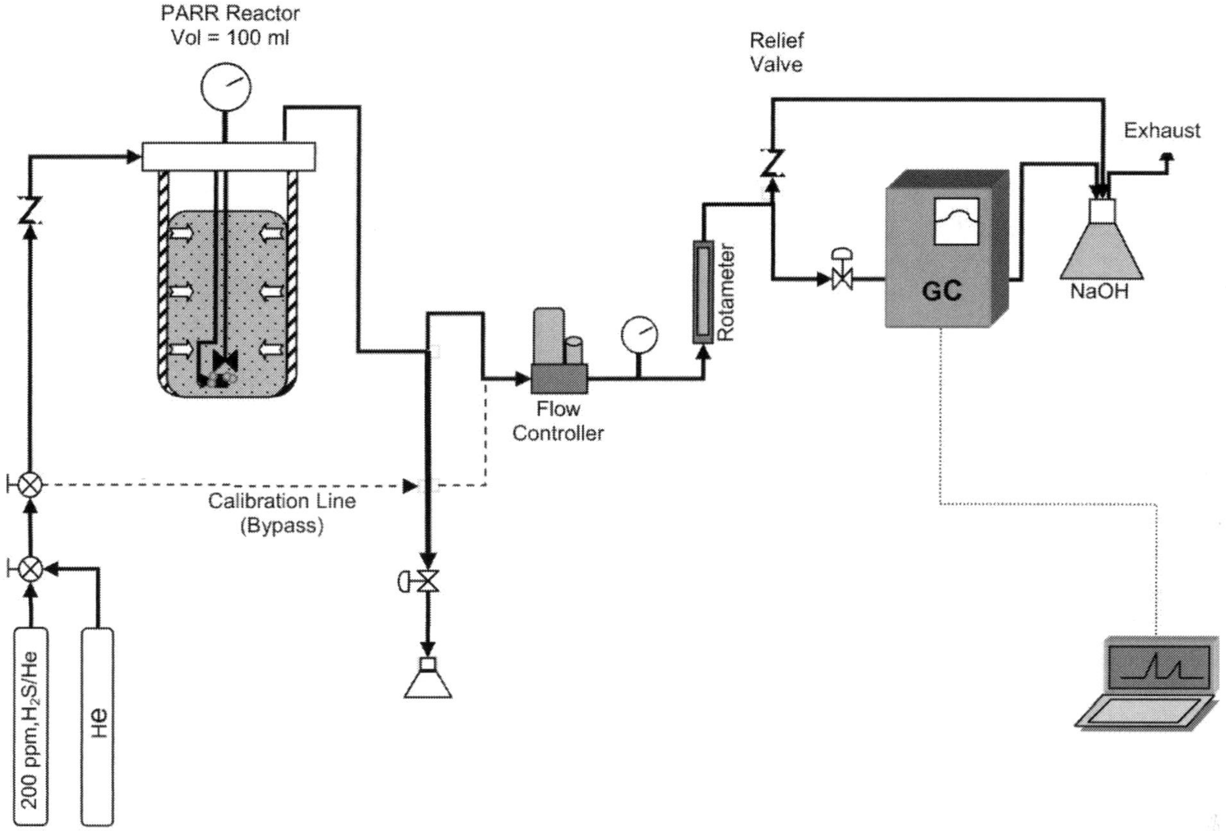

Figure 1: Experimental setup.

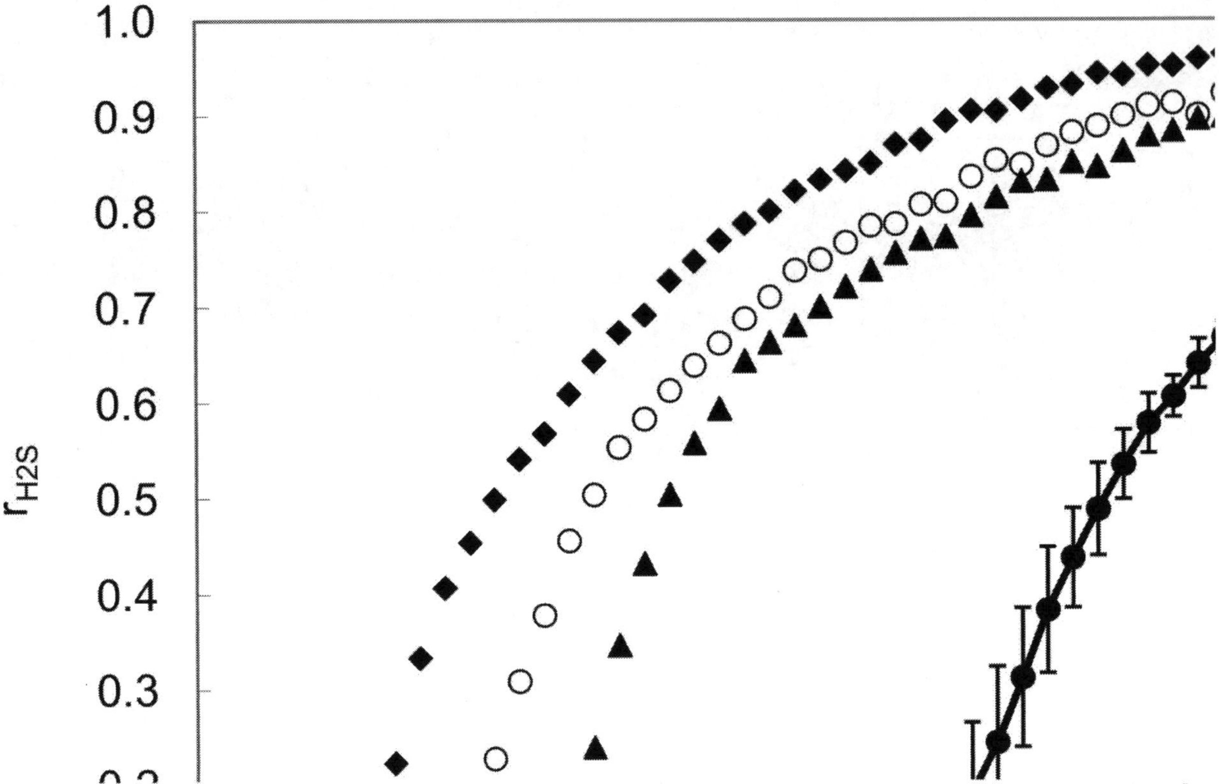

Figure 2: H_2S breakthrough curves obtained at 25 °C, 200 psi, 40 sccm and 60 rpm stirring for the (◆) oil phase, (○) the microemulsion, (▲) the microemulsion with the commercial iron oxide nanoparticles, and (●) the microemulsion with microemulsion-synthesized iron oxide/hydroxide nanoparticles

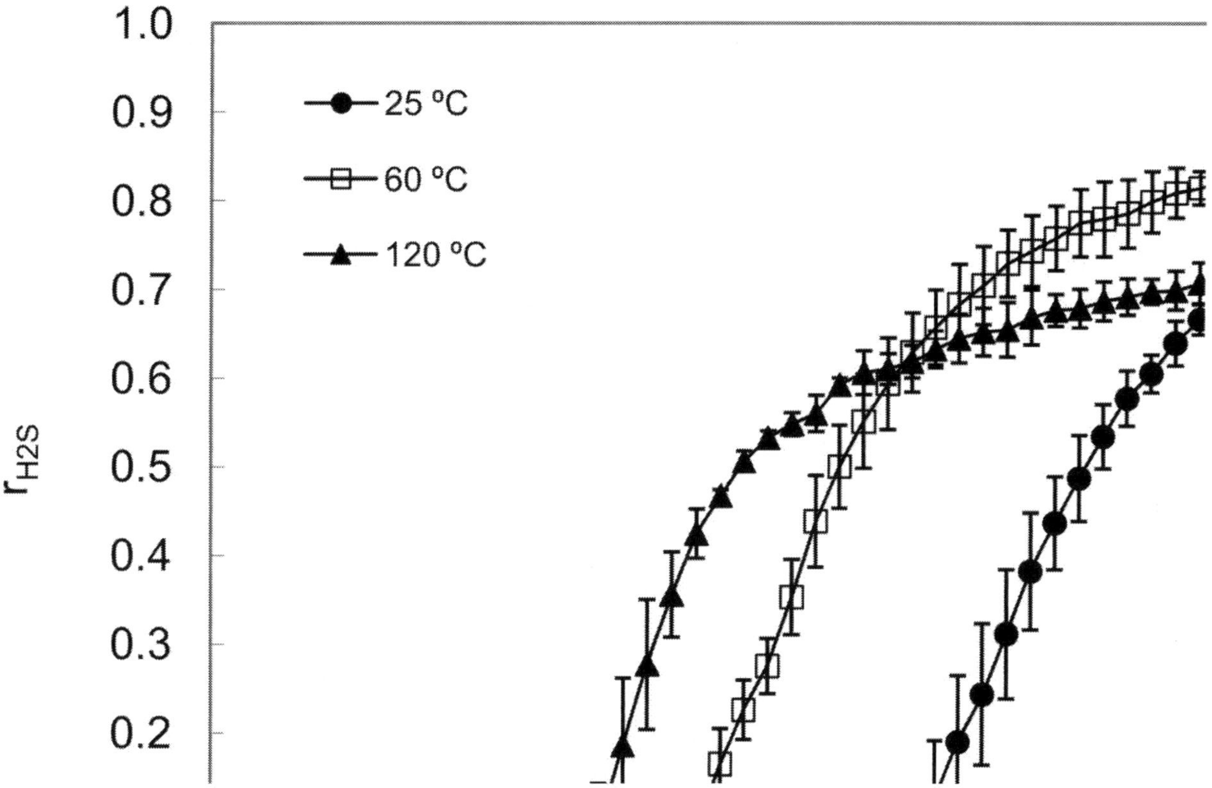

Figure 3: Effect of temperature on the H_2S breakthrough for the microemulsion with microemulsion-synthesized iron oxide/hydroxide nanoparticles at 200 psi, 40 sccm and 60 rpm stirring.

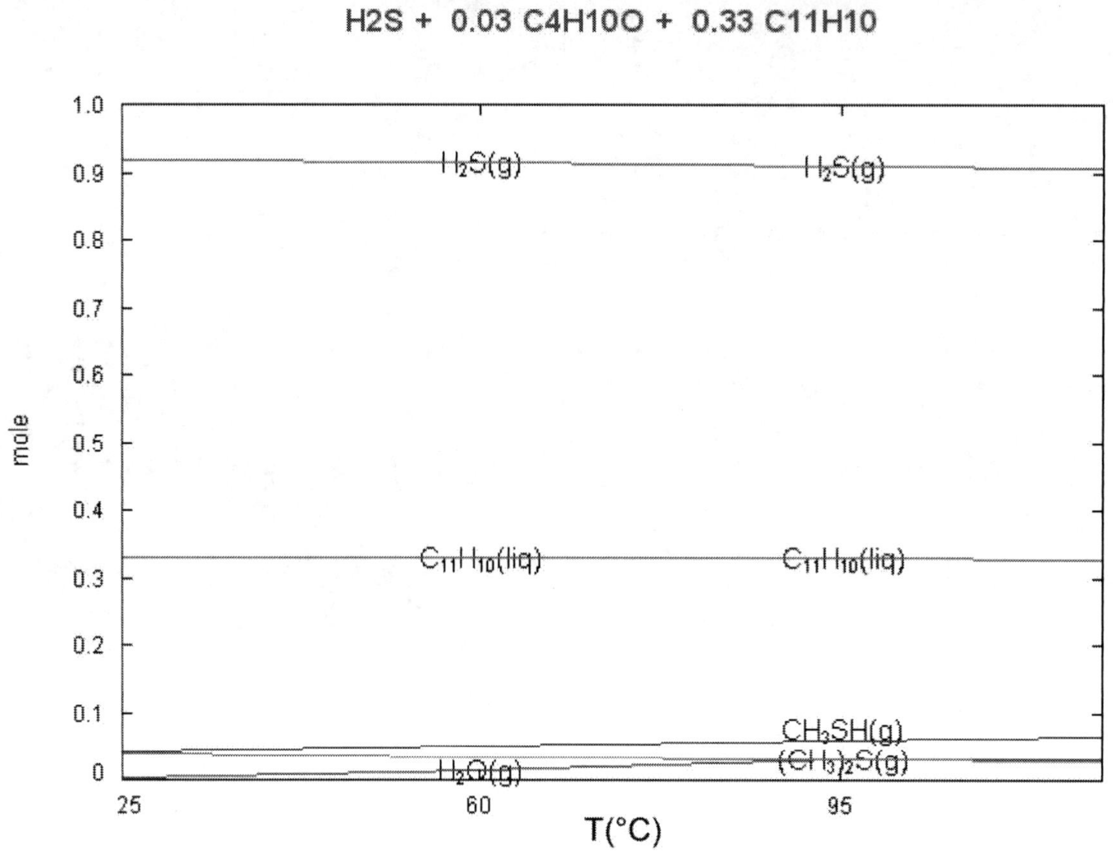

Figure 4: Most stable products from the reaction: H_2S + 0.03 $C_4H_{10}O$ + 0.33 $C_{11}H_{10}$ obtained by thermodynamic calculations using software FactSage®

328u Measurement of Colloidal Surface Charges by Second Harmonic Generation

Lars-Owe Schneider and Wolfgang Peukert. Institute of Particle Technology, University of Erlangen, Germany, Cauer Str. 4, Erlangen, 91058, Germany

Second Harmonic Generation (SHG) Spectroscopy is a well established technique for the investigation of planar interfaces. Recently it has also been applied to investigate the surface properties of colloidal interfaces [1]. In contrast to linear optical methods, where the separation of bulk and surface signals is often difficult if not impossible, second order nonlinear effects are exclusively surface sensitive. The signal generation depends on the nonlinear susceptibility ($\chi^{(2)}$) of the materials which vanishes for centrosymmetric materials. Interfaces between two such materials represent a symmetry break and do therefore possess a nonlinear surface susceptibility, which depends on the number of surface molecules, their orientation and present electro-magnetic fields [2,3]. The method can in general be applied to an interface which is formed by two centrosymmetric bulk media, and which is optically accessible. The poster will present some theoretical background of the method focusing on the application for particle surface charge measurements. Further, experimental results of surface charge measurements on silica particles will be presented. In this context the influence of particle concentration [4] as well as of the pH and the bulk electrolyte concentration will is illustrated. Electrophoretic zeta potential measurements will be presented as a reference. All experimental results will reveal that the SHG signal provides information on the potential at the Stern layer which is situated closer to the particle surface than the shear plane, where the zeta potential is measured. Therefore, SHG can be applied to directly monitor the ion displacements by adding ions to the suspension (with a higher affinity to the Silica surface than the already present ions) while the zeta-potential measurements do not reveal these phenomena. In this context we will investigate the Hofmeister series from a new perspective, a view which has not been experimentally accessible yet.

[1] H. Wang, E. Yan, E. Borguet, K. Eisenthal: Second harmonic generation from the surface of centrosymmetric particles in bulk solution Chem. Phys. Lett., 1996, 259, 15-20

[2] K. Eisenthal: Second Harmonic Spectroscopy of Aqueous Nano- and Microparticle Interfaces Chem. Rev., 2006, 106, 1462-1477

[3] L. Schneider, W. Peukert: Review: Second Harmonic Generation Spectroscopy as a Method for In-Situ and Online Characterization of Particle Surface Properties Particle and Particle Systems Characterization, 2007, 23, 351-359

[4] L. Schneider, H.-J. Schmid, W. Peukert: Influence of Particle Size and Concentration on the Second Harmonic Signal generated at Colloidal Surfaces Appl. Physics.B, 2007, 87, 333-339

Web Page: www.lfg.uni-erlangen.de

328v Thermal Transformation Of Metal Oxides Into Metal Nitrides Using Cyanamide Or Urea

Jelena Buha, Colloid Chemistry, Max Planck Institute of Colloids and Interfaces, Am Muehlenberg 1, Potsdam, Germany

Text Not Available.

328x Co-Formulation of Phenytoin and Pharmaceutical Polymers by Precipitation Using Compressed CO2 Antisolvent

Thakur Ranjit[1], Muhrer Gerhard[1], M. Mazotti[2], F. Fusaro[2], and F. Kluge[2]. (1) Novartis Pharmaceuticals, CH – 4002, Basel, Switzerland, (2) Institute of Process Engineering – ETH Swiss Federal Institute of Technology, Sonneggstrasse 3, CH – 8092, Zurich, Switzerland

The co-formulation of active pharmaceutical ingredients with polymeric excipients is a relevant way to achieve more favourable pharmacokinetics of the drug, e.g. enhancement of bioavailability, controlled release or drug targeting. Precipitation with compressed antisolvent (PCA) is a promising technique that enables the production of virtually solvent free drug-polymer co-formulations in a one step process. The PCA process is characterized by a number of advantages, namely mild operating conditions, high product purity and low solvent consumption, which satisfy precisely the demands for the manufacturing of pharmaceutically relevant drug-polymer co-formulations. Therefore PCA can be considered as a practical alternative to other processing techniques such as spray drying, lyophilization or melt extrusion.

The poorly water soluble anticonvulsant drug phenytoin and the biodegradable polymer PVP K30 have been selected as a model system to study the properties of co-precipitates formed by means of the PCA technique. The drug to polymer ratio was found to be the most influential parameter with respect to the obtained product morphology. Up to a relative drug content of 45 wt.%, fully amorphous co-precipitates are obtained, with the drug molecularly dispersed in the polymer matrix. At higher relative drug contents, products contain crystalline phenytoin besides the fully amorphous polymeric phase. The dissolution behaviour of the amorphous co-precipitates clearly outperforms the pure crystalline drug, as a determination of the intrinsic dissolution rate shows. Further, all fully amorphous co-formulations were stable and did not recrystallize during one year of storage at ambient conditions.

Co-precipitaion experiments with phenytoin have been extended to other water-soluble pharmaceutical excipients, namely hydroxypropyl methyl cellulose (HPMC), HPMC acetate succinate (HPMC-AS) and HPMC phthalate (HPMC-P). As compared to PVP, a similar behaviour has been observed for these polymers; however the maximal drug to polymer ratio up to which fully amorphous co-precipitates could be obtained was significantly lower. Currently, the four component system polymer, drug, solvent and CO2 anti-solvent, its thermodynamics and the possible mechanisms of particle formation for drug and polymer are under investigation. The long term stability of co-precipitation products indicates that the solubility of phenytoin in the polymers might play a key role for amorphous co-precipitates to be obtained. How this "solid solubility" is related to the structure of the polymer, e.g. to its ability to preferentially bond with the drug molecules, thereby hindering their organization into a crystal lattice, and whether drug-polymer solubility can be either predicted or measured, still remains to be fully understood.

IMPROVING CHEMICAL CONVERSION IN FLUIDIZED BEDS THROUGH SECONDARY INJECTION

D. Christensen, J. Nijenhuis, J.R. van Ommen, and M.-O. Coppens
Delft University of Technology - DelftChemTech
Julianalaan 136, 2628BL Delft, The Netherlands
E-mail: coppens@rpi.edu

ABSTRACT

The effect of secondary injection on the performance of a fluidised bed reactor has been studied. A modified Kunii and Levenspiel model was used to predict the effect on conversion and indicates that the reduction in bubble size caused by secondary injection results in a significant increase. This effect was confirmed experimentally, although the model appears to underestimate the experimental values.

INTRODUCTION

The mass transfer of a reactant from the bulk fluid to the catalyst in a bubbling fluidised bed is severely hindered because a large portion of the gas flow is in the form of bubbles. The reactant gas must diffuse through the bubbles and into the dense phase before it can reach the catalyst. As such, much of the gas bypasses the catalyst bed entirely. The result of the poor gas-solid contact throughout the bed is a low conversion and poor selectivity.

To improve the performance of a bubbling fluidised bed the size of the bubbles must be reduced. Using secondary gas injection, whereby a portion of the total gas flow is injected into the reactor via a fractal injector (Figure 1a), results in a reduction of the bubble size by 50% or more. Furthermore, the total volume of the bubbles is reduced (1, 2). These results indicate that a much better gas-solid contact is achieved, which should result in a higher conversion. The purpose of the current work is to directly study the effect of secondary injection on the conversion in a bubbling fluidised bed. Firstly, predictions of the effect using a reactor model are presented. Secondly, the results of preliminary conversion experiments using ozone decomposition (equation (1)) as a test reaction are reported. Ozone decomposition is often used as a test reaction because it has some favourable characteristics (3-5). Ozone is easy to detect at very low concentrations, thus temperature and volume changes due to reaction are negligible. It is simple to generate and the reaction kinetics are typically close to first-order.

$$2O_3 \rightarrow 3O_2 \qquad (1)$$

REACTOR MODELLING

The purpose of modelling the effect of secondary injection is to determine how the conversion is dependent on the bubble size. These results will be compared with the experimental data. The reactor model used is the fine-particle bubbling bed model proposed by Kunii and Levenspiel (6, 7). The assumptions inherent in this model are: bubbles, cloud, and emulsion can be modelled as separate phases; bubbles are spherical and follow the Davidson model for bubble behaviour (i.e., they travel faster than the emulsion gas and are surrounded by thin clouds); the wake of the bubbles is considered to be part of the cloud; bubble size is constant throughout the reactor; and the flow through the bubble phase is so much larger than the flows through the other phases that the latter flows can be ignored.

The model has been modified to include the flow through the emulsion phase because it is not negligible compared to the flow through the bubble phase for the total gas flow rate used in this study ($4 \times Q_{mf}$). The assumption of constant bubble size has also been removed by using the Darton equation (8) to predict the bubble size along the height of the reactor. This relation has been shown to be accurate in standard fluidisation columns (9). We must assume, however, that it will still be valid with secondary gas injection and the presence of internals.

It is not possible to implement the radial distribution of the secondary injection points in this one-dimensional axial model. Therefore, we assume that the fractal injector can be modelled as a series of levels at which the secondary gas is uniformly introduced into the bed. In addition, even though previous experiments have shown that secondary injection causes an increase in the flow through the emulsion phase (2), it was conservatively assumed that the emulsion phase remains at minimum fluidisation conditions throughout the reactor and that all of the secondary gas goes directly into the bubble phase. A schematic description of the model is shown in Figure 1b. Each phase also contains a reaction term (not shown). These modifications to the original K-L model result in a system of differential and algebraic equations that must be solved numerically. The general scheme is, firstly, to solve the system of equations up to the first injector level. Secondly, to adjust the boundary conditions to take into account the fresh secondary gas, and to continue the integration up to the next injector height, repeating this process until the final bed height is reached. Since the bed height depends on the bed porosity, which, in turn, depends on bubble size and flow rate (both of which are changing with height), iteration must be performed until the solution converges.

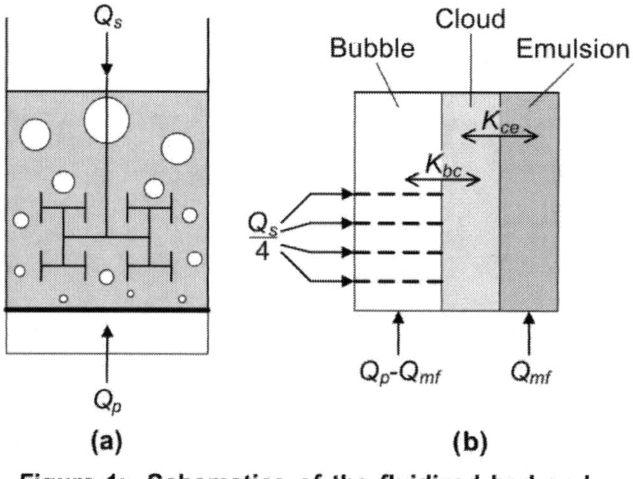

Figure 1: Schematics of the fluidized bed and fractal injector; (a) experimental setup; (b) model configuration

To take the effect of secondary injection on the bubble size into account, the Darton equation is multiplied by a bubble diameter reduction factor, β, as shown in equation (2). This factor is always between 0 and 1 and depends on the amount of injected secondary gas. Its value is determined from experiments, which will be discussed later. For example, if the bubble diameter with secondary injection is

25% smaller, then β is equal to 0.75. If β is known as a function of height then it can be replaced by $\beta(h)$ in equation (2). Here, $\beta(h)$ was not known, so it was assumed that the values measured halfway up the bed were the average values for the reactor. The parameter A_0, the area of the distributor plate per orifice, is unknown but is very small for porous distributors. Therefore, A_0 was neglected in this study.

$$d_b = \frac{0.54\left(U_0 - U_{mf}\right)^{0.4}\left(h + 4\sqrt{A_0}\right)^{0.8}}{g^{0.2}} \cdot \beta \tag{2}$$

A minimum bubble diameter, $d_{b,min}$ = 1.3 cm, had to be defined because the model would not converge at very small values. The model uses the result of equation (2) or $d_{b,min}$, whichever is greater at the current height in the bed. It should be noted that d_b is the average bubble size at that height. The rest of the model follows the development of Kunii and Levenspiel (6, 7). The steady-state material balances for the reactant in each of the phases, assuming first-order reaction, are:

$$\frac{dC_{A,b}}{dh} = \frac{-1}{\left(\delta \cdot U_b\right)} \cdot \left[\delta \cdot K_{bc} \cdot \left(C_{A,b} - C_{A,c}\right) + f_b \cdot k \cdot C_{A,b}\right] \tag{3}$$

$$0 = -\delta \cdot K_{bc}\left(C_{A,b} - C_{A,c}\right) + \delta \cdot K_{ce}\left(C_{A,c} - C_{A,e}\right) + f_c \cdot k \cdot C_{A,c} \tag{4}$$

$$\frac{dC_{A,e}}{dh} = \frac{-1}{\left(1-\delta\right) \cdot U_{mf}} \cdot \left[-\delta \cdot K_{ce}\left(C_{A,c} - C_{A,e}\right) + f_e \cdot k \cdot C_{A,e}\right] \tag{5}$$

This method of predicting the effect of a reduced bubble size on the conversion is general enough to be applied to modelling other techniques that decrease the bubble size throughout the reactor.

MODEL RESULTS & DISCUSSION

All parameters in the model, such as Q_0, Q_s, U_{mf}, β, bed mass, number and placement of injectors, etc., were chosen to match the experiments, which are described below. The reaction is assumed to be first order with the ozone concentration, while the reaction rate constant, k, is unknown and is used as an adjustable parameter. The modelling was carried out with and without secondary injection. An example of the calculated concentration profiles and bubble diameters for a total flow rate, Q_0, of $4\times Q_{mf}$ is given in Figure 2. In this particular case, the calculated conversions are 62.9% for the case without secondary injection and 68.7% for the case with $Q_s = 2\times Q_{mf}$ and β = 0.62. These results indicate that a strong relationship exists between the conversion and the bubble size in a fluidised bed reactor. It can be seen that, with secondary injection, the concentration in the bubble phase is lower, while the concentration in the emulsion phase is higher than in the normal bed. This change in concentration profiles indicates that mass transfer to the dense phase has increased, which results in a higher conversion for sufficiently high values of k.

The relationship between the decrease in bubble size (with increasing Q_s) and the relative increase in conversion (X_s/X_0) is confirmed when the case with β = 1 for all flow conditions is compared to the case with the experimentally determined β's (Figure 3). When β = 1 for all flow conditions and low to moderate values of k, the conversion changes very little. For high values of k the mass transfer is not fast enough to keep the catalyst supplied with reactant. The conversion decreases with increasing Q_s in this case because the reactant injected higher in the bed has less time to transfer to the emulsion phase. When the bubble size reduction is taken into

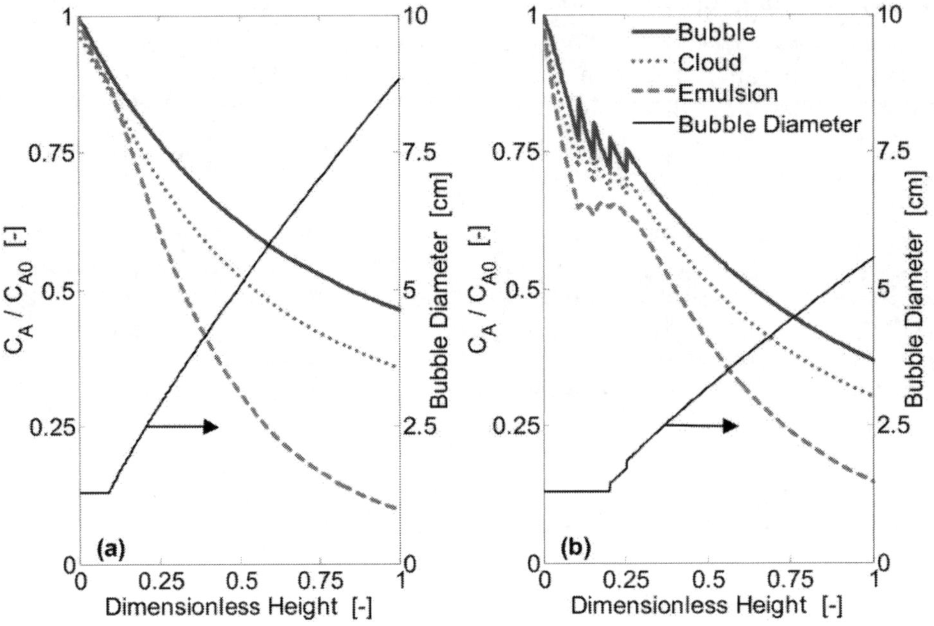

Figure 2: Concentration profiles and bubble diameters predicted by the modified K-L model with $Q_0 = 4\times Q_{mf}$, $k = 1$ $m^3_{gas}/(m^3_{solid}\cdot s)$, $d_{b,min} = 1.3$ cm; (a) no secondary injection; (b) $Q_s = 2.0\times Q_{mf}$, $\beta = 0.62$

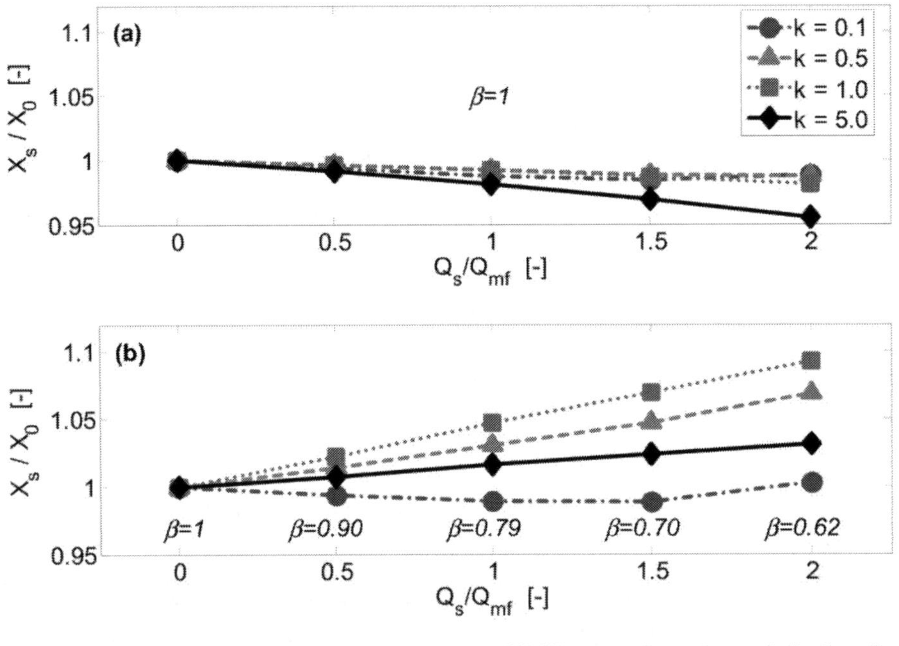

Figure 3: The relative change in conversion, X_s/X_0, as a function of Q_s for $Q_0 = 4\times Q_{mf}$ and various k; (a) with $\beta = 1$ for all flow conditions; (b) with β determined from experiments for each flow condition. k has units of $m^3_{gas}/(m^3_{solid}\cdot s)$.

account (β < 1) the conversion increases significantly (when k is greater than 0.1) with increasing Q_s and decreasing bubble size. The change in conversion is small when k is low because the reaction is kinetically limited; therefore an improvement in the mass transfer will have little effect. The increase in conversion when k is very high is also small because the absolute conversion at Q_s = 0 is already high, and is difficult to further increase. As a result, the biggest relative increases in conversion are seen at intermediate reaction rates. These results are only taking the reduction in bubble size into account but are neglecting the increase in emulsion phase flow that also occurs with secondary injection (2). Thus, an even higher conversion is expected in experiments.

EXPERIMENTAL

Experiments were conducted in a 20 cm wide × 1.5 cm deep quasi-2D column with a porous plate distributor. The lowest level of the fractal injector was at 6 cm and the highest at 14 cm. The injector configuration is as shown in Figure 1a. The bed mass consisted of 40 g of 1 wt% iron-impregnated Al_2O_3 particles diluted with 855 g of the same non-impregnated particles to give a settled bed height of 40 cm. The particle size was 250-300 μm and the particle density was approximately 1339 kg/m^3, which is in the A/B border region of the Geldart chart. Ozone was generated by an OAS Coolflow O_3 generator and mixed with air to produce the desired flow rate. The ozone concentrations were analyzed with an INUSA 2000 O_3 analyzer with a range of 0-100 ppm$_v$ and an accuracy of 0.1 ppm$_v$. The temperature of the feed gas and the cabinet in which the column was located were heated to 55°C. The relative humidity was monitored (but not regulated) and stayed within the range of 1-2% over the course of these experiments. A constant flow was drawn off the feed gas line to analyze the initial ozone concentration. By switching a set of valves the product concentration was sampled. Both sample lines had mass flow controllers to control the sample flow rates to 1.8 L/min. For the secondary gas injection experiments the feed gas was split into primary and secondary streams using needle valves and a calibrated rotameter to monitor the flow. The ozone concentration in the feed was in the range of 60-70 ppm$_v$, with the exact value depending on the back-pressure on the O_3 generator, which was typically less than 0.5 bar(g).

The method of determining bubble sizes from pressure fluctuations proposed by van der Schaaf *et al.* (10) and validated by Kleijn van Willigen *et al.* (11) was used to estimate the bubble size reduction caused by secondary injection. The standard deviation of the incoherent portion of a pressure fluctuation-time series measured in the bed at high frequency (in this case, 400 Hz) is directly related to the size of the bubbles passing the pressure sensor. Typically a calibration factor (usually obtained from video analysis) is required to determine an absolute bubble size with this technique, but for our purposes here, only the relative change in bubble size with respect to the case without secondary injection is required, so calibration was not needed. For more information regarding this technique, the reader is directed to (10) and (11). Kistler type 7261 piezoelectric pressure transducers and Kistler type 5011 amplifiers were used to measure the pressure fluctuations at three locations – the reference sensor located immediately above the distributor plate, and two others at 19 cm and 30 cm above the distributor. The pressure transducers were connected to sample ports in the side of the bed at these locations by 10 cm long ×

4 mm inner diameter copper tubes, well within the probe size recommendations proposed by van Ommen *et al.* (12).

The feed concentration and pressure fluctuations were measured simultaneously over a period of five minutes for each flow condition. The sampling valve was then switched, and the product concentration was measured simultaneously with the pressure fluctuations for an additional five minutes.

EXPERIMENTAL RESULTS AND DISCUSSION

The behaviour of β with increasing secondary injection for $Q_0 = 4\times Q_{mf}$ is shown in Figure 4 for two measurement positions. The 19 cm position is 5 cm above the topmost level of the fractal injector. It shows that β decreases significantly with increasing values of Q_s. The bubble size reduction at the 30 cm probe position, which is more than twice the height of the fractal injector, is still very large. The β's observed at 19 cm were used in the modelling discussed above.

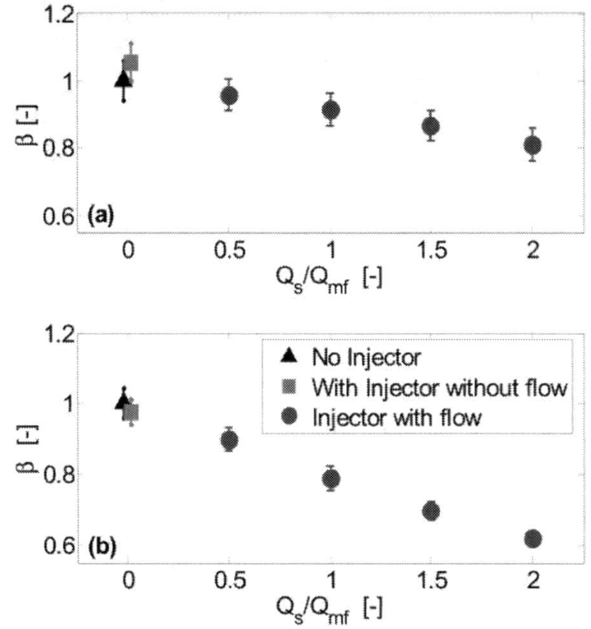

Figure 4: β as a function of Q_s at $Q_0 = 4\times Q_{mf}$; (a) measured at h = 30 cm; (b) measured at h = 19 cm. Error bars are 95% confidence intervals.

The conversion results are presented in Figure 5 for $Q_0 = 4\times Q_{mf}$. The error bars are the 95% confidence intervals. Although the uncertainty is quite large, there is a definite indication that conversion increases with increasing Q_s. This trend shows that secondary injection has a much greater effect than the presence of the internals. The relative increase in conversion is as high as 14%, which confirms that the conversion size depends on bubble size. These results need to be reproduced and the uncertainty in the data needs to be decreased. The fractal injector has not been optimized; future designs may yield even better results.

The model predictions are also presented in Figure 5. The value of k that gave the same conversion as the experimental result at Q_s = 0 was used (k = 0.29 $m^3_{gas}/(m^3_{solid}\cdot s)$). A large relative increase in conversion with decreasing bubble size was expected, given that this is an intermediate value of k. The model predicts higher conversions with increasing Q_s, but underestimates the absolute values and the rate of increase when compared to the experimental data. This is an indication that only taking the decrease in bubble size into account is not sufficient. Therefore the increase in emulsion phase flow caused by secondary injection should also be included in the model.

The benefit of secondary injection at an industrial scale is expected to be even larger than in this laboratory-scale reactor. The effect will be greater because the bubbles grow much larger, so the potential to increase the mass transfer between the phases is much higher.

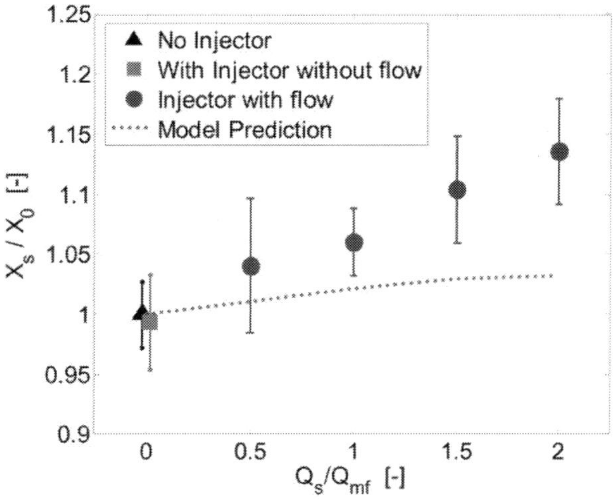

Figure 5: The relative change in conversion as a function of Q_s; $Q_0 = 4 \times Q_{mf}$. The model consistently underestimates the conversion. In the model, k = 0.29 $m^3_{gas}/(m^3_{catalyst} \cdot s)$. Error bars are 95% confidence intervals.

CONCLUSIONS

A modified version of the Kunii and Levenspiel model that incorporates emulsion phase flow and bubble growth has been used to predict the effect of secondary injection on the conversion in a bubbling fluidised bed reactor. The results indicate that the conversion increases because secondary injection reduces the bubble size, which results in an improved mass transfer. Experiments were also performed using ozone decomposition as a test reaction. These results also indicate that the conversion increases, while the bubble size decreases, with increasing secondary gas. The model predicts a positive trend in the conversion, but consistently underestimates the experimental values and rate of increase. The model should predict the data better by including an increase in the emulsion phase flow.

ACKNOWLEDGEMENTS

V. Twigt is gratefully acknowledged for his help with the experiments. We wish to thank the Dutch National Science Foundation, NWO, for financial support (Pionier/Jonge Chemici).

NOTATION

A_0	Area of distributor plate per orifice, [m^2]
$C_{A,i}$	Concentration of reactant in phase i, [mol/m^3]
C_{A0}	Initial concentration of reactant, [mol/m^3]
d_b	Bubble diameter, [cm]
$d_{b,min}$	Minimum bubble diameter allowed, [cm]
f_i	Solids volume fraction for phase i (bubble, cloud, or emulsion), [m^3_{solid}/m^3_{bed}]
g	Acceleration due to gravity, 9.8 [m/s^2]
h	Current height in the bed, [m]
k	First-order reaction rate constant, [$m^3_{gas}/(m^3_{solid} \cdot s)$]
K_{bc}	Interchange rate coefficient between bubble and cloud, [$m^3_{gas} / (m^3_{bubble} \cdot s)$]
K_{ce}	Interchange rate coefficient between cloud and emulsion, [$m^3_{gas} / (m^3_{bubble} \cdot s)$]

Q_{mf} Minimum fluidisation volumetric flow rate, [m^3/s]
Q_0 Total volumetric flow rate, [m^3/s]
Q_p Primary volumetric flow rate that enters the bed via the windbox, [m^3/s]
Q_s Secondary volumetric flow rate that enters via the fractal injector, [m^3/s]
U_b Bubble rise velocity in a bubbling fluidised bed, [m/s]
U_{mf} Minimum fluidisation velocity, [m/s]
X_0 Conversion of the reactant without secondary gas injection, [-]
X_s Conversion of the reactant with secondary gas injection, [-]

Greek

β Bubble diameter reduction factor, [-]
δ Bubble fraction, [m^3 of bubble / m^3 of bed]

REFERENCES

1. Y. Cheng, C.M. van den Bleek, M.-O. Coppens, "Hydrodynamics of Gas-Solid Fluidised Beds Using a Fractal Injector," Proceedings of the10th International Conference on Fluidisation, United Engineering Foundation, NY, 373–380 (2001).
2. F. Kleijn van Willigen, D. Christensen, J.R. van Ommen, M.-O. Coppens, "Imposing Dynamic Structures on Fluidised Beds," Cat. Today, Vol. 105, 560–568 (2005).
3. G. Sun, J.R. Grace, "The Effect of Particle Size Distribution on the Performance of a Catalytic Fluidised Bed Reactor," Chem. Eng. Sci., Vol. 45, 2187–2194 (1990).
4. C. Fryer, O.E. Potter, "Experimental Investigation of Models for Fluidised Bed Catalytic Reactors," AIChe J., Vol. 22, 38–47 (1976).
5. H. Bi, P. Jiang, R.-H. Jean, L.-S. Fan, "Coarse Particle Effects in a Multisolid Circulating Fluidised Bed Reactor For Catalytic Reactions," Chem. Eng. Sci., Vol. 47, 3113–3124 (1992).
6. D. Kunii, O. Levenspiel, "Fluidised Reactor Models. 1. For Bubbling Beds of Fine, Intermediate, and Large Particles. 2. For the Lean Phase: Freeboard and Fast Fluidisation," Ind. Eng. Chem. Res., Vol 29, 1226–1234 (1990).
7. D. Kunii, O. Levenspiel, "Fluidisation Engineering," 2nd Edition, Butterworth-Heinemann, Stoneham, Massachusetts (1991).
8. R.C. Darton, R.D. LaNauze, J.F. Davidson, D. Harrison, "Bubble Growth due to Coalescence in Fluidised Beds," Trans. I. Chem. Eng., Vol. 55, 274–280 (1977)
9. T. Baron, C.L. Briens, P. Galtier, M.A. Bergougnou, "Verification of Models and Correlations for Bubble Properties in Fluidised Beds," Chem. Eng. Sci., Vol. 45, 2227–2233 (1990).
10. J. van der Schaaf, J. C. Schouten, F. Johnsson, C.M. van den Bleek, "Non-Intrusive Determination of Bubble and Slug Length Scales in Fluidised Beds by Decomposition of the Power Spectral Density of Pressure Time Series," Int. J. Multiphas. Flow, Vol. 28, 865–880 (2002).
11. F. Kleijn van Willigen, J.R. van Ommen, J. van Turnhout, C.M. van den Bleek, "Bubble Size Reduction in a Fluidised Bed by Electric Fields," Int. J. Chem. React. Eng., Vol. 1, paper A21 (2003).
12. J.R. van Ommen, J.C. Schouten, M.L.M. vander Stappen, C.M. van den Bleek, "Response Characteristics of Probe-Transducer Systems for Pressure Measurements in Gas-Solid Fluidised Beds: How to Prevent Pitfalls in Dynamic Pressure Measurements," Powder Tech., Vol. 106, 199–218 (1999). Erratum: Powder Tech., Vol. 113, 217 (2000).

341b Dense Phase Discrete Particulate Flow Modeling

Srinivas Mohan, Applications, ANSYS-Fluent India Pvt. Ltd., Plot No. 34/1, Pune Infotech Park, Hinjewadi, Pune, 411057, India and Ahmad Haidari, Marketing, Ansys Inc, Fluent USA Inc, 10 Cavendish Court, Lebanon, NH 03766.

Modeling of multi-species, multi-sized particulates subject to fluidization is important in many technology applications such as coal and biomass combustion, gasification, granulation and many other processes. The modeling of these processes provides the engineer with the ability to predict and design equipment for optimum performance of these applications.

The model workhorse for multiphase flows involving particles is the Eulerian-granular model. The Euler-granular model has been used successfully to model various fluidized bed applications, it can prove costly when faced with a large size class of particles that have different hydrodynamic behavior. The Lagrangian modeling of particles does not suffer from this limitation.

In this work we describe how the DPM model in ANSYS Fluent, which is valid for low volume fractions of solids, has been extended to model dense flow. Validation of these model with experiments, as well as comparisons with the more established Euleriann-Granular models are also presented. The advantages and disadvantages of the modeling approaches are highlighted.

341c Downer-to-Riser Coupling Technique for Heavy Oil Catalytic Cracking

Changning Wu, **Yi Cheng**, and Yong Jin. Department of Chemical Engineering, Tsinghua University, Beijing, 100084, China

Fluid catalytic cracking (FCC) unit converts low value heavy hydrocarbons into a series of more valuable products such as gasoline and light olefinic compounds. Many recent developments in the FCC technique have focused on achieving cleaner gasoline with less content of olefins to meet the growing environmental concerns and maximizing some high-added value products such as ethylene and propylene. It is helpful to improve the product quality and product distribution by carrying out the main primary and secondary reactions in two different reaction zones with different hydrodynamics and process conditions. Consequently, the FCC reactions can be operated in coupled reactors, such as the riser-riser coupled reactor and the riser-downer coupled reactor, under a controllable and selective mode. A coupled high-density downer-to-riser reactor was proposed and demonstrated feasible operation at high flux conditions in a pilot-scale experimental setup. This coupled reactor aims to utilize both downer and riser advantages to improve the gasoline quality and efficiently control the distribution of FCC products. Downer accommodates high severity cracking reactions at the initial stage with the benefit of near plug flow operation; while riser is operated at a lower temperature with longer contacting time to promote hydrogen shift reactions. In this study, a 2D convection-diffusion-reaction model incorporated with hydrodynamics and a 14-lump kinetic model is established to predict the conversion and yields of different products in the coupled design. The kinetic model is based on a molecular description of cracking and hydrogen shift reactions. Promising results are achieved for the coupled downer-to-riser reactor compared with the conventional riser and other coupled reactors.

341d Experimental Investigation Of Rotating Fluidized Beds In A Static Geometry

Juray De Wilde[1], Ali Habibi[2], Guy B. Marin[3], Geraldine J. Heynderickx[4], Rahul Ekatpure[4], and Axel De Broqueville[5]. (1) Materials and Process Engineering Dept. (IMAP), Universite catholique de Louvain (UCL), Reaumur, Place Sainte Barbe, 2, Louvain-la-Neuve, B - 1348, Belgium, (2) Materials and Process Engineering, UCL, Place Sainte Barbe, 2, Louvain-la-Neuve, B - 1348, Belgium, (3) Laboratorium voor Petrochemische Techniek, Ghent University, Krijgslaan 281, S5, Gent, 9000, Belgium, (4) Laboratorium voor Petrochemische Techniek, Ghent University, Krijgslaan 281, S5, Gent, 9000, Belgium, (5) Grez-Doiceau, 1390, Belgium

Rotating fluidized beds in a static geometry are experimentally investigated. To introduce the rotating motion and the tangential fluidization of the particle bed, the fluidization gas is injected tangentially in the fluidization chamber via multiple gas inlet slots. The radial fluidization of the particle bed is introduced by forcing the gas to leave the fluidization chamber via a centrally positioned chimney. The fluidization behavior of both large diameter, low density polymer particles and small diameter, higher density salt particles is investigated at different solids loadings in a 24 cm diameter, 13.5 cm long non-optimized fluidization chamber. The behavior of the rotating fluidized bed is very different for the polymer particles and the salt particles. Whereas the polymer particles tend to form a dense and uniform bed, the salt particles tend to form a less dense and less uniform bed. Bubble formation is observed with the salt particles. A rotating fluidized bed and an acceptable gas-solid separation is obtained provided that the solids loading and the particle bed rotational speed (i.e. the gas flow rate) are sufficiently high. To allow a sufficiently high solids loading to build up, particles losses via the chimney have to be minimized. With both types of particles, slugging and channeling, that is, a non-uniform distribution of the gas over the gas inlet slots to the fluidization chamber may occur at too low rotational speeds or too low solids loadings. Static pressure profiles in the fluidization chamber are studied and shown to be a good indicator of channeling or slugging. To correctly balance the centrifugal force and the radial gas-solid drag force, an optimization of the fluidization chamber design and of the operating conditions for each given type of particles is shown to be necessary. Scale-up to a 36 cm diameter fluidization chamber is studied and scale-up rules are explained.

341e Modeling Of Three Phase Fluidization Using Cfd

Rajat Mittal, Applications, ANSYS-Fluent India Pvt. Ltd., Plot No, 34/1, Pune Infotech Park, Hinjewadi, Pune, 411057, India, **L. Srinivasa Mohan**, ANSYS Fluent India Pvt. Ltd., Plot no 34/1, Rajiv Gandhi Infotech Park, MIDC, Hinjewadi, Pune, India, and **Ahmad H. Haidari**, ANSYS Inc., 10 Cavendish Court, Lebanon, NH 03766.

Three phase fluidized bed reactors are a common occurrence in the process industries and are used in processes such as slurry reactors, and ebullent beds. In spite of the technological importance of these reactors our understanding of these reactor are limited due to the complicated fluid dynamics, and subsequently the heat and mass transfer characteristics of these rectors.

In this contribution we review recent efforts to understand three phase gas-liquid-solid fluidization from a fundamental viewpoint, and report on our efforts to model the fundamental problem of the motion of a bubble in slurries, and the solid phase mixing induced by the motion of the bubbles [1, 2]. By using the Eulerian multi fluid model in ANSYS CFX and ANSYS Fluent, we study the dependence of the bubble rise velocity on the solids concentration and the solid phase mixing induced by the motion of the bubble. We compare our results with the recent VOF-DPM studies reported in [2]. The implications of these results for reactor modeling, as well as closures for the coarser dispersed multi fluid models are also discussed.

References

[1]. Solid phase hydrodynamics of three-phase fluidized bed reactors – A convective/dispersive phenonmenon, International Journal of Chemical Reactor Engineering, Vol 5, Article A3, 2007.

[2]. Numerical simulation of gas-liquid-solid flows using combined front tracking and discrete particle method, Chemical Engineering Science

341f Dynamic Modeling Of Shallow Conical Spouted Beds

Jiandong Zhou[1], Duane D. Bruns[1], C. Stuart Daw[2], Charles E. A. Finney[3], and Sreekanth Pannala[4]. (1) Chemical Engineering, University of Tennessee, Knoxville, 419 Dougherty Hall, Knoxville, TN 37996, (2) Fuels, Engines, and Emissions Research Center, Oak Ridge National Laboratory, Oak Ridge, TN 37831-6472, (3) Engineering Science and Technology Division, Oak Ridge National Laboratory, Oak Ridge, TN 37831, (4) Computer Science and Mathematics Division, Oak Ridge National Laboratories, Oak Ridge, TN 37831

Spouted beds are widely used for gas-solids contacting in contexts where well-defined solids circulation is important. One particular application of growing interest is chemical vapor deposition (CVD) for particle coating. CVD of fuel particles is a critical fuel-preparation step for Advanced High-Temperature Gas-Cooled Nuclear Reactors. Nuclear fuel coaters typically involve CVD of U-containing kernels in shallow spouted bed. Controlling the detailed time-temperature-species history of particles is critical for meeting coating specifications Experimental measurement is greatly limited by operating environment, restricted physical access. Prior modeling has focused on prediction/correlation of time average bed properties (e.g., gas velocity profiles, spout height). Very little has been done to model solids and gas flow and mixing variations over time. Oak Ridge National Laboratory (ORNL) has developed detailed CFD models (based on MFIX) for spouted-bed CVD coating reactors. ORNL CFD models confirm the richness of spouted-bed dynamics (i.e., they are highly dynamic, not steady state) and capture the coater spouted-bed dynamics, but they are computationally intensive (e.g., several days to simulate 20s of dynamics).

Low-order dynamic models are needed to: i) rapidly sift through large sets of design and operating variations; ii) reduce complexity to understand fundamental patterns; iii)implement on-line model-based diagnostics and controls. This model simulates the dynamic variation of the gas inlet pressure across the shallow spouted bed since it is the most conveniently obtained variable during coating at high temperature. In the model, the spouted bed is divided into several different zones including entrainment zone where particles are picked up, spout zone where gas accelerates and carries particles from the bottom of the bed to the top solid surface, fountain zone and annular zone. In each zone, particle and gas behaviors are described based on mass balance and momentum balance. How dominant frequency of the gas pressure drop is analyzed, which helps to understand the frequency shift during coating process due to particle size and property changes and operation condition variation. It could be used to track and control the on-line coating process.

354a Mechanochemical Synthesis Of Blue Luminescent Functionalized Silicon Nanoparticles

Andrew S. Heintz, Chemical and Biomolecular Engineering, Tulane University, 300 Lindy Boggs, 6823 St. Charles Ave, New Orleans, LA 70118, Mark J. Fink, Chemistry, Tulane University, New Orleans, LA 70118, and Brian S. Mitchell, Chemical & Biomolecular Engineering, Tulane University, 300 Boggs Building, New Orleans, LA 70118.

A mechanochemical method for the synthesis of alkyl passivated blue photoluminescent silicon nanoparticles using high energy ball milling (HEBM) is described. The method provides for the simultaneous production of photoluminescent silicon nanoparticles and the passivation of the particle surface with alkyl groups covalently linked through Si-C bonds. As fresh silicon surface is formed during HEBM by particle fracture, the surface Si atoms react in-situ with liquid alkyls, such as 1-octyne and 1-hexyne. Functionalized silicon nanoparticles with physical dimensions less than the bulk Bohr radius of silicon display intense photoluminescence due to quantum size effects, and have exhibited a quantum yield of 0.60. The fluorescent nanoparticles are then characterized using spectroscopic techniques such as photoluminescence, NMR, and FTIR. The effects of mechanochemical processing parameters such as milling time and reactive media on nanoparticles properties and surface functionalization are described.

354b Molten Salt Synthesis And Luminescence Properties Of Rare-Earth Doped Yttrium Oxide And Rare-Earth Zirconates

Yuanbing Mao and Jane P. Chang. Department of Chemical and Biomolecular Engineering, University of California at Los Angeles, Los Angeles, CA 90095

A molten salt synthetic method is described in this work for synthesizing rare-earth doped yttrium oxide (RE:Y_2O_3) and rare-earth zirconates ($RE_2Zr_2O_7$) since it is one of the simplest, most versatile, and cost-effective approaches available for obtaining crystalline, chemically purified, single-phase powders at lower temperatures and often in overall shorter reaction times with little residual impurities as compared with conventional solid-state reactions. RE:Y_2O_3 are promising rare-earth-ion-activated phosphor materials with wide applications including amplifiers, lasers, waveguides, X-ray imaging, bioimaging, and field emission or electroluminescent displays due to their luminescent characteristics and stability in high vacuum. $RE_2Zr_2O_7$ are of great interest in recent years for their possible applications as high-temperature heating elements, oxidation catalysts, thermal barrier coatings, nuclear waste hosts, and host materials for luminescence centers. The as-prepared nanoparticles have sizes around 100 nm for RE: Y_2O_3 and around 30 nm for $RE_2Zr_2O_7$. They are thoroughly characterized by X-ray diffraction (XRD), scanning electron microscopy (SEM), transmission electron microsecopy (TEM), high resolution TEM (HRTEM), selected area electron diffraction (SAED), energy-dispersive X-ray spectroscopy (EDS), and UV-vis spectroscopy to delineate the structure and composition of these nanoparticles. Their room temperature luminescent properties (photoluminescence and catholuminescence) are measured as a function of the particle size and doping level. In order to process these materials for device fabrication, they are made water soluble by surface modification of these nanoparticles with DMSA (2,3-dimercaptosuccinic acid). We will discuss the effect of these nano-coatings on the optical properties of the nanoparticles.

354c Bifunctional Quantum Dot-Iron Oxide Nanoparticle Composites

Shuang Deng, Chemical and Biomolecular Engineering, Ohio State University, 140 W 19th Ave, Columbus, OH 43210 and Jessica O. Winter, Chemical and Biomolecular Engineering/Biomedical Engineering, Ohio State University, 140 W 19th Ave, Columbus, OH 43210.

Since their introduction, nanoparticles have made significant contributions to the field of biological imaging. However, nanoparticles also have interesting electromagnetic properties that may be exploited to directly manipulate cells. Nanocomposite materials containing both functionalities will provide enhanced functionality in contrast with their more limited single-component counterparts. We are investigating one class of nanocomposites, fluorescent-magnetic particles, which may allow for direct mechanical manipulation and simultaneous observation of subcellular features. These composites are composed of iron oxide and quantum dot materials. Quantum dots have been extensively studied for their unique optical and electronic prosperities, sharp emission band with broad excitation, and strong resistance to photobleaching. They are of great interest for their dimensional similarities to biomarcromolecules (e. g., DNA and protein), applications in biological imaging, and bioconjugation. Superparamagnetic iron oxide nanoparticles (SPIONs) have also been studied for biomedical applications, including MRI contrast enhancement, magnetic immobilization, and drug targeting. Here, we present synthesis and characterization of a new bifunctional nanocomposite: CdTe-Fe3O4@SiO2. CdTe is an ideal imaging material as the emission can be tuned to the near infrared where optical tissue absorption is at a minimum, whereas the silica shell can prevent the photooxidation and subsequent elution of toxic Cd2+ and biodegradation of iron oxide magnetic nanoparticles. Additionally, silica surfaces provide a mechanism for facile conjugation to targeting biomolecules (e.g., proteins). To create these composites, a two-step synthesis was employed. First, CdTe QDs were conjugated to magnetic Fe3O4 nanoparticles using a one-pot chemical method. Second, CdTe-Fe3O4 conjugates were modified with an outer silica shell. Particles were characterized using TEM, XRD, UV and fluorescence spectroscopy, and superconducting quantum interference device (SQUID). Result indicated that nanocomposites exhibit magnetic properties and fluorescence. These nanocomposites have potential in combined MRI/fluorescence imaging, diagnostics, circulating drug-delivery systems, and investigation of mechanotransduction in biological systems.

Web Page: www.chbmeng.ohio-state.edu/~winter/

354d Passivation Of Pigment-Grade TiO_2 By SiO_2 Atomic Layer Deposition In A Fluidized Bed Reactor

David M. King[1], Xinhua Liang[1], Kamal M. Akhtar[2], and Alan W. Weimer[1]. (1) Department of Chemical and Biological Engineering, University of Colorado, 1111 Engineering Dr., Campus Box 424, Boulder, CO 80309, (2) Research Center Baltimore, Millennium Chemicals Inc., 6752 Baymeadow Drive, Glen Burie, MD 21060

The pigment industry is a ~ $5 billion/year industry, of which a significant portion of process cost is to effectively passivate the highly photoactive and acid degradable TiO_2 substrate. Atomic layer deposition (ALD) of nanothick SiO_2 films is utilized here for the passivation of TiO_2 nanoparticle pigments. Conformal, pinhole-free films have been successfully deposited in a scalable fluidized bed reactor. Both anatase and rutile substrates were coated for this work. Films as thin as 2 nm, as verified by HRTEM, eliminated 98% of the photoactivity as compared to the uncoated substrate. H_2SO_4 studies were performed using ICP-AES to monitor the dissolution rate of the TiO_2 core. The 2 nm SiO2 films prevented > 99% of the TiO_2 cores from dissolving, which also verifies the conformal, pinhole-free nature of these films. In situ mass spectrometry was used to monitor reaction progress and to prevent excess precursor usage and process time. This residual gas analysis technique is advantageous to minimize process costs and is applicable to fluidized bed or similar flow-type reactor configurations at any scale. SiO_2 ALD films are very efficient and can be further optimized to meet the needs of all aspects of the pigment industry. Surface passivation can be extremely economical in large-scale batch processing using the process control measures presented in this work.

354e Polymer-Induced Single-Step Synthesis And Stabilization Of Noble Metal Nanoparticles In Water

Rajesh Sardar, Jong-Won Park, and Jennifer S. Shumaker-Parry. Department of Chemistry, University of Utah, 315 S 1400 E, Rm 2020, Salt Lake City, UT 84112

Metal nanoparticles, particularly gold, have received wide attention because of the unique optical properties of these particles for applications in photonics, detection systems, device fabrication, and catalysis. For this reason, different synthetic methods have been developed to generate monodisperse gold nanoparticles (AuNPs) either in aqueous or organic media. One of the most widely-used methods is the reduction of tetrachloroaurate ions (AuCl4-) in aqueous medium using sodium citrate.1 However, this method lacks good control over the synthesized particle size and the size dispersion is rather poor. The Brust method and various modifications are useful for the generation of AuNPs having core sizes ranging from 1 to 4 nm.2 In the Brust method, the transfer of AuCl4- into toluene or chloroform is performed using tetraalkylammonium bromide followed by reduction with sodium borohydride in the presence of alkylthiols. In addition to all of the above-mentioned methods, reduction of gold salt to AuNPs by amine-containing organic molecules also has been investigated and successfully accomplished.3 However, most of the amine compounds used in the AuNP synthesis are only soluble in organic solvents. As a result the reduction reaction must be performed in organic solvents, or in a biphasic system. Consequently, the applications of nanoparticles in biological system are unfavorable because the biomolecules are unstable in organic medium. We report a simple, inexpensive, single-step synthesis of gold and silver nanoparticles using poly(allylamine) (PAAm) as a reducing and stabilizing agent. The synthesis was carried out in aqueous solution, making the method versatile and environmentally friendly. The synthesized polymer-stabilized nanoparticles showed exceptional stability in water for months. The polymer-stabilized gold nanoparticles were allowed to undergo ligand exchange with a variety of ω-functionalized alkylthiols including acid-, alcohol-, amine-, and biotin-terminated thiols. The methodology also is applicable for the generation of silver nanoparticles.

1. Turkevich, J.; Stevenson, P. C.; Hillier, J. Discuss. Faraday. Soc. 1951, 55-75. 2. (a) Brust, M.; Walker , M.; Bethell, D.; Schiffrin, D. J.; Whyman, R. J. Chem. Soc., Chem. Commun. 1994, 801-802. (b) Collier, C. P.; Saykally, R. J.; Shiang, J. J.; Henrichs, S. E.; Heath, J. R. Science 1997, 277, 1978-1981. 3. (a) Selvakannan, P. R.; Mandal, S.; Pasricha, R.; Adyanthaya, S. D.; Sastry, M. Chem. Commun. 2002, 1334-1335. (b) Kumar, P. S.; More, A. S.; Shingte, R. D.; Wadgaonkar, P. P.; Sastry, M. Adv. Mater. 2004, 16, 966-971. (c) Hiramatsu, H.; Osterloh, F. E. Chem. Mater. 2004, 16, 2509-2511.

354f Supercritical Hydrothermal Synthesis Of Organic Inorganic Hybrid Nanoparticles And Perfect Dispersion

Tadafumi Adschiri, Institute of Multidisciplinary Research for Advanced Materials, Tohoku University, Katahira 2-1-1, Aoba-ku, Sendai, 980-8577, Japan

We propose a new method to synthesize organic-inorganic hybrid nanoparticles at supercritical hydrothermal conditions. By introducing organic species (aminoacids, carboxylic acids, amines, alcohols, aldehydes etc.) during supercritical hydrothermal synthesis, nanoparticles whose surface was modified with organic molecules were synthesized. This is due to the homogeneous phase formation for organic substance and aqueous solution at supercritical conditions. Particle size was in the range from 2.5 nm to 10 nm, and particle size dispersion is extremely narrow. Crystal shape can be controlled to be sphere, nano-cube, nano-ribbon etc. By selecting a proper modifier, particles could be dispersed perfectly in organic solvents or in aqueous solutions. By drying the colloid, we obtained self-assembly structure of nanocrystals. This implies a variety of applications of the nanoparticels including nanohybrid polymers, nano-ink, nano-paints, and bio-imaging probe.

Generalizing the Relationship between Inter-Particle Forces and Bulk Unconfined Yield Strength of Poly-Disperse Powders

By Kerry Johanson: Material Flow Solutions Inc. MatFlowSol@bellsouth.net

1.0 Introduction

Inter-particle forces are the foundation of bulk cohesive flow properties such as the unconfined yield strength. From a continuum point of view, unconfined yield strength is the major principal stress acting on a material that will cause that material to fail or yield in shear when it is in an unconfined state. It is the primary flow property that governs the development of hang-ups in process equipment. Yield strength is used to describe conditions where failure of free surfaces controls process behavior. All hang-ups in process equipment result in the formation of a free surface. By definition, the stress acting normal to any free surface is zero. However, stresses acting along the free surface may not be zero (see Figure 1). In a hang-up condition, the material on a free surface is supported by stresses that act along the free surface and are equal to the unconfined yield strength of the material. Both arching and rathole phenomena are examples of this type of free surface stress condition. However, free surfaces also occur at the surface of piles and agglomerates. Thus, unconfined yield strength is a key property in predicting arching, ratholing, pile formation, and agglomeration tendencies.

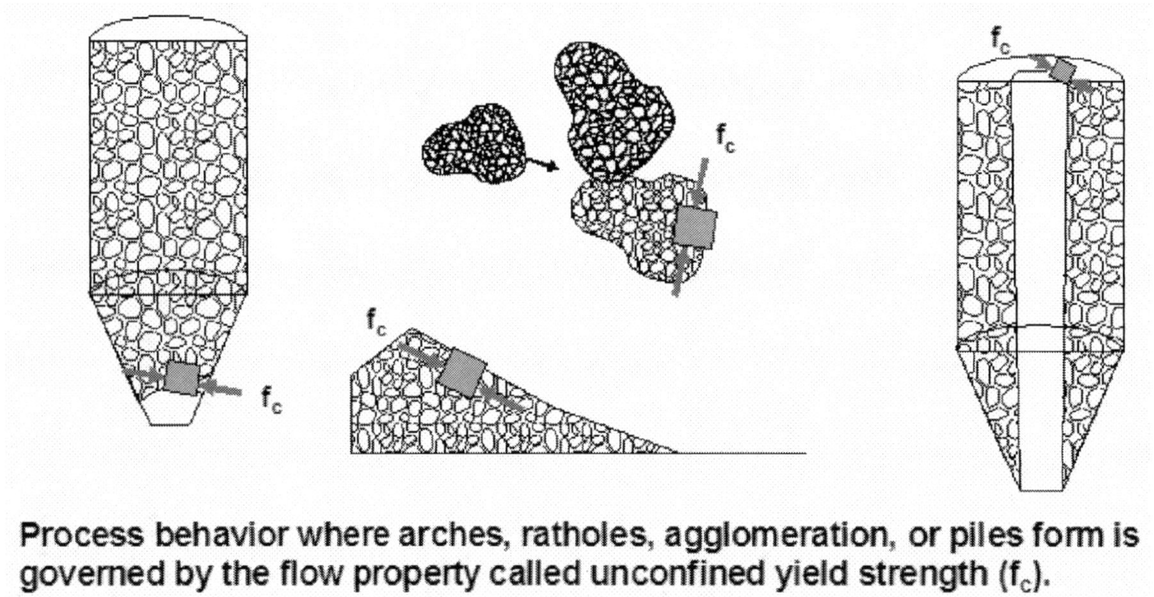

Figure 1. Process applications where bulk strength is important.

One can also look at unconfined yield strength from a particle assembly point of view. During a shearing event, particles move relative to one another. Some particles pull away from each other and some slide over adjacent particles. Thus, the bulk strength of an assembly of particles is related to the work required to pull adhering particles apart plus the work required to slide adjacent particles relative to each other. If one could determine the particles that were in the process of shear, then one could add up the incremental effect of each shearing particle or particle group and develop a method of estimating the bulk strength from a measure of inter-particle forces. The advantage of using this methodology over direct continuum measurement is that it allows for the incorporation of mechanistic models to predict bulk yield strength and, thus, allows engineers to answer questions regarding why a particular material possesses strength and what can be done to change the particle system to create a material with a prescribed strength. In the case of arching and ratholing, we want to design the material with the smallest strength. In the case of agglomeration, we need to find the material with the greatest yield strength. In the case of pile formation, we may require a median yield strength to optimize a particular process task such as mixing or blending.

Past researchers have developed relationships between yield strength and particle size based on the mode and mechanism of failure for mono-dispersed systems. The following Table summarizes some of these relationships.

Table I – Strength Models

Model	Mechanism	Source
$fc = \frac{K_1}{Dp^2}$	Van der Waals forces	Tardos [4]
$fc = \frac{K_2 \cdot \sqrt{C}}{Dp}$	Capillary forces	Rabinovich [2]
$fc = \frac{K_3}{Dp^{1/2}}$	Elastic fracture	Rumpf [1]
$fc = \frac{K_4}{Dp^n}$	Plastic-elastic fracture	Specht [3]

2.0 Model Development

It is important to stress that these relationships are only for systems where the particles have one unique size. Extending these models to poly-disperse conditions is a formidable task. This paper provides some simple scale laws based on the structure of the material to accomplish this task. Equation 1 provides a beginning framework to describe the additive effects of a system that exhibits the influence of multiple cohesive forces models.

$$fc = \frac{K_1 \cdot \sqrt{C}}{D_p} \cdot X_{capillary} + \frac{K_2}{D_p^{\,2}} \cdot X_{van} + \frac{K_3}{D_p^{\,n}} \cdot X_{fracture} \qquad (1)$$

Poly-disperse systems have the following behaviors that make direct extension of mono-dispersed equations difficult to use for poly-disperse prediction:

- Not all particles in the shear plane are in the act of shearing during a bulk shearing event.
- Moisture partition in the system is not uniform even in mono-dispersed systems.
- Shear planes may pass through soft agglomerates, thereby increasing bulk yield strengths.
- Shear planes may pass around agglomerates in the system, thereby effectively increasing particle size and reducing bulk strength.
- Soft agglomerates may be induced by shearing events.

However, one could break the particle assembly system into discrete volumes that characterize the structure of the particle bulk assembly. Each volume would contain a set of particles that, taken together, characterize a zone of the bulk assembly of particles (i.e. the smallest possible volume element exhibiting the same shear behavior). Simple poly-disperse laws can be made for each discrete structure and then combined to determine an overall strength relationship based on the statistical occurrence of each unique structure in the complete bulk assembly. Consider the case of this bimodal system of spherical particles. If capillary forces are the only forces acting on the system, then the following simple strength model can be used to estimate the strength of round particles of a given size. In case 1 the particles are all coarse and the strength is given by the equation in Figure 2. We used glass beads with a particle size of 468 micron as the coarse component for this research. The simple capillary model suggests that, the larger the particle size, the smaller the yield strength. Thus, coarse particles exhibited very little strength.

Case 1

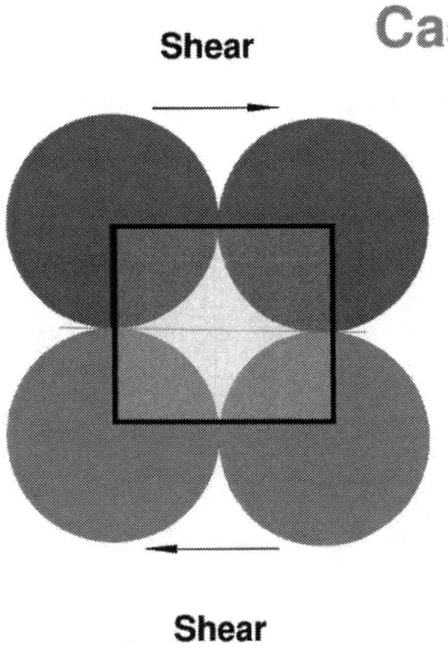

There are no small particles in the system. Shear takes place between adjacent large particles. The strength of this structure is found from the strength values for mono-dispersed systems for large particles.

$$fc = \frac{K}{Dp_A}$$

Figure 2. Estimated strength for just coarse particles

Adding fine particles to this coarse structure causes the voids between the coarse particles to fill with fine particles. In this case, a shear zone forming between two coarse particles will also shear though a section of fine particles within the bulk solid's pores. However, it should be pointed out that a portion of finer particles will not undergo a shear action, but are carried along for the ride within the pore structure. We used a discrete element analysis to determine the fraction of particles (fr_{void}) that occupied the shear plane. The total strength in this case will be due to breaking bonds between the coarse particle matrix as well as the fraction of fine particles in the shear plane between these coarse particles.

Case 2

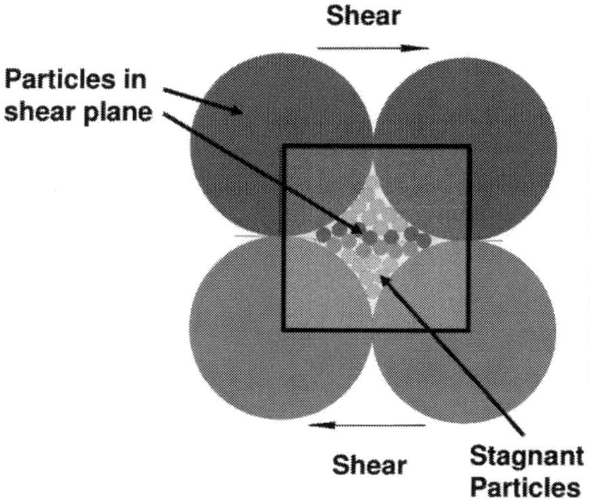

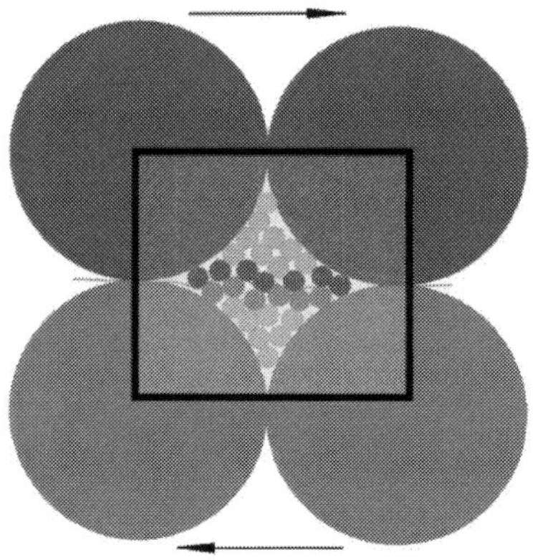

$$fc = \left[\frac{K}{Dp_B} \cdot \left(\frac{4}{\pi} - 1 \right) \cdot \left(\frac{Dp_A}{Dp_B} \right)^2 \cdot PF_B \cdot fr_{void} \right] + \frac{K}{Dp_A}$$

Number of small particle contacts per unit area of shear. Shear area defined by the distance between large particles.

Dp_A is the particle size of the large particles
Dp_B is the particle size of the small particles
PF_B is the packing factor of the small particle matrix
fr_{void} is the fraction of total small particles that are involved in shear
K is a function of surface tension and contact angle and moisture

Figure 3. Estimated strength for case where fine particles fill the coarse pore structure.

The final case occurs when the number of fine particles causes a separation of the coarse matrix. In this case, the unconfined yield strength approaches the strength of just the fine particles in the mixture.

Case 3

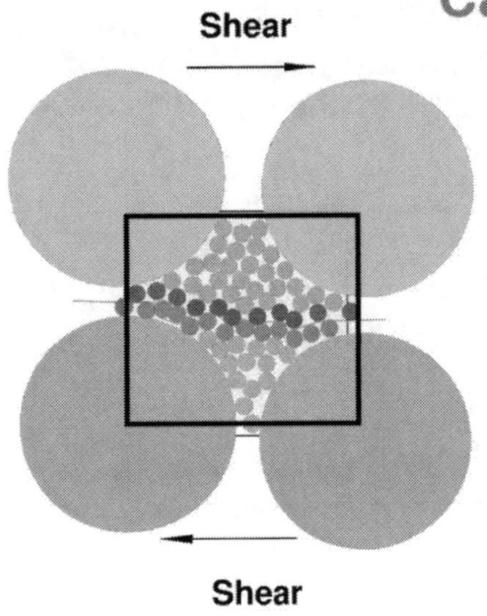

There are an excess of small particles in the system. Shear takes place between adjacent small particles. The strength of this structure is found from the strength values for mono-dispersed systems for small particles.

$$fc = \frac{K}{Dp_B}$$

Figure 4. ***Estimated strength for case where fine particles separate the coarse particles. Strength approaches that of just fine particles.***

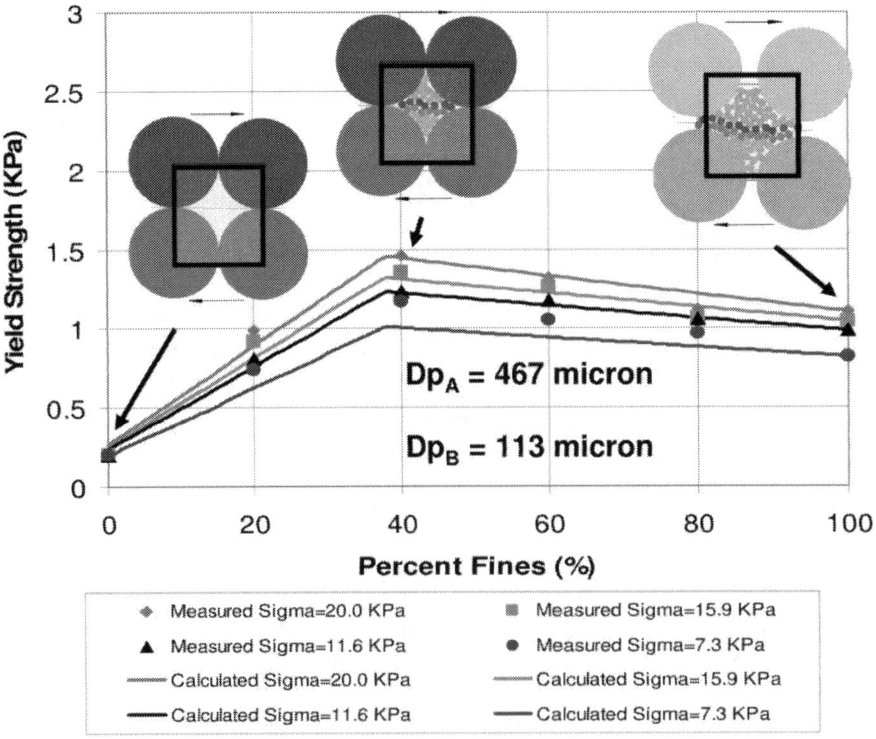

Figure 5. Calculated and estimated strength values for a mixture of two glass spheres of different size 467 micron and 113 micron.

$$fc = \frac{K}{Dp_B} \cdot X_1 + \frac{K}{Dp_A} \cdot X_2 + \left[\left[\frac{K}{Dp_B} \cdot \left(\frac{4}{\pi} - 1 \right) \cdot \left(\frac{Dp_A}{Dp_B} \right)^2 \cdot PF_B \cdot fr_{viod} \right] + \frac{K}{Dp_A} \right] \cdot X_3 \qquad (2)$$

We measured the strength of several bimodal mixtures of different particle sizes of glass beads. These mixtures show that maximum strength occurs when the voids are filled with fine particles (see Figure 5). The maximum strength obtained depends on the particle size of the fine component. The finer the components are, the greater the yield strength.

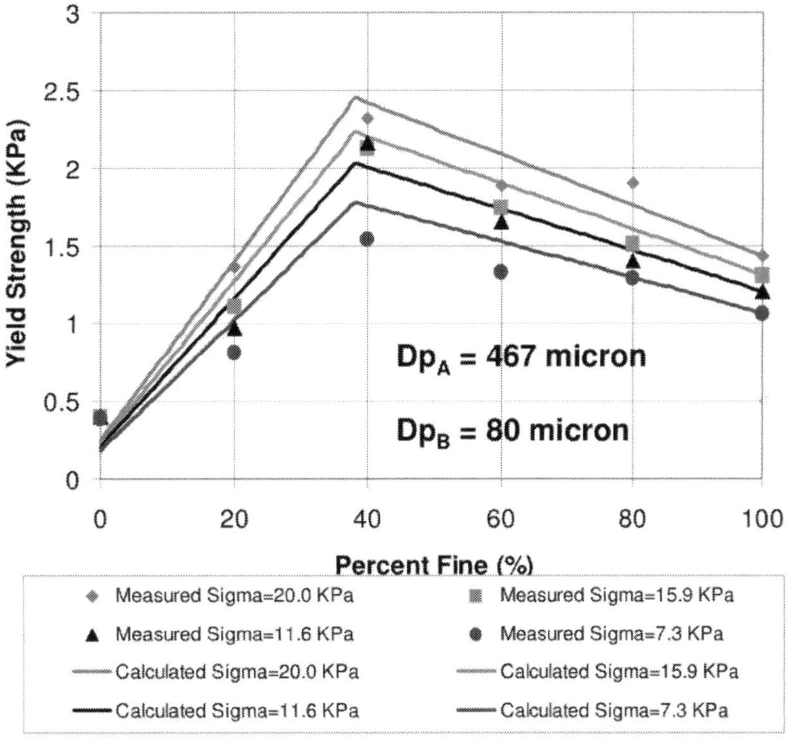

Figure 6. Calculated and estimated strength values for a mixture of two glass spheres of different size 467 micron and 80 micron.

3.0 Conclusions

Understanding the structure of the bulk assembly plays a primary role in developing models to predict the strength of bulk materials. One must understand the mechanisms acting within the bulk solid and the how the particle structure reacts to shear. Some particles are carried in solids voids and simply go along for the ride during a shear event. Future model refinement must include both mechanistic and structural effects to accurately predict real strengths of actual materials.

4.0 Nomenclature

C	Moisture content
Dp_A	Coarse particle size
Dp_B	Fine particle size
f_c	Unconfined yield strength
fr_{void}	Fraction of fine particles in shear zone
K	Capillary strength constant
K_1	General capillary strength constant
K_2	Van der Waals strength constant
K_3	Fracture strength constant
PF_B	Packing factor of fines
X_1	Fraction of unit volumes that are fine particles
X_2	Fraction of unit volumes that are coarse particles
X_3	Fraction of unit volumes that have fines in course voids
$X_{fracture}$	Fraction of material that fractures during shear
X_{van}	Fraction of material subject to Van der Waals forces during shear
X_{cap}	Fraction of material subject to capillary forces during shear

5.0 Acknowledgements

I would like to acknowledge the financial support of the Particle Engineering Research Center (PERC) at the University of Florida, The National Science Foundation (NSF Grant EEC-94-02989), Material Flow Solutions Inc. and the Industrial Partners of the PERC for support of this research.

6.0 References

1 H. Rumpf: *Particle Technology*. New York: Chapman and Hall, 1990.

2 Oil Mediated Particulate Adhesion And Mechanical Properties Of Powder Y. Rabinovich, M. S. Esayanur, K. Johanson, B. Moudgil, Proceedings World Congress on Particle Technology: paper 370. 2002

3 Caking of Granular Materials: An Experimental and Theoretical Study, Thesis, University of Florida, Dauntel Specht, 2006

4 Forces generated in solidifying liquid bridges between two small particles, Gabriel Tados, Rajan Gupta, Powder Technology 87 pg 175-180, 1996

FE prediction of particulate solids flow from silos fitted with a double cone insert

M. Jecmenica[1], S. Ding[2] and G. G. Enstad [1,2]
1. Telemark University College, P. O. Box 203, 3914 Porsgrunn, Norway
2. POSTEC dept. av TEL-TEK R&D Centre, 3917 Porsgrunn, Norway

Introduction

Investigations into the discharge of particulate solids from silos dated back to a time at the turning of 19th century (Janssen, 1895). However, the progress of knowledge has been rather difficult ever since; major progress in terminology and definition did not appear to have occurred until the work of Jenike and his colleagues (1961). Much progress in silo conceptual development may be credited to Jenike (1964); notable amongst these are the criteria for mass and funnel flow, which are widely used in silo functional design (Jenike, 1964, 1987; Drescher, 1991, 1992).

When mass flow is preferred, one usually designs a silo with a fairly steep hopper, or otherwise uses flow promoting devices such as inserts to obtain mass flow (Johanson , 1967; Enstad, 1998). A prediction of whether such inserts can convert a funnel flow silo into a mass flow silo is therefore desirable.

Decisions on whether inserts are suitable for a given application can be however fairly difficult. So far the theoretical methods to describe silo flow patterns are only suitable under certain circumstances. When the geometry of the silo becomes complex, for example, with an insert inside a silo, there is scarcely any analytical method available. Numerical modelling approaches are becoming powerful tools in the efforts recently in searching for solutions (Cundell, et al., 1979; Haussler, and Eibl, 1984; Rotter, 1998).

This paper presents the results of some numerical investigations of the flow patterns and velocity distributions in a silo, and the effects of a double cone insert on them with an in-house code developed.

Basic Feature in the Program

Collaborations were initiated by POSTEC to develop, based on the finite element method, a computational program capable of predicting flow behaviour around inserts (Karlsson et al., 1998, 1996; Karlsson, 1996). The program was written in FORTRAN with an application of Matlab for post processing.

In the program, the particulate solids were treated as continuum media. Assumptions needed to take advantage of a steady state silo discharge were also made. It greatly simplified the numerical treatment involved. Since the present work focuses on the investigations of solids flow around an insert, this steady state condition is acceptable. Under such an assumed discharging state, it is possible to use an Eulerian approach to set up a model to describe effects of the inserts on the particulate solids flow.

On the principle of plasticity, the process of a silo discharge was further approximated as a plastic flow. An elasto-plastic continuum model, using the Mohr-Coulomb yield condition, was adopted. Admittedly, it may be a very coarse approximation, but it has not yet been shown that a more complicated model is required.

The geometry of the silo under consideration was axi-symmetrical, 1970 mm in height and 750 mm in diameter, with a hopper of 470 mm in height and a 34.7° cone half angle. The outlet was 100 mm in diameter. This cone angle is relatively steep, so it can be expected to be not too far from producing mass flow conditions for some solids. The insert was an axi-symmetrical double cone; and was, aligning with the axis of the silo, fitted into the silo as shown in Fig. 1.

A second order triangular element was used to form the element domain. Meshes were first defined in a region 500 mm below the top of the silo. This region represented the material region confined by the silo and inserts. The governing equations and their boundary conditions were formulated and implemented in this domain accordingly. The reader is referred to Ding (2004) for the specific descriptions concerning the implementation and programming.

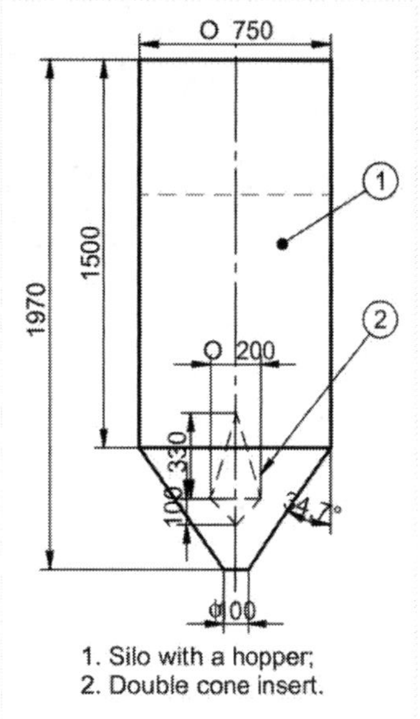

Fig. 1 An configuration of an insert within the model silo

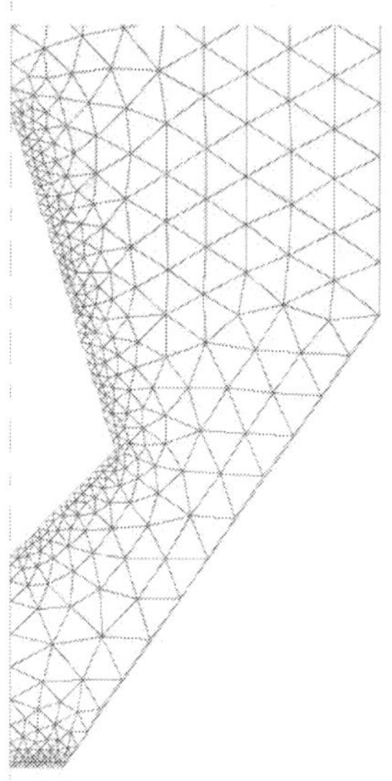

Fig. 2 A second-order triangular elements with finer meshes

The choice of values for the material parameters relates to the material model. The values of the parameters of shear viscosity (μ) and bulk viscosity (κ) as required in the model were rather arbitrarily set as $\mu = 1\times10^{5}$ Pa and $\kappa = 9\times10^{5}$ Pa respectively, though no tests are known from which they have been deduced. The other parameters were the bulk density ρ, the internal friction angle φ and the wall friction angle ϕ_w; they were all measurable, and were chosen as $\rho = 606$ kg/ m^3, $\varphi = 31.9°$ and $\phi_w = 21.8°$ in the implementation of model.

Numerical Results

Preliminary simulations were first carried out on the silo itself only. The program was created with an introduction of tracer particles in the continuum. Such tracer particles were seeded into the mesh to visualise the numerical results as paths and locations of material movement. Using this technique, the results are given in the form of tracer particle movements as shown in Fig. 3 a.

As seen, the dotted lines represent the original positions of the tracers assigned at three chosen levels, and the asterisks, also forming three lines accordingly, represent the tracer locations in the last step of the simulation. It is evident that the particles near the centre and close to the wall at the upper level moved down, but the particles close to the wall at the transition level did not.

The velocity distributions in Fig. 3 b show further that the material in the centre region moves faster than that close to the wall region, where no material movement occurs around the transition level. Obviously, a stagnant zone developed around the transition area in the silo, especially in the hopper.

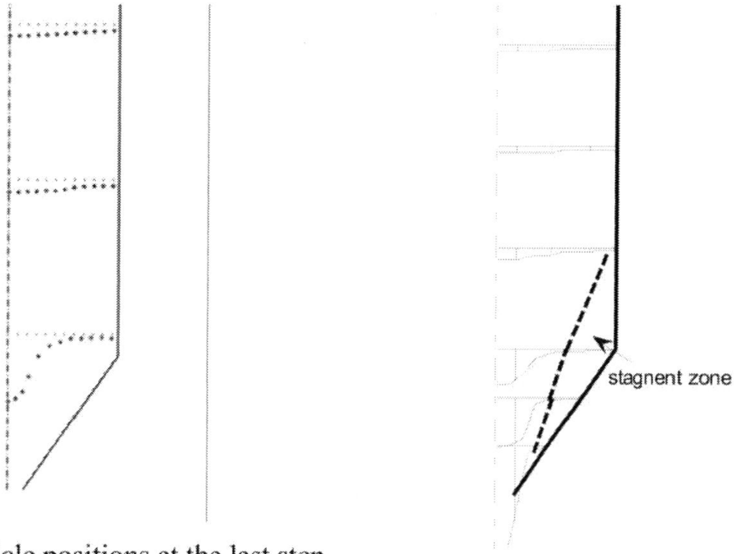

(a) Tracer particle positions at the last step of simulation

(b) Velocity distributions

Fig. 3 Funnel flow silo and its stagnant zone

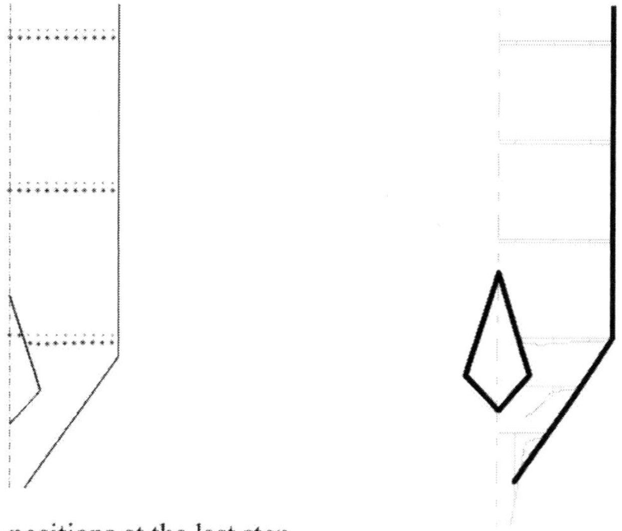

(a) Tracer particle positions at the last step of simulation

(b) Velocity distributions

Fig. 4 Influence of the double cone insert on flow pattern

In this funnel flow silo, the double cone insert was modelled within the silo. The lower tip of the cone was 120 mm above the outlet. Finer meshes were designed in the region around the inserts and the outlet. An example was given in Fig. 2 as a configuration of the double cone and silo.

Simulations of the effects of the double cone insert on the flow were carried out. They are shown in Fig. 4 in the form of tracer locations and velocity distributions, similar to those in as Fig. 3. One can see that the double cone insert overall improved the evenness of the flow pattern. They caused the solids to be discharged quite evenly in the cylinder section, and made the solids in the hopper move in a manner similar to a mass flow.

In the next step, the silo to be investigated is getting to be rather shallow with a cone angle of a 45° to the vertical. The other dimensions remained the same. The modelling of the flow was first attempted again without the insert; the result is shown in Fig. 5 a, where the asterisks represent the tracer particles in the last step of the simulation. As shown, the particles near the centre and close to the wall at the upper level moved down, but the particles close to the wall at the transition level did not. The velocity distributions in Fig. 5 b also show that the material in the central region moved faster than that close to the wall region, where no material movement occurred around the transition level. Clearly, a larger zone of stationary solids (compared with that as in Fig. 3) developed around the transition area in the silo, especially in the hopper.

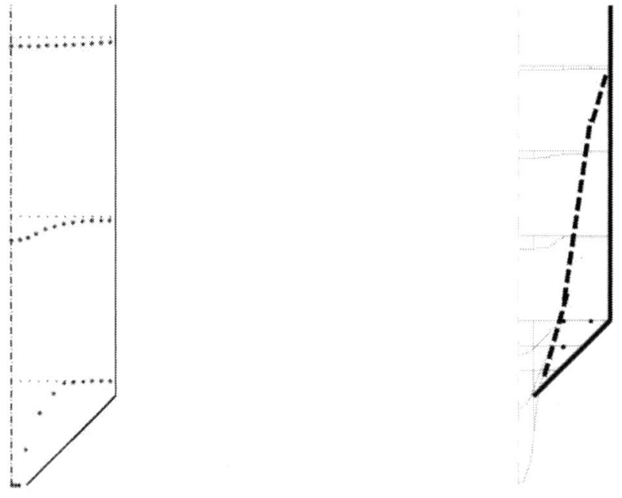

(a) Tracer particle position at the last step of simulation.

(b) Velocity distributions

Fig. 5 Development of a channel flow in a silo with a 45° hopper

The same insert was fitted into the silo, aligning its axis and that of silo; the base of the double cone insert was adjusted to the level of transition of the silo. Keeping the other conditions exactly the same as those in the previous calculations, simulations were carried out and the results are shown in Fig. 6

The results show that the installation of the double cone insert at this level improved the material flow pattern, with a very even movement of solid in the whole cylinder section. But below the transition level, the discharge velocity distribution was very uneven, with solids almost stagnant close to the hopper wall. To reduce the unevenness of flow pattern in the hopper, further efforts were made to focus on this portion of the silo.

It was presumably understood by the inventors of such insert that the length of the lower part of the cone may play an important role in the flow pattern. Modification was then carried out by extending the lower tip of the double cone to the outlet of the silo, while keeping the widest part of the cone at the same level, and its diameter unchanged.

The results show that the installation of the double cone insert at this level improved the material flow pattern, with a very even movement of solid in the whole cylinder section. But below the transition level, the discharge velocity distribution was very uneven, with solids almost stagnant close to the hopper wall. To reduce the unevenness of flow pattern in the hopper, further efforts were made to focus on this portion of the silo.

It was presumably understood by the inventors of such insert that the length of the lower part of the cone may play an important role in the flow pattern. Modification was then carried out by extending the lower tip of the double cone to the outlet of the silo, while keeping the widest part of the cone at the same level, and its diameter unchanged.

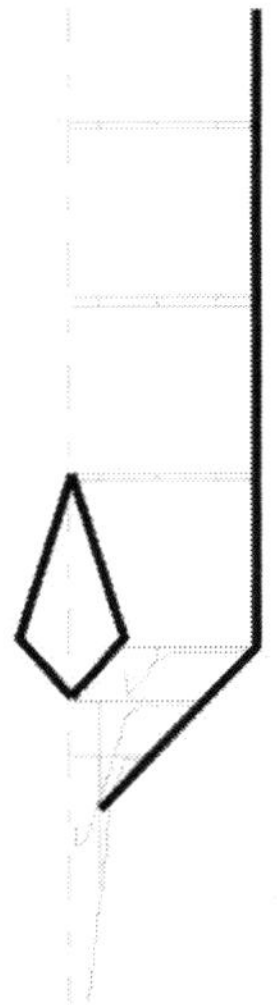

Fig. 6 Flow in a silo with a 45° hopper with a double cone insert

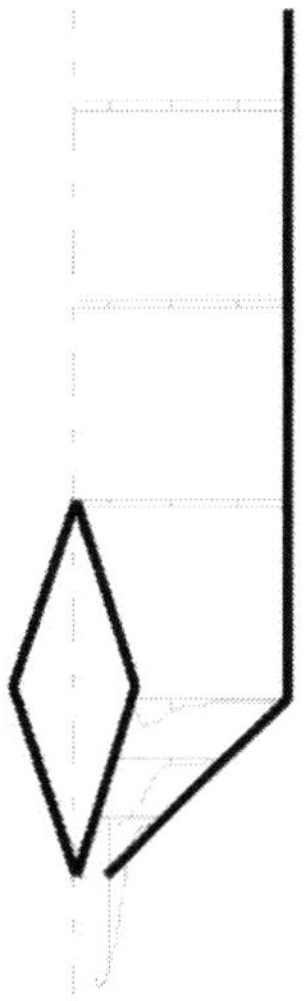

Fig. 7 Flow in a silo with a 45° hopper having an extended double cone°

The results are shown in Fig. 7. Compared with the results shown in Fig. 6, the simulations show that the extension of the lower cone led to somewhat reduced unevenness of discharge in the hopper, whilst very even movement in the whole cylinder section was unaffected. The height of the lower cone might play a role to improve discharge; however, the improvements were very limited under the current conditions. Other methods of flow improvement, which might include reducing the hopper wall friction, should also be considered.

An investigation into the effect of wall friction on the flow was carried out next. In the zone just above the hopper, it was evident that some material remained stationary. There is therefore no obvious reason to decrease the friction along the wall of the insert and cylinder. A reduction of the wall friction between the material and the wall of the hopper alone was therefore implemented. Simulations were carried out. Fig. 8 shows the numerical results after resetting the friction between material and hopper wall to 15° and to 18° (in place for the previous 21.8°).

Comparing the results in Fig. 8 with those in Fig. 7, one can see that after a decrease of the wall friction on the hopper wall, the unevenness of discharge in the hopper was reduced and the velocity distribution was improved. It is also evident from the three different wall friction conditions calculated, as the hopper became smoother, the flow pattern became progressively better. The hopper wall friction is very important when inserts are used to improve the flow pattern.

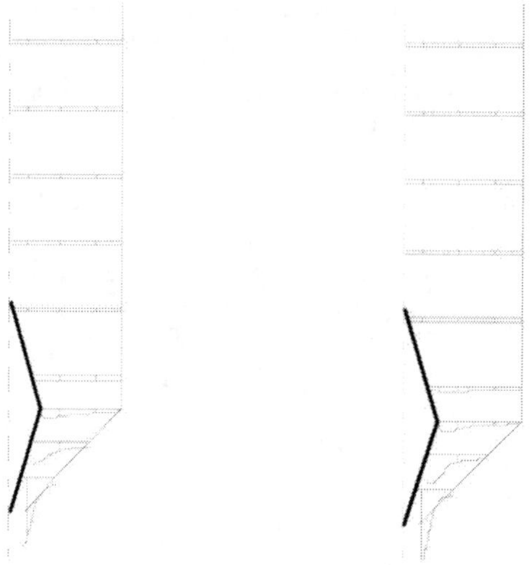

(a) ϕ_w=15° (b) ϕ_w=18°

Fig. 8 Velocity distributions with different hopper wall friction

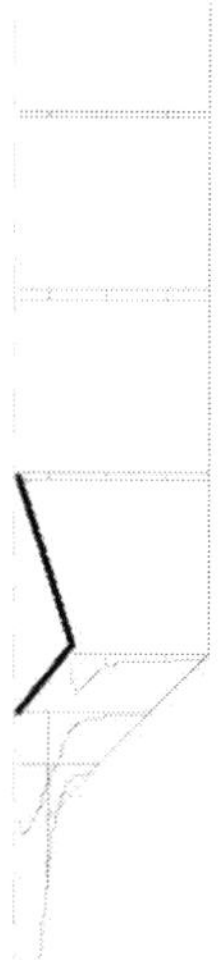

Fig. 9 The effects of the lower cone extension (ϕ_w=15°)

One question remained concerning with the role played by the extension of the lower cone of an insert when the hopper wall friction is low. To explore this question, a further calculation was carried out in which the lower cone was returned to its original geometry, but the hopper wall friction angle was kept at 15°. The numerical results are shown in Fig. 9.

It is evident in Fig. 9 that the unevenness of the flow velocities on discharge was similar to that found when the double cone insert was not extended to the outlet (Fig. 6). Comparing the result again with that for the smoothest hopper wall (Fig. 8), it is clear that the extension of the lower cone to the outlet is a valuable change, but it is important that the wall friction is also made low at the same time; perhaps at 18° for the present example case.

Concluding remarks

Simulation results show that placing an insert in a certain position in a silo is a method to influence the discharging flow pattern. The double cone inserts as described were able to change the material flow pattern from funnel flow to mass flow in a silo whose hopper has an inclination angle larger than that of the critical value for mass flow. Beyond this point, mass flow could not be obtained.

It has been also shown that, in a silo with a hopper inclination angle below a certain value, the vertical extension of the lower cone of a double-cone insert would improve the silo discharge pattern, but its effect is hardly noticeable if the hopper wall friction is high.

A combination of this extension of the lower cone and a decrease of friction between the solids and walls of hopper would eventually improve the silo discharge mode, and permit an increase of the hopper half angle.

References

1. Cundell, P. A. and Strack, O. D. L., (1979), "A Discrete Numerical Model for Granular Assemblies", *Geotechnique*, Vol. 29, No. 1, p.47-65.
2. Ding S., (2004), "Investigation of flow and pressure in silos during filling and discharging in presence of inserts", PhD thesis, University of Edinburgh, Dec.
3. Drescher A., (1992), "On the criteria for mass flow in hoppers", *Powder Technology*, 73, p. 251-260.
4. Drescher A., (1991), "Analytical Methods in Bin-Load Analysis", Elsevier Science Publishers B.V.
5. Enstad G. G., (1998), "Use of Inverted Cones and Double Cones as Inserts for Obtaining Mass Flow", POSTEC Newsletter No 17.
6. Johanson J. R., 1967 - 68, "The placement of inserts to correct flow in bins", *Powder Technology* 1, 328-333.
7. Haussler U., and Eibl, J., (1984), "Numerical Investigation of Discharging Silos", *Journal of Engineering Mechanics*, 110(6), P. 957-971.
8. Janssen H. J., (1895), "Verssuche uber Getreidedruck in Silozellen", Zeitschrift des Verein Deutscher Ingenieure, 39(35),1045-1049.
9. Jenike A. W., (1961), "Gravity Flow of Bulk Solids", Bulletin 108, Utah Engineering Experiment Station, University of Utah,.
10. Jenike A. W., (1964), "Storage and Flow of Solids", Bulletin 123, Utah Engineering Experiment Station, University of Utah,.
11. Jenike A. W., (1987), "A Theory of Flow of Particulate Solids in Converging and Diverging Channels based on a Conical Yield Function", *Powder Technology*, 50(3), 229-236.
12. Karlsson T., Kilsinski M. and Runesson K., (1998), "Finite Element Simulation of Granular Material Flow in Plane Silos with Complicated Geometry", Powder Technology.
13. Karlsson T., (1996), "Finite Element Simulation of Flow In Granular Materials", Licentiate Thesis 1996:8L, Luleå University of Technology,
14 Rotter J. M., (1998), "Challenges for the future in numerical simulation", Chapter 36 in *Silos: Fundamentals of Theory, Behaviour and Design*, (Eds C.J. Brown and J. Nielsen), E & FN Spon, London, pp 584-604.

379c Rheological Behavior of Flowing Powders: Use of a Couette Cell to Validate Constitutive Equations for Different Regimes

Mehrdad Kheiripour Langroudi, Chemical Engineering, City College of the City University of New York, 140th Street and Convent Avenue, New York, NY 10031 and Gabriel I. Tardos, Department of Chemical Engineering, The City College of the City University of New York, 140th Street & Convent Ave, New York, NY 10031.

The rheological behavior of slowly moving dense granular materials was studied in a cylindrical Couette-type cell by measuring the shear stress-shear rate dependence. The aim of this work is to validate constitutive equations by measuring torque on the rotating cylinder as well as normal stresses on the outer (stationary) wall. We improved the traditional Couette device (that does not allow for axial flow) by providing continuous powder feed and discharge through a conical hopper attached to the lower part of the bin. The axial flow increases the porosity of the bed and destroys the chains formed by particle-particle interaction in the dense bed. Experiments performed with only the top and the bottom part of the inner cylinder showed that a comparatively large portion of the torque is generated by these parts. Using a large length-to-diameter ratio for the inner cylinder assures that the "end" effects are small.

We found that a static pile of grains does not transmit normal stresses unaltered and to get correct values the layer has to be continuously sheared. These tests were performed using a traditional Jenike shear cell. Materials with different stiffness such as glass particles (ranges from 0.1 mm to 2 mm), polyethylene granules and cast steel shots were tested and all of them proved the above behavior. We used these results for the design of the Couette device.

In the "no-axial-flow" (closed) high solid concentration bed, shear stresses increase due to packing of particles and formation of chains that result in stress build-up that eventually reaches equilibrium. To support this, we observed the free surface of granules on the top of the bed that were horizontal at the onset of the experiment and turned to be concave thereafter. Under these conditions, increasing the shear rate provides some space for the particles to move and cause the decrease in shear stress. Hence, there is a range in which shear stress decreases by increasing shear rate when there is no axial flow. At lower shear rates, the shear stresses is independent of shear rate, the movement is very slow and the forces are due to friction: the powder moves in the quasi-static regime.

In the "axial-flow" (open) bed, creating a continuous axial flow significantly decreases the torque (shear stress on rotating cylinder) and normal stresses on the wall. Providing enough space for the granules to move results in breaking the chains. This small porosity change provides space for the particles to change their position randomly and collide causing the bed to move into the "Intermediate regime" of flow where the powder shows some "fluid like" behavior. This is characterized by a power-law relationship between the shear stress and shear rate [1]. Experiments were performed on a variety of particles including glass particles (0.1 mm to 2 mm), Grace catalyst (0.1 mm) and Acrylic polymers (0.1 mm) and a characteristic power index in the range between n=0.3 ~ 0.6 was determined.

Reference

1. G. I. Tardos, S. McNamara, I. Talu, Slow and intermediate flow of a frictional bulk powder in the Couette geometry, Powder Technol. 131 (2003) 23-39

379d Caking of Binary Powder Mixtures

Elsa Schreyer and Stefan Palzer. Technology Group, Nestlé Product and Technology Centre Singen, Lange Str. 21, Singen (Hohentwiel), Germany

In the food and pharmaceutical industries, water soluble amorphous components - in particular various carbohydrates - are ubiquitous. Water acts as a plasticizer on water soluble amorphous material. Plasticizers are compounds that increase the free volume (i.e. the volume that is not occupied by molecules) and therefore decrease the glass transition temperature. The molecular mobility increases and the material soften. As the water content of an amorphous matrix increases, the glass transition temperature decreases. A dramatic decrease of viscosity at temperature above the glass transition temperature can be observed. Indeed the amorphous matrix softens and becomes more and more plastically deformable at temperatures above the glass transition temperature, influencing the stability of a powder bulk. As the viscosity declines viscous flow can develop. If two amorphous particles are in contact, they can stick together and build viscous bridges at their contact point. This is the sintering phenomenon which provokes undesired agglomeration and caking, i.e. the change from a bulk of flowable particles to a sticky mass. Sintering and caking are not only depending on temperature and water content. Indeed mechanical stress applied on a bulk can lead to viscoelastic deformation and amplifies the caking if the particles are sticky. And as sintering is a kinetic phenomenon, storage time has also a great influence on the degradation of flowability. Crystalline particles may also form interparticle bridges but the mechanism of bridge formation differs from sintering. Practically crystalline particles have to be exposed to relative humidity higher than their critical humidity (i.e. the relative humidity at which the crystalline substance starts to take up moisture from the surrounding atmosphere and to dissolve without any viscosity development) to cake. This phenomenon leads to liquid bridges formation between particles that solidify after drying. The critical humidity of crystalline substance being often higher compared to the relative humidity responsible of softening of amorphous substance, they are consequently less prone to caking than the latter. Caking mechanisms of pure amorphous and pure crystalline products are known. However the caking behaviour of powder mixes remains difficult to predict. In case of a powder mix, the properties of each single component have to be taken into account: particle size, particle shape and concentration of each component in the mix. In the food industry, many recipes and products contain mixtures of crystalline and amorphous substances in their powdered form in various particle size distributions. Therefore, knowledge of the caking behaviour of mixtures is important. Actually the most sensitive component of a mix will lead the mixture to cake. Hence it is useful to know which amount of sensitive components, like amorphous ingredients, can be added to a mix without impairing its properties during storage or processing. In the current study, binary mixes of crystalline and amorphous substances in their powdered form and at different concentrations were prepared. The caking behaviour of the mixtures was investigated using time-consolidation measurements with a ring shear tester. The mixes were stored at controlled temperature, relative humidity and consolidation stress. The flowability after storage was analysed as a function of the components' particle size distribution and concentration. This study presents the relationship between the internal factors (mix components characteristics), the external factors (relative humidity, temperature and consolidation stress) and the resulting caking behaviour of binary mixtures.

Granular flow through an orifice – effect of granule properties on the onset of jamming.

Derek Geiger, Mark Wandstrat and Paul Mort
Procter & Gamble Co., ITC, 5299 Spring Grove Ave., Cincinnati, OH, USA

Abstract

This paper describes methods to measure flow rate and jamming probability of granules discharged through a flat-bottom cylindrical hopper with a circular orifice. The intrinsic jamming onset for ideal particles (spherical, monodisperse, smooth) is experimentally measured by two independent methods, with good agreement. For non-ideal particles, the jamming probability increases as a function of granule shape, size distribution and frictional properties. An empirical model of the jamming onset is introduced to quantify these effects over the range of materials investigated. The jamming probability can be taken as a measure of differentiation between relatively free-flowing granules.

Background

While shear cell systems are regarded as the most quantitative test measures of incipient flow of bulk cohesive powders and granules, shear cells are relatively insensitive to differences in very free flowing materials.[1] According to the continuum model, a "cohesionless" material measured using a shear cell (i.e., a yield locus passing through the origin) implies flow through a hopper cone opening of infinitesimal diameter; at this limit, the continuum model fails because the orifice must be at least as large as the particle diameter. Further, local packing and jamming effects require that the opening be significantly larger than a single particle.[2] A theoretical analysis of intrinsic cohesion suggests a boundary layer effect of 3-5 particle diameters;[3] as a starting point, this can be taken as a theoretical minimum dimensionless orifice size for an otherwise free-flowing granular material. Experimental studies on the physics of jamming in granular flows report jamming transitions on the basis of a dimensionless hopper opening at the onset of the jamming transition,[4, 5, 6] where the dimensionless hopper opening (B/D) is defined by the ratio of the opening size (B) relative to a monodisperse particle size (D). In the case of granular materials that are not monodisperse spheres, an empirical funnel flow measurement is often used, for example in the industrial evaluation of flow aides. To the knowledge of the authors, there is no current theoretical basis for determining intrinsic cohesion of polydisperse and/or non-spherical granules. The current work is an empirical investigation, seeking to elucidate the effects of both size distribution and shape factors on the jamming probability of free-flowing granules.

Experimental

The experimental device used in this work is a flat-bottom cylindrical hopper with a set of interchangeable disks with a range of discrete orifice sizes; it is available as a commercial flow testing system, Flodex™ (Hanson Research, Chatsworth, CA, USA), with additional disks machined as required to extend the range of orifice diameters. The range of orifice sizes used with the Flodex includes: 2.0, 2.5, 3.0, 3.5, 4, 5, 6, 7, 8, 9, 10, 12, 14, 16, 18 and 20 mm. Experiments were run using common materials including narrowly-classified glass beads, Ottawa sand, and a variety of granular detergent samples. In all cases, these samples were relatively free-flowing, i.e., as measured using shear cells.

A schematic of the orifice flow instrument is shown in Figure 1. The stainless steel hopper is filled by pouring the granular sample through the loading funnel. Approximately 100 ml of material is used in the test, filling the 5.7 cm diameter hopper to a height of about 4 cm. After the sample settles, the spring-loaded discharge gate is opened and the sample is allowed to drain through the orifice into a receiving cup below. Once the flow stops and remains stopped for 30 seconds, the mass of discharged material is measured. For each measurement, the mass% discharged is calculated according to the formula: (mass% discharged) = 100 * (mass discharged) / (sample mass). The average of the three mass% discharge calculations is plotted as a function of the dimensionless orifice size (B/D), with the mass% discharged on the ordinate and the dimensionless orifice size on the abscissa. The orifice size is made dimensionless by scaling the orifice size (B) to a characteristic particle size (D). The Sauter mean (D_{32}) is used as the characteristic size. Note the Sauter mean is weighted toward the surface area of the granules; the surface area is relevant to frictional interactions in the flow.

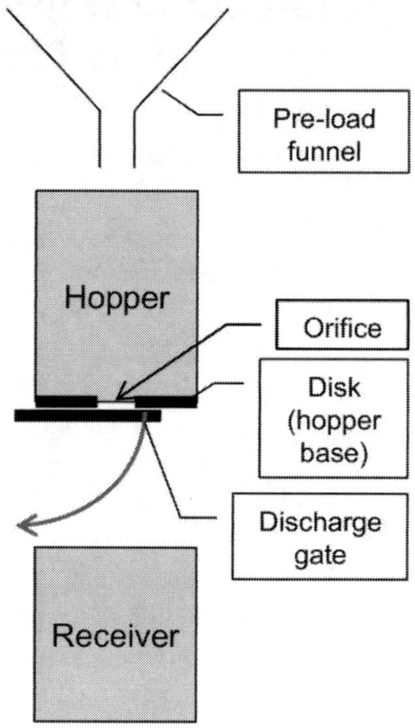

Figure 1. Schematic of flat-bottom cylindrical hopper device (Flodex).

This procedure is repeated using incrementally larger orifice sizes until the hopper discharges without jamming for three consecutive trials. The averaged data are linearly interpolated to find the Jamming Onset (J), defined here as the value of the dimensionless orifice size at the point of 25 mass% average discharge (Figure 2).

In addition, the drained angle of repose, $\phi_{drained}$, was calculated using equation 1 for experiments where the hopper discharged without jamming the orifice, i.e., where B/D > J. In this case, $M_{retained}$ is the mass of material retained in the hopper, D is the diameter of the hopper, B is the orifice diameter and ρ_{bulk} is the repour (loose) bulk density of the material.

$$\phi_{drain} = \arctan\left(\frac{24 \cdot M_{retained}}{\pi \cdot \rho_{bulk} \cdot \left(2 \cdot D^3 - 3 \cdot B \cdot D^2 + B^3\right)}\right) \quad (1)$$

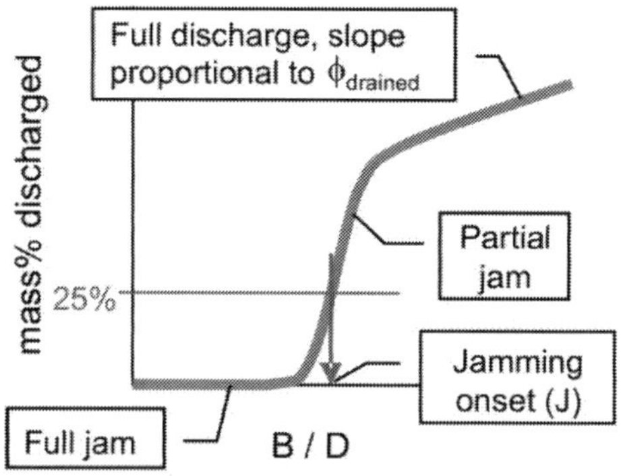

Figure 2. Interpolation of Jamming Onset (J).

The cumulative granular discharge was measured by placing the receiver cup on an electronic balance connected to a data acquisition program. Again, this experiment was done using orifice sizes above the jamming onset. The flow rate was determined by taking the time derivative (slope) of the cumulative discharge. The mass flow rate is converted to a volumetric rate using the repour bulk density.

Results and discussion

Jamming measurements and multivariate model

Results for nearly "ideal" materials including tightly classified glass beads and washed Ottawa sand are shown on Figure 3. These samples are nearly mono-disperse. The glass beads are nearly spherical, while the sand is slightly irregular in shape. Both are relatively smooth-surfaces and have a low drained friction angle. The jamming onset for the glass beads is in the range of 3 to 5 particle diameters. The washed sand has a slightly higher jamming onset of about 7 diameters. Other granular samples characterized in this work included some with considerably broader size distributions and irregular shapes, with relative jamming onsets up to 30 and drained friction angles in excess of 60°.

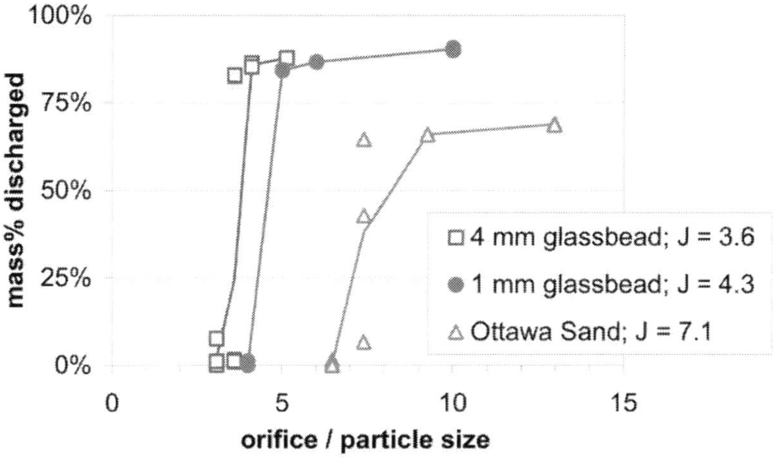

Sample	$\phi_{drained}$ (°)
4 mm glassbead	25.5
1 mm glassbead	19.0
Ottawa sand	38.6

Figure 3. Cylindrical hopper discharge data, jamming onset calculations and drained angle of repose for glass bead and washed sand samples.

Data from 22 samples were used to generate a multivariate model for the jamming onset (J) as a function of particle parameters (eq. 2).

$$ln(J) = ln(k_j) + a\, ln(\alpha) + b\, ln(\sigma_g) + c\, ln(\chi) \qquad (2)$$

Significant parameters include particle shape, size distribution and roughness or other excess friction: α is the median aspect ratio, σ_g is the geometric standard deviation of the particle size distribution, and χ is an excess friction factor defined as χ = max [1, tan(ϕ_{drain})]. The aspect ratio is the ratio of major to minor axes, measured using an automated image analysis system, Solids Sizer (JM Canty, Buffalo, NY). The excess friction factor is based on the drained friction angle measurement (eq. 1). The data used in the multivariate regression are shown in Figure 4; the results are:

k_j = 2.78, a = 2.7, b = 1.4, c = 1.2.

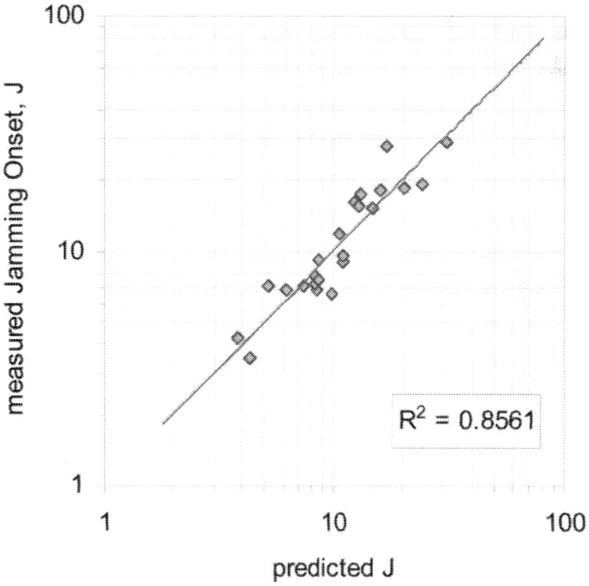

Figure 4. Multivariate model of jamming onset (J) as a function of particle aspect ratio, size distribution, and excess friction factor (eq. 2).

Note that all three of model parameters (α, σ_g, χ) have a logical minimum value of 1.0. Thus, extrapolation to the power-law intercept, k_j, represents the jamming onset of an idealized sample (mono-disperse, spherical, smooth, low friction). Note the high value of the exponent particle aspect ratio (a=2.7) implies a strong effect of shape on elevated jamming probability.

Flow rate measurements and Beverloo analysis

Flow rate data were analyzed using the Beverloo equation.[7] It is shown here in volumetric form (eq. 3), where V' is the volumetric feed rate, B is the orifice diameter, D is the characteristic particle size (Sauter Mean), g is the gravitational constant and C and k_f are fit parameters; $B-k_fD$ represents a reduced orifice size for active flow caused by a boundary layer that scales with particle size. For the purpose of regression analysis as a function of relative orifice size (B/D), the Beverloo equation can be written in linearized form, eq. 4, where C is solved using the regression slope and k_f using the intercept.

$$V' = C\sqrt{g}(B - k_f D)^{5/2} \quad (3)$$

$$\frac{V'^{2/5}}{D} = C^{2/5} g^{1/5}\left(\frac{B}{D} - k_f\right) \quad (4)$$

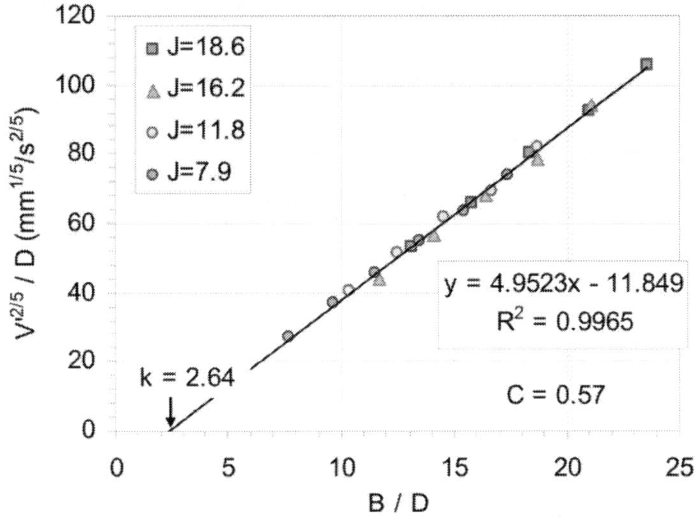

Figure 5. Regression of flow rate data using equation 4.

Several flow rate data sets are shown in Figure 5. These data sets represent similar materials, but with different shape and size distribution characteristics. The jamming onsets (J) of these samples span a significant range from 8 to 20; however, the Beverloo regressions are remarkably similar. All lie on a common slope (C ~ 0.57) and extrapolate to a similar zero-flow intercept (k_f ~ 2.64). In other words, the granular flow behavoir is consistent above the jamming onset, but subtle differences in the granular characteristics and frictional properties have a significant effect on the onset of jamming.

Consider k_f to represent an "intrinsic jamming onset" at which the flow rate goes to zero. Note the extrapolated value for intrinsic jamming (i.e., no flow) is only slightly lower than the power-law intercept of the multivariate model (Figure 4). Recall, the measurement of the jamming onset, J, is based on the interpolated orifice size for an average 25% sample discharge from the test hopper. Thus, it is logical that k_j, which is based on multivariate analysis of J, may slightly exceed k_f, which implies zero flow. On the other hand, the slope of the jamming probability function tends to be quite steep in the range from zero to 25% discharge. So, for the purpose of this paper, we can merge the interpretation of k_j and k_f – both represent an intrinsic jamming limit. Indeed, even though each was measured independently using different methods, their values are nearly identical.

An interpretation of this result is that the "intrinsic jamming onset" is characterisitic of orifice flow for an ideal particle (monodisperse, spherical, no excess friction), and that differences between the intrinsic onset and the actual measured jamming onset is due, in large part, to irregulatities in the size distribution, shape and frictional roughness of the granules.

Conclusions

The statistical analysis of jamming is a promising method for characterization of relatively free-flowing granular materials. In this work, we evaluated the effect of dimensionless orifice size on jamming probability and flow rate. A multivariate analysis of the jamming probability suggests an intrinsic jamming probability of about 2.8 particles diameters for an idealized sample; irregularities such as size distribution, shape or excess friction may cause an increase in the jamming probability. While all three parameters are significant in the regression analysis, the result suggests that shape (aspect ratio) is an especially important contributor to jamming. Analysis of the flow rate data using the Beverloo equation extrapolates to a similar value for the intrinsic jamming onset, providing good experimental agreement among two independent methods.

The experimental methods described herein are relatively simple. Continued work covering a wider range of materials can help to build a stronger empirical model for the effects of shape, size distribution and other frictional properties on jamming probability. In addition, first principle modeling of dense flows may help to elucidate the underlying mechanisms of jamming and the real effects of non-ideal particle characteristics, and perhaps help to build a more theoretical model.

Lastly, the reader should temper this report with the caveat that all of the samples used herein are relatively free-flowing granular materials. As materials become more cohesive and/or compressible, it is not clear that the flow rate and jamming onset experiments will continue to converge on a common intrinsic jamming onset.

References

1 de Jong, J.A.H., Hoffmann, A.C. and Finkers, H.J., Properly determine powder flowability to maximize plant output, Chemical Engineering Progress, 95, No. 4, 25-34, (1999).

2 Birks, A. H., Bradley, M.S.A. and Farnish, R., The conversion of the analytical simple shear model for the Jenike failure locus into principle stress space and implication of the model for hopper design, in Handbook of Conveying and Handling of Particulate Solids, A. Levy, H. Kalman, Eds., Elsevier Science, 95-105, (2001).

3 Weir, G. J., The intrinsic cohesion of granular materials, Powder Technology, 104, 29-36, (1999).

4 I. Zuriguel, A. Garcimartín, D. Maza, L. Pugnaloni and J. M. Pastor, "Jamming during the discharge of granular matter from a silo," Phys. Rev. E 71, 051303 (2005).

5 K. To, P-Y. Lai and H.K Pak, "Jamming of Granular Flow in a Two-Dimensional Hopper," Phys. Rev. Letters, 86, 71-74 (2001).

6 P. Mort, K. McKenzie, J. Wambaugh and R. Behringer, "Granular flow through an orifice – effect of granule size and shape distributions," Proceedings of the World Congress of Particle Technology 5, Orlando, FL, USA, 2006.

7 W.A. Beverloo, H.A. Leniger and J. Van de Velde, *Chem Eng. Sci.*, 15 (1961).

379f Power Consumption And Granular Flow In A Vibro-Fluidized Stirred Granular Bed

Kenneth J. Ford, James F. Gilchrist, and Hugo S. Caram. Lehigh University, Department of Chemical Engineering, Bethlehem, PA 18015

This work examines the behavior of a deep granular bed subject to simultaneous vertical vibration and stirring. The power necessary to stir a 5.6 cm diameter and 20 cm deep granular bed of 150 micron glass beads using a 4-blade vane impeller was measured over a wide range of forcing conditions. Impeller rotation rates from 0 – 1000 rpms and vibration accelerations in the range $0 \leq \Gamma = \omega^2 a/g \leq 4.0$ were explored. Both the power required for stirring and vibration were recorded. Sharp changes in vane power draw indicate flow transitions from dense granular flow to vibro-fluidized flow at a critical acceleration (Γ_c). The total power profile (vane plus vibration power) demonstrates a minimum just above the critical acceleration. Progressive increases or decreases in both vane speed and Γ independently show hysteresis as the flow bifurcates between two primary states of dense granular flow and loose-packed, vibrofluidized behavior. These observations are compared to those found in fluidized systems and flows generated in high-shear granulators.

A Visual Comparison of Pneumatic Conveying of Soft and Hard Materials

Néstor Vásquez [a], Karl Jacob [a], Shrikant Dhodapkar [b]

[a] The Dow Chemical Company, Midland, MI 48667, USA
[b] The Dow Chemical Company, Freeport, TX 77541, USA

Abstract

Dilute phase pneumatic has been widely studied due to its importance in the transport of bulk solids, and its numerous applications in the industry. In general, most of the studies focus on the measurement and prediction of the conveying design parameters such us gas velocity and pressure drop, while visual analysis of particles dynamics is used only in limited experiments in which specific phenomenon like particle collisions are studied in a controlled manner. In this work, the particles dynamics were analyzed during normal operation of a 2 inch pneumatic conveying system, considering two testing materials: hard pellets and soft pellets. The visual analysis was done on straight sections and on two of the more common elbows: the long radius bend (LRB) and the blind "T" (BT). Videos of particle movement during conveying were recorded with a high speed video camera in conjunction with pressure drop measurements. A number of conveying trials were performed in order to include high and low superficial gas velocities and several loadings (μ) for each material.

The experiments performed with the hard pellets showed the normal differences in pressure drop between the elbow and the straight sections, with the elbows as a significant component of the higher pressure drop. The difference between the LRB and the BT was also as expected, with higher pressure drop for the latter. Pressure drop oscillations were detected during these trials, which were found to be related to the formation of particles clusters as shown by the high speed videos. The clusters originated from the rotary valve during the discharge of its pockets into the conveying line. As the hard pellets move through the LRB, they follow a trajectory along the outer wall of the elbow, showing little, if any bouncing. In this type of elbow the particle velocity does not decrease substantially, which would explain the low pressure drop. Through the BT, the hard pellets showed a significant change in particle velocity as the pellets impacted the stationary particles in the BT pocket. The required reacceleration of the particles should explain the higher pressure drop produced by this type of elbow.

The experiments with the soft pellets showed unexpected results. In this case, the bouncing of the particles was so intense that the particles did not significantly change their average axial velocity through the elbow. It appears that the elbow affects the conveying velocity of the particles upstream, so by the time the pellets get into the elbow, they have already reduced their velocity. Thus, the pressure differential per unit of length ($\Delta P/L$) for the elbow and the straight section upstream was not different. It has to be noted that the ΔP across the elbows was measured by placing the pressure taps at sufficient distance downstream the elbow to permit the reacceleration of the particles. Through the LRB, the soft pellets bounced intensely so there was not a higher particle concentration close to the outlet wall of the elbow curvature. Similarly, the soft pellets did not show a significant change in velocity when passing through the BT. For both elbows, the particles bounced creating a homogeneous distribution of particles inside the line. When comparing the LRB against the BT using the soft pellets, the LRB showed lower pressure drop than the shorter elbow. However, its $\Delta P/L$ was similar to the straight section upstream for the same trial. This type of pellet did not show the particle clusters or the pressure drop fluctuations. In this case the bouncing of the particles increased the pressure drop due to the reduction of the conveying velocity, which increased the drag force.

From the visual analysis, it can be concluded that:

- When conveying hard pellets, the rotary valve creates clusters of particles that produce pressure oscillations. This phenomenon was not detected when conveying soft pellets.
- The hard pellets concentrate at the outer wall of the elbow when passing through the LRB. In this case, their velocity does not change substantially compared to the straight sections, which would explain the low pressure drop of this type of elbow.
- When passing through a BT, the hard pellets significantly change their conveying velocity after impacting with the stationary particles in the BT pocket. This change in velocity and the consequent reacceleration helps to explain the higher pressure drop developed by this type of elbow.
- When conveying soft pellets, the $\Delta P/L$ is higher than that for hard pellets due to the intense bouncing, which reduces the conveying velocity (the particles must be continuously reaccelerated), increasing the drag force.
- The intense bouncing of the soft pellets reduces the $\Delta P/L$ difference between the elbows and the straight sections and homogenizes the particle distribution inside the line during conveying.

INTRODUCTION

Dilute phase is the most widely used method of pneumatic conveying and it has been the subject of many research studies. Researchers have focused their work in several areas; from conveying behavior and state diagram[1] to pressure drop estimation and scale-up analysis. Many of these studies have been developed by the industrial requirements and the developing of new materials that need to be transported. Additionally, theoretical developments have been done in order to improve the current models, allowing more accurate predictions of solids conveyability. Nevertheless, the wide variety of material properties that affect the pneumatic transport, restricts the applicability of the conveying models making the experimental an important, if not indispensable, part of the design process. As many advantages, the pneumatic conveying also generates many challenges not only on terms of operation and setting, but also in terms of understanding the mechanisms that affect the particles dynamics and the pressure drop in the line[2,3]. In general most of the research has focused on the measurement and prediction of the conveying design parameters such us gas velocity and pressure drop, while visual analysis of particles dynamics is used only in limited experiments, in which specific phenomenon like particle collisions are studied in a controlled manner. In this work, the particles dynamics were analyzed during normal operation of a 2 inch pneumatic conveying system, focusing the observations on the straight sections and through long radius bend (LRB) and blind "T" (BT) type elbows. Two plastic pellets were used as testing materials in which the modulus of elasticity was the differencing property.

EXPERIMENTAL PROCEDURES

The conveying tests were performed in a 0.052 m I.D. aluminum pipe conveying system approximately 35 m long. Along the line, several pressure transmitter measured the pressure differential ΔP. Thus, pressure transmitter PDT 403 measures ΔP in horizontal direction, PDT 404 measures ΔP across the elbow under analysis, and PDT 401 measures the total ΔP. The airflow was measured by a Vortex type flow meter located downstream the filter receiver. Figure 1 shows the conveying system layout.

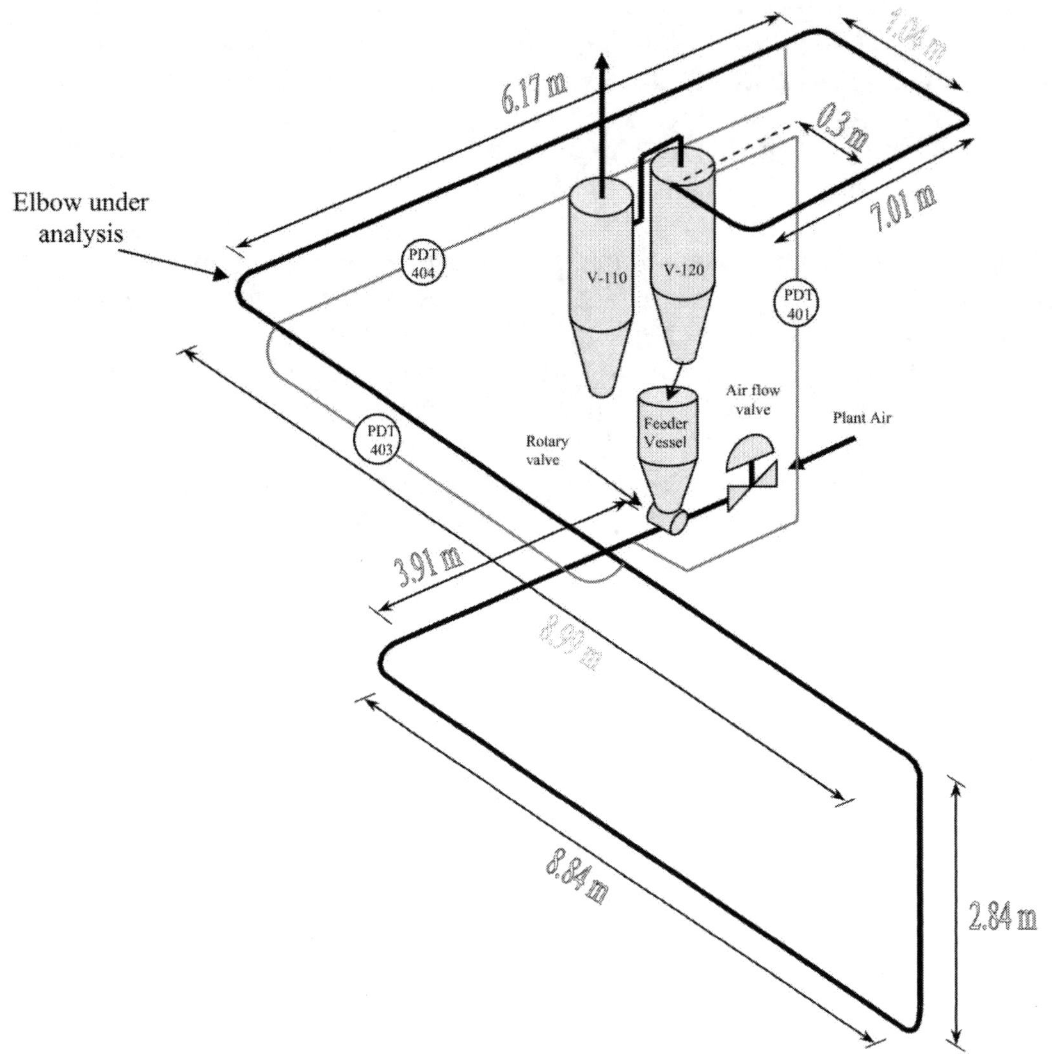

Figure 1: Conveying system layout

The two materials for testing were selected based on the requirements of minimum particle degradation during conveying and similar size and density, being their modulus of elasticity their only significant difference. Table 1 shows the principal characteristics of the materials tested:

Table 1: Material properties

Name	Hard Pellets	Soft Pellets
Particle shape	Ellipsoid	Ellipsoid
Equivalent spherical diameter, mm	4	4
Particle density, g/cc	0.889	0.861
Bulk density, kg/m^3	556	542
Tensile Modulus, 100% Secant, MPa	91.7 (100% secant)	6.6 (2% secant)

The long radius bend (LRB) and the Blind T (BT) elbows were selected for this study due to their wide application in pneumatic conveying systems. In both cases, the elbow under analysis was placed in horizontal direction, so gravity effect was neglected. Because of the reacceleration that the solids experience after passing through an elbow, the downstream tap for the ΔP across the elbow was placed 5.56 m downstream the elbow, allowing the pellets to recover their initial velocity[4].

A series of conveying tests were performed with the purpose of analyzing the particle dynamics and pressure drop at different solids loading and two superficial air velocities: 15 m/s and 30 m/s.

RESULTS AND DISCUSSION

For comparison purposes, the pressure drop across the elbows under study was expressed as ΔP/L, where L is the distance between the pressure taps. Thus a direct comparison between the pressure drop of the horizontal straight section and the elbows could be done.

Hard Pellets

The first set of experiments performed with the hard pellets was done using the LRB as the elbow under analysis. Then, the LRB was replaced by the BT and the tests were repeated using fresh conveying material. Figures 2, 3, and 4 show the pressure drop comparison for straight section vs. LRB, straight section vs. BT, and LRB vs. BT.

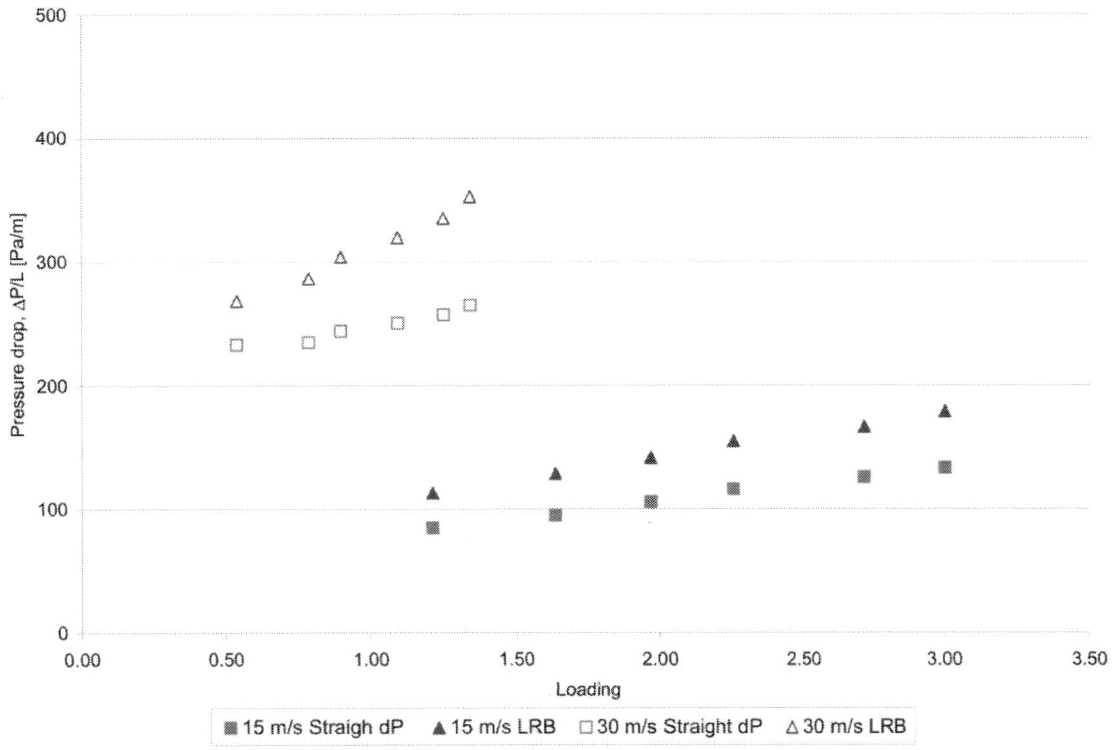

Figure 2: Straight section ΔP/L vs. LRB ΔP/L for hard pellets

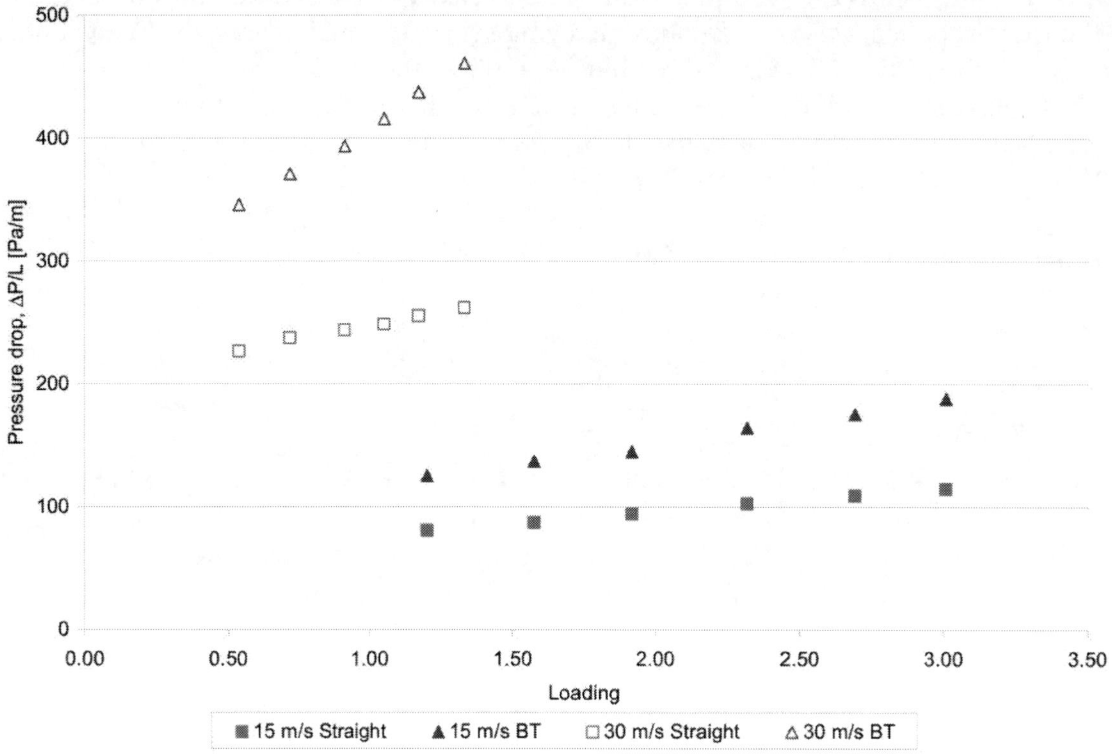

Figure 3: Straight section ΔP/L vs. BT ΔP/L for hard pellets

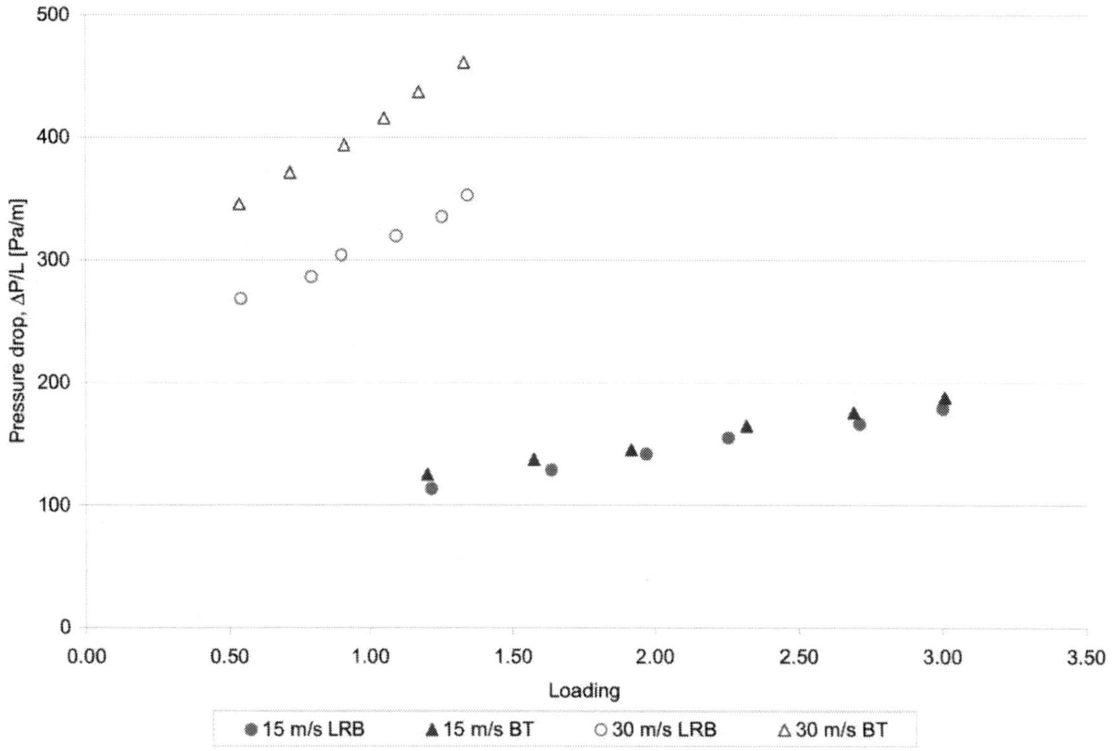

Figure 4: LRB ΔP/L vs. BT ΔP/L for hard pellets

As expected, the pressure drop produced by either elbow was greater than the pressure drop along straight sections. This is principally due to the change in conveying velocity that the solids experience when passing through an elbow and the posterior reacceleration. Note that the conveying velocity is defined as the axial component of the linear particle velocity. When comparing the $\Delta P/L$ between the elbows the LRB shows lower pressure drop than the BT although, as shown by Figure 4, its advantage is only significant at high air velocity.

Pressure oscillations

During the tests and specially at lower solids loading (μ), pressure oscillations were detected by the pressure transmitters. These oscillations showed constant frequency during the entire test and it coincided with the pulsing noise produced by the solids during conveying. The video analysis showed that the pressure oscillations were produced by particles clusters during the conveying process. Further analysis determined that the clusters originated from the rotary valve during the discharge of its pockets into the conveying line. At lower solids loadings, the slow rotation of the rotary valve feeds the line discontinuously producing the clusters shown by the high speed videos. On the other side, at higher solids loadings, the rotary valve empties its pockets at a higher frequency into the line, which increases the solids concentration in the pipe and homogenizes the solids distribution. This explained why the pressure oscillations were detected only at lower μ.

Visual analysis

High speed videos were recorded during the conveying trials with the purpose of analyzing the particles dynamics on both: straight sections and through the elbows under study. The videos recorded on hard pellets showed relatively low bouncing, with most of the pellets moving principally in the axial direction in the straight section. On the other side, when the hard pellets move along the LRB, they follow a trajectory along the outer wall of the elbow, showing little, if any bouncing. In this type of elbow the particle velocity does not decrease substantially, which would explain the low pressure drop. Figure 5 shows how the hard pellets concentrate on the outer wall of the LRB.

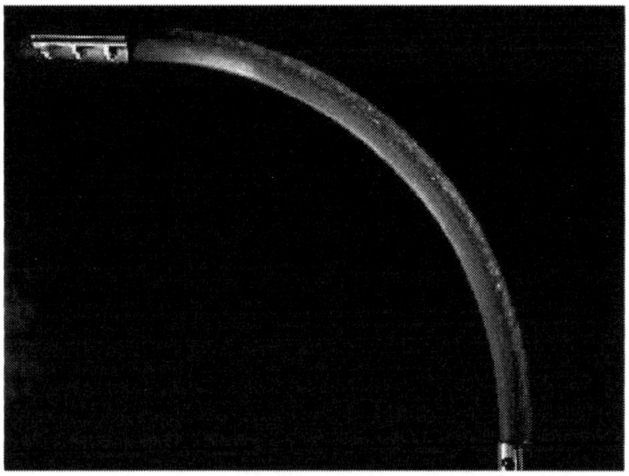

Figure 5: Hard pellets moving through the LRB

Another important consequent of this behavior is that the concentration of the particles in the outer wall intensifies the contact between the solids and the wall, which under certain conditions, may be the cause of particle attrition, floss generation or elbow wear. Through the BT, the hard pellets showed a significant change in particle velocity as the pellets impacted the stationary particles in the BT pocket. The required reacceleration of the pellets should explain the higher pressure drop produced by this type of elbow. Additionally, the BT seems to reduce the clusters downstream by homogenizing the flow since it works as a capacitor. Figure 6 shows a picture of the hard pellets passing through the BT.

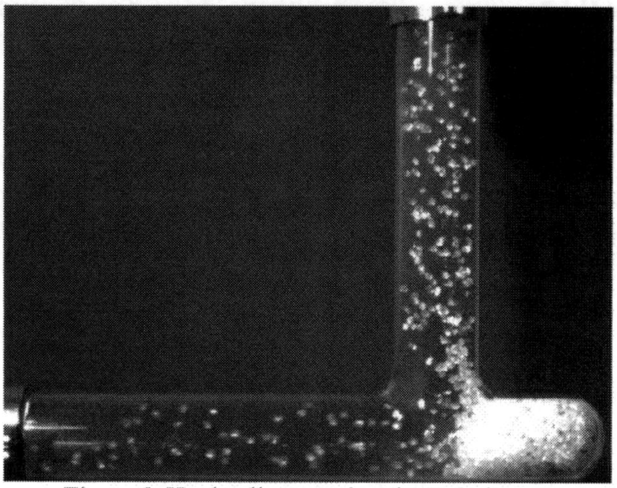

Figure 6: Hard pellets moving through the BT

Soft Pellets

The conveying trials with the soft pellets were performed using the same procedure and operational conditions as with the hard pellets. However, with this material the results show significant differences with the hard pellets. Figures 7, 8, and 9 show the pressure drop comparison for straight section vs. LRB, straight section vs. BT, and LRB vs. BT respectively.

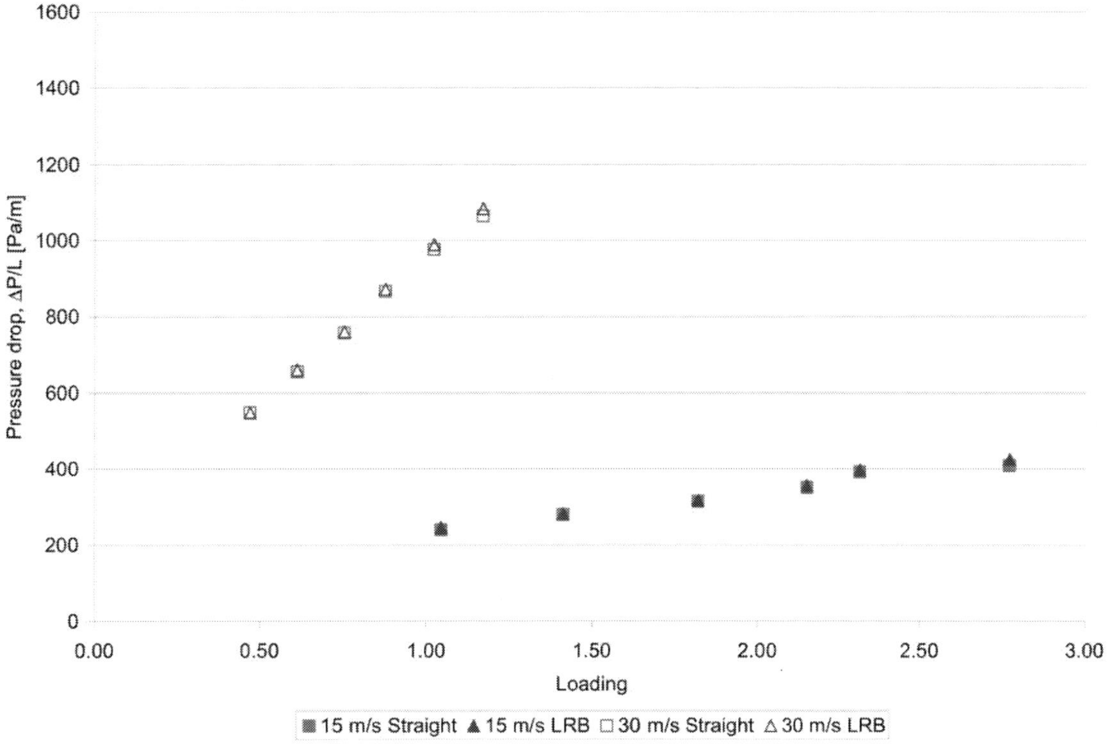

Figure 7: Straight section ΔP/L vs. LRB ΔP/L for soft pellets

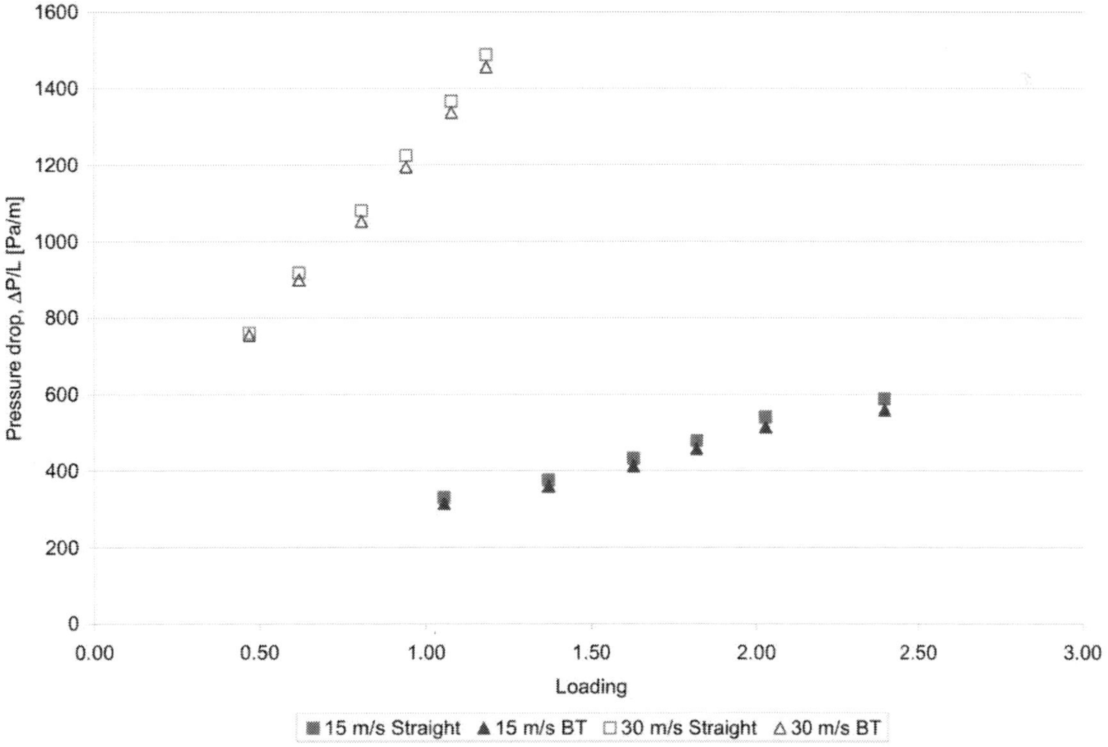

Figure 8: Straight section ΔP/L vs. BT ΔP/L for soft pellets

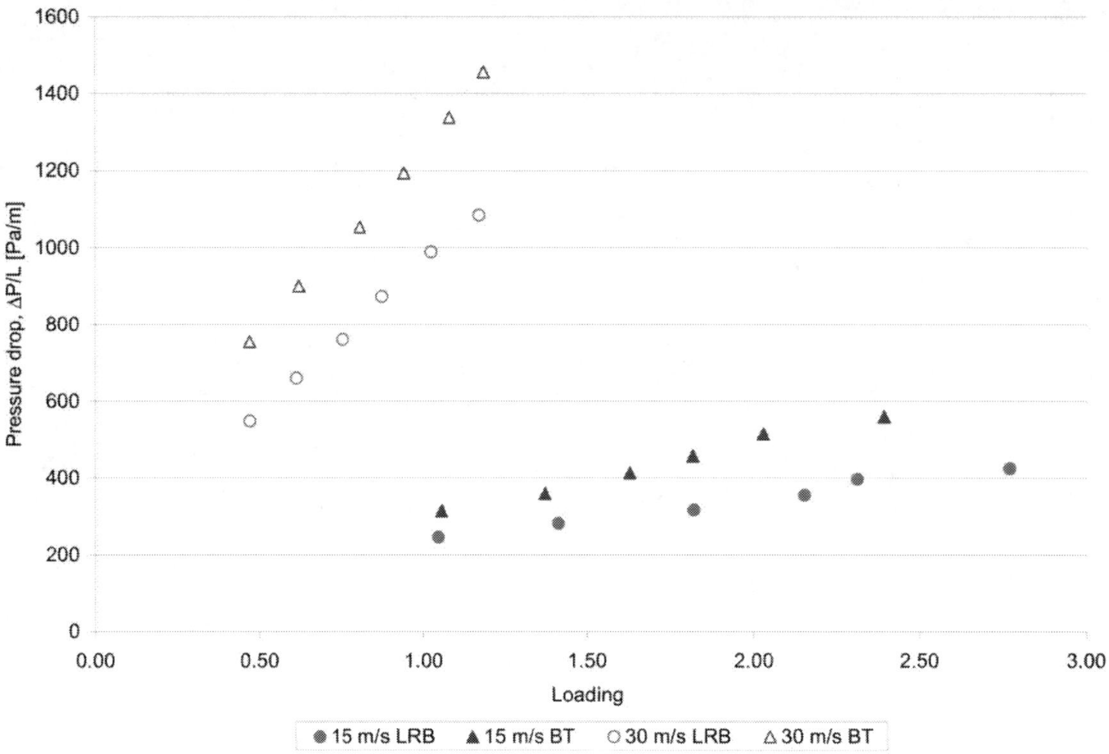

Figure 9: LRB ΔP/L vs. BT ΔP/L for soft pellets

As Figures 7 and 8 show, the pressure drops per unit of length are basically the same for straight sections and elbows. These values contradict the expected results since the change in particles direction along the elbows should reduce the conveying velocity, requiring additional energy to reaccelerate the pellets downstream. In addition to this significant difference compared to the hard pellets, the soft particles produce much higher pressure drop for similar μ and air velocity. In fact this difference, which has been reported for straight sections[5], it is also developed by the elbows. However, with the soft pellets the elbow pressure drops were always higher for the BT, even at low air velocity as shown by Figure 9.

Visual analysis

Due to the unexpected results from the pressure drop measurements, the visual analysis became a very important source of information, providing visual evidence of the reasons behind the similar pressure drops in straight sections and along the elbows. Thus, the videos of the soft pellets moving along the LRB showed very different behavior compared to the hard pellets. In this case, the particles did not concentrate in the outer wall of the elbow but they were homogeneously distributed as in the straight sections. Figure 10 shows the soft pellets passing through the LRB.

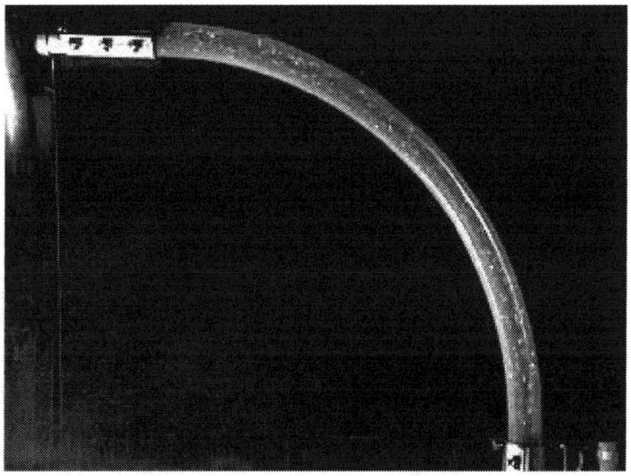

Figure 10: Soft pellets moving through the LRB

This behavior seems to be the results of the intense bouncing that this type of material produces[2]. Additionally, the tests with the soft pellets did not show the pressure oscillations or particles clusters as seen with the hard material. It seems that the bouncing of the soft pellets reduces the conveying velocity of the particles and it homogenizes their distribution inside the line. All this minimizes the cluster formation with the consequent reduction in pressure oscillations.

Through the BT, the soft pellets showed similar intense bouncing with no indication of a significant reacceleration of the particles after changing conveying direction in the elbow. These observations indicate that the soft pellets have similar conveying velocities before and after the elbows. This would explain why the straight sections and the elbows showed similar $\Delta P/L$ in the trials with this type of pellets. Figure 11 shows the soft pellets moving through the BT.

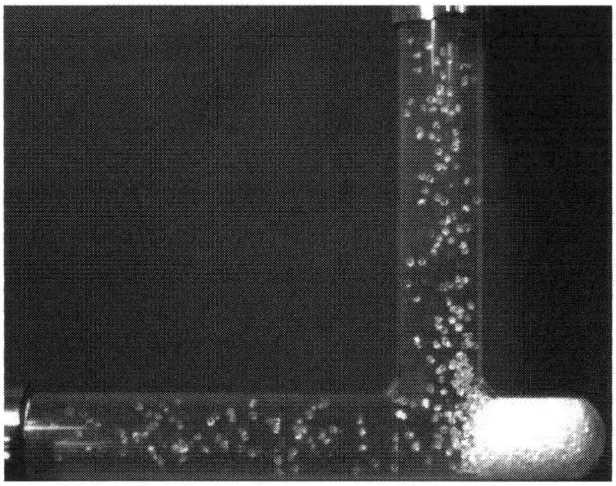

Figure 11: Soft pellets moving through the BT

COMMENTS AND CONCLUSIONS

The hard pellets showed the pressure drops that were expected, i.e. higher $\Delta P/L$ for the elbows over the straight sections and for the BT over the LRB. However, the soft pellets showed that the pressure drop per unit of length across the elbow is practically the same as for the straight sections. These results show the importance of the particles bouncing and its effect on the pressure drop. The observations also indicate that the soft pellets do not reduce their conveying velocity when passing through the elbows. The conveying velocity of the soft pellets is mainly affected by the bouncing of the particles, which does not seem to be affected by the change in conveying direction.

From the visual analysis, it can be concluded that:

- When conveying hard pellets, the rotary valve creates clusters of particles that produce pressure oscillations. This phenomenon was not detected when conveying soft pellets.
- The hard pellets concentrate at the outer wall of the elbow when passing through the LRB. In this case, their velocity does not change substantially compared to the straight sections, which would explain the low pressure drop of this type of elbow.
- When passing through a BT, the hard pellets significantly change their conveying velocity after impacting with the stationary particles in the BT pocket. This change in velocity and the consequent reacceleration explains the higher pressure drop developed by this type of elbow.
- When conveying soft pellets, the $\Delta P/L$ is higher than that for hard pellets due to the intense bouncing, which reduces the conveying velocity (the particles must be continuously reaccelerated), increasing the drag force.
- The bouncing of the soft pellets reduces the $\Delta P/L$ difference between the elbows and the straight sections and homogenizes the particle distribution inside the line during conveying.

BIBLIOGRAPHY

[1] Zenz, F.A. and Othmer, D.F. (1960) *Fluidization and fluid particles systems,* Reinhod, New York

[2] Luis Sanchez, Nestor A. Vasquez, George E. Klinzing and Shrikant Dhodapkar (2005) *Evaluation of models and correlations for pressure drop estimation in dense phase pneumatic conveying and an experimental analysis.* Powder Technology, Volume 153, Issue 3, 1 June 2005, Pages 142-147

[3] O. Molerus (1981*) Prediction of pressure drop with steady state pneumatic conveying of solids in horizontal pipes.* Chemical Engineering Science, Volume 36, Issue 12, 1981, Pages 1977-1984

[4] Stephen Tallon, Clive E. Davies and Bernard Barry (1998) *Slip velocity and axial dispersion measurements in a gas-solid pipeline using particle tracer analysis.* Powder Technology, Volume 99, Issue 2, 15 September 1998, Pages 125-131

[5] Vásquez, N. A., Jacob, K. V., Cocco, R. A., Dhodapkar, S. S., and Klinzing, G. E (2007) *Visual analysis of particle bouncing and its effect on pressure drop in dilute phase pneumatic conveying.* Powder Technology

379h Design, Simulation, And Performance Of An Aerogel Bead Coating System

Joel, L. Plawsky, The Isermann Department of Chemical and Biological Engineering, Rensselaer Polytechnic Institute, Ricketts Building, 110, 8th Street, Troy, NY 12180 and Howard Littman, Isermann Department of Chemical and Biological Engineering, Rensselaer Polytechnic Institute, Ricketts Building, 110, 8th Street, Troy, NY 12180.

A spout fluid bed coating system was designed to coat porous particles in a size range from 0.1 – 2 mm in diameter. The goal of the equipment was to process the particles and to insure that just their outer surface gets coated. The inner surface area of the particles needed to remain open to preserve their functionality. Due to the low density and fragility of the porous particles, it was important to design the coater such that the particles made as little contact with the walls of the coater as possible and that the coating dried before the particles touched either the walls of the equipment or each other. Particle coating for these types of particles has never been attempted and so a finite element package, COMSOL, was used to help design the system and then to simulate the fluid flow within the system. Key elements of the simulation were to determine the flow streamlines for the air, the pressures in the various components of the system, and the particle trajectories. Simulations were initially done in two-dimensions followed by a full three-dimensional simulation. The paper will discuss the details of the design, the simulation, and experimental results we obtained on the actual coating of the particles and their performance following coating. A conventional, pharmaceutical coating, Surelease, from Colorcon, was used as the coating material and the apparatus can be used to coat low density as well as high density materials. It is well suited for many coating applications including those in the pharmaceutical industry.

379i Pharmaceutical Tablet Coating Variability: Variation Of Tablet Residence Times And Coating Optimization

Arjun V. Kalbag[1], Carl R. Wassgren Jr.[1], Sai Sumana Penumetcha[2], and Jose D. Perez-Ramos[3]. (1) School of Mechanical Engineering, Purdue University, 585 Purdue Mall, West Lafayette, IN 47907-2088, (2) Industrial and Physical Pharmacy, Purdue University, West Lafayette, IN 47907, (3) GlaxoSmithKline, Research Triangle Park, NC 27709

Pharmaceutical tablet coating variability: Variation of tablet residence times and coating optimization.

A successfully coated batch of pharmaceutical tablets will have a prescribed coating thickness on each tablet's surface with little inter- and intra-tablet variability. Variability needs to be minimized for functional coated tablets, since the amount of coating controls the release profile and bioavailability of the drug. Large variations in coating thickness can adversely influence the effectiveness of the tablet. Uncoated tablet batches are placed within horizontal rotating pans with the coating solution sprayed onto the surface of the rolling tablet bed. The coating process is terminated when the prescribed amount of coating is sprayed onto the tablet bed. For minimum inter-tablet coating variability, the variation in tablet-to-tablet spray zone residence time should be reduced, since the amount of coating deposited on the tablets is directly related to the residence times. The current work investigates inter-tablet coating variability of spherical tablets in a horizontal axis pan coater using the discrete element method. Measurements of the fractional residence time, defined as the ratio of time spent by a tablet in the spray zone to the total coating time, are made for five tablet batch sizes and seven pan rotational speeds. The discrete element computer simulations suggest that *average* fractional residence time is independent of pan speed and total coating time for spherical tablets and is equal to the ratio of the number of tablets exposed to the spray to the total number of tablets, referred to as the coating fraction. Independently deduced analysis and video experiments supports the discrete element computer simulation results. While the average fractional residence time determines the average amount of coating received by a tablet, it is the variability in residence times that controls the amount of variability in the coated batch of tablets. Analytically, coefficient of variation of fractional residence time is proportional to the -1/2 power of the number of coating "trials" in a randomly coated batch of tablets. A coating trial occurs over an interval of time where the tablets in the spray zone and the tablets outside the spray zone appear to be segregated, *i.e.* not mixed. For extremely small intervals of time the tablet bed appears static and unmixed, as the tablets have not had sufficient time to move into new positions. For very large intervals of time the tablet bed appears to be well mixed and tablet positions appear to be randomly distributed in space with respect to the original reference state. The critical interval of time is the largest time interval where the tablets in the spray zone can be considered to be segregated from the tablets outside the spray zone. Physically this is approximately equal to the time taken for the spray zone to be populated with a new set of tablets that previously lay outside the spray zone and is equal to the time per coating trial.

Decreasing the time per coating trial reduces inter-tablet variability as it increases the number of coating trials. The time per coating trial can be decreased by increasing pan speed, increasing the spray zone aspect ratio (spanwise-to-streamwise spray zone dimension), choosing an optimal tablet load that minimizes slumping motion (rigid body motion) in the tablet bed and increasing tablet-pan and tablet-tablet friction coefficients. Analysis independent of the discrete element simulations suggests a ways to optimize inter-tablet coating variability by controlling average fractional residence time, and the time per coating trial, ensuring only an allowable fraction of tablets is over- or undercoated.

416a Crosslinked Polymeric Nanoparticles Via Brush Copolymers

Chong Cheng[1], Kai Qi[1], David S. Germack[1], Ezat Khoshdel[2], and Karen L. Wooley[1]. (1) Center for Materials Innovation, Department of Chemistry and Department of Radiology, Washington University in St. Louis, One Brookings Drive, Saint Louis, MO 63130-4899, (2) Unilever Research, Port Sunlight, Quarry Road East, Bebington CH63 3JW, United Kingdom

Well-defined core-shell brush block copolymers were prepared by tandem ring-opening metathesis polymerization (ROMP) and living radical polymerization (LRP), and used as unimolecular templates for the synthesis of polymer-based nanoparticles. Several bifunctional compounds, containing a norbornene functionality and an alkoxyamine, haloester, or trithiocarbonate functionality, were synthesized. ROMP of these compounds gave well-defined polyfunctional macroinitiators or macro-chain transfer agents for LRP. Sequential LRP of different monomers using these macroinitiators or macro-chain transfer agents yielded core-shell brush block copolymers with controlled macromolecular architectures. The entire synthetic procedure of core-shell brush block copolymers could be simplified by one-pot sequential ROMP and LRP based on their excellent compatibility. Shell-crosslinked nanoparticles were prepared by selectively crosslinking the shell domains of core-shell brush block copolymers, and nanocages were further obtained by degradation of the core domains of these nanoparticles. Core-crosslinked nanoparticles with controlled shape and size were also synthesized by selectively crosslinking the core domains of core-shell brush block copolymers.

References: 1). Cheng, C.; Qi, K.; Germack, D. G.; Khoshdel, E.; Wooley, K. L. Adv. Mater. 2007, 19, accepted. 2). Cheng, C.; Khoshdel, E.; Wooley, K. L. Macromolecules 2007, 40, 2289-2292. 3). Cheng, C.; Khoshdel, E.; Wooley, K. L. Nano Letters 2006, 6, 1741-1746. 4). Cheng, C.; Qi, K.; Khoshdel, E.; Wooley, K. L. J. Am. Chem. Soc. 2006, 128, 6808-6809. 5). Cheng, C.; Khoshdel, E.; Wooley, K. L. Macromolecules 2005, 38, 9455-9465.

416c Reactive Nanoparticles For Biomedical Cements And Construction Materials

Samuel C. Halim, Tobias J. Brunner, Neil Osterwalder, and Wendelin J. Stark. Chemistry and Applied Biosciences, Swiss Federal Institute of Technology (ETH Zurich), Wolfgang-Pauli-Str. 10, ETH Hönggerberg, Zurich, 8093, Switzerland

In solid/gas- or solid/liquid-reactions reactive particles strongly influence important parameters such as reaction rate, mechanism and product stability (e.g. fuels, cements, catalysis). The reactivity of particles strongly depends on the availability of surface area of the starting material, which ideally consists of well characterized homogenous nanoparticles. The present contribution investigates how such materials can be manufactured and explores their use as rapidly binding biomedical cements. Recent advances in the synthesis of nanoparticles by flame spray synthesis made not only a wide variety of oxides accessible to the flame process, but also reactive salts (calcium phosphates, gypsum, cement, limestone) [1] and even metal nanoparticles [2]. Tricalcium phosphate (TCP) is a very promising candidate to find application as a bone substitute material. We show that nanoparticulate amorphous TCP produced in the flame exhibit extremely fast reaction times, as required for a clinical application [3]. Furthermore, a similar reactivity was found for both, nanoparticulate gypsum (anhydrous calcium sulphate) and Portland cement derived from flame processing. Thereby, the gypsum showed a strongly increased hardness when compared to commercially available Alabaster [4]. We compare the gain in reactivity of biomedical cements, gypsum and Portland cement if nanoparticles are used instead of conventional micron-sized starting materials. The use of online monitoring systems such as isothermal calorimetry allowed studying the reactivity in detail. Based on these experimental finding, we developed a physical model describing the cement hardening process.

References: [1] S. Loher, W.J. Stark, M. Maciejewski, A. Baiker, S.E. Pratsinis, D. Reichardt, F. Maspero, D. Günther, Fluoro-apatite and calcium phosphate nanoparticles by flame synthesis, *Chem. Mater.* 17 (1), 36-42 **(2005)**.

[2] R. N. Grass, W.J. Stark, Gas phase synthesis of fcc-cobalt nanoparticles, *J. Mater. Chem.*, 16, 1825-30 **(2006)**.

[3] T.J. Brunner, M. Bohner, C. Dora, C. Gerber, W.J. Stark, Comparison of amorphous TCP nanoparticles to micron-sized alpha-TCP as starting materials for calcium phosphate cements, *J. Biomed. Mater. Res. Part B: Appl Biomater*, published online **(2007)**.

[4] N. Osterwalder, S. Loher, R. N. Grass, T.J. Brunner, L. K. Limbach, S. C. Halim, W. J. Stark, Preparation of nano-gypsum from anhydrite nanoparticles: Strongly increased Vickers hardness and formation of calcium sulfate nano-needles, *J. Nanopart. Res.*, 9, 275-81 **(2007)**.

416d Molecular Dynamics Studies on the State of Dispersion of Silica Nanoparticles in a Polyethylene Melt in the Presence of Nonionic Surfactants

Yangyang Shen, Rutgers University, 98 Brett Road, Piscataway, NJ 08854, Aleksey Vishnyakov, Department of Chemical and Biochemical Engineering, Rutgers, the State University of New Jersey, 98 Brett rd, Piscataway, NJ 08854, and M. Silvina Tomassone, Chemical and Biochemical Engineering, Rutgers University, 98 Brett Road, Piscataway, NJ 08854.

The addition of nanoparticles to polymer composites has been shown to influence significantly their mechanical, optical, and electrical properties. However, nanoparticle aggregation often leads to a breakdown of the composite nanostructure and properties. For example, a dielectric breakdown was observed in polyethylene impregnated by silica nanoparticles when the filling fraction exceeded 3 vol %. One of the ways to prevent nanoparticle agglomeration is the addition of surfactants to the composites in order to increase their affinity to the polymer. Here we study the factors that affect silica nanoparticle dispersion by means of molecular dynamics (MD) simulations. All the molecular models used in this work are described by the united-atom approach, with the exception of the chemical modifications performed on the silica nanoparticle surface to include a hydroxylated layer, for which atomistic models are used. We examine both the hydroxylated surfaces (i.e. surfaces contain a high number of silanols) and the dehydroxylated surfaces (i.e. surfaces have a low coverage of hydroxyl groups).

Our results show that the binding energy between the nanoparticles decreased by 37.5% by adding the nonionic surfactant oleyl alcohol ($C18H36O$) to a concentration of 3.3%, which indicates much less overall interparticle agglomeration. To further quantify the results the specific heat (Cv) was calculated during the simulations. By studying the specific heat the optimal conditions for nanoparticle dispersion could be determined. It was concluded that for a filling fraction of 1.15% or less, the particles remain in a homogeneously dispersed state that confirms experimental findings. The transitional boundary between dispersed and clustered states is shifted to a larger filling fraction, showing that the range of dispersed states expands. The width of the transition region becomes narrower, indicating lower fluctuation level when the surfactant is present. In addition, the effects of nanoparticle filling fraction, surface silanol (Si-O-H) group coverage, and surfactant concentration on nanoparticle dispersion are also investigated.

416e Gas Binding Studies With Nanoparticulate Metal Complexes Prepared With Compressed Carbon Dioxide

Chad A. Johnson, Chemical Engineering, University of Kansas, 1530 W 15th Street, 4120 Learned Hall, Lawrence, KS 66045, Joseph Nguyen, Chemistry, University of Kansas, 2010 Malott Hall, 1251 Wescoe Dr, Lawrence, KS 66045, A.s Borovik, Chemistry, University of California-Irvine, Irvine, CA 92697, and Bala Subramaniam, Center for Environmentally Beneficial Catalysis/Chemical Engineering, University of Kansas, 1501 Wakarusa Dr., Suite A-110, Lawrence, KS 66047.

Nanoparticles of metal complexes are rare because of the limited methods for their preparation. We have used precipitation with compressed antisolvent (PCA) to process nanoparticles of neutral metal-salen complexes (N,N'-salicylidiene)-1,2-ethylene-di-aminato(2-)). Current work is focused on comparing the gas binding and catalytic properties of precipitated nanoparticles against the starting material. Preliminary gas uptake studies with a suite of techniques such as Tapered Element Oscillating Microbalance (TEOM), Thermogravimetric Analysis (TGA), and Rubotherm magnetic suspension balance have demonstrated that nanoparticles in the solid state bind both nitric oxide and dioxygen while the starting material shows no measurable binding. Research is also focusing on coating the metal complex nanoparticles on solid supports followed by testing the ability of the coated supports to bind different permanent gases. These results will be presented and discussed.

Synthesis of Unique Ag Nano Plate for Printed Electronics

Young Jin Kim, Chang gun Lee, Sangho Kim, Woojin Lee, †Won Hui Lee, †Insoo Kim

Tokusen U.S.A., Inc. 1500 Amity road P.O. BOX 1150, Conway, Arkansas 72033

ykim@tokusenusa.com

†Advanced Materials Science & Engineering, Dong-A University 840 Hadan2-dong, Saha-gu, Busan 604-714, Korea

iskim@dau.ac.kr

1. Introduction

FPCBs(Flexible Printed Circuit Boards) and PCBs(Printed Circuit Boards) used for mechanical support and electrical connections to conventional electronic components have conductive pathways etched from a Cu sheet laminated onto a nonconductive substrate.

The conventional process, which uses etched Cu for the conductive pathways, is not environment friendly or cost effective due to the requirement for hazardous etchant materials and a multi-step manufacturing process. Additionally, compact electronic devices require FPCBs & PCBs with narrow and accurate conductor patterning. Precious metals with very low electrical resistivity, such as Ag, Au, and Pt are used for electrical components, on finished boards. However, the conventional etching process can't employ precious metals for conductive pathways due to high material costs and excessive waste of circuit material.

Recently, advances in printing technology and circuit materials have allowed direct printing of conductive pathways. The circuit materials most commonly used in this type of printing are Ag paste made using either Ag flake powder (under 10um size) or globular Ag powder. Ag flake powder is inexpensive and readily available, but is difficult to print into accurate circuit patterning with low resistivity. To get lower resistivity, the curing temperature must be higher than 500°C for adequate fusion of the individual Ag particles. Thus, Ag flake is limited to heat resistant substrates. Globular Ag powder, on the other hand, can be successfully cured at temperatures under 200°C but it is more expensive and difficult to handle.

In order to overcome the aforementioned problems, a thin & flat Ag plate powder was developed. The Ag plate powder consists of planar silver crystals that are under 200nm thick and 2 to 7um in width. This particular morphology is very well suited for accurate circuit patterning with high conductivity even at low curing temperatures.

2. Experimental

2-1. Sample preparation

The Ag nano plates were synthesized through a thermal decomposition method. The Ag nano plates were then ultrasonically dispersed in ethanol and agitated for 5 min. before being coated on a glass substrate by spin coating for 10sec. at 1500rpm. After deposition, heat treatments were carried out at various temperatures on hot plate. In order to evaluate the suitability of the Ag nano plates for printing, an Ag paste was formulated using the Ag nano plates and additives.

2-2. Properties test

Electrical resistivity of the heat-treated Ag paste after curing was tested by using a 4-point probe (CMT-2000, AIT) and its particle shape and cross section were observed by using a FE-

SEM (Field Emission Scanning Electron Microscope, JEOL)

3. Results and Discussion

3-1. Size & Shape

Fig. 1 shows particle morphologies of the conventional Ag flake and the new Ag nano plate. As can be seen in the pictures, the conventional Ag flakes (A) are irregularly shaped due to their production process; a ball mill process that makes the flakes from globular Ag powder. This irregular shape results in low tap density and requires extensive use of binders when used in Ag paste. Ag nano plates (B) are formed by a continuous arrangement of atomic units during the thermal decomposition process. This manufacturing process produces thin, planar Ag crystals with exceptional packing properties, which are ideal for printing.

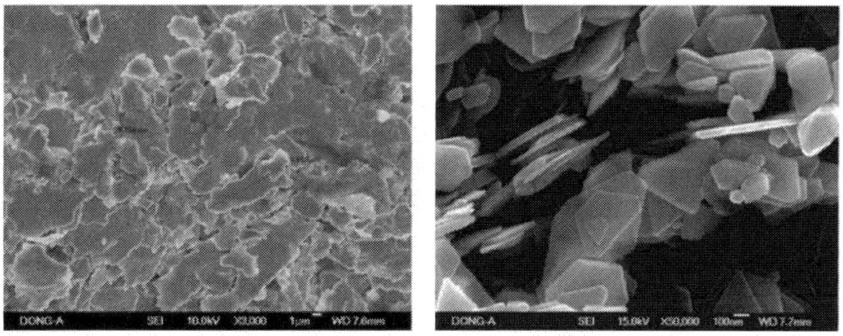

(A) (B)

Fig.1 Morphology comparisons between the conventional Ag flake (A) and Ag nano plate (B).

3-2. Heat treatment & Electrical resistivity

60 wt% of Ag nano plate was dispersed in H2O without any additives. Ag nano plate was ultrasonic-treated for 5 min, and, after then, Ag nano plate was coated on the glass slide by a spin coating method. Coating surface was uniform and very flat due to original characteristics of

Ag nano plate. The electrical resistivity and cross-section was investigated for samples treated under various curing conditions. The curing was performed at hot plate of 100, 150, 200 and 250℃ for 15 min.

The test results are shown in Fig. 2 and Fig. 3. As shown in the Fig. 2, the Ag nano plates were completely fused at 200°C, although melting point of Ag bulk is 960℃. It was confirmed that the plates melted at much lower temperature due to large contact area by flat shape and a nano size effect caused by thin thickness of the Ag nano plate. As can be seen in Fig 3, the electrical resistivity of the fused Ag plates is similar to that of Ag bulk (1.6 μΩ·cm). This is very attractive phenomena, considering many companies want to have both low process temperature and low electrical resistivity (several μΩ·cm)

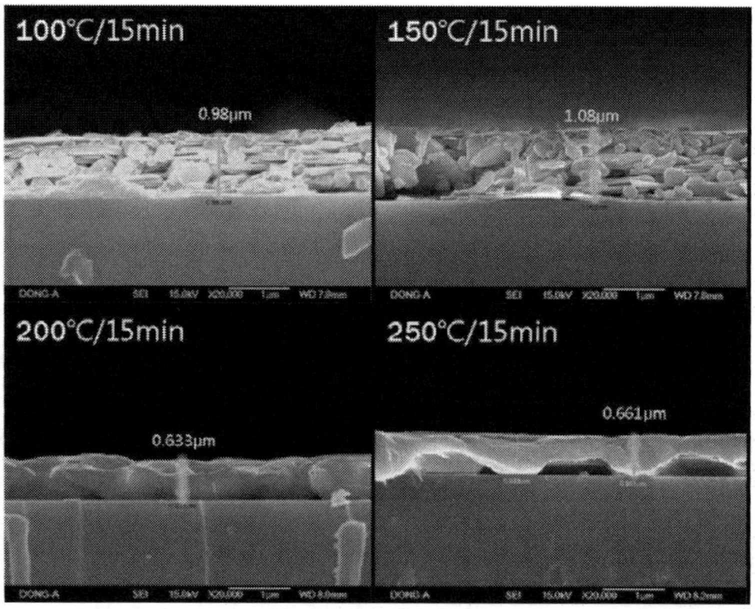

Fig. 2. Cross-sections of the Ag nano plate paste melted at various temperatures.

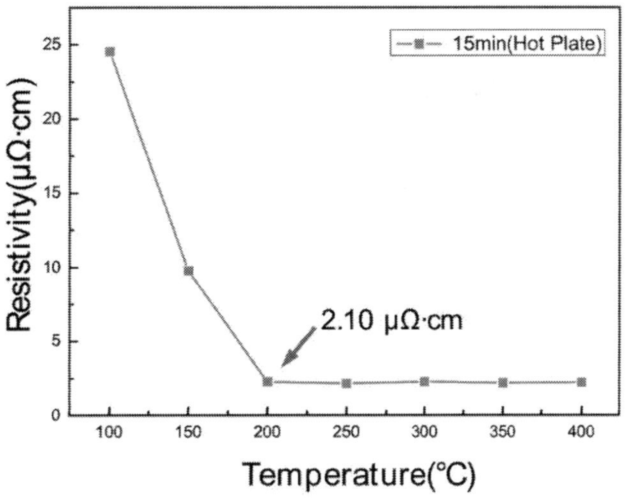

Fig.3. Electrical resistivity of the Ag nano plate paste melted at various temperatures.

3-3. Gravure printing test

Ag paste was formulated using nano-plate, resin, solvent and some additives. Line was formed using a gravure printing method (350 line, 30 m/min) and the surface was observed by FESEM.

Fig. 4 shows line printed using conventional Ag flake and Ag nano plate by a gravure printing method. As can be seen in the pictures, many pinholes were observed on the line formed by conventional flake. On the other hand, the Ag nano plate formed very dense and compact line. This appears to be attributed to flat shape and uniform size distribution of Ag nano plate. This indicates that Ag nano plate can form high resolution line without pinhole.

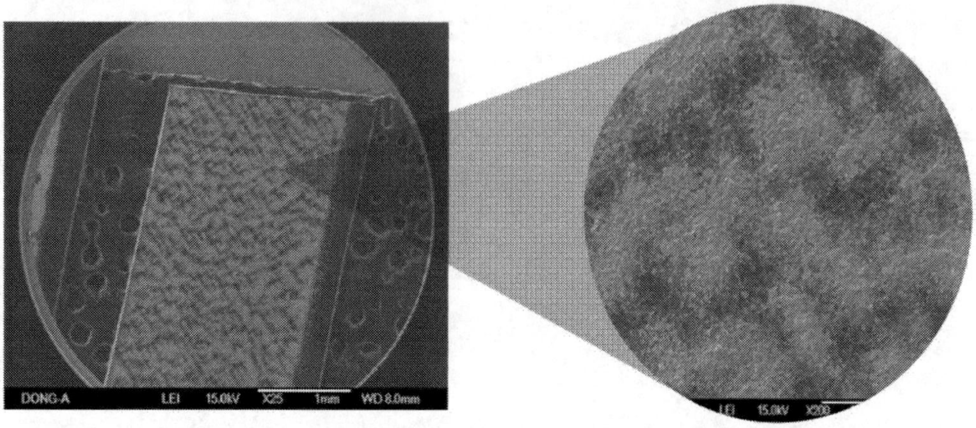

(A) Gravure pattern with conventional Ag flake paste - 350line, 30m/min

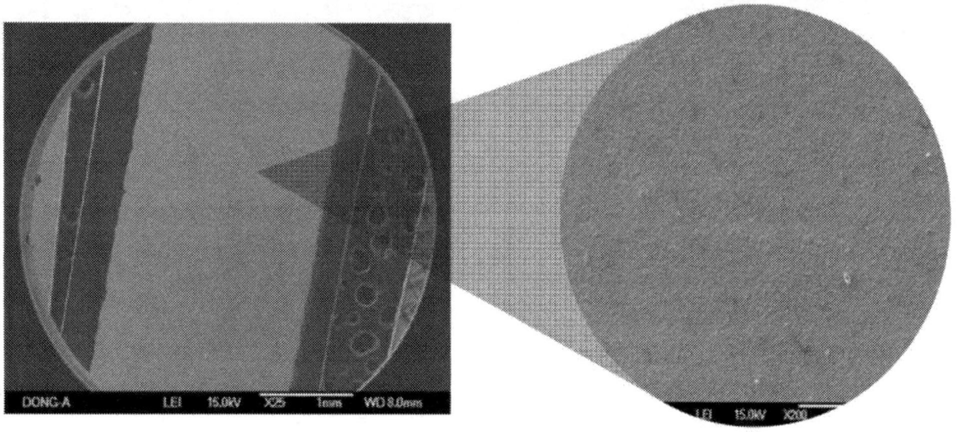

(B) Gravure pattern with Ag nano plate paste - 350line, 30m/min

Fig. 4. Gravure pattern comparisons between conventional Ag flake paste and Ag nano plate paste.

4. Conclusions

New types of flakes were introduced in the paper. This flake has thickness of <200 nm, length of 2-7μm and completely flat surface. Due to these properties, many attractive phenomena were observed.

1) Low Electrical Resistivity:

The coating layer formed by pure Ag nano plate without any additives was completely fused at hot plate of 200°C, and thus lower electrical resistivity was observed. (2.1 μΩ·cm)

2) Formation of Good Coating Layer

Compact and dense line was formed without pinhole when gravure printing method (350 line, 30m/min) was employed. This indicates that Ag nano plate can be used for printing of high-resolution line in mass production.

Characterization of the flow and fluidization properties of ultrafine powders

Qing Huang, Hui Zhang and Jesse Zhu
Particle Technology Research Centre
University of Western Ontario

1. Introduction

In 1973, Dr. Geldart classified powders into 4 groups according to their different fluidization behaviors. Group C powders, which are very fine (generally smaller than 25-30 microns), are very cohesive and thus often referred as fine powders or cohesive powders. It was found that the cohesive nature of fine powders comes from the relatively larger interparticle forces among particles in relation to the gravity (2).

However, because of many unique properties, fine powders are more and more applied in, for example, pharmaceutical, cosmetic and powder coating industries. Therefore, various flow aids - methods to assist powders to flow and be fluidized smoothly - have been studied. Adding certain nanoparticles is one of the effective ways (3).

Powder coating, distinguished from traditional liquid coating, is a method of coating dry powders directly onto the surface of substrates without using any organic solvents. The general procedure is: fluidized particles are first carried by compressed air through a spray gun and charged thereof at first. With the hydrodynamic force imposed by the compressed air and the electrical force caused by the electrical potential between the gun and the grounded substrate, particles are coated onto the surface of substrate. At last, the coated substrate is cured in an oven during which the polymer resin materials are cross-linked and a layer of firm coating is formed (3). However, the current powder coating technology has only been applied to low-end products such as automobile's underhood parts due to the low cosmetic quality, which is caused by the large particle size. Ultrafine powder coating (less than 20 microns) instead, could solve the cosmetic problem, and make it possible to apply to high end products including automobile exterior surface. Generally speaking, the smaller the particles, the more uniform surface can be achieved, but, the more difficult to be fluidized. However, ultrafine powder coating is able to achieve as good coating quality as liquid coating only if the good fluidization quality of ultrafine powders is insured.

In this study, several kinds of nanoparticles were mixed into paint powders to improve the flow and fluidization qualities. To further exam and understand the influence of these nanoparticles, 13 different formulae in combination were made. To determine the flowability of powders, the widely used flow index - angle of repose - was measured. Besides, shear stress of powders, which is also related to their flow property, was measured with an automated shear cell. Furthermore, to determine the fluidization property of powder samples, fluidized bed tests were carried out in a fluidized bed, where the pressure drop and bed expansion were measured.

2. Experiments

2.1 Experimental Materials

The experimental powders were polyurethane based and directly from powder coating industry. The density is 1.57 g/cm^3 , and the average diameter is 16 μm. The nanoparticles adopted were SiO_2, Al_2O_3, and TiO_2 respectively. Their properties are shown in the Table 1.

Besides the impact of each individual type of nanoparticles, the effects of the combination of these nanoparticles were also studied. Table 2 lists all the formulae of nanoparticles employed in this study.

Table 1. Properties of nanoparticles

Name	Diameter (nm)	Density (ρ_s)(g/cm^3)	Specific surface (BET) (m^2/g)	Tap density (g/l)	Remark
SiO_2	12	2.0	110±20	50	Silica_R972
Al_2O_3	13	3.3	100±15	50	Al_2O_3_AluC
TiO_2	21	3.8	50±15	130	TiO_2_P25

Table 2. formulae of combination of flow conditioners

Combination Formulae	Composition (weight ratio)
No.1 (TiO_2)	TiO_2
No.2 (Al_2O_3)	Al_2O_3
No.3 (SiO_2)	SiO_2
No. 4 (TiAl)	$TiO_2 : Al_2O_3=1 : 1$
No. 5 (TiSi)	$TiO_2 : SiO_2 = 1 : 1$
No. 6 (AlSi)	$Al_2O_3 : SiO_2 =1 : 1$
No. 7 (Ti2Al)	$TiO_2 : Al_2O_3 = 2 : 1$
No. 8 (TiAl2)	$TiO_2 : Al_2O_3 = 1 : 2$
No. 9 (Ti2Si)	$TiO_2 : SiO_2= 2 : 1$
No.10 (TiSi2)	$TiO_2 : SiO_2 = 1 : 2$
No.11 (Al2Si)	$Al_2O_3 : SiO_2 =2 : 1$
No.12 (AlSi2)	$Al_2O_3 : SiO_2 =1 : 2$
No.13 (TiSiAl)	$TiO_2 : Al_2O_3 : SiO_2 =1 : 1 : 1$

2.2 Experimental methods

Angle of repose values were measured by a Hosokawa Powder tester in this study.

The shear stress tests were carried out in the shear cell accessory of a FT4 powder rheometer. Following a universal shear stress test procedure, a yield locus for each powder sample would be obtained. Its offset, called cohesion, at the Y axis indicates the maximum shear stress the powder sample can sustain without any normal stress.

In a fluidized bed, the powder is first fluidized for a certain period of time to erase its memories. Then gas velocity was adjusted to a lower point, and the bed height and pressure were measured simultaneously when the powder bed gets stabilized.

2 Results and analysis

The nanoparticles were mixed in paint powders with the method patented by our research group (3). SEM images disclosed that the nanoparticles distributed on the surface of paint powder in the form of small agglomerates. The weight percentages of nanoparticles varied from 0 to 3.0%, which covered most of the range our technology would use. Figure 1 shows the bed expansion data of powders mixed with the combination of nanoparticles Al_2O_3 and SiO_2 with ratio of ½. It is clear that the powder could not be fluidized without nanoparticles; 0.1wt% nanoparticles made some improvement but it was far from enough to make the powder fluidized; 0.3wt% -1.0wt% of nanoparticles largely improved the fluidization quality and as a result, the bed expansion increased dramatically and whole bed height could finally doubled. A comparison of the angle of repose and cohesion value with regard to weight percentage of nanoparticle SiO_2 is shown in Figure 2. In the figure, the angle of repose was plotted as black squares, and the one on the right hand is for cohesion shown as red circles. Similar trends were observed: both angle of repose and cohesion decreased with the increasing amount of SiO_2 and reached the minimum at around 1.0 wt% - 1.2 wt%, and then increase gradually. Therefore, one could conclude that the nanoparticles SiO_2, used in this study could improve the flowability of polyurethane powder effectively, and the improvement increased with the amount of SiO2 till around 1.0 wt% - 1.2 wt%, the improvement slightly decreased afterwards.

The comprehensive study on other formulae of nanoparticles indicated that all the formulae of nanoparticles adopted in the study could improve the flow and fluidization quality to certain degrees. Other formulae except TiO_2 had their effective nanoparticles amount very close to SiO_2: 0.3wt%-1.2wt% nanoparticles could effectively improve the flow and fluidization qualities of paint powders. The influence of TiO_2 showed up only when the weight percentage reached 1.2%.

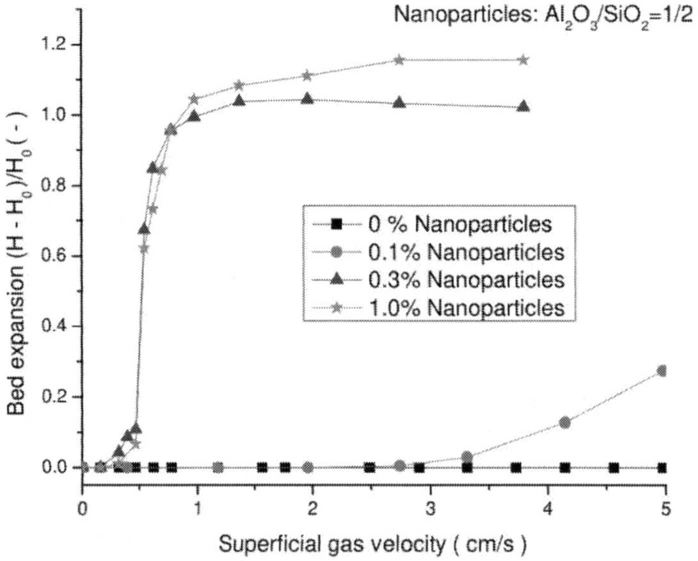

Figure 1. Bed expansion of ultrafine paint powder with nanoparticles

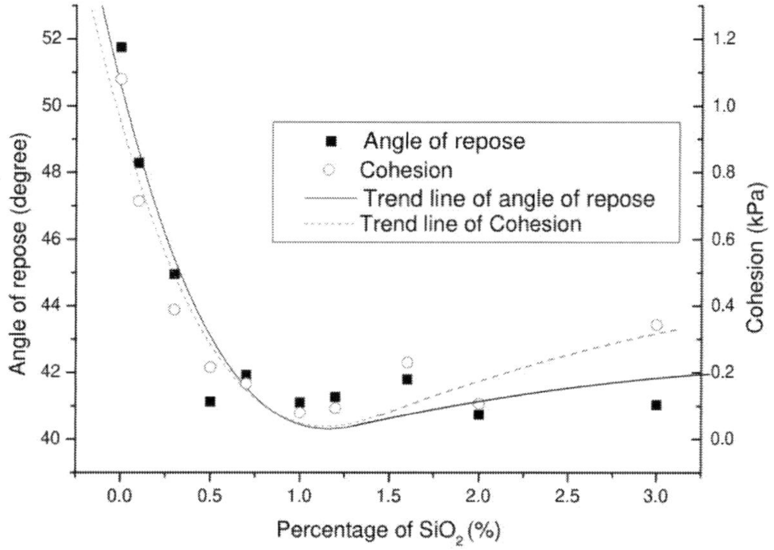

Figure 2. Comparison of Angle of repose and cohesion

4 Conclusions

A comprehensive study concerning the influence of 3 kinds of nanoparticles and their combination on the flow and fluidization qualities of ultrafine paint powders were carried out. It was found that:

1) With suitable amount of nanoparticles, ultrafine paint powders could be fluidized and the bed expansion could reach as high as more than 1. Similarly, the flow index test results (angle of repose and shear stress) showed that the flowability of powders with nanoparticles was improved significantly.

2) All the formulae used in this study could effectively improve the flowability and fluidizability of ultrafine polyurethane powders to certain degrees. An amount of 0.3wt%-1.2wt% nanoparticles is recommended for all the formulae except TiO_2. And more than 1.2w% of TiO_2 would be preferred in practical operations.

References

1. Geldart D. Types of gas fluidization, Powder Technol. 7 (1973) 285-297.

2. Zhu J-X “Fluidization of Fine Powders”, Chapter 10 in Advances in Granular Materials: Fundamentals and Applications, eds. S. J. Antony, W. Hoyle and Y. Ding, Royal Society of Chemistry, London, 2003, pp 270-295.

3. Zhu J and Zhang H, "Fluidization Additives to Fine Powders", U.S. Patent 6,833,185, December 2004

461b Understanding Ion-Mobility Of Aerosol Nanowires And Its Application To The Measuement Of Length

Soo Kim, George Mulholland, and **Michael R. Zachariah**. University of Maryland and NIST, 2125 Glenn L Martin Hall, College Park, MD 20742

In a previous experimental study, we demonstrated that gas phase electrophoretic mobility separation enables one to classify diameter-selected carbon nanotubes (CNTs) by length [Kim, S. H. and Zachariah, M. R. Nanotechnology 16 (2005) 2149-2152] and have subsequently used this capability to track the growth rate of CNT's in free-flight [Kim, S. H. and Zachariah, M. R., J. Phys. Chem. B 110 (2006) 4555-4562]. In this talk, we also develop a theoretical model to describe the behavior of nanotubes (or nanowires) undergoing Brownian rotation in an electric field, and provide a more rigorous interpretation of the experimental results. The Bolzmann expression for the orientation probability includes both the free charge energy as well as the polarizability energy. In the theoretical model, we computed the orientation-averaged electrical mobility and the precipitating time of a rotating nanowire in DC electric field in the free-molecular limit. This analysis was used to obtain the precipitation time of a nanowire in a differential mobility analyzer (DMA). Based on the theoretical model and its comparison with experimental measurement, we found that (i) a stronger electric field was required to select longer nanowires for a given diameter and (ii) shorter nanowires with the aspect ratio of •<~30 (df=15 nm) freely rotate for applied electric field up to ~1kV/cm, while longer nanowires (•>~30) were aligned by the electric field. The experimentally determined nanowire length was in fair agreement with that computed by our theoretical model, in which the effect of the nanowire's alignment on mobility classification in various electric fields was accounted for. The over prediction of the model for longer nanowires is shown to result from the bent structure of the longer nanowires. The methodology should be generic to other nanostructures with a high aspect ratio.

Web Page: www.enme.umd.edu/~mrz

Effect of Primary Particle Size on Steady-State Aggregates Size and Structure under Turbulent Conditions in Stirred Tank

Lyonel Ehrl, Miroslav Soos, Amgad Salah Moussa, Massimo Morbidelli

Institute for Chemical and Bioengineering, Department of Chemistry and Applied Biosciences, ETH Zurich, 8093 Zurich, Switzerland

Abstract

The evolution of the steady-state size and structure of aggregates produced under turbulent conditions in stirred tank for primary particle diameters equal to $120\,\text{nm}$, $420\,\text{nm}$, and $810\,\text{nm}$ was studied experimentally for various values of the volume average shear rate and solid volume fraction. The steady-state aggregate structure was characterized by image analysis of two-dimensional pictures of the aggregates obtained by confocal microscopy. The mass fractal dimension showed to be invariant to primary particle size and applied shear rate with a value equal to 2.64 ± 0.18. The absolute values of the root-mean-square radius of gyration obtained from small-angle static light scattering measured for similar solid volume fraction and hydrodynamic conditions proved to be independent from the primary particle size. The root-mean-square radius of gyration for all studied conditions follows a power-law scaling with the volume average shear rate with an exponent equal to $-1/2$, which is in agreement with our previous data and data published in the literature. Using these values combined with a force balance approach leads to an aggregate strength independent from the aggregate size and the obtained values agree well with values published in the literature. Relating the two measured moments of the cluster mass distribution, namely the root-mean-square radius of gyration and the zero-angle intensity, with the independently measured aggregates structure can be used to evaluate the effect of multiple scattering within the aggregates.

Keywords

Aggregation, Breakage, Primary Particle Size, Small-Angle Static Light Scattering (SASLS), Image Analysis, Aggregate Structure, Dense Clusters, Rayleigh-Debye-Gans (RDG), Steady-State Reversibility, Maximum Stable Aggregate Size, Aggregate Strength

1 Introduction

Many solid-liquid separation processes, such as settling, flotation, filtration, or dewatering are preceded by an aggregation step to improve the solid phase characteristics for better removal. Depending on the physical properties of the dispersed particles and the suspending fluid, the aggregation process can be driven by several mechanisms, such as Brownian motion[1, 2], shear[1, 3], or differential settling[4]. Shear aggregation can be used to produce aggregates with well-defined size distribution and morphology, whereas, to properly control the process it is necessary to understand and quantify the physical mechanisms which control the aggregation process.

Shear driven aggregation typically results in a sigmoid-like evolution of the aggregates size, due to a self-accelerating growth, followed by a steadily slowing relaxation to a steady-state aggregate size[5-7]. It was found that the initial time evolution of the cluster mass distribution (CMD), if presented in terms of a certain characteristic size and plotted in appropriate dimensionless time, follows a single curve[8, 9]. After this initial phase of self-accelerating aggregate growth, the kinetics gradually slows down as breakage sets in and eventually reaches steady state. This steady state predominantly depends on the size and concentration as well as surface and bulk properties of the primary particles, composition of the liquid phase and type of coagulant, and average value and distribution of the shear rate in the vessel. For systems in which non-precipitating coagulant is used, it was shown that the steady state is controlled by a balance between aggregation and breakage, with $d_f \approx 2.6$, independent of the applied history of solid volume fraction[8], or shear rate[6, 9].

There exists only little work on the effect of primary particle size on the aggregates characteristics[6, 10, 11], whereas the change in particle size is often accompanied by a change of the surface chemistry and/or solid volume fraction. The goal of this work is to investigate the sole effect of primary particle size on the steady-state aggregate size and structure for various values of the shear rate[9] and solid volume fraction[8, 9]. Therefore, three polystyrene latexes with similar surface chemistry, so that the primary particle diameter, d_p, is the only changing parameter, were aggregated under turbulent conditions in stirred tank.

2 Materials, Methodology, and Measurement Methods

In the experiments two white sulfate polystyrene latexes were used, both supplied by *Interfacial Dynamics Corporation* (Portland, OR, USA) (product-no: 1-400, $cv = 4.0\%$, batch no: 1034,2, $wt\% = 8.0$, surface charge density: $4.8\mu C/cm^2$; product-no: 1-100, $cv = 8.4\%$, batch no: 1614,1, $wt\% = 4.1$, surface charge density: $6.7\mu C/cm^2$). The mean diameter of the latex particles were equal to $420nm$ and $120nm$, as measured by SASLS, in agreement with the particle size declared by the producer. The measured particle size distributions are very narrow, and the latexes can be considered as mono-disperse. The latexes are chosen in a way to ensure similar surface chemistry (nature and number of surface charge groups per surface area) to that of the latex used in our previous work[9] (product-no: 1-800, $d_p = 810nm$, $cv = 2.0\%$, batch no: 642,4, $wt\% = 8.1$, surface charge density: $5.2\mu C/cm^2$). Hence, the primary particle diameter is the only changing parameter and, therefore, its effect can be studied independently from other phenomena that might be induced by a variation in the surface or bulk chemistry of the particles.

All experiments were performed in a $2.5L$ stirred tank coagulator, as shown in Figure 1, operated with stirring speeds covering a range from 200 to $1073rpm$[9, 12].

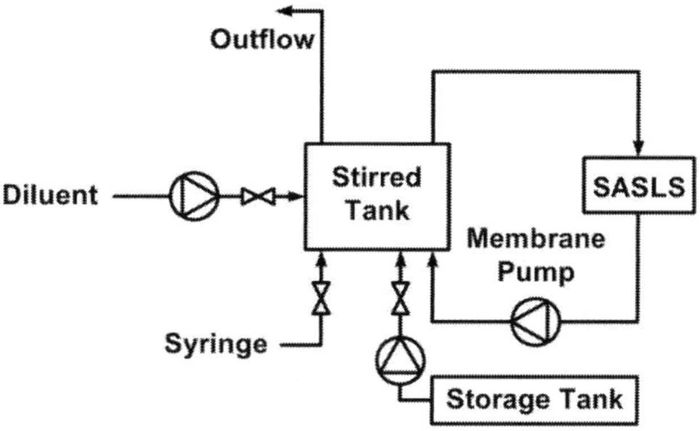

Figure 1: Schematic drawing of the experimental setup.

Each experiment consisted of two parts. In the first part, a suspension of primary particles was destabilized by a salt solution and aggregated under turbulent conditions until the system reached steady state. In the second part the dynamic response of the system on either a variation of the shear rate or on a variation of the solid volume fraction, φ, via dilution was investigated. For dilution experiments, a particle-free salt solution with the same salt concentration as that in the tank was pumped at a constant flow rate into the tank. In this way the systems steady-state response was investigated for solid volume fractions ranging from 2×10^{-5} down to below 1×10^{-6}.

A SASLS instrument, *Mastersizer 2000* (*Malvern*, U.K.), was used in all experiments for on-line characterization of the CMDs in the stirred tank in terms of angle-dependent intensity of scattered light, $I(q)$. Analysis of $I(q)$ in the Guinier region[13] (for qR_g up to about unity) allows one to extract certain characteristic quantities of the CMD, namely the root-mean-square (rms) radius of gyration, $\langle R_g \rangle$, and the zero-angle intensity, $I(0)$. Within the limits of the Rayleigh-Debye-Gans (RDG) theory[14, 15] $I(0)$ scales with the second power of the scatterers mass and constitutes a second order moment of the CMD. Outside the limits of the RDG theory, i.e., for primary particles of a size comparable to or larger than the laser wavelength[16-18], for small non-fractal aggregates[19], or in the case of very dense aggregates, where multiple light scattering within the aggregate is present[16, 20] the intensity of the forward scattered light will scale with the mass of the scatterer to a power smaller than two and, therefore, $I(0)$ will not anymore correspond to the second order moment of the CMD.

If the primary particle diameter is within the limits of the RDG theory, and the obtained aggregates are large enough to exhibit a well developed fractal scaling, the mass fractal dimension of the aggregates, d_f, is an additional quantity, which can be obtained from the slope of the power-law region of the log-log plot of $S(q)$ vs. q. However, this is only explicitly valid within the limits of the RDG theory[13-15, 17], and for many systems, such as those studied in this work, with primary particle diameters comparable to or larger than the laser wavelength[16-18], with dense aggregates, where multiple light scattering within the aggregates occurs[16, 20], with small relative aggregate sizes[19], or with a certain polydispersity of the CMD[21], the slope of the power-law region of the log-log plot of $S(q)$ vs. q cannot be directly interpreted as d_f.

Therefore, we will denote this slope as the scaling exponent, SE, which is not necessarily equal to d_f. According to this the aggregate structure was independently characterized using image analysis of two-dimensional images of aggregates, taken from samples withdrawn for off-line measurements at various operating conditions. In this way, the structure of aggregates can be characterized by the perimeter fractal dimension, d_{pf}, obtained from the scaling of the projected surface area, A, vs. the perimeter, P, of the binary images of aggregates[22], $A \propto P^{2/d_{pf}}$. Subsequently, the perimeter fractal dimension was used to evaluate the mass fractal dimension according to a correlation presented by Lee and Kramer[23].

Detailed description of the experimental methodology, the used measurement techniques, and measurement data analysis can be found in Moussa et al.[8] (dilution experiment) and Soos et al.[9] (shear experiment, image analysis).

3 Evaluation of Aggregate Strength

Assuming that the maximum stable aggregate size results from a balance of the hydrodynamic force exerted on the aggregate, F_{hyd}, and the cohesive force, F_{coh}, holding the aggregate together (hereafter called the aggregate strength) the criterion for breakage reads as $F_{hyd} > F_{coh}$[24, 25]. Since the aggregate sizes in all our experiments are comparable to or smaller than the Kolmogorov micro-scale[8, 9], we assume that breakage occurs within the viscous sub-range. Further due to a very compact aggregate structure[9] the hydrodynamic force can be approximated as the hydrodynamic force acting on the hemisphere of a sphere of comparable size in a simple shear flow[26] with $F_{hyd}^{vs} = (5/8)\pi\eta r^2 G$, where η is the dynamic viscosity, r is the characteristic size (which is assumed to be adequately represented by $2\langle R_g \rangle$ in the case of an aggregate), and G is the local shear rate. In the limits of observable aggregation kinetics the steady-state CMD, and with it also the maximum aggregate size, is a result of equilibrium between aggregation and breakage and, therefore, depends on the solid volume fraction[8]. For this reason, the aggregate strength will be extracted from the critical aggregate size for breakage obtained under infinite dilution conditions, at which aggregation can be safely neglected and the aggregate size is controlled by breakage only[9].

4 Results

The steady-state size and structure of aggregates produced from polystyrene latexes with primary particle diameters equal to $120\,nm$, $420\,nm$, and $810\,nm$ under turbulent conditions in stirred tank was obtained for various values of the volume average shear rate and solid volume fraction. Applying step changes in stirring speed it was found that the steady-state CMD is completely reversible with no dependency on shear rate history. An example of an experiment aiming to investigate this issue is shown in Figure 2, for a primary particle diameter equal to $420\,nm$. As can be seen, complete reversibility of the steady-state values, characterized by $\langle R_g \rangle$ and $I(0)$, with respect to a variation of the shear rate was observed. Similar findings were observed for the $120\,nm$ as well as $810\,nm$[9] primary particles.

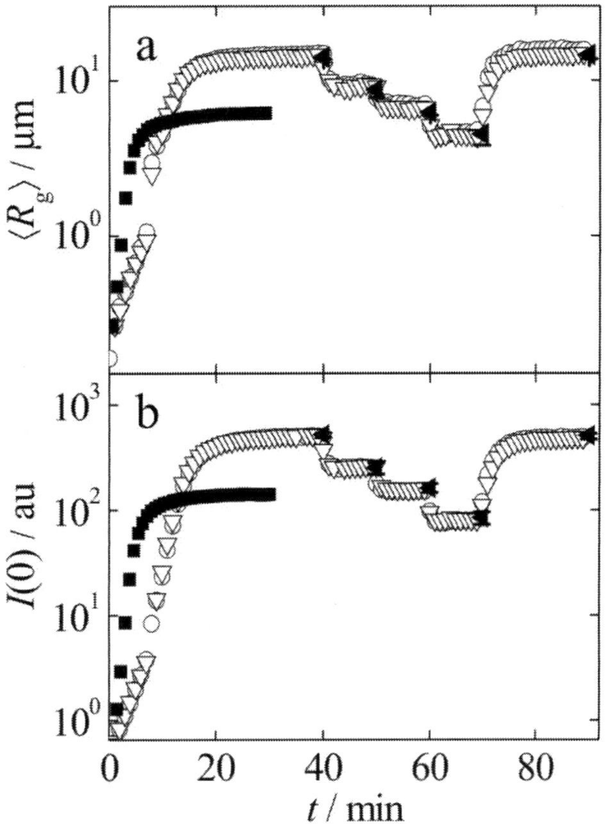

Figure 2: Examples of the time evolution of (a) the rms radius of gyration, $\langle R_g \rangle$, and (b) the absolute zero-angle intensity, $I(0)$ (in arbitrary units, au), during step changes of the volume average shear rate, $\langle G \rangle$, through a change in rotation speed, measured for a solid volume fraction equal to 2×10^{-5} and a primary particle diameter of 420 nm; (○,▽) the volume average shear rate was changed in the following manner: $\langle G \rangle = 106 - 287 - 517 - 1097 - 106\,\mathrm{s}^{-1}$; (■) $\langle G \rangle = 517\,\mathrm{s}^{-1}$; (○,▽,■) on-line measurement, (◀) off-line measurements.

In Figure 3 are presented images of aggregates produced under similar experimental conditions (200 rpm, $\varphi = 2\times10^{-5}$) for different sizes of primary particles. The effect of rotation speed for all investigated primary particle sizes is summarized in Figure 4. As can be seen, the steady-state aggregate structure obtained via image analysis provided a mass fractal dimension equal to 2.64 ± 0.18 (hatched area), independent of primary particle diameter and shear rate. However, this does not agree with the lower values of the scaling exponent obtained from light scattering data, as can be seen from the comparison of the values of SE obtained from $S(q)$ with d_f obtained from image analysis[23]. This is caused by the non-validity of the RDG theory combined with the presence of a large portion of small non-fractal aggregates, especially for the case of large primary particles and highest stirring speeds.

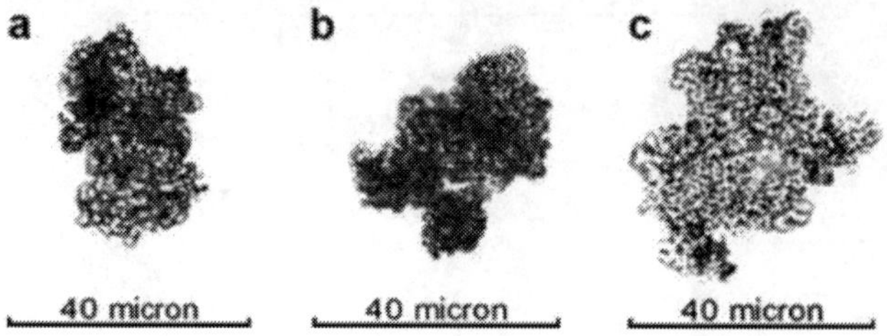

Figure 3: Images of aggregates composed of primary particles with different diameters (a) 810nm[9], (b) 420nm, and (c) 120nm for a volume average shear rate equal to $106\,s^{-1}$ obtained by confocal microscopy from samples taken at steady state with $\varphi = 2\times10^{-5}$.

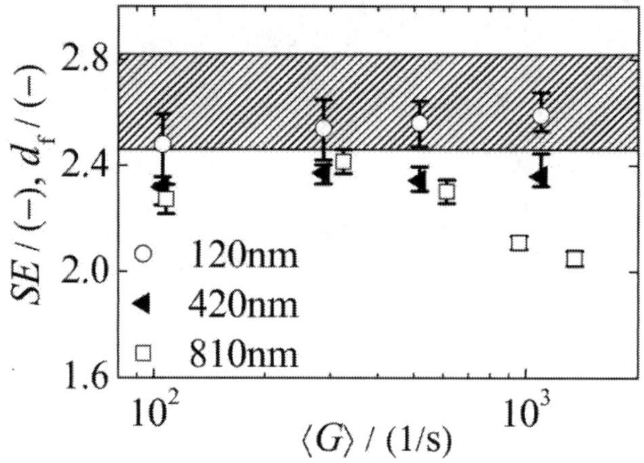

Figure 4: Comparison between values of the scaling exponent, SE (symbols) together with the range of fractal dimension, d_f (hatched bar), obtained by image analysis[23].

The absolute values of rms radius of gyration, presented in Figure 5, obtained under similar solid volume fraction and hydrodynamic conditions proved to be independent from the primary particle diameter. Moreover, the rms radius of gyration for both $\varphi = 2\times10^{-5}$ as well as $\varphi \rightarrow 0$ followed a power-law scaling with the volume average shear rate according to $\langle R_g \rangle = \langle G \rangle^{-1/2}$, which is in agreement with data published in the literature. Due to the scaling behavior of $\langle R_g \rangle$ versus $\langle G \rangle$, the constant force approach[27] is assumed to be the most appropriate way to evaluate aggregate strength, which was evaluated from the critical aggregate size for breakage, obtained at low solid volume fraction, where aggregation can be safely neglected. It was found to be equal to 6.18 ± 0.91nN, 3.53 ± 0.77nN, and 6.12 ± 1.54nN for primary particle diameters equal to 810nm[9], 420nm, and 120nm, respectively. Comparing this values to available data in the literature[28-32] we found good agreement. The observation of an aggregate strength independent from the primary particle diameter is explained by the non-ideality of the particles surface, in particular the effect of the surface roughness is hold responsible in agreement with AFM pull-off force measurements on micron sized polystyrene particles[30, 31, 33] providing absolute bonding forces of the same order.

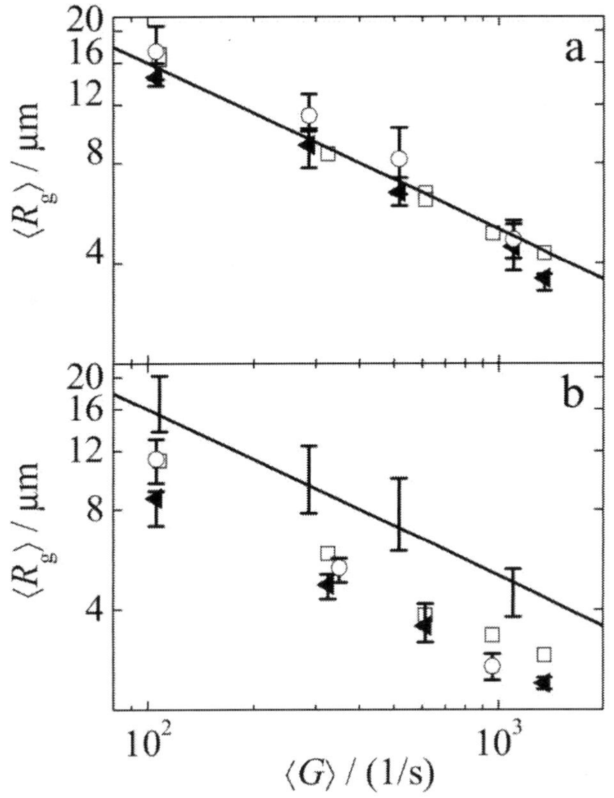

Figure 5: Scaling of the rms radius of gyration, $\langle R_g \rangle$, (a) at steady state for a solid volume fraction equal to 2×10^{-5} and (b) at infinite dilution conditions for different primary particle diameters, i.e., (○) $120\,\text{nm}$, (◀) $420\,\text{nm}$, and (□) $810\,\text{nm}$[9], as a function of the volume average shear rate, $\langle G \rangle$. The solid line indicates a scaling with a slope equal to $-1/2$.

The second measured moment of the CMD, the zero-angle intensity of scattered light, showed a power-law scaling with the rms radius of gyration according to $I(0) \propto \langle R_g \rangle^a$, with the exponent depending on the primary particle diameter (values for a equal to 2.36, 1.64, and 1.38 for d_p equal to $120\,\text{nm}$, $420\,\text{nm}$, and $810\,\text{nm}$, respectively), valid both at finite volume fractions as well as at infinite dilution conditions. These slopes do not agree with the expected scaling, within the limits of RDG theory for a mono-sized or self-preserving system of scatterers with constant mass fractal dimension $a = d_f$. This deviation is explained by three reasons, first, the primary particle diameters become comparable or larger than the laser wavelength of the scattering device and, therefore, the RDG theory is not anymore valid, second, due to a very dense aggregate structure multiple light scattering within the aggregate occurs, and third, for larger primary particles the relative aggregate size decreases, leaving a large portion of small non-fractal aggregates. Using an independently measured mass fractal dimension, e.g., obtained by image analysis, the relation between $I(0)$ and $\langle R_g \rangle$ can serve to evaluate the effect of multiple light scattering within the aggregates.

5 Acknowledgement

The authors thank Matthäus Ulrich Bäbler, Marco Lattuada, Amgad Salah Moussa, and Hua Wu, for useful discussions and their helpful suggestions. This work was financially supported by the Swiss National Science Foundation (Grant No. 200020-113805/1).

6 References

1. von Smoluchowski, M. Versuch einer mathematischen Theorie der Koagulationskinetik kolloider Lösungen. Zeitschrift für Physikalische Chemie. 1917; 92:129-168.
2. Lin, M.; Lindsay, H. M.; Weitz, D. A.; Ball, R. C.; Klein, R.; Meakin, P. Universal diffusion-limited colloid aggregation. Journal of Physics: Condensed Matter. 1990; 2:3093-3113.
3. Saffman, P. G.; Turner, J. S. On the collision of drops in turbulent clouds. Journal of Fluid Mechanics. 1956; 1:16-30.
4. Friedlander, S. K. Smoke, Dust and Haze (First Edition edition). New York: John Wiley & Sons: 1977.
5. Reich, I.; Vold, R. D. Flocculation-Deflocculation in Agitated Suspensions - I. Carbon and Ferric Oxide in Water. Journal of Physical Chemistry. 1959; 63:1497-1501.
6. Kusters, K. A. The Influence of Turbulence on Aggregation of Small Particles in Agitated Vessel. PhD Thesis, Eindhoven University of Technology, Netherlands, Eindhoven, 1991.
7. Flesch, J. C.; Spicer, P. T.; Pratsinis, S. E. Laminar and Turbulent Shear-Induced Flocculation of Fractal Aggregates. A.I.Ch.E. Journal. 1999; 45:1114-1124.
8. Moussa, A. S.; Soos, M.; Sefcik, J.; Morbidelli, M. Effect of Solid Volume Fraction on Aggregation and Breakage in Colloidal Suspensions in Batch and Continuous Stirred Tank. Langmuir. 2007; 23:1664-1673.
9. Soos, M.; Moussa, A. S.; Ehrl, L.; Sefcik, J.; Wu, H.; Morbidelli, M. Effect of shear rate on the aggregates size and morphology investigated under turbulent conditions in stirred tank. journal not decided. to be published;
10. Serra, T.; Colomer, J.; Casamitjana, X. Aggregation and Breakage of Particles in a Shear Flow. Journal of Colloid and Interface Science. 1997; 187:466-473.
11. Selomulya, C.; Bushell, G.; Amal, R.; Waite, T. D. Aggregation Mechanisms of Latex of Different Particle Sizes in a Controlled Shear Environment. Langmuir. 2002; 18:1974-1984.
12. Waldner, M. H.; Sefcik, J.; Soos, M.; Morbidelli, M. Initial growth kinetics of aggregates in turbulent coagulator. Powder Technology. 2005; 156:226-234.
13. Sorensen, C. M. Light Scattering by Fractal Aggregates: A Review. Aerosol Science and Technology. 2001; 35:648-687.
14. Kerker, M. The Scattering of Light and Other Electromagnetic Radiation New York: Academic Press: 1969.
15. Jones, A. R. Light scattering for particle characterization. Progress in Energy and Combustion Science. 1999; 25:1-53.
16. Lambert, S.; Thill, A.; Ginestet, P.; Audic, J. M.; Bottero, J. Y. Structural Interpretations of Static Light Scattering Patterns of Fractal Aggregates - I. Introduction of a Mean Optical Index: Numerical simulations. Journal of Colloid and Interface Science. 2000; 228:379-385.

17. Farias, T. L.; Köylü, Ö. Ü.; Carvalho, M. G. Range of validity of the Rayleigh-Debye-Gans theory for optics of fractal aggregates. Applied Optics. 1996; 35:6560-6567.
18. Tishkovets, V. P.; Petrova, E. V.; Jockers, K. Optical properties of aggregate particles comparable in size to the wavelength. Journal of Quantitative Spectroscopy & Radiative Transfer. 2004; 86:241-265.
19. Lattuada, M.; Wu, H.; Morbidelli, M. A simple model for the structure of fractal aggregates. Journal of Colloid and Interface Science. 2003; 268:106-120.
20. Petrova, E. V.; Tishkovets, V. P.; Jockers, K. Polarization of Light Scattered by Solar System Bodies and the Aggregate Model of Dust Particles. Solar System Research. 2004; 38:309-324.
21. Soos, M.; Sefcik, J.; Morbidelli, M. Master Curves for Aggregation and Gelation: Effects of Cluster Structure and Polydispersity. Industrial & Engineering Chemistry Research. 2007; 46:1709-1720.
22. Mandelbrot, B. B.; Passoja, D. E.; Paullay, A. J. Fractal character of fracture surfaces of metals. Nature. 1984; 308:721-722.
23. Lee, C.; Kramer, T. A. Prediction of three-dimensional fractal dimensions using the two-dimensional properties of fractal aggregates. Advances in Colloid and Interface Science. 2004; 112:49-57.
24. Tambo, N.; Hozumi, H. Physical Characteristics of Flocs - II. Strength of Floc. Water Research. 1979; 13:421-427.
25. Higashitani, K.; Inada, N.; Ochi, T. Floc breakup along centerline of contractile flow to orifice. Colloids and Surfaces. 1991; 56:13-23.
26. Coufort, C.; Liné, A. Forces on Spherical Particles in Terms of Upstream Flow Characteristics. Chemical Engineering Research & Design. 2003; 81:1206-1211.
27. Potanin, A. A. On the Computer Simulation of the Deformation and Breakup of Colloidal Aggregates in Shear Flow. Journal of Colloid and Interface Science. 1993; 157:399-410.
28. Yeung, A. K. C.; Pelton, R. Micromechanics: A New Approach to Studying the Strength and Breakup of Flocs. Journal of Colloid Interface Science. 1996; 184:579-585.
29. Blaser, S. Flocs in Shear and Strain Flows. Journal of Colloid and Interface Science. 2000; 225:273-284.
30. Hodges, C. S.; Cleaver, J. A. S.; Ghadiri, M.; Jones, R.; Pollock, H. M. Forces between Polystyrene Particles in Water using the AFM: Pull-Off Force vs Particle Size. Langmuir. 2002; 18:5741-5748.
31. Hodges, C. S.; Looi, L.; Cleaver, J. A. S.; Ghadiri, M. Use of the JKR Model for Calculating Adhesion between Rough Surfaces. Langmuir. 2004; 20:9571-9576.
32. Kobayashi, M. Breakup and strength of polystyrene latex flocs subjected to a converging flow. Colloids and Surfaces A. 2004; 235:73-78.
33. Heim, L. O.; Ecke, S.; Preuss, M.; Butt, H. J. Adhesion forces between individual gold and polystyrene particles. Journal of Adhesion Science and Technology. 2002; 16:829-843.

461d Surface Coverage On A Nanoparticle

Leonard F. Pease III[1], De-Hao Tsai[2], Rebecca A. Zangmeister[1], Michael R. Zachariah[2], and Michael J. Tarlov[1]. (1) Process Measurements Division, National Institute of Standards and Technology, 100 Bureau Drive, Gaithersburg, MD 20899, (2) University of Maryland and NIST, 2125 Glenn L Martin Hall, College Park, MD 20742

Here we present electrospray-differential mobility analysis (ES-DMA) as a tool to determine the conformation and surface coverage of long, flexible molecules on engineered nanoparticles. As an initial model system, we examine gold nanoparticles coated with single-stranded DNA (ssDNA), which hold potential for a variety of uses ranging from cancer treatments to DNA sequencing probes. Key to each of these applications is the determination of the coverage and configuration of the strands on the surface of the particle, as they affect the stability, properties, and utility of the suspension. We employ ES-DMA to determine the size distribution of both coated and bare particles, and thus the apparent coating thickness, based on their electrical-to-drag force ratio. We then determine the configuration and surface coverage with a simple analytical model. We find the ssDNA adopts a random coil configuration and the surface coverages determined by this method, $2.5*10^{12}$-$8.7*10^{13}$ molecules/cm^2, are in accord with those reported by other techniques. These results indicate the value of ES-DMA within the repertoire of tools for the characterization of engineered nanoparticles.

461e Electrokinetic Properties And Stability Of Engineered Fullerene (C_{60}) Nanoparticles In Aqueous Solutions

Kai Loon Chen and Menachem Elimelech. Department of Chemical Engineering, Environmental Engineering Program, Yale University, New Haven, CT 06520-8286

Buckminsterfullerene C_{60} is an emerging nanomaterial that has attracted interest for its diverse potential applications. Since fullerene C_{60} is only soluble in some organic solvents and is almost insoluble in water, recent efforts have been made to synthesize fullerene nanoparticles suspended in water suitable for use in aqueous and biological systems. Currently, there are two techniques employed to synthesize these nanoparticles. One technique involves first dissolving fullerene in an organic solvent before ultrasonicating the mixture in water to remove the organic solvent. Another technique simply requires the prolonged stirring of fullerene in water. Past studies have shown that fullerene nanoparticles synthesized through both techniques exhibit a negative electrophoretic mobility, the source of which is still unclear. In this study, we compare the electrokinetic properties and stability ratios of fullerene nanoparticles synthesized through both techniques with the objective of understanding the source of their surface charge. Electrophoretic mobility measurements showed that both nanoparticles remained negatively charged over the pH range of 2–12, and that the electrophoretic mobility became less negative towards the lower pH. The aggregation kinetics of these nanoparticles were derived through time-resolved dynamic light scattering over a range of sodium and calcium chloride concentrations. The nanoparticles synthesized through prolonged stirring proved to be more stable to aggregation, implying that the synthesis technique plays an important role over their surface characteristics. The electrokinetic properties of the fullerene nanoparticles were further investigated in the presence of cationic and anionic surfactants.

DEVELOPMENT OF A NEW METHOD TO GET A RELIABLE POWDER FLOW CHARACTERISTICS USING ONLY ONE TO TWO GRAMS OF POWDER

Kari Seppälä[1], Jouko Yliruusi[2] and Jyrki Heinämäki[2]

[1]Orion Corporation, Espoo, Finland
[2]Division of Pharmaceutical Technology, Faculty of Pharmacy, P.O.Box 56, FI-00014 University of Helsinki, Finland

INTRODUCTION

In pharmaceutical technology, the flowability of powder is an important characteristic that affects in many pharmaceutical systems. For example in predicting the weight variation of tablets in fast operating tablet presses it is necessary to know the flowing properties of the powder accurately.

To date, only few reports are available on development of a novel flowability testing method for poorly flowing powders in a small scale. Some of the proposed methods are based on direct measuring of powder flow. The reason why direct methods intended for the measuring flowability of poorly flowing powders have not worked is because the powder forms a vault structure. When a vault structure is formed it doesn't break just by force or vibrating it. This is the reason why there isn't any measurement device that determinates accurately the flowability of poorly flowing powders using small sample amounts. There are a couple of indirect methods to study flowability, for example determination a ratio of bulk and tap density (i.e. Carr's and Hausner indexes) which has a correlation with flowability, These approaches, however, can not provide accurate information on flowability since in determination of bulk and tap density, the shape of the particles affects more than electrical forces.

The basic idea in our new flowability testing method is to break the vault structure with a single upward motion, and consequently, let the powder flow through an orifice freely until it forms a vault structure again. This cycle is repeated until the powder has flowed completely. Hereby, the device described in this paper measures the powders ability to form a vault structure and the strength of that vault structure. This ability is directly proportional to the flowability of a powder. The resolution of the present flowability device has also turned out to be high.

MATERIALS AND METHODS

Materials

The materials used were glucose anhydrate (Ph. Eur.), sucrose (Dan Sukker / Suomen sokeri Oy, Food Grade), partially pregelatinized maize starch (Starch 1500, Colorcon Ltd., U.K.), microcrystalline cellulose, MCC (Avicel PH-101, Avicel PH-102 and Avicel PH-200, FMC Corporation, U.S.A), magnesium stearate (Ph. Eur.), caffeine (Ph. Eur.), carbamazepine (Hawkins Inc, USP), and paracetamol (Ph. Eur.).

Methods

All materials were manually screened before testing by using a standard sieve (750 µm) to remove lumps. If the powder sample is composed of lumps, the real flowability value can not be determined by this method or any method based on direct flowing of the powder through an orifice. Bulk and tap densities were determined by using an established Ph.Eur. method and device (Erweka SVM tapping device, Erweka Apparatebau GmbH, Germany). For preparing binary powder mixtures of MCC and magnesium stearate (0.5% w/w), the powders were mixed for 2 min at 22 rpm in a glass jar using a laboratory-scale Turbula mixer (Turbula T2F, Willy A. Bachofen AG Machinenfabrik, Switzerland).

The schematic diagram of the present novel flowability testing device designed for rapid flowability determinations of powders in a small scale, is presented in Figure 1. The system includes a frame, motor, sample holder, funnel, orifice and analytical scale. The volume of the funnel is 5.96 ml and the diameter of orifice is 3.0 mm. Motor stirs the funnel vertically which makes the powder flow through an orifice and the breaking of the vault structure possible. With analytical scale connected to a computer it is possible to calculate the flow rate (mg/s) and investigate the shape of the mass function. To minimize the non-linearity of the mass flow, 5% in the beginning and at the end of the mass flow function were not taken into count while calculating the flow rate. In flowability determinations, the

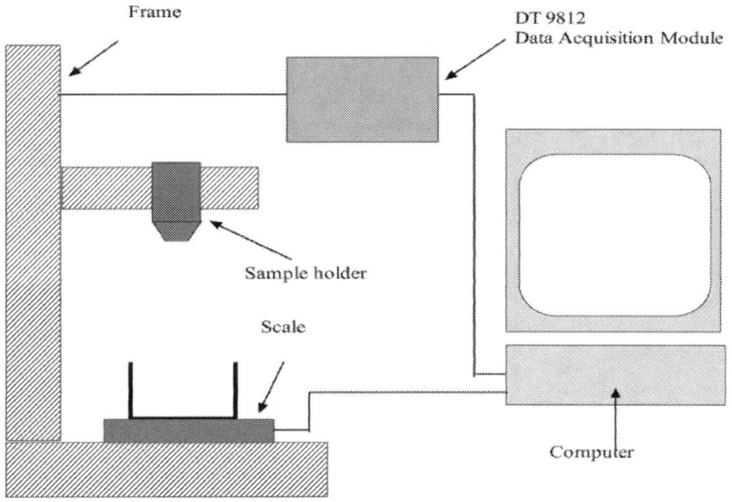

Figure 1. Block diagram of the measuring arrangement.

device was at a lower position 80% of the time and at a upper position 20% of the testing time. This makes it possible that the upward movement is powerful enough to break the vault structure of the sample. Furthermore, the potential flowing time (down position) was long enough to let the powder flow freely until it forms a vault structure again with the frequency of 1 Hz.

RESULTS AND DISCUSSION

All powders have numerous different attributes that affect their flowability rate. Such attributes are for example particle shape, density, surface roughness and static electricity. Every device that is designed to measure flowability, emphasizes the other of these attributes more. Therefore, not a single device available to date can properly measure flowability of powders from small powder samples. Compared to the other flowability testing methods, the present testing device directly measures the flowing of the powder sample. Therefore, it is evident that the method can value both the shape of the particles and electric charge of the particles fairly the same amount.

Scanning electron micrographs of the materials tested are illustrated in Figures 2-4. Sugar excipients (i.e. glucose anhydrate, sucrose and partially pregelatinized maize starch) exhibited particles with a relatively round and spherical shape. Particle size of these excipients ranges from almost micronized scale of 20 micrometers (pregelatinized starch) to 400 micrometers (sucrose). The three MCC grades (Avicel) tested exhibited particles with a more elongated and fiber like shape. Particle sizes of MCCs were approximately 50 micrometers (Avicel PH-101), 100 micrometers (Avicel PH-102) and 200 micrometers (Avicel PH-200). Caffeine, carbamazepine and paracetamol were found to have clearly the smallest particle size among the test materials ranging from approximately 5 micrometers to 50 micrometers.

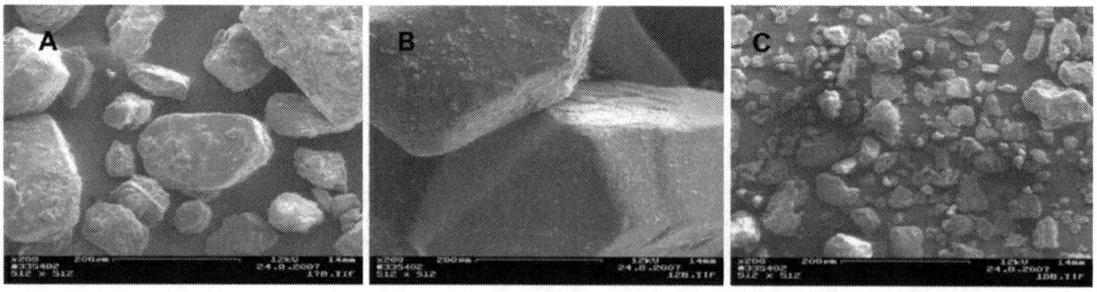

Figure 2. Scanning electron micrographs of sugar excipients glucose anhydrate (A), sucrose (B) and pregelatinized starch (Starch 1500) (C).

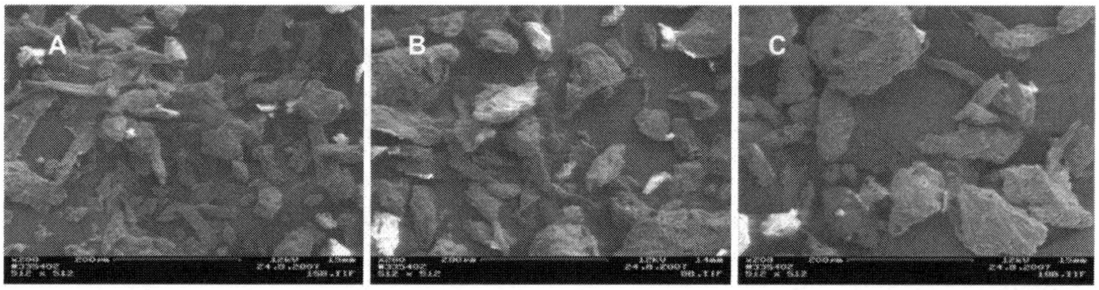

Figure 3. Scanning electron micrographs of microcrystalline cellulose grades Avicel PH-101 (A) Avicel PH-102 (B) and Avicel PH-200 (C).

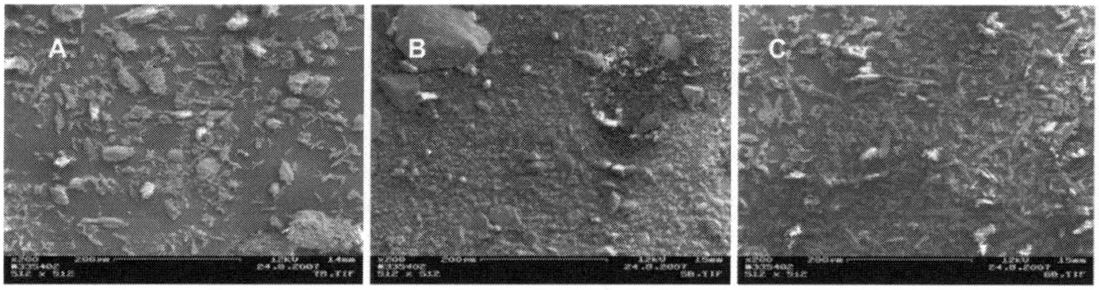

Figure 4. Scanning electron micrographs of model drugs caffeine (A), carbamazepine (B) and paracetamol (C).

Table 1 summarized comparison of flow rates (in mg/s and ml/s) and Carr's indexes of the materials tested. The results on flowability obtained with the present novel method were in accordance with the calculated Carr's indexes. As regards with sugar excipients, sucrose exhibited the best flowing characteristics and partially pregelatinized maize starch the worst flowing capacity. Flowability of MCC (Avicel) was greatly dependent on the grade of it. With respect to flowability, the MCC grades tested were ranked in the following decreasing order: Avicel PH-200, Avicel PH-102 and Avicel PH-101. The flowability results obtained with the present novel method with drugs were also in accordance with those obtained with the reference methods. With respect to flowability, drugs tested were ranked in the following decreasing order: caffeine, carbamazepine and paracetamol. With drugs, variation of the flowability results obtained with the present novel method was found to be very low.

Table 1. Flow rates (in mg/s and ml/s) and Carr's indexes of the studied materials.

Flow rate (mg/s)	Glucose	Sucrose	Starch 1500
AVG	**100,8**	**195,6**	**137,6**
SD	1,4	26,2	5,0
RH%	56 %	56 %	46 %
Flow rate (ml/s)			
AVG	**0,131**	**0,228**	**0,239**
SD	0,000	0,032	0,014
RH%	56 %	56 %	46 %
Carr's index (%)			
AVG	**14,3**	**7,9**	**18,0**
SD	0,8	1,2	2,3
RH%	57 %	57 %	59 %

Flow rate (mg/s)	Avicel PH 101	Avicel 102	Avicel PH 200
AVG	**22,9**	**66,3**	**55,4**
SD	0,6	2,3	1,3
RH%	56 %	57 %	56 %
Flow rate (ml/s)			
AVG	**18,9**	**0,195**	**0,166**
SD	1,4	0,031	0,004
RH%	63 %	57 %	56 %
Carr's index (%)			
AVG	**18,9**	**16,3**	**14,8**
SD	1,4	1,4	1,2
RH%	63 %	60 %	59 %

Flow rate (mg/s)	Caffein	Carbamazepine	Paracetamol
AVG	**9,0**	**6,7**	**1,7**
SD	0,2	0,8	0,2
RH%	49 %	47 %	52 %
Flow rate (ml/s)			
AVG	**0,051**	**0,028**	**0,025**
SD	0,001	0,003	0,002
RH%	49 %	47 %	52 %
Carr's index (%)			
AVG	**17,4**	**26,0**	**25,0**
SD	0,3	1,1	1,6
RH%	52 %	55 %	56 %

As seen in Figure 5, mixing time of the tableting lubricant (magnesium stearate) slightly increased the flow rate of both Avicel PH-101 and PH-102. This is obviously due to the formation of an uniform lubricant film around the MCC particles which enhances flowing properties of the binary mixture.

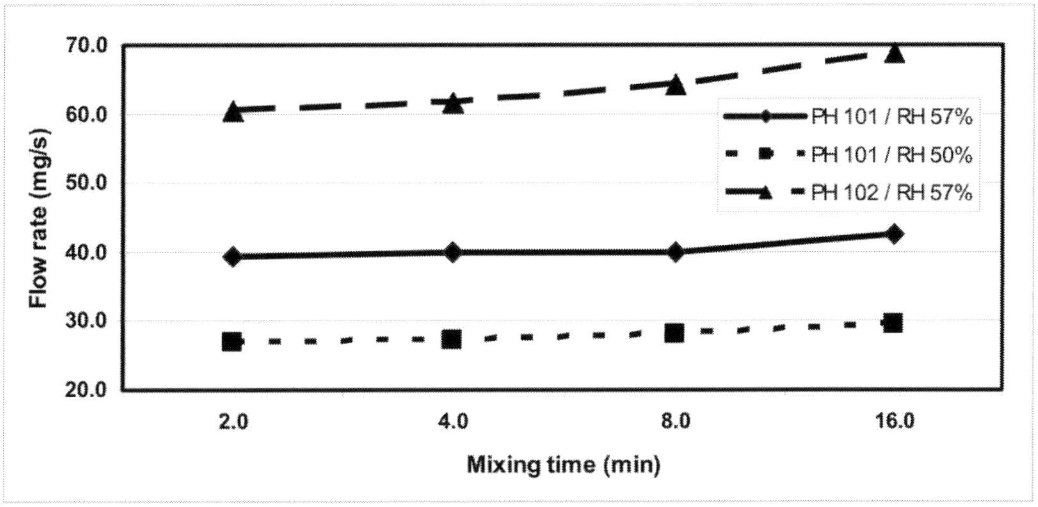

Figure 5. Mixing time of magnesium stearate slightly increases the flow rate of Avicel PH-101 and PH-102.

The effects of seasonal humidity fluctuation in the flow rate of Avicel PH-101 are shown in Figure 6.

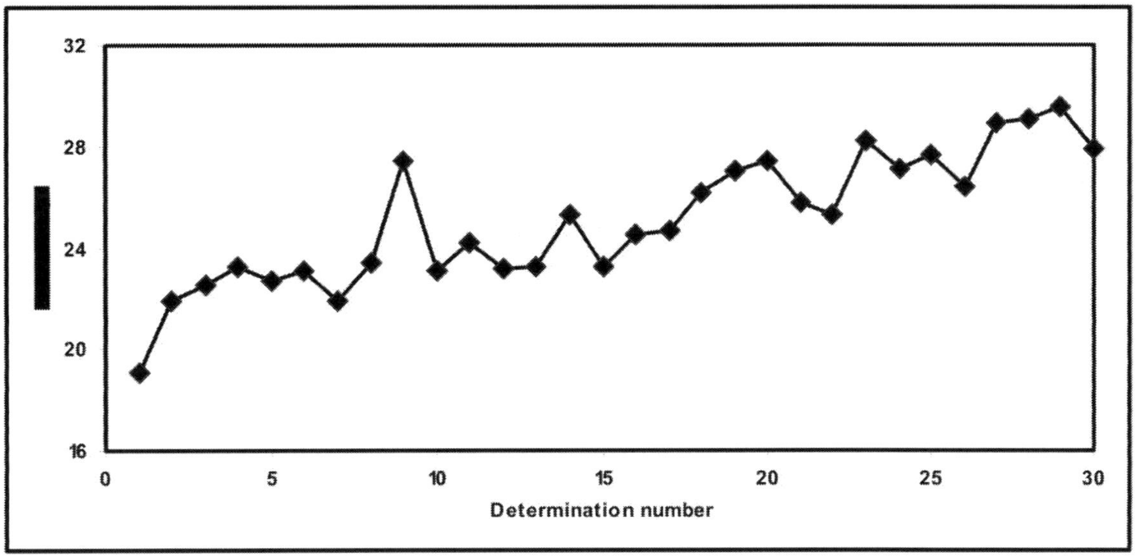

Figure 6. Effect of seasonal humidity fluctuation in the flow rate of Avicel PH-101.

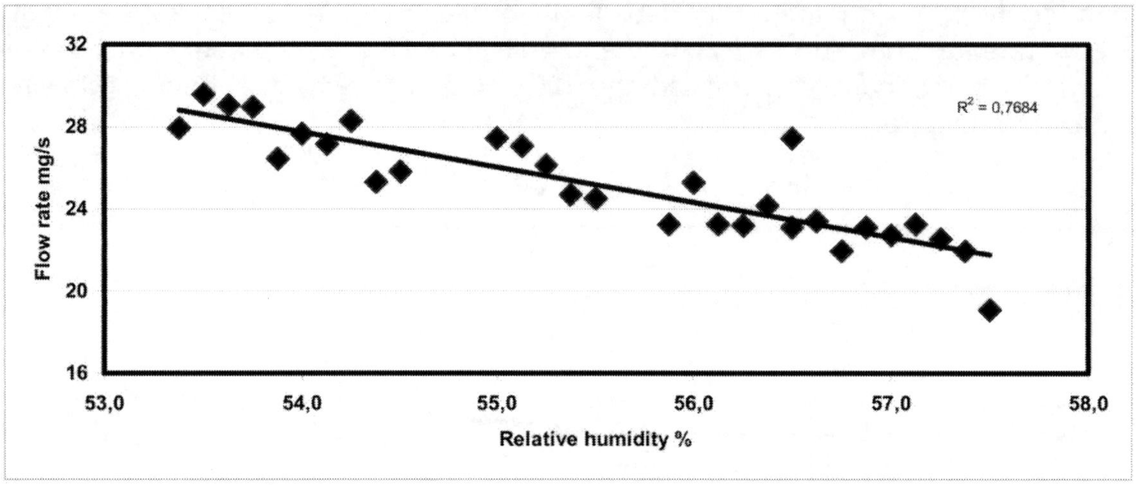

Figure 7. Linear correlation between relative humidity of air and the flow rate of Avicel PH-101.

CONCLUSIONS

In conclusion, the present novel flowability testing method provides a new tool for a rapid determination of flowing characteristics of powders and granules at a small scale.

REFERENCES

Rios, M. Developments in powder flow testing. Pharm. Technology Eur. 30 (2) 38-49 2006.

Antikainen, O.K., Rantanen, J.T. and Yliruusi, J.K. Use of the Kohonen self-organizing map to predict the flowability of powders. S.T.P. Pharma Sci 10 (5) 349-354 2000.

European Pharmacopoeia 5th Ed., Council of Europe Strasbourg, ISBN:92-871-5281-0, 2004.

461g Controlled Particle Deposition by Design of An Electrochemical Adsorption Cell

Ehsan Bakhshi, R&D Center of National Petrochemical Company, Tehran, Iran, Fatemeh Abniki, National Iranian Petrochemical Company, Ghadir Group, Tehran, Iran, and Majid Mosalla, Metallurgy & Material Science Department of Shiraz University, Tehran, Iran.

Deposition of colloidal particles on to a substrate takes place in two stages : (1) transport stage wherein the particles are brought into the vicinity of the collector by bulk fluid flow or diffusion; (2) attachment process wherein the particle is physically deposited. The attachment stage is governed by colloidal interaction forces such as London van der Waals, electrical double layer, Born repulsion , hydration hydrophobic interactions . In the theoretical section of this work it will be assumed that the main energies contributing to adsorption interaction between the surfaces will be due to electrical double layer and the van der Waals attraction . The sum of these energies provides the total interaction energy between surfaces and this constitutes the underlying assumption for the classical D.L.V.O. theory of colloidal stability.Predictions based on the D.L.V.O. theory suggest the occurrence of favorable and unfavorable deposition conditions when overall attractive and repulsive energies exist between surfaces respectively . Indeed many experimental results in the literature confirm the occurrence of favorable and unfavorable deposition conditions. However , The D.L.V.O. theory suffers from a few shortcomings discussed by Elimelech et .al . discuss the deposition studies in the literature and point out that ideal collectors and particles are usually used for investigation of deposition processes . Ideal particles are smooth, spherical , mono disperse with narrow size distribution . Ideal collectors used are parallel plate, stagnation point flow, bed of glass beads and these provide hydrodynamics and interaction energies that can be described within the existing framework . It was mentioned that the deposition process consists of two main stages i.e. transport and deposition stages. The transport stage may occur by bulk fluid flow , interception, sedimentation under the . influence of gravity, inertial impaction and diffusion. For non – Brownian particles the diffusion process becomes negligible compared to the bulk fluid flow . The polystyrene latex particles used in this study were 5.4 Micrometer and hence effect of diffusion becomes negligible. For reasonably low flow rates, Re Conclusion For the present experimental conditions the main forces contributing to the deposition process are the surface forces and hydrodynamics. At the highest flow rate of 34 cm3 .min1 the colloidal forces were shown to be negligible compared to the hydrodynamics. Indeed the effect of surface forces at the KCI concentration of 0.001 mol/ dm3 apparently only became significant at flow rate of 1.7 cm3/min . At this flow rate and KCI concentration of 0.001 mol/dm3 the experimental results show clear evidence of conditions governing the occurrence of favorable and unfavorable deposition conditions. The variation of collector potential under unfavorable deposition condition dos not reduce large changes in the deposition rate whereas variation of collector potential under favorable conditions produce significant changes in the deposition rate. At collector potentials of -300 mV vs. SCE hydrogen evolved and this was accompanied by a rise in the pH of the solution. This rise in pH affected the zeta potential of the particles though slight but significant.

Web Page: www.nipc.net

461h Pull-off Force of Dry Coated Fine Powders Under Small Consolidation

Yuhua Chen[1], Miguel A. S. Quintanilla[1], Jun Yang[1], Rajesh Dave[1], and Jose-Manuel Valverde[2]. (1) New Jersey Center for Engineered Particulates, New Jersey Institute of Technology, 138 Warren Street, YCEES 208, Newark, NJ 07102, (2) University of Seville, Seville, Spain

Dry Coating is a novel technique to prepare engineered particles and it helps reduce the adhesion force of very fine powders and improve their flowability and fluidizability. However, at present, the mechanism of the fluidizability improvement due to nano-additives is not well understood. Since particles are compressed at least by the weight of particles above themselves when stored, transported and processed, it is important to account for such low consolidation when proposing a mechanism of the improvement in fluidizability or flow. In this work, a three dimensional model taking into account the contact deformation and Surface Area Coverage of nano-additives is proposed to predict the pull-off force, the force required to separate two contact particles, under consolidation stress up to 10,000 Pa for sparsely and densely dry coated group C powder cornstarch. Sparse coating demonstrates that two dry coated particles contact each other through contact of silica and cornstarch, while dense coating illustrates that two dry coated particles contact each other by silica-silica contact. The pull-off force measurement is conducted in a Seville Powder Tester (SPT), which uses gas flow either to fluidize to erase memory of the powder or to compress the bed in order to subject it to a controllable consolidation stress. The analytical results have good agreement with the experiment results indicating: (1) the pull-off force of sparsely coated cornstarch is larger than that of densely coated cornstarch due to the greater hardness of fumed silica; (2) the pull-off force increases with increasing the consolidation force due to larger deformation in contact area; (3) under consolidation stress less than 10,000 Pa the nanosized silica is not fully embedded in the cornstarch because the deformation of cornstarch is not big enough though it increases with raising the consolidation force and decreasing the surface area coverage (SAC).

Problems of Scale-up of Solids Handling Processes[1]

by

John Carson[2], Thomas Troxel, and Eric Bengtson

Jenike & Johanson, Inc.

Abstract

Scale-up of solids handling processes can be very troublesome, particularly if new equipment or process steps are being incorporated. Unfortunately, if scale-up is not handled correctly, commercial operation suffers in terms of the time and expense it takes to correct problems, time to market, and perhaps derating of the equipment or process. If a sound theory has been (or can be) developed that accurately models the process and equipment independent of scale, bench scale tests can be run and the results used as input data in the mathematical model. If such a theory is not available, the only safe approach is usually to model some or all of the critical elements of the system at or close to full scale.

Examples of systems that can and cannot be scaled-up are presented, along with scaling rules. The focus is on storage and processing vessels and does not include other types of bulk solids handling systems such as conveyors. Case histories of successful scale-up are presented.

Introduction

Edward Merrow, author of the famous RAND study in the mid-80's concerning problems of start-up and operation of solids processing plants, wrote an update article several years ago [1]. He identified and examined over 500 recently completed, fully operational projects that met the following criteria:

- Technology must involve chemical processing
- Facility must be commercial (not pilot or semi works)
- Capital expenditure must be more than $7 million
- Operational history must be at least one year long
- Process must be continuous or semi-continuous (not batch)

The sample comprised 36 process industries and companies around the world with projects averaging $66 million, with a median execution time for engineering and construction of 19 months and median total cycle time from conceptual design to steady-state operation of 28 months.

As in his previous report, Merrow found a clear relationship between increased particle processing difficulty and plant start-up times. The typical liquid- or gas-process required about 2 months to complete start-up. Liquid- and gas-fed facilities with a solid intermediate or product required slightly more time. Facilities that consume a previously processed - and therefore reasonably homogenous – feedstock required an average of a little more than 4 months, whereas "raw" solid feedstock facilities required an average of more than 7 months.

---

[1] Submitted for presentation at 2007 AIChE Annual Meeting, Salt Lake City

[2] Author contact information: Jenike & Johanson, Inc., 400 Business Park Dr., Tyngsboro MA 01879; phone: 978/649-3300; email: jwcarson@jenike.com

The cause of problems appears to be lack of stability in basic technical data which results in schedule slip and cost growth. Merrow believes that the problem stems from lack of fundamental knowledge of solids processing technology and difficulty in extrapolating results from one facility to another when the facility designs are not identical.

Merrow investigated innovation difficulties and concluded, "Innovation in solids processing has been difficult, slow and expensive". He measured process innovation in terms of "new process steps" which entailed the use of commercially new chemistry, commercially unproven major equipment, or first-ever match of feed and equipment. In all plants, operability (defined as percent of design after start-up) decreased as the number of process steps increased. However, the change was most dramatic for solids-processing plants using solid feed. For example, a plant using a refined solid feed with no new process steps averaged 80% operability, whereas one with four new process steps averaged only 65%. For plants using raw solid feed, the corresponding numbers were 70% and 40%. He stated, "It is extremely imprudent to begin commercial design of a second version of a process before the pioneering design has been proven. A highly innovative solids-processing facility commonly requires 24 to 60 months to reach complete steady-state operation at the original design rates. It is not unusual to see pioneer solids processing facilities significantly modified, returned to an older, more established technology, or even abandoned without having achieved beneficial operation."

One of the reasons that solids processing innovation is so difficult is the inability to extrapolate from small pilot plants -- which is not the case with liquid- and gas-processing facilities because reaction conditions can be adequately simulated at a small scale. Another reason for innovation difficulty is that pilot plants for solids processing must be not only on a much larger scale but also fully or nearly fully integrated to provide reliable data. This results in a very expensive non-commercial facility, so it is tempting to bypass this stage. Merrow stated, "Bypassing an integrated pilot plant stage under these circumstances is almost always an economic mistake. The added cost of the pilot plant will be more than offset by improvements in the first few years of commercial plant operability".

Our firm's 40+ years of consulting engineering experience on industrial bulk solids projects confirms virtually all of Merrow's observations and conclusions. We have found that if scale-up is not done properly, the resulting problems can be expensive both in cost as well as time. These problems include:

- Time and expense to correct problems. The time it takes to correct a problem, particularly the implementation of a correction, is likely to cut into the time that the plant is available for production. Correcting problems with existing equipment often involves access restrictions, headroom availability, etc., which both limits the options available as well as drives up the cost. Furthermore, the cost often comes from the plant's operating or maintenance budget as opposed to capital budget.

- Time to market. Plant downtime to address scale-up problems may mean a delay in getting new product to market, thereby potentially affecting competitive positioning.

- Customer complaints. Perhaps because of scale-up problems the equipment works but does not produce the desired uniformity of product quality. This may result in customer complaints and loss of market.

- Possible derating of equipment or process. Sometimes scale-up problems are so difficult if not impossible to correct it is necessary to limp along at a less-than-desired production rate. Obviously this can affect the economics of a plant.

It is commonly understood that if a sound theory has been (or can be) developed that accurately models the process and equipment independent of scale, bench scale tests can be run and the results used as input data in the mathematical model. If such a theory is not available, the only safe approach is usually to model some or all of the critical elements of the system at or close to full scale. The key is determining which processes and equipment have a sound theory of operation so that scale-up can be done effectively and correctly.

Solids handling processes with good theoretical basis or model

One area of solids handling in which there is a sound theoretical basis involves the flow of bulk solids through a converging hopper. The pioneering work of Drs. Andrew Jenike and Jerry Johanson at the University of Utah [2-4] involved development of stress and velocity fields during discharge from converging hoppers by gravity. They postulated, and later proved by mathematical and experimental models, that both stress and velocity fields develop in a radial manner emanating from the apex of the hopper. This is perhaps more obvious for velocity fields than for stress fields, since if mass flow develops, particle movement will vary from being vertical at the axis of symmetry to being parallel to the hopper walls in the region just inside the hopper. For stress fields Jenike postulated the following based on an analogy from soil mechanics attributed to Sokolovski [5]:

$$\sigma = \gamma \; r \, s(\Theta) \qquad (1)$$

$$\psi = \psi(\Theta) \qquad (2)$$

where:

σ	=	mean normal stress
γ	=	bulk density
r	=	radial distance from hopper apex
s (Θ)	=	stress function dependent on angle measured from vertical
ψ	=	angle between direction of major principal stress σ_1, and coordinate ray

This theory led to the development of design procedures for achieving mass flow while avoiding problems of cohesive arch formation. Over the last nearly 50 years since this theory was first developed, literally thousands of bins and silos have been designed and built using this as the design basis, and almost without exception they have worked as designed. This involves containers varying in size from small press feed hoppers and portable bins all the way up to multi-thousand ton capacity silos. Split-model tests (described below) showing displacement fields in mass flow hoppers confirm the assumption of a radial velocity field. Also, hopper wall pressure measurements confirm the development of a radial stress field.

The second area in which a proven model has been developed involves fluid flow through a packed bed of bulk solids. The rate that interstitial gas (fluid) migrates through a granular material under the influence of a pressure gradient depends on permeability, and may be written using Darcy's Law as:

$$u = -K \, dp/dx \, / \, \gamma \qquad (3)$$

where: u = superficial gas velocity through solids bed

K = permeability coefficient
dp/dx = gas pressure gradient across the bed
γ = bulk density

More detailed forms have followed, including the well-known forms of Ergun [6] and Leva [7]. The latter's form is:

$$-dp/dx = \frac{k(1-\varepsilon)^{(3-n)}}{\varepsilon^3} \frac{\rho^{(n-1)}u^n}{D_\rho^{(3-n)}} \mu^{(2-n)} \tag{4}$$

where:
dp/dx = gas pressure gradient across the bed
k = empirically determined permeability coefficient
ρ = density of the gas
ε = void ratio
u = superficial gas velocity through solids bed
n = gas state (n=1 for laminar, n=2 for turbulent)
D_p = particle diameter assuming uniform spheres
μ = viscosity of gas

This can be simplified for laminar flow (*n*=1) to:

$$-dp/dx = \frac{k(1-\varepsilon)^2}{\varepsilon^3} \frac{u}{D_\rho^2} \mu \tag{5}$$

Because of the difficulty in determining an appropriate particle diameter, and since *k* must be empirically determined, this equation may be simplified by lumping the terms together into one overall material dependent coefficient, called permeability factor *K*. The result is the same as Darcy's Law, eqn. (1). Permeability factor *K* can be determined empirically usually in the following form, where K_o, γ_o and *a* are empirically determined constants.

$$K = K_0(\gamma/\gamma_o)^a \tag{6}$$

Bench scale tests used to provide input values to theories

In order to design bins, silos, hoppers, and other types of storage and processing vessels for bulk solids, several key parameters are required. These include:

- Cohesive strength, internal friction and wall friction. Using a representative sample of the bulk solid, these parameters are measured and are a function of moisture/volatiles content, time of storage at rest, temperature, consolidating pressure, and particle size and shape. The most common devices used for these measurements are the Jenike Shear Tester [3,8] and the Schulze Ring Shear Tester [9].

- Compressibility. The bulk density of a bulk solid is not single or dual valued, but instead varies as a function of the applied consolidation stress. A test to measure this parameter can be found in [10].

- Permeability. The ability of air (or other process gas) to flow through a packed bed of solid varies strongly with the bulk density of the material. A permeability tester has been developed to measure this parameter.

- Abrasive wear. An abrasive wear tester has been developed [11] which allows measurement of the absolute wear rate as a function of applied pressure using a sample of bulk solid and a coupon of the wall material upon which the bulk solid will be sliding. It should be noted that this test is limited to wear due to sliding abrasion, not impact. Combining this data with radial stress and velocity field assumptions, one can accurately estimate the wear life of bins and other similar storage and process equipment. [12]

Scale modeling

Sometimes physical scale models are required when the complexities of a solids handling system cannot be handled by theory and bench scale tests alone. The validity of the scale-up, however, in many cases depends on utilizing underlying theory of solids flow.

One example where this is useful involves scale modeling of vessels that contain inserts and/or gas introduction. One consideration is the variation of certain material properties with consolidation pressure, in particular internal and wall friction as well as bulk density [13].

Deaeration of fine powders as well as the effects of body forces due, for example, to gas pressure gradients or applied vibration, are sometimes more pronounced in small vessels than in larger ones. As long as the effects of body forces are properly taken into account, scale-up can proceed with confidence.

If the solids flow pattern is of interest, care must be taken in constructing scale models so as to properly observe the behavior. In particular, observing flow pattern only at bin walls, by, for example, using transparent material to build the bin, provides only very limited information as to flow behavior. Splitting a bin and replacing one half with a vertical transparent wall provides a means of observing the flow pattern; however, this flow pattern will be strongly affected by the presence of the vertical, stationary wall. A better approach is to use split 3-dimensional models as described in [14].

Scale-up of some types of blenders is also possible. If the blending mechanism involves in-bin, gravity flow, the scaling rules are the same as for any other bin configuration [15]. If the blender operates by tumbling, it is essential that geometric, dynamic and kinematic similarity be maintained [16]. Geometric similarity requires that there be a constant ratio of linear dimensions between the test unit and full scale. This includes not only the relative dimensions of the container and axis of rotation, but also such items as the percentage of the container that is full as well as the sequence of filling individual ingredients into the container. Dynamic similarity requires that there be constant dynamic forces at corresponding points at both units. The appropriate term is the Froude number. The manner in which kinematic similarity can be obtained is less certain than the previous two. Some authors [17] state that the requirement is for a constant ratio of velocities at corresponding points. Empirical data is available for scaling of rotating drum, twin shell (V-type), and double cone mixers. An alternative means of achieving kinematic similarity would be to maintain a constant number of revolutions of the blender. This has been found to be appropriate for tumble blenders, mostly for materials with low cohesive strength. [18]

Examples where full-scale tests are required

Sometimes neither a well developed theory nor scaling rules have been developed that allow confident design at full scale. Examples of such situations are as follows:

- Transmission of vibrations. The ability of an external or internal vibrator to activate material within a storage vessel is not at all well understood. As a result, sizing and placement of vibrators is essentially trial-and-error, hopefully based to some extent on past experience of what works and what doesn't.

Sometimes vessels vibrate on their own due to movement of the bulk solid within. The resulting vibrations can either be high frequency, low amplitude (a humming sound), or low frequency and high amplitude (sometimes called silo quaking). There are a number of mechanisms by which such vibrations can occur, many of which are described in [19, 20]. Sometimes scale models are effective in pointing in the direction of changes that will affect the severity of the vibrations, while at other times scale models give a false impression of changes that will be effective. The bottom line is that the only sure way of knowing whether or not a change will be effective in a large vessel is to implement the change and observe the result.

- Feed rate uniformity. While there is a good basis for estimating critical discharge rates from an orifice, the uniformity by which material will flow - particularly over short time intervals - can only really be understood in a full-scale model.

- Segregation problems. There are many mechanisms of segregation [21] and some bench scale tests by which some of these mechanisms can be investigated. [22, 23] However, for many segregation problems the only way to assess the full magnitude of the problem as well as the effectiveness of modifications is to run a full-scale test.

- Anisotropic materials. Such materials are difficult to test in most shear testers, particularly if the anisotrophy is due to particle shape effects. Sometimes scale model tests provide insight into this behavior, but often full scale tests are required.

Case histories

Purge column

BP Amoco Chemical (now Ineos) built a 200,000-metric ton per year polypropylene (PP) plant in 1995 and 1996 at Geel, Belgium, that uses the BP Amoco gas phase process and BP Amoco's high-activity support catalyst. The plant produces random and blocked copolymers and a wide range of homopolymers.

The design included a purge column following the reactors. The purge column uses nitrogen to reduce residual volatiles, deactivates the trace catalyst residues using a small quantity of steam and provides surge capacity between the reaction and pelletization sections of the plant.

Engineers at BP Amoco specified the minimum and maximum purge duration (the exposure time to purge gas). They also specified the required solids and gas flow rates as well as the gas temperature and pressure at the top of the vessel. Mass flow is an important design parameter for the purge column because product transitions are done "on the fly".

Operating in mass flow minimizes the possibility of cross-contamination between product grades. Short-circuiting of product that could result in high volatiles is also avoided by operating in mass flow. The combination of low volatiles, no cross-contamination and deactivated catalyst residues are important requirements for a high-quality PP product.

Based on flow property data for the materials being handled, engineers developed the functional design for the purge column [24]. It's a tall, narrow cylinder section (H/D = 7.3), below which is a steep-sided, conical stainless-steel hopper. Purge gas is introduced just above the top of the conical hopper section.

The purge column has been operating for nearly ten years, and has performed to meet the design basis. According to Jim Lee, research associate for BP Amoco Chemical, "The performance of the purge column has exceeded our expectations. In fact, it did not require any modification when the result of the PP unit was debottle-necked by 40% in 1998".

This technology has been introduced at several other plants that use the Ineos gas phase process, as well as a number of other PE and PP production facilities.

This is an excellent example of scale-up based on using sound theoretical models to predict solids stresses and permeability, combined with bench scale tests that provided input values to the theories.

Blending ABS pellets

A new compounding plant was being designed for ABS plastics. The plant would produce a wide range of ABS pellets compounded to customer specifications. Production runs would be relatively short with frequent formula changes. Based on experience at smaller plants, engineers for the project settled on using a V-type rotating blender. Batch ingredients would be fed into a weigh hopper and then discharged into the blender. The blended batch would then be discharged into a surge bin feeding the extruder. While this process had worked successfully at other compounding plants, this plant would blend ingredients with a wider size range. The most extreme case involved blending two materials which consisted of cylindrical pellets with a nominal size of about 3mm and nearly spherical particles (beads) with a size range from about 1mm down to 150 micrometers.

Engineers evaluated V-type rotating blenders by sampling a scale-model blender. [18] They found that significant segregation could occur, and maintaining a blend in the discharged material would be difficult. Results from a typical blending test showed only about 9% of the blender batch fell within the acceptable range of 15% +/- 2% beads blended with pellets. A number of modifications were made to the model blender to try to improve discharge uniformity with only moderate success.

The cause of the problem was a combination of inadequate blending and segregation that occurred during discharge. When the blended material flowed out of the V-blender, it flowed in a funnel flow pattern with a fast moving central flow channel and no flow along the walls of the blender. The resulting shear and central depression allowed sifting and rearrangement of the smaller particles in the blend, resulting in non-uniform discharge.

Using a closed insert in the scale-model V-Blender made a significant improvement in blending, but still only about 72% of the batch was within the desired blend specification.

An alternate geometry using a conical blend vessel with an insert was designed based on the flow properties of the materials. The vessel geometry was selected to ensure uniform mass flow during discharge to minimize segregation. Tests with this configuration in a scale model showed much better performance than with the V-type blender of similar scale, with over 90% of the discharge being within the acceptable range.

Three full-scale conical blenders (8,000 lb. capacity each) were designed and built. Before they were shipped to the plant, blend tests were run on one of these blenders. Results were consistent with those from the scale model, providing proof that the scale-up had been done properly.

This is a good example of scale-up where, although theoretical models don't exist, the scaling parameters are understood.

Air-assisted discharger

A power plant was switching to a new coal supply to meet Clean Air Act requirements. Due to this switch, it was determined that fluxing with magnetite would be required to maintain suitable slag melting temperatures for boiler tapping. Reliable feed of magnetite from two 8,000 cu. ft. (600 ton) silos was critical, and the stored magnetite had to be equally split to five separate batch weigh hoppers for pneumatic conveying.

One approach that was considered was to use a mass flow bin with a belt feeder to control discharge. In order to minimize the chances of the fine magnetite flooding uncontrollably from the 600T storage silo it would be necessary to close the slide gate when initially filling the silo from empty, and the material would have to be left for a minimum of 24 hours prior to the slide gate being opened and the feeder activated. It would also be necessary to keep a 200T level of material in the silo at all times prior to re-filling, and the silo would have to be filled on its symmetry axis at a maximum fill rate of 500 TPH.

An alternate approach was to partially fluidize the material in the silo via an air-assisted discharger. This solution would have the benefit of requiring no minimum head of material in the silo, thus allowing a live capacity of 600T to be realized.

In an air-assisted discharger, small amounts of air are injected into the material via a porous membrane. The air acts to fluidize a small boundary layer of material, thus reducing the boundary friction and enabling the material to discharge at higher rates than would be the case without air injection. The discharger is comprised of a number of different gas plenums, each designed to allow, nominally, different amounts of gas to be injected into the material. Control of the air-assisted discharger is critical from the standpoint of obtaining mass flow in the silo above. The flow rate from the discharger cannot be regulated, but must be on or off.

Bench-scale fluidization tests were carried out on the magnetite at several moisture levels to determine if it would be feasible to discharge the magnetite via an air-assisted discharger. Scale-model air-assisted discharger tests were also run. Both sets of tests showed that the magnetite was a good candidate for this type of feed system.

Two air-assisted dischargers (one for each silo) were designed, build and installed. This system has been working reliably for nearly seven years.

This is an example of scale-up using a combination of theoretical models with inputs from bench scale tests, in conjunction with proper scaling approaches.

Abrasive wear

A manufacturing facility stores and handles a hard abrasive bulk solid. Since significant abrasive wear of the silo wall was observed in several silos, abrasive wear tests [11] were run to investigate alternative materials of construction including hard liners. These tests were conducted using the bulk materials handled at the plant and at pressures that simulate bulk solid flow in the existing silos, assuming that mass flow develops.

A wear analysis of the silos was performed using a simulation tool [12]. The existing silo design was analyzed and the results compared to measurements made in the transition region of two silos.

Accurate records had been kept of the total throughput of each silo. Wall thickness measurements were made at 6 to 12" increments around the circumference at each of eight elevations, then averaged and compared to original thickness.

Comparisons of predicted vs. average total wear in the two silos showed excellent agreement in the region of highest wear, namely just below the cylinder-cone transition where the normal pressure against the wall is highest.

This project provided increased confidence that the theoretical models which use bench-scale test results as inputs are valid. As with any theoretical model, having continued feedback as to its accuracy and limitations is important so the user applies it properly the next time.

References

1. Merrow, Edward, "Problems and Progress in Particle Processing", Chemical Innovation, published by The American Chemical Society, January 2000
2. Jenike, A.W., *Gravity Flow of Bulk Solids*, University of Utah Engineering Experiment Station, Bulletin No. 108, Oct. 1961.
3. Jenike, A.W., *Storage and Flow of Solids*, University of Utah Engineering Experiment Station, Bulletin No. 123, Nov. 1964.
4. Johanson, J.R., "Stress and Velocity Fields in the Gravity Flow of Bulk Solids", J. Appl. Mech. Series E 31:499-506, 1964
5. Sokolovski, V.V., *Statics of Soil Media*, Butterworths Scientific Publications, 1960
6. Ergun, S., "Fluid Flow through Packed Columns", Chem. Eng. Prog., Vol. 48, No. 2, pp. 89-94, 1952
7. Leva, Max, *Fluidization*, McGraw-Hill Book Co., New York, 1959
8. ASTM D6182-06 "Standard test method for shear testing of bulk solids using the Jenike shear cell", Amer. Soc. For Testing and Materials Intl., West Conshohocken, PA
9. ASTM D6773-02 "Standard shear testing method for bulk solids using Schulze ring shear tester", Amer. Soc. For Testing and Materials Intl., West Conshohocken, PA
10. ASTM D6683-01 "Standard test method for measuring bulk density values of powders and other bulk solids", Amer. Soc. For Testing and Materials Intl., West Conshohocken, PA
11. Johanson, J.R. and Royal, T.A., *Abrasive Wear Tester*, U.S. Patent 4,446,717, May 8, 1984
12. Johanson, J.R. and Royal, T.A., "Measuring and Use of Wear Properties for Predicting Life of Bulk Materials Handling Equipment", Bulk Solids Handling 2:517-523, 1982
13. Carson, J. W., "Addressing Critical Solids Handling Aspects at the Pilot Scale", Presented at AIChE Spring 1988 Meeting, New Orleans, LA, March 1988.
14. Johanson, J.R., "The Use of Flow-Corrective Inserts in Bins", J. Eng. Industry, Series B 88, 1966, pp 224-230
15. Carson, J. W. and Royal, T. A., "In-Bin Blending Improves Process Control", Powder Handling & Processing, Vol. 4 No. 3, Sept. 92, pp 301-307

16. Carson, J. W., T. A. Royal and R. J. Hossfeld, "Tumble Blending with Mass Flow Containers Improves Productivity and Quality", Powder Handling & Processing, Vol. 6 No. 4, Oct./Dec. 1994, pp 413-416

17. Wang, R.H. and Fan, L.T., "Methods for Scaling up Tumble Mixers", *Solids Handling*, ed. by Chem. Eng., pp. 245-251, 1981.

18. Troxel, Thomas G., "Modeling and Scale-Up of Tumble Blenders for High Segregating Materials", presented at 5th World Congress on Particle Technology, Orlando FL, April 2006.

19. Purutyan, H., K. E. Bengtson, and J. W. Carson, "Identifying and Controlling Silo Vibration Mechanisms: Part I", Powder and Bulk Engineering, Vol. 8, No. 11, Nov. 1994, pp 58-65

20. Purutyan, H., K. E. Bengtson, and J. W. Carson, "Identifying and Controlling Silo Vibration Mechanisms: Part II", Powder and Bulk Engineering, Vol. 8, No. 12, Dec. 1994, pp 19-28

21. Carson, J.W., Royal, T.A., and Goodwill, D.J., "Understanding and Eliminating Particle Segregation Problems", Bulk Solids Handling, Vol. 6, Feb. 1986, pp 139-144

22. ASTM D6940-04 "Standard Practice for Measuring Sifting Segregation Tendencies of Bulk Solids", Amer. Soc. For Testing and Materials Intl., West Conshohocken, PA

23. ASTM D6941-05 "Standard Practice for Measuring Fluidization Segregation Tendencies of Powders", Amer. Soc. For Testing and Materials Intl., West Conshohocken, PA

24. Carson, J., Lee, J.H. and Pittenger, B. H., "Gravity Flow Processing Vessels", Chemical Processing, 8/99, pp. 49-53

494b Pilot Plant Scale-down to Troubleshoot a Commercial Crystallization Process

Joseph B. Powell, Chemical Process Technology, Shell Global Solutions US Inc., Westhollow Technology Center, P. O. Box 1380, Houston, TX 77251-1380

A case study of commercialization of a new process involving production of a solids product via intermediate crystallization is presented, in which small-scale, multi-throughput reactions, crystallizations, and separations were conducted in vials at the 1 – 5 gram level, to create a virtual, multi-throughput pilot plant to diagnose off-spec performance in the commercial unit. Confidence in the mini multi-pilot approach was gained via correlation with performance observed during earlier unit operations at the 1 – 4 liter scale, piloted during initial process development. Scale-down to allow rapid response, side-by-side testing to elucidate impacts of trace chemistry was paramount to successful post startup troubleshooting.

Industrial Aspects of Fouling Associated with Polymerization

Istvan Lengyel, Debashis Chakraborty, Paul Chauvel and Chris Christenson
The Dow Chemical Company, Freeport, TX

Oliver Reipschläger, Matthias Schäfer,
Dow Olefinverbund GmbH, Schkopau, Germany

Abstract

This paper describes a general method of solving difficult fouling problems in the process industries. The fouling specifically is organic/polymer in nature. Specific tools are also described. In many cases, problems that had existed in our processes for more than 25 years were solved in <6 weeks. The approach requires writing a clear definition of success. Then, it is important to write a list of the things that must be true for success to be true. Recent new tools that have proven very useful include computational chemistry and its interface to engineering modeling tools.

Introduction

Unplanned polymerizations in chemical and polymer processes are, unfortunately, a relatively common fact of life in industrial-scale chemical processes. The products of these reactions have a tendency to foul surfaces in critical unit operations, resulting in down time, unnecessary capital expenditures for redundant capabilities, increased maintenance costs and potential chemical exposure of the maintenance personnel. Consequently, speedy recovery of these kinds of foulings, and to design sustainable procedures for avoiding them is critical for profitable plant operations and process safety. This paper describes a general approach that has been used successfully in Dow to understand the root cause of these types of problems and to minimize or eliminate them. Figure 1 shows a few examples of the problem, which are quite familiar to experienced process engineers.

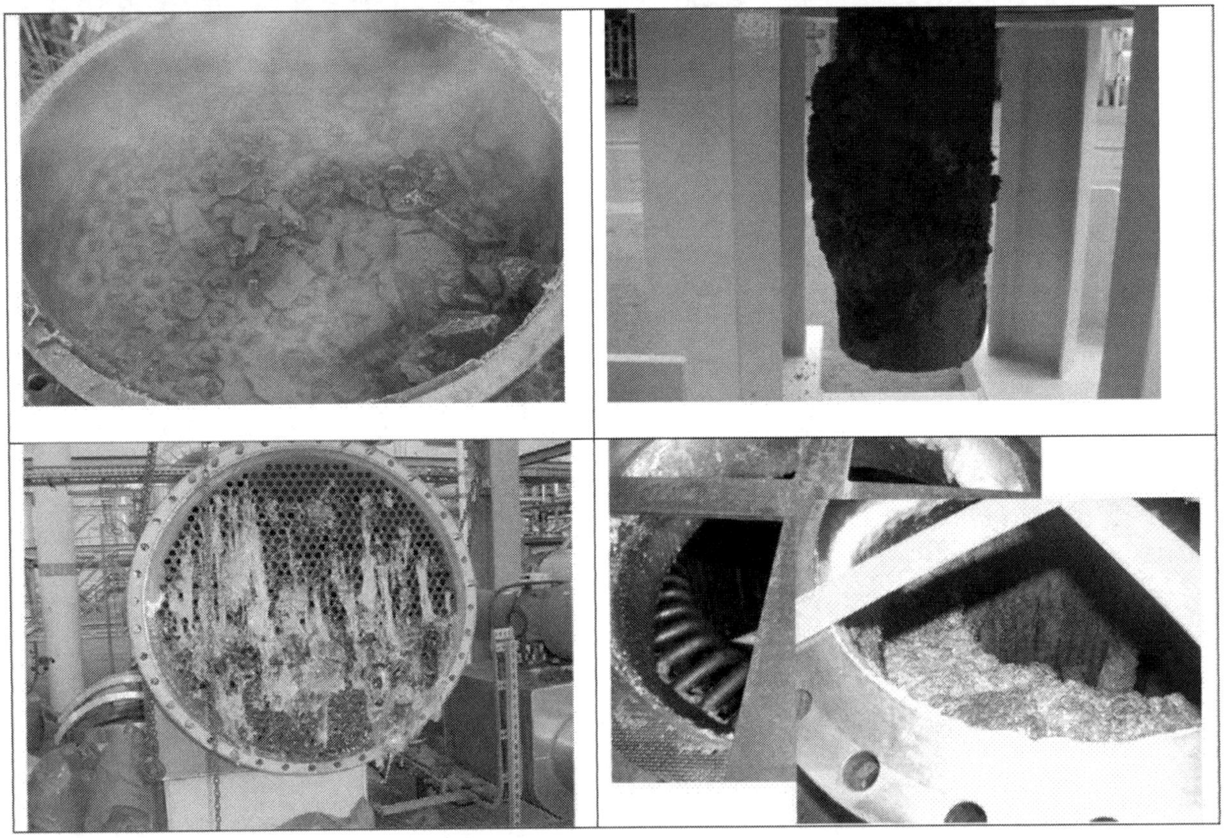

Figure 1: Images of recent fouling problems at Dow Chemical that have been solved using the methods described in this paper. Top left) Top of a falling film evaporator that plugged with tars every 8-10 days; Top right) Carbon deposit on the feed nozzle of a gas phase thermal chlorinator, Bottom left) Polymer deposit on the tube surface of a heat exchanger, Bottom right) A stirred tank reactor with internal cooling and polymer fouling.

In two top images in Figure 1, the plants that experienced this fouling lived with the problems for more than 25 years. These types of problems often share some similar characteristics. First, they are typically the result of some kind of trace chemistry and not the intended chemical reaction or polymerization. This means that they occur on a significantly longer time scale than the desired reaction. Long time scales are typically present in continuous processes, where the goal is often to run for years between shutdowns. Recycle streams can build-up concentrations of impurities and provide a reservoir for potential undesired chemistries to occur. Additionally, these impurities' concentration may exceed the solubility limit, and precipitate at a critical unit operation. The locations are often on heat exchangers, filters, catalyst beds or evaporators. Corrosion products such as rust can serve as Lewis-acid catalyst for many polymerization reactions. One final characteristic is that the fouling problem will tend to come and go – that is it will be "solved" and then reappear some weeks or months later.

The Methodology for Solving Fouling Problems in Process Industries

The general method for permanently solving fouling problems is at first inspection so simple that it is interesting that most of the problems have not been solved previously to our work in Dow. The detective work needs rigorous, objective analysis that relies heavily on data (experimental or computational), developing multiple hypothesis, testing and elimination of those, followed by designing, and implementing well-tested, sustainable solutions. The problem solving work is guided by the following principles.

Definition of Success

The first step is to write the definition of success in one to two sentences. For example:

"Success is not having to shut the plant down due to fouling of the head of the evaporator more often than every 6 months".

Sometimes there is an interim success statement and a long term success statement. In the example above, the unit was fouling about every 10 days. Long term success was "we never have to shut the plant down due to fouling of the head of the evaporator".

What has to be true for success to be true?

List what has to be true for success to be true. In general, this includes the following:

- First, the chemical structure of the fouling material, or at least its general molecular framework, needs to be determined.
- Then, there needs to be a fundamental mechanistic understanding of the chemical reactions that produce the fouling material.
- The physical properties of the fouling material and the media around it need to be determined.

- The reaction rates to make it need to be estimated. If the proposed reaction rate approximately balances with the fouling rate, in the context of a mass balance of the system and the appropriate physical properties, then the problem is probably pretty well understood.
- Once the fundamentals are understood, they can be integrated into an engineering model of the process to determine where to intervene.
- The models can be tested with either process or experimental data. The authors have a general reluctance to go to the lab and perform experiments first, as these can slow progress towards solutions. Often the needed information can be taken from the literature or the process itself.

The Tools Used for Solving Fouling Problems in Process Industries

The rate limiting step to solving fouling problems is to understand fundamentally how the fouling material is made and deposited within the constraints of the process. Again, these are often trace side reactions that create and accumulate material with time.

Use of the Analytical Lab

Many problem solving teams make use of the analytical lab at their company to gain some understanding of the molecular structure of the fouling material. The authors do this as well. Very briefly, the use of GC-MS, LC-MS and pyrolysis GC-MS, preferably with an exact mass detector, have proven to be very useful. NMR compliments MS very well. Simple techniques such as elemental analysis are helpful (CHN, XRF, ICP). Scanning electron microscopy coupled with energy dispersive x-ray gives elemental maps of the fouling materials' microstructure. There are numerous other techniques that have proven to be helpful that cannot be detailed here.

Use of Computational Chemistry for Thermodynamics and Kinetics

In our recent problem solving exercises the fastest way to evaluate hypothesis was to complement experimentation with computations. The accelerated problem solving always accompanied with maximizing the value of a reduced number of experiments with complementary computations. The literature is abundant today and can provide many plausible reaction pathways once the general structure of the fouling material is understood. Because of recent advances in both hardware and software, computational chemistry has become one of the fastest tools to evaluate multiple possible scenarios for the formation of materials.

Classical, and more advanced computational chemistry techniques implemented in various software packages (Gaussian[1], Spartan[2], Jaguar[3], Turbomole[4]) are suitable to perform geometry optimizations, and additional computations for thermochemistry (H^{298}, S^{298}, and Cp). As single molecule computations approximate primarily gas-phase thermochemistry; additional techniques are necessary for liquid and solid phase properties computations such as vapor-liquid equilibria, miscibility, and solubility. One of the most useful, and accurate methods is the CosmoTherm[5] package developed by A. Klamt. Calculation of mixture properties is as straightforward as single compounds'. Solubility and miscibility between gas-liquid, liquid-liquid and liquid-solid can be addressed. While the method can treat charge-neutral systems very well, and can deal with ionic liquids or infinite dilution ionic solutions; it cannot estimate the properties of intermediate dilution ionic solutions, here further basic scientific development is necessary. In case spectral properties are required for deducing possible compositions of a mixture, UV-VIS, IR, and NMR computations are accurate enough to guide analytical work. The thermodynamic calculations can frequently be performed within hours. Thermochemistry is suitable only for very rough screening of possible scenarios; most of the time estimation or direct measurements of kinetics is necessary. As many classical theories of reaction rates exist for organic reactions, they may be used in combination with modern computational chemistry. Some of these theories are linear free-energy relationships, transition state theory, and collision rate theory. This is the area where computations can help most in reducing the number of experiments, as estimation of relative rates are reasonably accurate. Consequently, minimal experiments for calibration of absolute rates are required only.

As we mentioned before, these are often trace chemistry side reactions with very low yields: low enough that they are not industrially interesting but high enough to cause problems in the process industries. They are often briefly mentioned in most organic chemistry texts or not at all because the yield is so low. Consequently, general principles of organic chemistry are always useful for enumerating all possible scenarios and test each of them. Frequently it is useful to understand electron distributions, reactivity of functional groups. Since in an industrial setting speed is critical, there is always a compromise between speed and accuracy while applying computational chemistry techniques. The authors most often prefer ab-initio hybrid density functional theory, which is fast and accurate enough for trends. When it's affordable and necessary, higher level electron correlation methods are used.

While ab-initio quantum chemistry can deal with a great variety of molecular structures – including organo-metallic compounds, its speed is not fast enough for computing large number of compounds within minutes. Classical group contribution methods are very fast; however, the classes of compounds that they can handle is much more limited. To broaden the classes of compounds, high level quantum chemistry maybe used to obtain "virtual" experimental data for finding thermochemistry parameters of experimentally unknown group values[6]. Additivity and transferability of groups' reactivity is also a well-known phenomenon[7] that is automated and utilized for rapid kinetics estimations.

Connecting Chemistry Understanding to Engineering Simulations

Once the underlying chemistry is qualitatively understood, and quantitative thermochemistry and kinetics is developed, there is a great need to translate this knowledge to engineering simulations. Engineering analysis most of the time requires additional knowledge of physical properties, such as vapor-liquid equilibria, solubility, miscibility, viscosity, diffusivity, heat transfer coefficients, which are generally not primary importance to chemistry understanding. In our recent trouble shooting successes it was a critical part to understand the differences between the two fields, and intentionally concentrate on connecting them with providing the suitable pieces of information to each. In this process, Cosmo-RS and Cosmotherm package was a great help to turn single-molecule gas phase simulation results to fluid phase property computations. These steps are largely automated and can be routinely used by the experts.

Once the thermochemistry, kinetics, and necessary physical properties are obtained engineering simulation packages, such as ASPEN or gPROMS are used for process simulations to account for the effect of transport, process controls and recycle streams. The impact of the existing process control methods can be determined. The extremes of the process conditions that can drive trace chemistry, such as skin temperatures in direct fired heat exchangers or other heated equipment, as well as the impact of residence time, and thus the yields of the undesired foulants can be estimated.

An Example of Successful Application of Methodology and Tools

Formation of Polymeric Tar on Top of a Catalyst Bed

Hydrogenation processes are very common in the chemical and refining industries. Dow has one such process where a mixture of olefins and olefinic aromatics are being hydrogenated to saturate the olefins. One of the hydrogenation reactors is a packed bed of catalyst that experienced an abrupt increase in pressure drop across the reactor about every 90 days. When the reactor was opened, about 1 meter of polymeric tar was found on top of the reactor. Unfortunately, the plant became very experienced at removing this tar and replacing the catalyst, optimizing the turn around process so that only slightly more than 2 weeks of shut down were required per event. This problem existed for the life of the plant (more than 20 years). The loss of production and loss of economic profit became very significant as the market price of the product greatly increased.

Success was defined as: "The plant can run for 3 years without a shut down due to an increase in pressure drop across the hydrogenation reactor in question".

What was necessary for success was:

- We need to know what the macromolecular structure of the polymeric tar is.
- We need to know what the raw materials are that are available to produce it.
- We need to write out the chemical reactions that could lead to its production under the conditions available in the process and estimate their thermochemistry, kinetics.
- We need to predict the yield of the material based on this fundamental understanding.
- Then, we need to translate this understanding to engineering interventions that have a high probability of success while living within the economic constraints that exist for the business.

Use of the Analytical Lab

The polymeric tar was analyzed in the analytical lab using elemental analysis (CHN), XRF, pyrolysis GC-MS and solid state NMR. The empirical formula of the tar was found to be about C1H1. Few if any trace elements were present (very little oxygen, low trace amounts of catalyst dust). Solid state NMR revealed that there were equal amounts of sp2 and sp3 hybridization carbon. Pyrolysis GC-MS found many single aromatic fragments with short aliphatic side chains. Thus, the tar appeared to be a network of single aromatic rings connected with a lattice of short aliphatic side chains.

Use of Computational Chemistry for Thermodynamics and Kinetics

The feedstocks in the process included a mixture of styrene, ethyl benzene, xylenes and some di-olefins. The reactions that are possible for such mixtures included Diels-Alder cyclo addition reactions, polymerization of styrene monomer to make polystyrene, and perhaps free radical reactions. At first, the last reactions were deemed to be unlikely as the process conditions were between 200-300 ^{o}C at about 40 bar total pressure.

The Diels-Alder reactions were examined using the literature[8] and quantum chemistry. The findings were that Diels-Alder cyclo additions do occur in the process. However, as the molecules move through the process the temperature of the process increases markedly (from about 200 to about 300 C) and the reverse reaction occurs (the retrograde Diels-Alder reaction) as the negative entropy of the process overcomes exothermicity of the reaction and drives the products back to the starting materials. An example of the equilibrium constants for this reaction is shown in Figure 2.

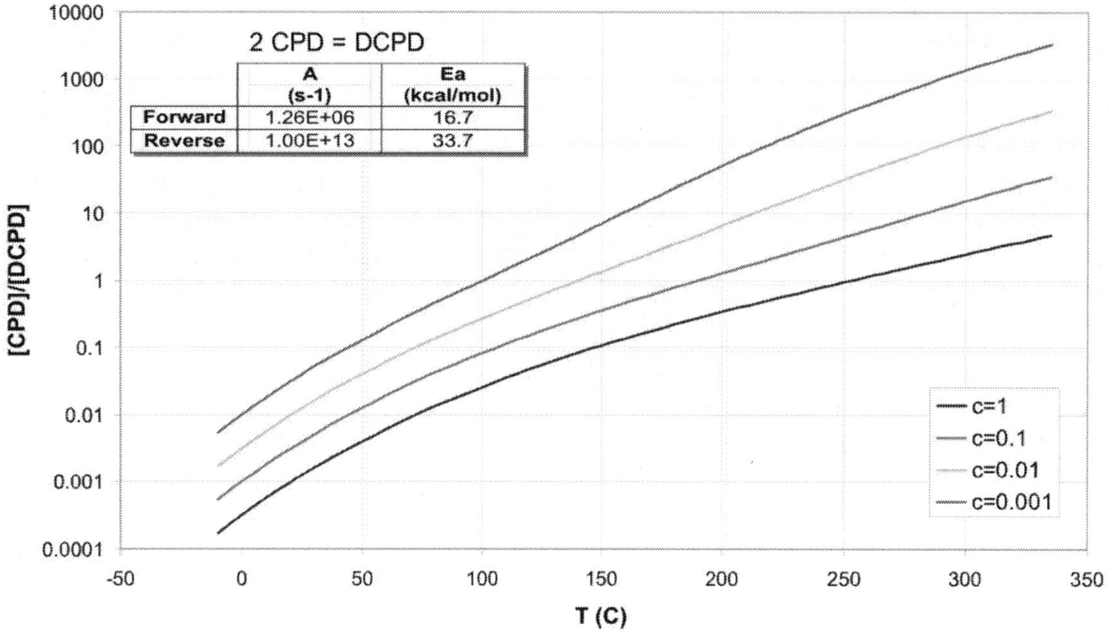

Figure 2. Ratio of cyclopentadiene and dicyclopentadiene at equilibrium as a function of temperature and initial concentration in moles/liter (c=0.001, c=0.01, c=0.1 and c=1.0). The Arrhenius pre-exponential factor and activation energies are also listed. The equilibrium constant is independent of the media as is approximately true for all Diels-Alder reactions in which apolar molecules react[9].

There was evidence of small amounts of polymerization of polystyrene from data collected from liquids in an intermediate knock out pot. Pyrolysis GC-MS data from this liquid show evidence of low molecular weight polystyrene that has incorporated other monomers into its structure. This is shown in Figure 3. However, from the literature it is unlikely that polystyrene is the primary cause of the tar formation, as this reaction too is reversible. The process conditions exceed the so called "ceiling" temperature of polystyrene. Exceeding this temperature causes the polymer to revert back to its monomers.

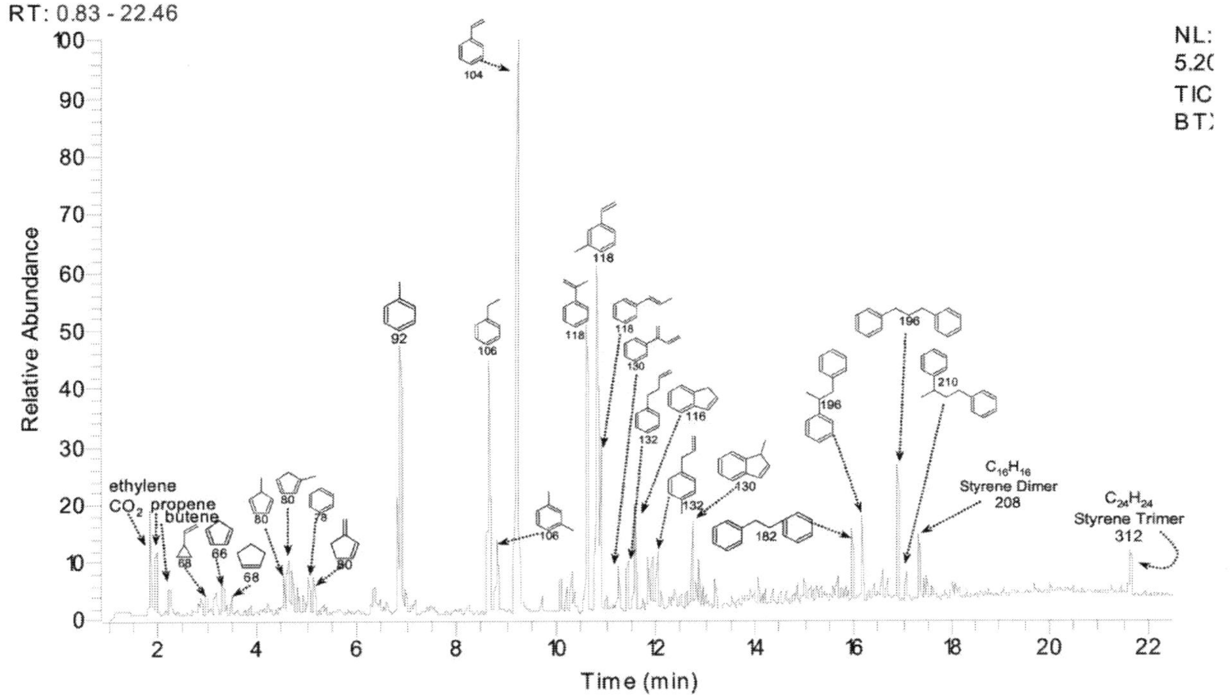

Figure 3. Pyrolysis GC-MS data of liquid from intermediate knock out pot. Fragmentation pattern shows styrene monomer, dimer and trimer, which is typical of polystyrene.

This leaves free radical polymerization as the only plausible reaction mechanism available to produce the tars. The process conditions of about 40 bar and 300 C at first make this mechanism seem unlikely. The necessary initiation reaction step is shown in Figure 4. As we see, breaking the carbon-carbon bond in the benzylic position has at least about 70 kcal/mol activation barrier. The first question that arises is how can this reaction occur at 300 °C, as this temperature is not enough with physically pre-exponential values to proceed with a reasonable rate? Examination of the process provides an answer. There is an old direct fired furnace in the process that is being run at rates far past its initial design capacity. A brief description of the furnace is show in Figure 5. The skin temperature of this furnace can be estimated using standard engineering methods. The results are that the skin temperature is between 550 and 600 °C.

There is a laminar sublayer near the surface of the tubes in the furnace. This is shown in Figure 6. Molecules such as ethyl benzene will experience temperatures of about 550-600 °C in this sub layer. A fraction of them, about 1 in 10^8, will become free radicals and react with the fluid around them. Collision theory can be used to estimate the number of reactive collisions. The ultimate size of the tar molecules can also be estimated to be about 8-10 monomer units. The estimated physical properties of these tars are above the condensation point in the process that exists on top of the catalyst bed (about 300 C). The ultimate proof is that this analysis provided a correct estimate of the yield of tars. Experimentally, the plant digs out about 7 parts in 1 million of its feed (the mass of the polymeric tar) on shutdown. The fundamental estimate produces a yield of about 5 parts per million.

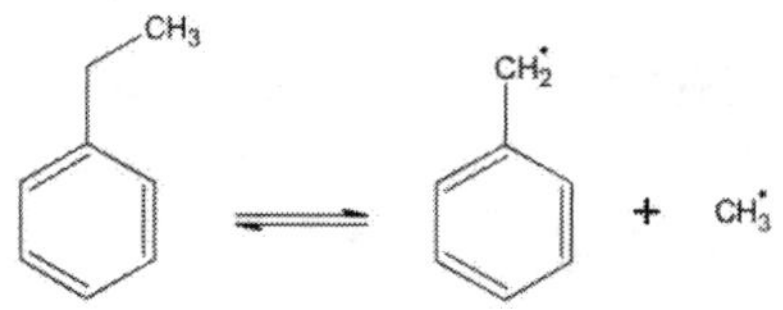

$$\Delta H^{298} = 70.5 \text{ kcal/mol}$$

$$\Delta S^{298} = 39.2 \text{ cal/mol/K}$$

Figure 4. The lowest barrier reactions to produce reactive free radicals in this system are those that produce benzylic radicals, which are stabilized by resonance with the benzene ring.

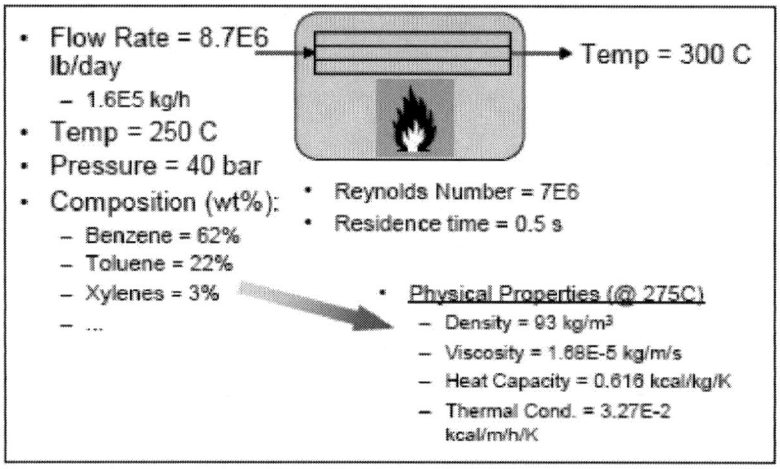

Figure 5: Brief description of the furnace prior to the hydrogenation reactor. Standard engineering calculations produce an estimated skin temperature of between 550 and 600 °C.

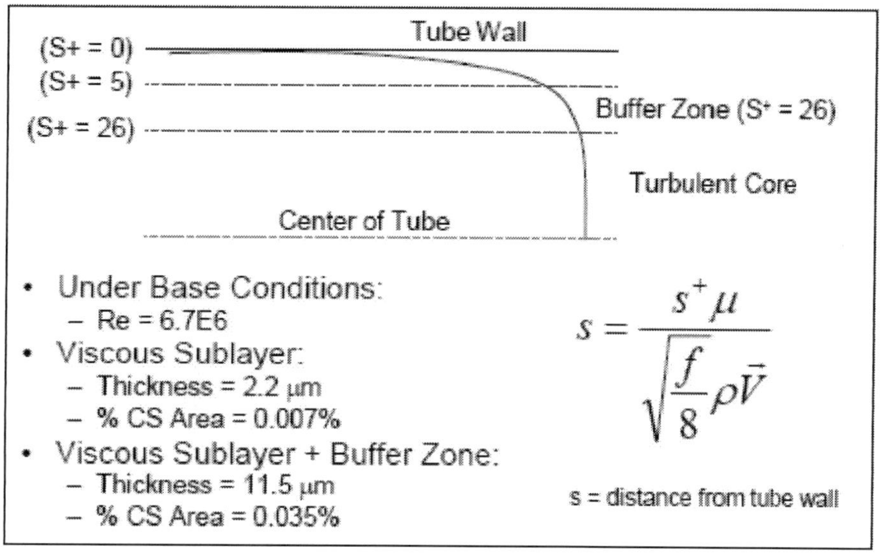

Figure 6: Depiction of laminar sub layer near the skin of a tube in the direct fired furnace shown in Figure 5.

In terms of an engineering intervention, a response surface specific to this particular furnace can be constructed. It is shown in Figure 7. The relative amount of polymer is depicted in the contours on the plot – with red being higher and blue being lower. The plant can use this guide to move away from regions of higher tar formation to lower tar formation, all within the constraints imposed by the existing heat balance in the process.

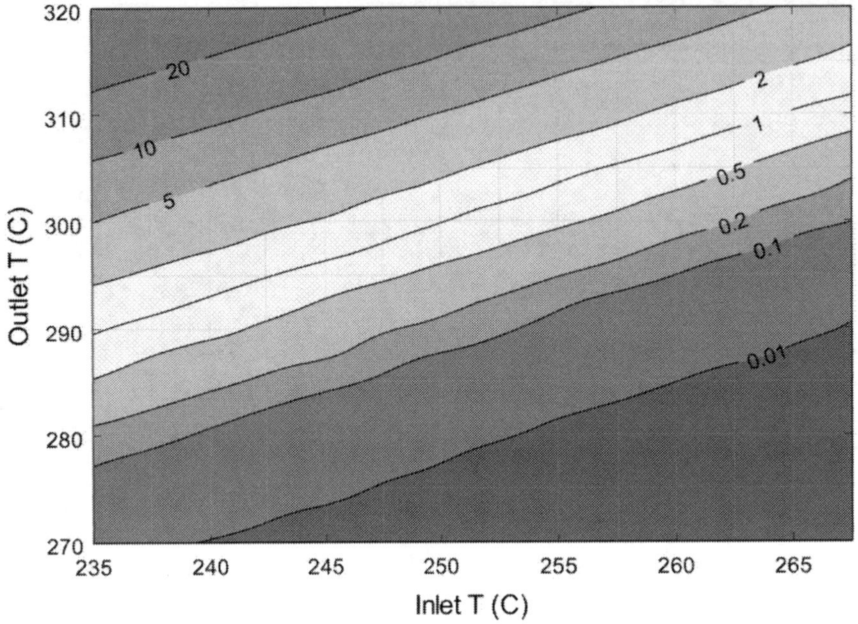

Figure 7. This is a diagram of the engineering intervention for tar formation in hydrogenation process. The furnace inlet and outlet temperatures are the x and y axes. The contours represent the relative rates of polymer formation for this particular furnace.

The technique shown above helped to design not only process conditions to avoid coke formation but also sizing of a replacement furnace.

Summary

The methods described in this report have been used to solve more than ½ dozen fouling problems at Dow in the last 3 years. In several cases, the problems had existed for more than 25 years. In these cases, multiple attempts had been made to solve them that were unsuccessful. The authors would appreciate comments from readers of this paper about how these methods fit with their experiences.

---

[1] *Gaussian 03*, M. J. Frisch, G. W. Trucks, H. B. Schlegel, G. E. Scuseria et al., Gaussian, Inc., Pittsburgh PA, 2003. www.gaussian.com
[2] Spartan '06, Wavefunction Inc., Ivine CA. www.wavefun.com
[3] Jaguar, Schrödinger LLC, www.schrodinger.com
[4] Turbomole, University of Karlsruhe, www.turbomole.com
[5] A. Klamt, *COSMO-RS, From Quantum Chemistry to Fluid Phase Thermodynamics and Drug Design* (Elsevier, Amsterdam) 2005. www.cosmologic.de

[6] D.J. Frurip, N.G. Rondan, J.W. Storer "Implementation and Application of Computational Thermochemistry to Industrial Process Design at the Dow Chemical Company" in *Computational Thermochemistry; Prediction and Estimation of Molecular Thermodynamics*, eds. K.K. Irikura, D.J. Frurip (ACS Publications, 1998).

[7] R. Atkinson *Atmospheric Oxidation* in "Handbook of Property Estimation Methods for Chemicals" Eds. R. S. Boethling, D. Mackay (Lewis Publishers, 2000)

[8] A. Wassermann: *Diels-Alder Reactions, Organic Background and Physico-Chemical Aspects*, Elsevier, 1965; K. N. Houk, L. J. Luskus: *J. Org. Chem., 38(22),* 3836, **1973.**

[9] Herndon, W.C., Grayson, C.R., Manion, J.M., Retro-Diels-Alder Reactions. III. Kinetics of Thermal Decompositions of exo and endo Dicyclepentadiene, *J. Org. Chem.*, Vol 32, pp. 526, **1966.**

494d Particle Information Required For Effective Piloting Or Scaling Of Particle Processes

Bruce D. Hook Sr., Performance Chemicals Process Research, The Dow Chemical Co., 2301 N. Brazosport Blvd, Building B4810, Freeport, TX 77541

this paper will review the particle properties important to understand to scale up from laboratory or piloted processes to full scale processes.

494e A New Type Of Liquid-Solids Draft Tube Spout Fluid Bed-Hydraulic Transport Of 1.94 Mm Glass Particles In Water

Howard Littman, Isermann Department of Chemical and Biological Engineering, Rensselaer Polytechnic Institute, Ricketts Building, 110, 8th Street, Troy, NY 12180, John Paccione, Environmental Health Sciences, SUNY at Albany, School of Public Health, 547 River Street, Flanigan Square, Troy, NY 12180, Lealon L. Martin, Chemical & Biological Engineering, Rensselaer Polytechnic Institute, 110 8th Street, Ricketts Building, Troy, NY 12180, and Joel L. Plawsky, Rensselaer Polytechnic Institute, Chemical Engineering, 110 Eighth St RI-122, Troy, NY 12180-3590.

A new type of draft tube spout-fluid bed (DTSFB) is described in which the draft tube physically separates the particle feeder section from the particle separation section. This arrangement allows the fluid mass flowrate in the draft tube and the dynamic pressure drop across it to be independently specified. As a result, the solids mass flowrate can be varied by changing the solids fraction, the particle velocity or both. Equations are presented to show how this new DTSFB works and how the solids fraction and particle velocity in the draft tube are set and varied. The data of Grbavcic et al (Can. J. Chem. Eng. Vol. 71, pp. 895-905, 1992 are used to flesh out the design of a water driven bed.

Experimental Investigation into Tailings as an Ingredient Material for Belite Mineral

Guoxian Xiao[1], S. Ding[2], Guoqiang Zhang[1], Ci Zhou[1] and Zhihai Gao[1]
1. Dept. of Material Sci. & Eng., Soochow Univ., Suzhou, China
2. Postec dept. av Tel-Tek, Kjølnesring, Porsgrunn, Norway

Introduction

The particulate solids known as tailings after the extraction of gold accumulated in the mining fields gradually, and were usually regarded as a waste, imposing great challenges to the environment in surroundings. However, analysis found that the tailings mainly consist of SiO_2 and CaO, along with some small amounts of other constituents, and could be therefore used as raw meal materials for the products in constructions.

It has been understood that the dicalcium silicate mineral (known as C_2S, belite) has advantages over the general silicate cement in, for example, heat hydration and soundness (lower), with a smaller quantity of CO_2 release while retaining the silicate cement's good durability strength [1, 2]. It attracted lots of attentions recently; and many activities are being conducted to produce belite C_2S with different raw materials [4 - 6]; however, the tailings were rarely used as a material in such efforts. In the present study, investigations were carried out in order to find an operation condition, under which the belite C_2S could be produced with the tailings. The possibility of producing new brand cements was also to be explored by blending the belite C_2S with silicate cements.

Composition and Properties of the tailings

The tailings are a kind of yellowish powders, with a moisture content of 1.8%. It has a relative narrow spectrum in the particle size distribution, with a mean size about 150 μm. The chemical compositions of the tailings are given as in Table 1. As seen, the tailings consist mainly of SiO_2, CaO, and a small quantity of Al_2O_3 and Fe_2O_3 with a high ignition loss. The contents of Alkalies (Ka_2O, Na_2O, MgO) are however also rather high. Never the less, the tailings appear to be a good candidate as a raw material for a product of belite C_2S of cement.

Table 1. Chemical composition of the tailings and limestone (%)

Constituents	SiO_2	CaO	Fe_2O_3	Al_2O_3	K_2O	Na_2O	MgO	TiO_2	Ignition loss	Σ
Tailings	30.24	26.48	6.09	6.66	1.13	1.28	3.02	0.57	24.04	99.51
Limestone	3.87	52.40	0.34	1.20	0.10	0.05	0.42	-	41.54	99.92

X-ray diffraction examination as shown in Fig. 1 revealed that the tailings contain mainly in mineral phases of Calcite, Quartz and Dolomite, with the Calcite as the largest in quantities among the three. The quartz exists in a crystal-phase with a α-SiO_2 modification.

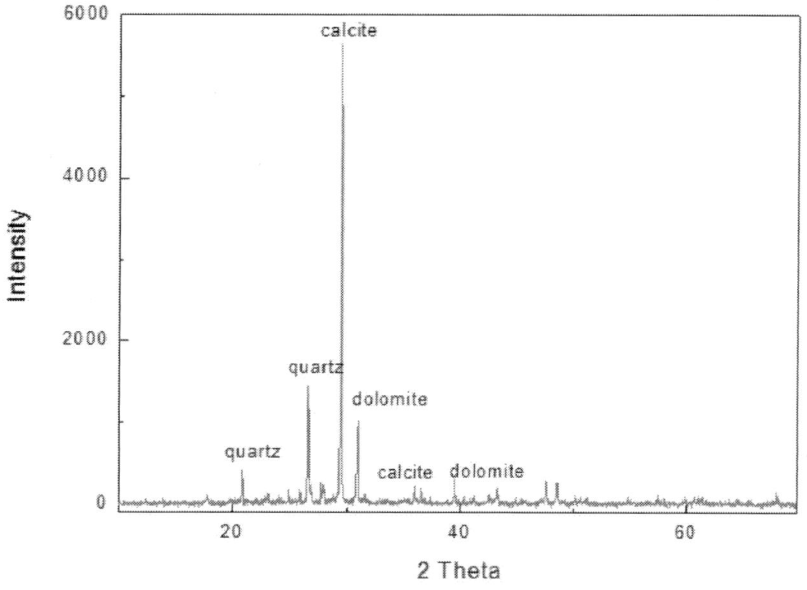

Fig. 1 XRD analysis

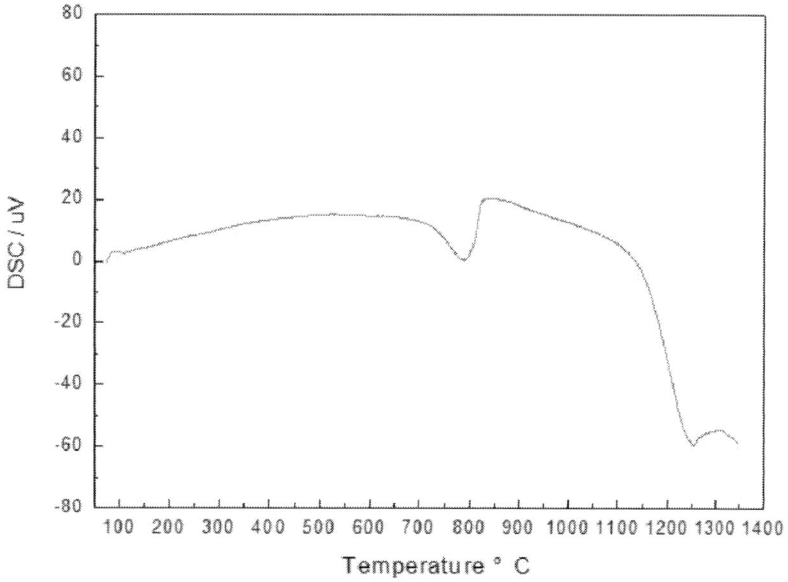

Fig. 2 DTA analysis

Figure 2 shows the results of analyses in a DTA diagram. It is easy to note that a typical endothermic peak occurred when the tailings were heated to a temperature at 787.27°C. This was caused by the decomposition of the Calcite and Dolomite into CaO and CO_2.

The fusion point of the tailings is 1255 °C due to the presence of various foreign constituents known as Alkalies (K_2O, Na_2O, TiO_2, Au etc.). The impurities of foreign compositions, incorporated as solid solutions, transfer the tailings to form a multi-phase mixed crystal, and bring down its solidus (eutectic) temperature. It was in fact observed that the tailings were totally melted down in whitewares when being heated to a temperature of 1250 °C; and was hard to be taken out from the appliance.

Calculation of raw mix composition and formation of Belite

As shown in Table1, the tailings are quite rich both in SiO_2 and CaO; still, the tailings alone contain not sufficient CaO to be a raw mix meal for the product of belite mineral. It is planned that the extra CaO required would be compensated with limestone, which is mainly comprised of CaO. The chemical compositions of the limestone used in the present experiment are also given in Table 1 (the lower row). Referred to Table 1, it is likely that the raw mix composition be prepared by blending the tailings and the limestone with a proper ratio. Based on the lime formulas, i.e., the conservation of the CaO, it was found that a mixture of the tailings and the limestone in a ratio of 1:1 (in weight) meets the requirement for chemical ingredients of Belite C_2S mineral cement product. Table 2 shows the chemical composition of such a mixture.

Table 2. Chemical composition of raw meal

SiO_2	CaO	Fe_2O_3	Al_2O_3	K_2O	Na_2O	MgO	TiO_2	Loss	Σ
17.06	39.44	3.22	3.93	0.62	0.67	1.72	0.29	32.79	99.74

The tailings and the limestone were ground into powders respectively in a ball mill, with a specific surface of a $420m^2/kg$ for the tailings and a $345m^2/kg$ for the limestone. These two powders were then blended in a ratio of 1:1 (in weight); the mixed powders were made into small cylindrical columns (Φ25mm×35mm) with a small amount of water as an agent. The cylindrical columns were then put into a high-temperature furnace after being dried in an oven dryer.

The column was fired with a designated temperature and time in the furnace. Liquid phases (melt) were observed in the column when it was heated to 1250°C or 1350°C, indicating that sintering (clinkering) occurred in the mixture.

Afterwards it was taken out from the furnace, and cooled down rapidly to a ambient temperature with a fan. The column solidified, accompanying with observable shrinkage in volume; and turned out in the end to be rather grey in colour. Such a rapid cooling is necessary to prevent the column from pulverizing, or transforming /converting C_2S from β-C_2S modification into γ - C_2S modification, i.e., leading a decrease of its strength [8].

X-ray diffraction analysis and free-CaO in Belite mineral

The cooled columns were ground into powders. The content of free lime (uncombined CaO) was determined according to the chemical analysis of cement and the X-ray diffractometry (diffraction analysis) as well [8]. The measurement results are given in Table 3, and Fig. 3(a - d) for different samples obtained / produced under different designated conditions (namely various fired temperature and time).

It is evident from Table 3 that: 1. the content of free-CaO in samples decreased considerably when the temperature increased from 1250 °C to 1350 °C; 2. the content of f-CaO also decreased if calcining time prolonged in a settled fired temperature (1350°C). Overall, the content of free-CaO in the belite mineral was however rather high if the temperature was lower than 1350°C, and was regarded as undesirable, and might cause liability [7, 8].

Table 3. The contents of free-lime CaO

Burning (Sintering) Condition	1150 °C	1250°C	1350°C	
	1h	1h	0.5h	1h
Free-CaO (%)	13.14	5.64	1.13	0.84

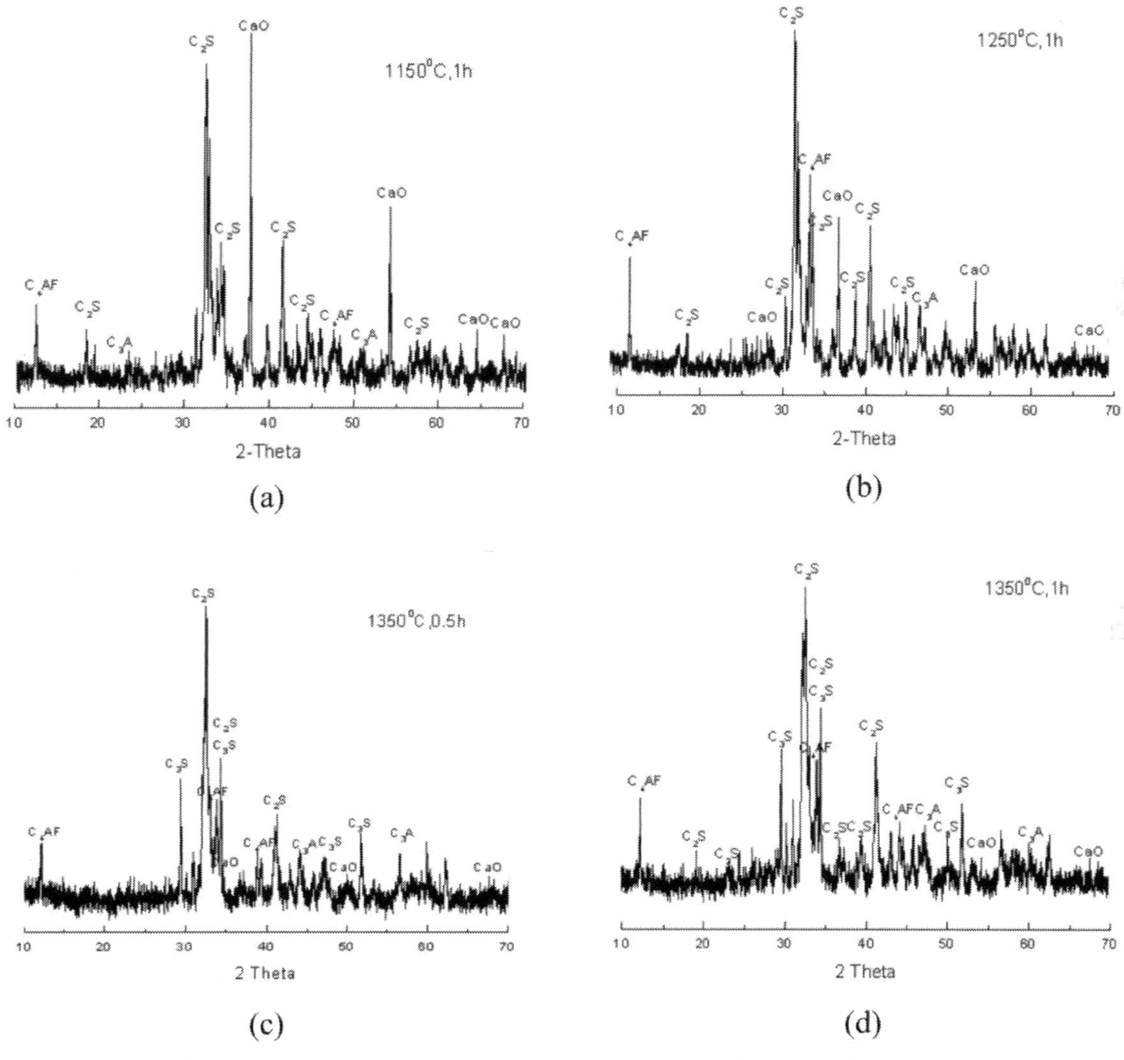

Fig. 3. XRD measurements for the content of C_2S and Free- CaO

Diagrams in Fig. 3(a - d) showed that the belite (in β - C_2S modification) was produced as prevalent mineral, along with the formation of small quantity of C_3A and C_4AF minerals when the tailings were fired with limestone at high temperatures. As seen, the peaks of belite phase hoisted as the temperature rose, indicating the belite mineral phase growing with the rise of temperature. The diagram of XRD (X-ray diffraction) also illustrated that the content of the free-CaO decreased with the time at a designated temperature.

Looking closer to the processes of the formation the belite C_2S, it might be found that: 1. the belite C_2S mineral started to form at a temperature of 1150 °C; the reaction was however not complete and the content of free CaO was therefore very high (13.14%); 2. the chemical reaction continued, the belite C_2S (in β - C_2S modification) increased while the free-CaO content decreased considerably (to 5.64%) at a temperature of 1250 °C; 3. when the temperature rose to 1350 °C, reaction to form belite C_2S mineral completed, a small quantity of C_3S also formed at the same time, as a result, the content of free-CaO dropped dramatically to 1.13% (0.84%).

Applications of belite mineral

To assess the properties of belite C_2S minerals formed under different burning conditions, belite mineral samples named HJ-A and HJ-B were prepared by firing the mixture of the tailings and the limestone (1:1 in weight) for one hour at a temperature of 1250°C or 1350°C respectively. The minerals thereby made were then ground with a ball mill into powders, with a specific surface of 425m^2/kg for HJ-A, and of 447m^2/kg for HJ-B. They were then respectively used to blend with the general silicate cement in ratios of 30:70, 45:55 to make different HJ- cements, which are denoted as given as in Table 4.

Table 4. HJ- cements by mixing silicate cement with HJ-A or HJ-B

HJ Cement	Ratios in weights (%)		
	Silicate cement	HJ-A	HJ-B
HJ-A-1	70	30	N/A
HJ-A-2	55	45	N/A
HJ-B-1	70	N/A	30
HJ-B-2	55	N/A	45

Table 5. The property of the HJ-cements

Test No.	H_2O (ml)	Setting times h:min		3d/MPa		28d/MPa	
		Initial	Final	Bending	Compression	Bending	Compression
HJ-A-1	121	2:22	3:37	4.2	17.7	7.6	37.9
HJ-A-2	128	2:29	3:46	3.4	13.6	7.0	35.1
HJ-B-1	124	1:34	2:29	5.5	22.3	8.1	45.2
HJ-B-2	129	1:37	2:39	4.6	19.2	7.4	39.4

The determinations of HJ- cement properties, compared with those of pozzolanic cements, was carried out based on the requirement as specified in National Standards. The results are given in Table 5. One can see in Table 5 that the properties of the cements made by blending silicate cement with HJ-A or HJ-B all appeared to be normal in the initial / final setting times and the comprehensive strength etc.; the consumptions of the water in hydration were also normal.

The strength development of bending / compression of the HJ- cements were also measured and are shown in Fig. 4 and Fig. 5. It is evident there that all HJ-cements showed similar trends of strength development (increase) over a period from 3 days to 28 days.

As shown in Fig. 4, one can see that the bending strength of all HJ- cements (mixed with 70 % silicate cement) is greater than that of Grade 42.5 pozzolanic cement; the same is for the bending strength of HJ-B cements even mixed only with 55 % silicate cement, whereas the bending strength of HJ-A cement could only march that of Grade 32.5 pozzolanic cement if doped with a less than 55% of silicate cement.

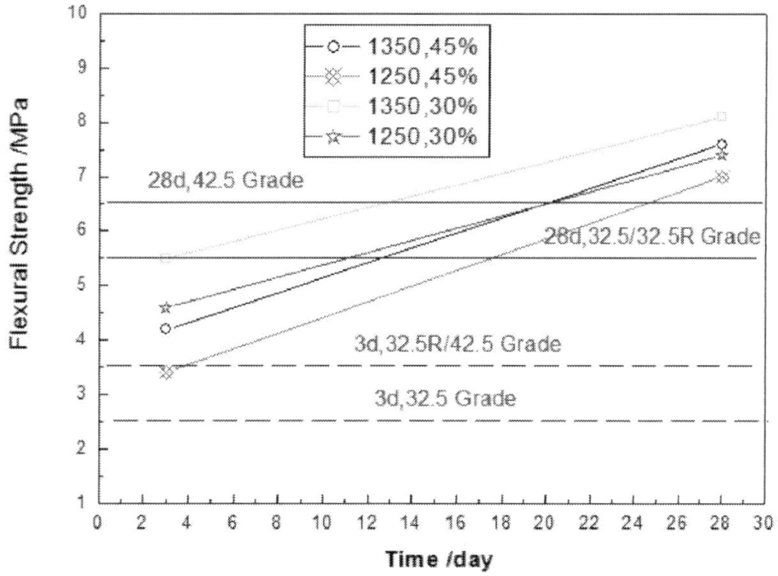

Fig. 4. Development of bending strength

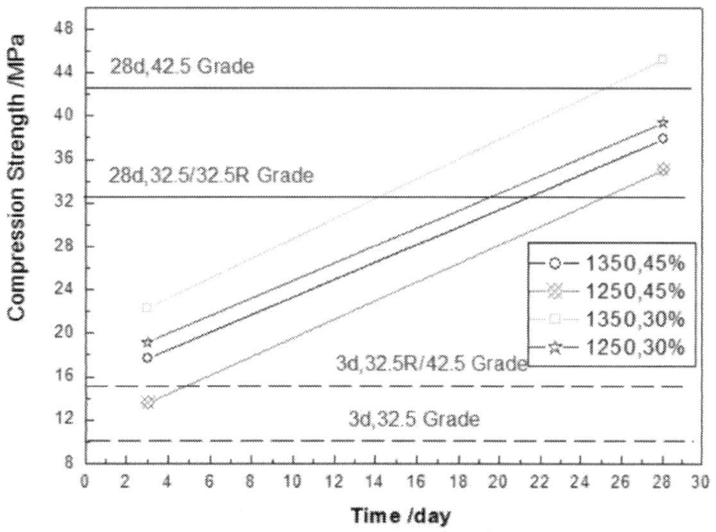

Fig. 5 Development of compression strength

One may also see in Fig. 5 that HJ-B cement (with 30% HJ-B) is greater than the Grade 42.5 pozzolanic cement in the compressive strength; HJ-B cement (with 45% HJ-B) than the Grade C32.5R pozzolana cement. While for the HJ-A cements, its compressive strength (with 30 %HJ-A) could reach that of Grade R32.5 pozzolanic cement, however when its portion increased in the mixture, HJ-A Cement (with 45% HJ-A) could only match that of Grade 32.5 pozzolanic cement in the compressive strength.

Conclusion

As having been shown, the tailings are rich in SiO_2 and CaO, with small quantity constituents Al_2O_3 and Fe_2O_3 etc; and could be used as a raw meal for product of cement when mixed with limestone in a ratio of one to one in weights.

A proper condition was found to sinter the raw meal made by mixing tailing and limestone, that is, the mixture should be fired / burned at temperature around 1350 °C for 0.5 to 1 hour. Under such a condition, the HJ-belite mineral produced was, as revealed with X-ray diffraction analysis, very good in quality with a very low free-CaO.

Investigations into the properties of the belite mineral were also carried out. The HJ-cements (made by blending the HJ- belite mineral and silicate cement) appeared to be normal in the initial / final setting times and the comprehensive strength. Further examinations of the strength of HJ-cements confirmed that it is feasible to produce different kinds of HJ- cement by mixing HJ belite C_2S mineral with general silicate cement in a proper ratio.

References

1. Fatang Li and Qingzhi Wang, Desong Wang, Zhimin Sun and Xuewang Zhu (2003), "Advantages and Research Progress of Belite Cement", *Hebei Industrial Technogy*, 20(4), pp. 57-61

2 Fengqin Xie, Lijiu Wang, Liping Wang and Ping Qu(1998), "Investigation into the Belite Cement-Pozzolanic Cement", *Journal of Dalian Sience and Technology University*, 38(5), pp.571-574

3. Hui Lu, Jingyu Zhong and Yueming Fan(1996), "Properties of Belite Produced at Different Temperatures", *Journal Sience and Technology University of South China (Natural Science Version),* 24(2), pp.124-128

4. Jiang Fenghua and Xu Delong(2006), " Sintering Condition and Compression Strengh of Q-Phase Belite Cement", *Journal of Chinese Ceramic Society*, 34(6), pp.749-752

5. Jiang Fenghua, Xu Delong and Zheng Shaohua(2004), "Sintering Condition of New Q-Phase Belite Cement", *Journal of Chinese Ceramic Society*, 32(10), pp.1289-1292

6.Guo Suihua, Zhang Hongtao, Lin Zhen, Wensheng Zhang, and Chen Yimin (2002), "Relationship between Agglomeration and Microstructure of High Belite Clinker and Performance of Cement, "Journal of Chinese Ceramic Society", 30(4), pp. 417-422

7. W. H. Duda, Cement data-book, International process engineering in the cement industry, Macdonald & Evans London, 1977.

8. L. Otta, Cement engineers' handbook, Bauverlag Gmbh, Wiesbaden and Berlin, 1983.

503a Cluster Formation And Particle Dynamic In Zinc Sulfide Cvd Reactor

Yousef Sharifi, Chemical, Materials, and Biomolecular Engineering, University of Connecticut, 191 Auditorium Rd Unit 3222, Room 204, Storrs, CT 06268

Zinc sulfide has a large direct band gap and is widely used in optical devices. A popular method to produce high quality zinc sulfide is chemical vapor deposition (CVD). Operating conditions (e.g. high temperature) in a CVD are usually suitable to enhance adduct formation. The clusters formed may leave the reactor, settle, or deposit on the substrate creating defects on the deposited film. In this presentation we will discuss a two dimensional CFD model which predicts the particle growth and distribution inside the CVD reactor. We employed density functional theory to estimate unknown kinetic parameters (such as nucleation, and surface growth rates) and possible crystal morphology in the gas phase. Our results show that (a) particle diameter will increase with an increase in temperature, and (b) the average particle diameter is higher at the reactor exit. The axial distribution of particles shows larger particle formation rate closer to the reactor inlet which decreases towards the reactor outlet. The clusters morphology varies from bubble-like for small particles to BCT structure for large particles.

503b The Axial Solids Hold-up Profile in the Riser of a Cfb

Manon Van de Velden, The University of Birmingham, Birmingahm, United Kingdom, Xianfeng Fan, School of Physics and Astronomy, University of Birmingham, Edgbaston, Birmingham, B15 2TT, United Kingdom, Andy Ingram, Department of Chemical Engineering, University of Birmingham, Edgbaston, Birmingahm, United Kingdom, and **Jan Baeyens**, Department of Chemical Engineering, The University of Birmingham, Edgbaston, Birmingham, B15 2TT, United Kingdom.

There is a literature discrepancy about the hydrodynamic fluidisation mode in the riser as function of the superficial gas velocity (U) and the solid circulation flux (G), with the possible existence of three types of hold-up profiles: -a constant dilute profile in the dilute regime -an S-shaped profile in the regime often referred as "fast fluidisation" -a constant and dense profile in the dense regime. Continuous transitions between these three regimes can be found by varying U and or G. As the research concentrates on gas/solid reactions, operating at low to moderate values of G compared with catalytic gas reactions, the dense regime is of less importance and was not experimentally investigated.

In order to define the transition between the dilute and S-shaped profile, solids hold-up profiles were measured at combined (U,G)-values. The zone of recycling bed material from the L-valve was moreover investigated using the PEPT-technique with 18F-labelled tracers.

Experimental data will be illustrated and correlated together with literature data. The transition from dilute to S-shaped profile is obtained at G > 3.9 U1.8.

The recycle loop can impose a high pressure drop on the riser, even at small G-values. In order to satisfy the pressure balance of the CFB, an S-shaped profile has to exist even at small values of G. This effect of the re-entry zone was measured by PEPT and expressed in an acceleration length, i.e. the height above the L-valve needed by the particle to achieve its stable particle velocity. Results are again Expressed as height vs. (U,G), and compared with a theoretical model to predict the acceleration zone. Experimental and model results are in very good agreement. In general, the re-entering particles will spend a time less than 5 to 7.5 % of their residence time in the riser, hence without major importance for calculating conversions in function of this residence time.

Literature data moreover enable us to define the transition between the S-shaped and dense profiles, obtained at G > 11 U1.8.

The simple equations proposed by the research prove of better accuracy than the more complicated correlations found in literature.

*Corresponding author: Professor Jan Baejens; e-mail: J.Baejens@bham.ac.uk; tel. +44 121 4145343; fax. +44 121 414 5323

503c The Effect of Horizontal Baffles on Gas Bypassing in Deep Fluidized Beds of Group a Particles

A. Issangya, S. B. Reddy Karri, and T. M. Knowlton. Particulate Solid Research, Inc., 4201 West 36th Street, Suite 200, Chicago, IL 60632

ABSTRACT

Solids mixing and bubble characteristics studies in smaller laboratory-scale test units have shown a fluidized bed to be well mixed. Recent work has shown that deep fluidized beds of Group A particles in the bubbling and near-turbulent bed modes can exhibit severe gas bypassing. The gas bypassing (also called jet streaming) occurs in the form of a high velocity air jet or rapidly moving series of bubbles that preferentially flow up near the column wall. The rest of the bed stays defluidized or poorly fluidized. The jet generally rotates around the column wall, but there are instances when it anchors at one position (i.e. at cyclone dipleg return locations). Multiple gas bypass streams can also be present. If gas bypassing occurs in industrial beds it can result in afterburning, poor gas/solids contacting and poorly fluidized entrances to standpipes or discharge regions of cyclone diplegs.

Baffles or tubes have traditionally been added to fluidized beds to improve gas-solid contact and/or to enhance heat and mass transfer. The objective of this work was to study how horizontal baffles can be used to reduce or eliminate gas bypassing in deep beds of Group A particles. Tests were conducted in a 0.3-m-diameter test unit using 76-micron FCC catalyst particles. Bed heights from 0.76 to 3.05 m and gas velocities of up to 0.9 m/s were investigated. Visual observation, measurement of differential pressure fluctuations and radial bubble void fraction profiles were used to determine how effective steel grating baffles and horizontal rods were in reducing gas bypassing. It was found that for a given operating condition, the ability of the internals to eliminate gas bypassing depended on vertical baffle spacing, open area, and grid (or mesh) spacing.

FOUR WAYS TO INTRODUCE STRUCTURE IN FLUIDIZED BED REACTORS

John Nijenhuis[1], J. Ruud van Ommen[1,*], and Marc-Olivier Coppens[1,2]

[1]Delft University of Technology - DelftChemTech
Julianalaan 136, 2628 BL Delft, The Netherlands
[2]Howard P. Isermann Department of Chemical and Biological Engineering
Rensselaer Polytechnic Institute, 110 8th Street, Troy, NY 12180, USA
*T: +3115 278 2133; F: +3115 278 5006; E: j.r.vanommen@tudelft.nl

ABSTRACT
Structuring fluidized beds can facilitate scale-up and increase conversion and selectivity by controlling the bubble size. We present four approaches to structure fluidized beds: oscillating the gas flow, distributing the gas injection, imposing an electric field to induce interparticle forces, and optimising distributed particle properties such as the size.

INTRODUCTION
Fluidized beds are used for several catalysed gas-phase reactions, especially for highly exothermal reactions and processes with regular catalyst replacement. However, they have serious drawbacks, the two most important being the difficult scale-up and the inefficient use of reactant gas due to the formation of bubbles. The rather small particles that are used in fluidized beds almost always ensure the absence of mass transfer limitations on the particle scale, but on a larger scale mass transfer limitations are present in most fluidized bed processes, especially in the bubbling regime but also in the turbulent regime. This leads to large differences between reactant concentrations in the dense and dilute phases, resulting in a lower conversion and selectivity.

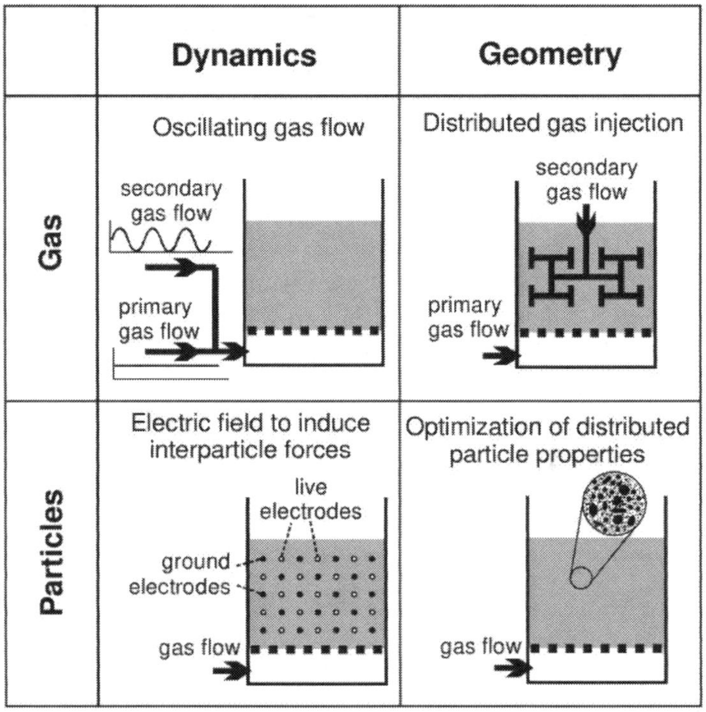

Figure 1. Overview of four approaches to introduce structure in gas-solid fluidized beds.

In this paper, we present four ways to manipulate the hydrodynamics of fluidized bed in order to introduce more structure and to reduce the bubble size. This will lead to a simplified scale-up and an increased interphase mass transfer, typically leading to better conversion and selectivity. This can either be done by (1) modifying the gas supply or (2) interfering in the particle phase. In both cases, either the dynamics can be changed or the configuration can be altered. This gives in total four different possibilities (see Fig. 1), which will be illustrated in this paper:

1.a. Oscillating the gas supply.
1.b. Distributing the gas supply over the height of the bed.

2.a. Varying the interparticle forces using electric fields.
2.b. Varying the particle size distribution and other distributed particle properties.
For a more extensive description, the reader is referred to van Ommen et al. (1).

DYNAMIC OPERATION OF THE GAS SUPPLY

Conventionally, the gas flow to a fluidized bed is kept at a constant value; this value is only adjusted now and then, *e.g.*, to change the production level. However, by continuously varying the instantaneous gas flow rate around a given average value additional degrees of freedom are obtained. One approach is to apply feedback control: based on one or more measurements of the fluidized bed behaviour, the gas flow rate could be adapted in order to maintain a desired state. Some first steps in this direction have been made (2,3,4), but it seems to be a long way before this approach could be applied to large-scale installations. An alternative approach is to apply a periodically varying gas flow without feedback control. It is known that pulsing the gas (5,6) – as well as oscillating the distributor plate (7) – can cause considerable changes in the fluid bed hydrodynamics and significantly improve reactor performance (8). Furthermore, chaotic and other strongly non-linear systems may turn periodic by oscillating a characteristic (order) parameter. This explains ripples on sand beaches, and many other regular patterns seen in nature.

Coppens *et al.* (9) showed that by oscillating the gas flow introduced through the porous bottom distributor plate, bubble patterns in fluidized beds may indeed become ordered and periodic. Experiments were first carried out in a quasi-two-dimensional bed. A sinusoidally oscillating gas flow was added on to a constant primary gas flow above minimum fluidization, so that the total gas flow would remain above minimum fluidization. For an air-sand system, regular bubble patterns were observed within a broad range of frequencies (from 2.5 to 7 Hz) and amplitudes (an oscillating component of 0.2 to 0.7 times the gas flow required for minimum fluidization). These patterns are hexagonal: bubbles rise in ordered rows with constant inter-bubble distance, with each row horizontally shifted with respect to the previous row by half the inter-bubble distance (Fig. 2). Above a certain height, the regularity of the patterns is destroyed, as fluctuations in the system start to dominate the hydrodynamics.

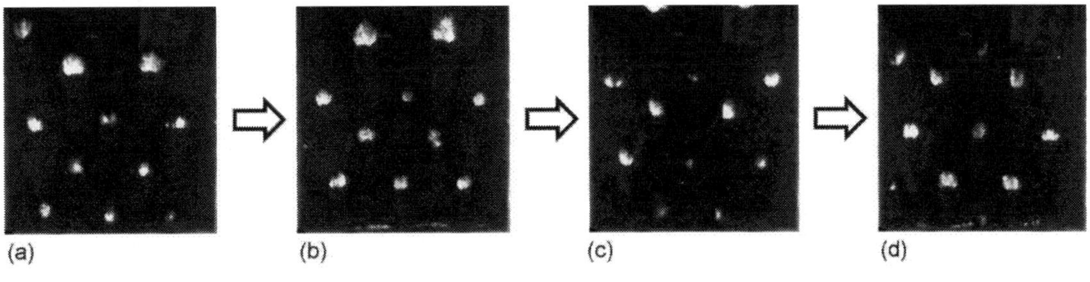

Figure 2. Regular bubble patterns obtained in a quasi-2D bed of 40 cm wide and 43 cm high. The airflow is oscillated at a frequency $f = 3.5$ Hz. The constant component of the gas flow is equal to the minimum fluidization gas flow; the amplitude of the oscillating component is half of that. The sequence shows 4 snapshots out of one period of the bubble pattern (frequency 3.5/2=1.75 Hz).

For the above quasi-2D air-fluidized bed of sand particles, the bubble pattern is regular to a height that is approximately equal to the bed width; the wider the bed, the less the left and right wall influence the pattern, and the greater the height over which the pattern formation persists.

The large front and back wall help in stabilizing the patterns in quasi-2D fluidized beds. Wang and Rhodes (10) also observed regular bubble pattern in discrete particle simulations, but not as clear as in the experimental systems. Coppens *et al.* (9) observed regular bubble patterns in experiments in 3D cylindrical beds too, but only for bed heights of a few centimetres. These patterns are similar to the patterns seen in even shallower, vibrated granular layers (11). Up to now, we have not yet succeeded to stabilise such patterns for higher beds.

It is important to note that both for 2D and 3D systems the waves are not a simple linear resonance phenomenon: the pattern wavelength is not inversely proportional to the driving frequency, while the pattern is formed in a range of frequencies and not at specific frequencies. Also, the ordered patterns in fluidized beds are propagated upwards via the rising gas flow, which differentiates these patterns from those observed in vibrated granular layers, where all the energy is transmitted to the particles via the moving bottom plate. As a result, dissipation is much stronger in vibrated granular matter than it is in gas-solid fluidized beds, where it is possible to influence the entire bed dynamics via a change in inlet gas dynamics. Further research is needed to apply this approach to deep 3D beds and to optimise the dynamic structuring such that a significant reduction of bubble size is reached.

DISTRIBUTING THE GAS SUPPLY OVER THE HEIGHT OF THE BED

An alternative way to obtain more degrees of freedom in operating a fluidized bed is to vary the gas injection in space instead of in time. In a fluidized bed, the gas is normally fed via the bottom, but examples of distributed gas supply do exist. In combustion, staged injection of the air can be used to reduce NO_x emission (*e.g.*, 12,13). Moreover, the application of membranes in fluidized beds to supply or remove gas has been studied (*e.g.*, 14,15).

To distribute the gas over the bed, Coppens (16) proposed to connect all injection points by a hierarchical, tree-like fractal structure. Gas flows from the stem of this tree to all branch tips, spread out over the reactor at optimised locations, where it exits. Via the bottom plate, enough gas is fed to ensure at least minimum fluidization throughout the bed. One important reason to use a fractal design is its intrinsic scalability, mimicking nature. When scaling up the tree-like injector, new branching generations are added to serve larger reactor volumes. The fractal injector branches in such a way that the length and diameter of branches of a given generation is the same. In this way, fluid leaves all outlets at the same flow rate, because the hydraulic path lengths or pressure drops from the inlet to all outlets are all the same. For outlets lying in the same horizontal plane, this avoids radial non-uniformity. In the vertical direction, by spacing the outlets according to a designated pattern, one could compensate for the axial gradients in gas amounts and reactant concentrations. Moreover, the bubble size can be controlled, as less primary gas leads to smaller bubbles initially, while fresh feed blown into the reactor at various locations tends to break up existing bubbles or blow particles apart, leading to an emulsion phase of higher void fraction (17).

Experiments to study the effect of secondary injection using a fractal injector are carried out in a quasi-2D Plexiglas column, 20x1.5x80 cm. The column was operated within a cabinet controlled at 30ºC. The bed material consisted of mono-disperse glass beads, with a diameter of 550 μm and a density of 2400kg/m^3 (Geldart B; u_{mf}=0.21 m/s). The settled bed height was 40 cm. The fractal injector has 16 injection points (similar to the schematic in Fig. 1); the lowest row at 6 cm above the bottom plate and the highest row at 14 cm. Pressure fluctuations were measured using piezo-electric pressure transducers, Kistler type 7261, at 20 cm above the sintered metal porous distributor and in the plenum. The pressure fluctuations were measured with a frequency of 200 Hz. The bubble size was derived from the pressure fluctuations using a technique described by van der Schaaf *et al.* (18). The so-called 'incoherent standard deviation' obtained in this way is a measure for the bubble diameter. The incoherent standard deviation has been shown to be a good quantitative descriptor of the average bubble diameter at a certain height in a fluidized bed, although a calibration of this value (for example, using video analysis or optical probes) is required to determine absolute bubble diameters (19). In this study, we are interested in the reduction of average bubble diameter and therefore only consider the relative value of the incoherent standard deviation.

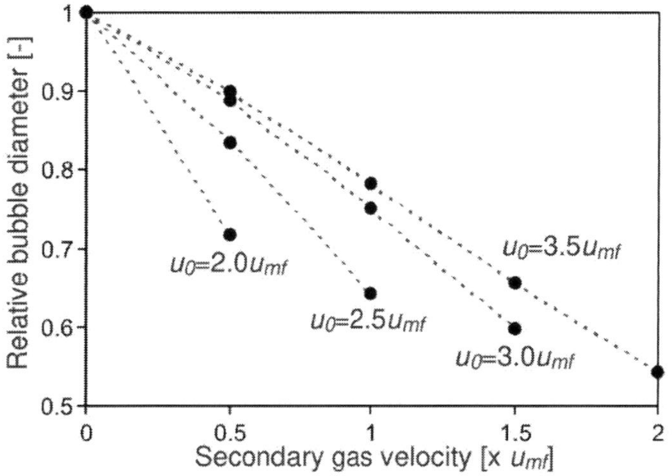

Figure 3. Relative bubble diameter, 20 cm above the distributor, as a function of the secondary gas velocity through the fractal injector. For every total gas velocity, the bubble diameter in the absence of secondary injection is defined as unity.

Figure 3 shows that for every total flow rate studied the bubble diameter is significantly reduced by feeding part of the gas (the secondary gas flow) via the fractal injector. The larger the fraction of the gas that is introduced via the fractal injector, the smaller the bubble size. At a higher measuring position the effect is somewhat smaller, since this position is further away from the upper level of the fractal injector. Further research is aimed at elucidating the mechanism behind the effects of distributed injection to allow a meaningful and directed approach to optimisation.

ELECTRIC FIELDS TO INDUCE INTERPARTICLE FORCES

By definition, a state of fluidization exists when the force of gravity on a set of particles is balanced by the drag arising from the flow of the fluidizing gas. It follows, therefore, that small interparticle forces, which may not be noticeable in other circumstances, may have a significant influence in a fluidized bed (20). Pandit *et al.* (21) recently showed via simulations that by introducing artificial interparticle forces, the behaviour in a fluidized can be moved from the Geldart B regime to the Geldart A regime. Since the bubble size in the Geldart A regime is smaller, imposing additional interparticle forces may be a way to reduce the bubble size.

By imposing an electric field on a fluidized bed of semi-insulating particles, the particles become polarised (*i.e.*, dipoles are induced on the particles); the net electric charge on each particle remains zero. These dipoles lead to an interparticle force, which strongly depends on the particle separation distance and the relative orientation of the particles in the electric field, both in magnitude and direction (22,23). Particles with the centre-to-centre axis aligned to the electric field will attract each other, while particles adjacent to each other in the field will repel one another. Particles at an angle to each other will experience a torque that attempts to align them to the field. Due to the electric field induced interparticle force, the particles in the bed will tend to form strings in the direction of the field. However, the electric field strength should be small enough to allow the free movement of particles; that is, the fluidity of the system must be preserved. By using an AC field instead of a constant (DC) electric field, this fluidity criterion is met.

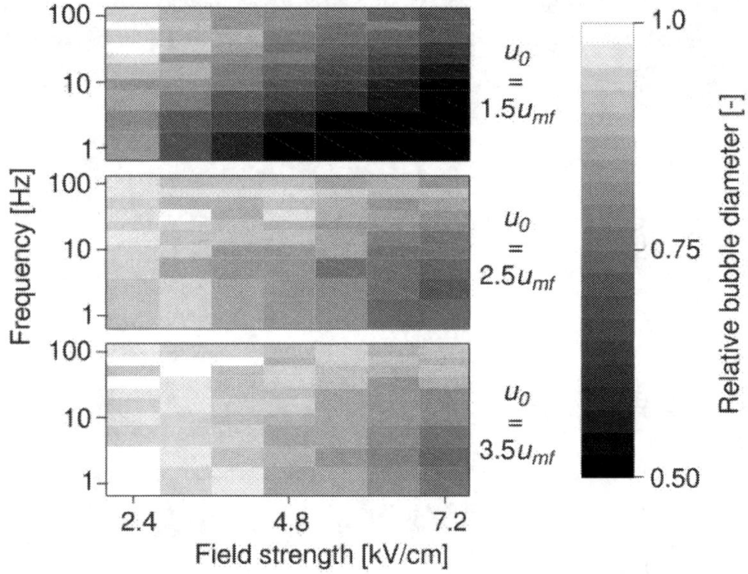

Figure 4. Ratio of bubble diameter with and without electric field, as determined by pressure fluctuation analysis at 20 cm above the distributor at 40% relative humidity, for three different gas velocities. Each subplot shows the bubble reduction ratio (*cf.* the grey scale) as a function of field strength (linear scale) and frequency (logarithmic scale).

In our experiments, the electric fields are introduced in the bed by stringing thin wires through the column, and alternately, both horizontally and vertically, driving these with an AC potential or grounding them (*cf.* Fig. 1). This creates a strongly inhomogeneous field in both the horizontal and vertical directions in the column. When no field is applied, the influence of these wires on the bubble behaviour is so small as to not be measurable. Oscillating (AC) electric fields with a frequency of 1 – 200 Hz and a strength up to 8 kV/cm were applied. The 2D set-up used is similar to the one used for the experiments with the fractal injector.

Figure 4 shows the ratio of the bubble diameter as a function of field strength and frequency to a base measurement without the presence of an electric field. The bubble size is derived from pressure fluctuations as described in the previous section; the pressure fluctuations are measured at 20 cm above the distributor plate. The figure shows that a large decrease in bubble diameter is obtained at higher field strengths (around 5 kV/cm and up). However, one should be careful: DC fields, or very low frequency AC fields, may gridlock the particles and lead to defluidization. With increasing flow rate, the net effect of the electric fields becomes smaller. Further research will be aimed at, among other things, increasing the effect at higher gas velocities and developing acceptable electrode configurations for industrial applications.

OPTIMISING DISTRIBUTED PARTICLE PROPERTIES

It is a well-known fact in fluidization technology that the diameter and density of particles determine the type of fluidization: Geldart defined four particle types, each with a distinct fluidization behaviour (24). Not only the size of the particles, but also their size distribution strongly influences the fluidization behaviour. For example, the addition of fines (particles with a diameter < 45 μm (25)) improves the fluidization behaviour and leads to a better mass transfer. Sun and Grace (26) showed experimentally that a wider particle size distribution leads to higher conversions for ozone decomposition and suggested that this is due to the disproportionate amount of fines in the dilute phase (27). The current practice is, however, that fluidized bed particles used for catalysed gas-phase reactions are mainly optimised on the scale of a single particle. Most attention is given to their pore size distribution such that a high surface area is achieved and that the active sites are easily accessible by the gaseous components. Little attention is paid to the earlier mentioned mass transfer from gas in the dilute phase to the particles in the dense phase, essential to practical fluid bed operation.

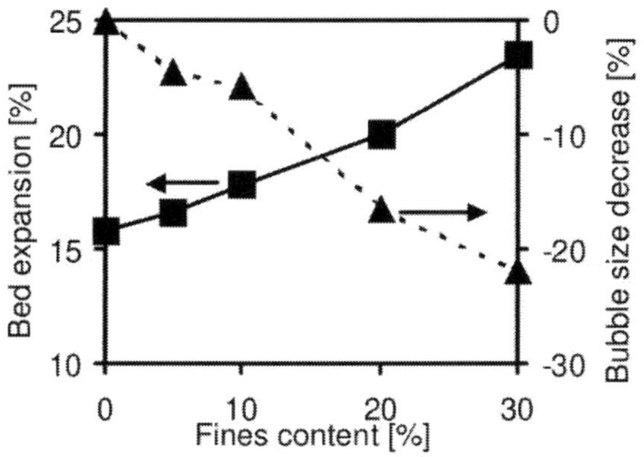

Figure 5. The influence of the fraction of fines in the bed on the bed expansion (compared to the bed height at minimum fluidization) and the decrease of the bubble diameter (compared to 0% fines). The bed is fluidized at 4.3 cm/s.

We aim at improving the conversion and selectivity of gas-solid fluidized bed reactors by designing mixtures of particles with optimal properties (size distribution, density, shape, elasticity), aided by high-throughput experimentation, a novel approach for hydrodynamics research. The automated set-up we use allows access to quantitative information from large numbers of particle mixtures. Catalyst carrier materials such as silica and alumina are used as particles. Experiments are carried out in two industrially relevant fluidization regimes: bubbling fluidization and turbulent fluidization. In the experiments, pressure measurements, optical probes, and video analysis are used to assess the hydrodynamics.

Some preliminary results were carried out in a quasi-2D Plexiglas column (30x2.0x80 cm), operated within a cabinet controlled at 25ºC. The bed material consisted of alumina Geldart A particles, of which the fines concentration was increased from 0 to 30%; the median particle diameter is 75 μm at 0% fines. The increase in fines resulted in a decreased minimum fluidization velocity from 4.3 to 2.9 mm/s, and a decrease in voidage at minimum fluidization due to a better fit of small particles between the larger ones. Figure 5 shows the results for fluidization at 4.3 cm/s. An increasing amount of fines leads to an increased bed expansion. Moreover, the addition of fines results in a decrease of the bubble diameter and a smaller total amount of gas in the bubbles. This will enhance the mass transfer of components in the gas phase to the particle surface. Compared to the previous three approaches, our work on this fourth way of manipulating the fluidized bed hydrodynamics is still in the start-up phase. We will work on further development of the automated screening set-up, production of tailored particle mixtures, and scanning of a large number of these mixtures. Moreover, computational fluid dynamics will be used to obtain more insight into the interplay between particles with different properties.

CONCLUSIONS

We presented four approaches for controlling fluidized bed hydrodynamics. In this way, fluidized beds can be structured, which facilitates scaling-up. Moreover, the bubble size can be decreased which, for most catalytic fluidized bed processes, leads to increased conversion and selectivity. Better hydrodynamic control is achieved by introducing additional degrees of freedom. This can be done either by manipulating (1) the gas or (2) the particles:

1.a. Varying the gas supply in time: we showed that an oscillating gas flow leads to regular bubble patterns.

1.b Distributing the gas supply over space: using a fractal injector, a much more even distribution of the gas is obtained leading to a strong reduction in bubble size.

2.a. Using an electric field, semi-insulating particles can be polarised. This induces interparticle forces, resulting in a significant decrease of the bubble size.

2.b. By tailoring the size distribution and other distributed properties of the particles, the bubble size can be decreased and the bed voidage increased.

Our group continues to investigate and optimise these four approaches.

ACKNOWLEDGEMENTS
We would like to thank D. Christensen, F. Kleijn van Willigen, and other current and former group members for their contributions to this research.

REFERENCES

1. van Ommen, J.R., Nijenhuis, J., van den Bleek, C.M., Coppens, M.-O., 'Four ways to introduce structure in fluidized bed reactors', Ind. Eng. Chem. Res. 46, 4236-4244, 2007.
2. Kaart, S., Schouten, J.C., van den Bleek, C.M., 'Improving conversion and selectivity of catalytic reactions in bubbling gas-solid fluidized bed reactors by control of the nonlinear bubble dynamics', Catal. Today 48, 185-194, 1999.
3. de Korte, R.J., Schouten, J.C., van den Bleek, C.M., 'Controlling bubble coalescence in a fluidized-bed model using bubble injection', AIChE J. 47, 851-860, 2001.
4. Harrison, A.J.L., Lim, C.N., Gilbertson, M.A., 'Modelling and control of the dynamics of bubbling fluidized beds', Proc. IMechE Part E: J. Process Mechanical Engineering, 220, 43-53, 2006.
5. Massimilla, L., Volpicelli, G., Raso, G., 'A study on pulsating gas fluidization of beds of particles', Chem. Eng. Progr. Symp. Ser. 62, 63-70, 1966.
6. Wong, H.W., Baird, M.H.I., 'Fluidization in a pulsed gas flow', Chem. Eng. J. 2, 104-113, 1971.
7. Köksal, M., Vural, H., 'Bubble size control in a two-dimensional fluidized bed using a moving double plate distributor', Powder Technol., 95, 205-213, 1998.
8. Pence, D.V., Beasley, D.E., 'Chaos suppression in gas-solid fluidization', Chaos 8, 514-519, 1998.
9. Coppens, M.-O., Regelink, M.A., van den Bleek, C.M., 'Pulsation induced transition from chaos to periodically ordered patterns in fluidised beds', in: Proceedings of the 4th World Conference on Particle Technology, Paper 355, 2002.
10. Wang, X.S., Rhodes, M.J., 'Pulsed fluidization - a DEM study of a fascinating phenomenon', Powder Technol., 159, 142-149, 2005.
11. Melo, F., Umbanhowar, P., Swinney, H.L., 'Transition to parametric wave patterns in a vertically oscillated granular layer', Phys. Rev. Lett. 72, 172-175, 1994.
12. Jensen, A., Johnsson, J.E., Andries, J., Laughlin, K., Read, G., Mayer, M., Baumann, H., Bonn, B., 'Formation and reduction of NOx in pressurized fluidized-bed combustion of coal', Fuel 74, 1555-1569, 1995.
13. Lyngfelt, A., Amand, L.E., Gustavsson, L., Leckner, B., 'Methods for reducing the emission of nitrous oxide from fluidized bed combustion', Energy Convers. Manage. 37, 1297-1302, 1996.
14. Adris, A.E.M., Grace, J.R., 'Characteristics of fluidized-bed membrane reactors: Scale-up and practical issues', Ind. Eng. Chem. Res., 36, 4549-4556, 1997.
15. Patil, C.S., van Sint Annaland, M., Kuipers, J.A.M., 'Experimental study of a membrane assisted fluidized bed reactor for H2 production by steam reforming of CH4', Trans. IChemE Part A Chem. Eng. Res. Des., 84, 399-404 MAY 2006
16. Coppens, M.-O., 'Structuring fluidized bed operation in a nature inspired way', in: Arena, U., Chirone, R., Miccio, M., Salatino, P. (eds.), Fluidization XI, Engineering Conferences International, New York, USA, 83-90, 2004.
17. Kleijn van Willigen, F., Christensen, D., van Ommen, J.R., Coppens, M.-O., 'Imposing dynamic structures on fluidised beds', Cat. Today 105, 560-568, 2005.

18. van der Schaaf, J., Schouten, J.C., Johnson, F., van den Bleek, C.M., 'Non-intrusive determination of bubble and slug length scales in fluidized beds by decomposition of the power spectral density of pressure time series', Int. J. of Multiphase Flow 28, 865-880, 2002.
19. Kleijn van Willigen, F., van Ommen, J.R., van Turnhout, J., van den Bleek, C.M., 'Bubble size reduction in a fluidized bed by electric fields', Int. J. of Chem. Reactor Eng. 1, paper A21, 2003. http://www.bepress.com/ijcre/vol1/A21
20. Seville, J.P.K., Willett, C.D., Knight, P.C., 'Inter-particle forces in fluidisation: a review', Powder Technol. 113, 261-268, 2000.
21. Wang, X.S., Rhodes, M.J., 'Pulsed fluidization - a DEM study of a fascinating phenomenon', Powder Technol., 159, 142-149, 2005.
22. Colver, G.M., 'An interparticle force model for ac-dc electric fields in powders', Powder Technol. 112, 126-136, 2000.
23. Kleijn van Willigen, F., van Ommen, J.R., van Turnhout, J., van den Bleek, C.M., 'Bubble size reduction in electric-field-enhanced fluidized beds', J. Electrostat. 63, 943-948, 2005.
24. Geldart , D., 'Types of gas fluidization', Powder Technol., 7, 285-292, 1973.
25. Yates, J.G., Newton, D., 'Fine particle effects in a fluidized-bed reactor', Chem. Eng. Sci., 41, 801-806, 1986.
26. Sun, G., Grace, J.R., 'The effect of particle size distribution on the performance of a catalytic fluidized bed reactor', Chem. Eng. Sci. 45, 2187-2194, 1990.
27. Sun, G., Grace, J.R., 'Experimental determination of particle dispersion in voids in a fluidized bed', Powder Technol. 80, 29-34, 1994.

Biological Nutrients Removal from Municipal Wastewater using a Novel Circulating Fluidized Bed-Bioreactor

Nabin Chowdhury, Jesse Zhu, George Nakhla, and Ajay Patel
Powder Technology Research Centre, University of Western Ontario

Abstract: The objective of this study was to examine the feasibility of a novel liquid-solid circulating fluidized bed bioreactor (LS-CFBBR) in municipal wastewater treatment. The CFBBR employs attached microbial films for biodegradation of both organics and nutrients, in both soluble and particulate forms, within a single circulating fluidized bed unit. This new technology combines the more compact and efficient fixed-film process with the recently developed biological nutrient removal (BNR) process that provide the additional removal of nitrogen and phosphorous. A lab-scale CFBBR has been demonstrated to be able to treat degritted municipal wastewater The CFBBR was operated at an empty bed contact time (EBCT) of 0.82 hours with and without particle recirculation. More than 80% organic removal, 70% nitrogen removal and 30% phosphorus removal were observed without particle recirculation. With particle recirculation the system was able to achieve effluent quality characterized by BOD_5 <5 mg.L^{-1}, TSS of <5 mg.L^{-1}, ammonia-nitrogen <1 mg.L^{-1}, total nitrogen <7 mg.L^{-1}, phosphorus <1 mg L^{-1} and UV transmissivity of 85% at an overall HRT of 2 hours. A high transmissivity of the effluent makes it ready for ultra-violet (UV) disinfection and reuse. Combination of nitrification, denitrification and biological phosphorus removal in one unit saves space, reduces energy consumption and also produces less sludge. Remarkably lower sludge yield of 0.13 g VSS.g^{-1}COD, approximately one-third of the conventional treatment processes, of this system is able to reduce overall cost of wastewater treatment process. To commercialize this novel technology a pilot-scale CFBBR at a capacity of 6000 L d^{-1} has been established and commissioned at the Adelaide Pollution Control Plant in London, Ontario for the treatment of degritted municipal wastewater, for pre-commercialization confirmation.

1. Background

Water sustainability is one of the most essential things to the survival of modern human society. Fixed-film systems or biofilm technology is perfectly suited to enhance sustainability, because microbial communities are uniquely able to convert complex substrates into simple forms and purify water [1]. A biofilm can be defined as a complex coherent structure of cells and cellular products, like extra-cellular polymers, which either form spontaneously as large, dense granules, or grow attached on a solid surface (static biofilms) or on suspended carriers (particulate supported biofilms) [2].

Deterioration of water bodies as a result of eutrophication has incited regulatory agencies to more tightly regulate nitrogen and phosphorus discharges. Biological removal of nitrogen and phosphorus (BNR) is usually accomplished by hybrid systems consisting both aerobic (aerated) and anaerobic (unaerated) bioreactors. These BNR processes employ a combination of anaerobic, anoxic, and aerobic suspended growth biological reactors with or without primary clarification. BNR processes are known to offer several advantages over the more conventional activated sludge processes, namely superior effluent quality, a significant reduction in aeration energy requirements due to utilization of formed nitrates to reduce organic matter, improved sludge settling characteristics, a reduction in sludge quantities due to lower bacterial yields in the anoxic tanks, and the elimination/minimization of chemical sludge. Laboratory and full-sale evaluations have indicated that nitrification and denitrification coupled with enhanced biological phosphorus removal (EBPR) are efficient in removing nutrients from wastewater [3].

Due to stringent environmental regulations and space constraints, the suspended growth activated sludge processes are less attractive than fixed film processes, and hence bioreactor innovations in wastewater treatment systems have included the fluidized bed, moving bed, and biological aerated filter reactors designs [4]. Fluidized beds are known for their excellent interfacial contact and mass transfer properties. Although the fluidized bed bioreactors (FBBRs) are very efficient in removing organics, separate stage nitrification, and denitrification [2,5,6] very few studies [7-12] have attempted to integrate biological nutrient removal (BNR) processes with the FBBR.

This research work was conducted to study: (a) removal of organic, nitrogen, and phosphorus from municipal wastewater without particle recirculation and (b) compare the nutrient removal efficiencies with particle recirculation using a laboratory-scale novel gas-liquid-solid circulating fluidized bed bioreactor.

2. Materials and methods

In this study a laboratory scale CFBBR was established and operated at room temperature (22±2°C) to treat municipal wastewater collected regularly from grit-chamber of the Adelaide Wastewater Treatment Plant, London, ON.

2.1 Reactor description

A schematic of CFBBR shown in Figure 1 was fabricated in the laboratory using plexiglass. The system was consisted of two columns, the riser column (0.77 L) and the downer column (3.3 L). The riser column was operated as an anoxic reactor and the downer column was operated as an aerobic reactor. Liquid and solid circulation between riser and downer was maintained using the riser-downer and downer-riser connecting lines. Reactors arrangement and recirculation connections were fabricated as mentioned by Cui [12]. Lava rock particles with an average diameter of 670 μm (300 to 1000 μm) were used as a carrier media for biofilm attachment in the anoxic and aerobic columns. The particle porosity was about 33% and the total porosity was 61%.

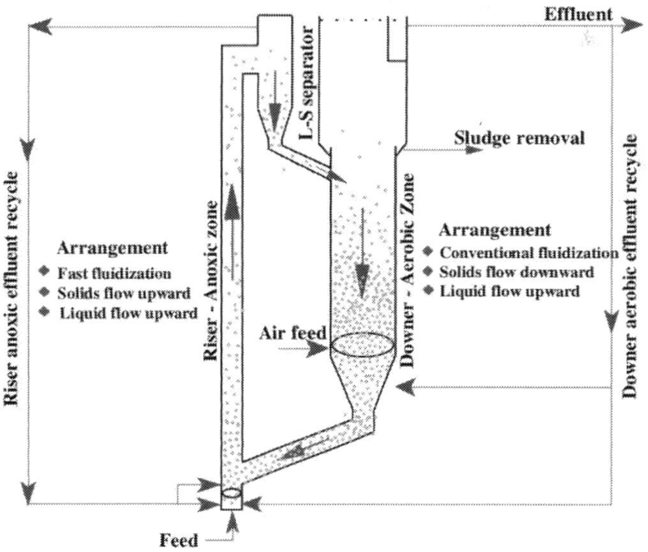

Figure 1: Schematic of the CFBBR

The CFBBR was operated at EBCT of 0.82 h with and without particle recirculation, denoted henceforth as Mode I, and Mode II respectively and corresponding nominal hydraulic retention times (nHRT) of the system were 0.32 h, calculated as a product of EBCT×(1-

porosity). The air was fed at the bottom of the downer column. Both anoxic and aerobic beds were fluidized by recirculation of liquid from the solid-liquid separator of riser column and downer column. Superficial liquid velocities of 3.6±0.2 $cm.s^{-1}$ and 0.12±0.2 $cm.s^{-1}$ were maintained in the riser column and downer column respectively by controlling recirculation flow rates. An airflow rate of 0.315±0.1 $m^3.d^{-1}$ was maintained in the downer column.

2.3 Analytical methods

Total suspended solids (TSS), volatile suspended solids (VSS), biological oxygen demand (BOD_5), and total Kjeldahl nitrogen (TKN) were analyzed according to the standard methods of the American Public Health Association [13]. Dissolved oxygen (DO) was measured using Thermo Orion (810 A+) meter. HACH methods and testing kits (HACH Odyssey DR/2500 spectrophotometer and COD digestion unit) were used to measure total chemical oxygen demand (TCOD), soluble chemical oxygen demand (SCOD) (Method 8000), and total phosphorus (TP) (Method 10127). Ammonium nitrogen (NH_4-N), nitrite nitrogen (NO_2-N), nitrate nitrogen (NO_3-N), and soluble phosphorus (PO_4-P) were measured using ion chromatography (IC, Dionex 600, USA) equipped with CS16-HC and AS9-HC columns respectively.

2.4 Reactor start-up

The system was inoculated with 1 L of return activated sludge (RAS) collected from Adelaide Wastewater Treatment Plant, London, ON, which contained mixed liquor total suspended solid (MLTSS) and mixed liquor volatile suspended solid (MLVSS) of 3511 and 2810 $mg.L^{-1}$ respectively.

3. Results and discussion

The CFBBR was operated without recirculation of bioparticles, lava rock particles with attached biomass, (Mode-I) and with bioparticles recirculation (Mode-II) for 260 and 110 days respectively. Table 1 shows the pseudo steady state effluent characteristics of the system at Modes I and II.

Table 1: Summary of Steady-state Effluent Parameters during Mode-I and II

Parameter	**Mode-I (EBCT = 0.82 h)**		**Mode-II (EBCT = 0.82 h)**	
mg/L	**Range**	**Average ± SD (n)**	**Range**	**Average ± SD (n)**
TSS	2 - 10	6 ± 2.9(10)	2 - 6	4.4 ± 1.5(10)
VSS	2 – 8	5 ± 2.3(10)	2 - 5	3.2 ± 1.1(10)
COD	15 - 24	18 ± 3.2(10)	19 - 35	26 ± 4.8(10)
SCOD	13 - 22	15.8 ± 3.8(10)	16 - 28	20.6 ± 4.1(10)
BOD	5 - 10	7.5 ± 1.8(10)	8 - 12	9.2 ± 2.1(10)
SBOD	4 - 8	5.2 ± 1.6(10)	5 - 8	6 ± 1.2(7)
TKN	1.4 – 2.1	1.7 ± 0.2(10)	1.5 – 1.85	1.8 ± 0.09(10)
NH_4-N	0 – 0.9	0.35 ± 0.4(10)	0.4 – 0.9	0.67 ± 0.17(10)
NO_3-N	3.1 – 8.8	5.8 ± 1.7(10)	5.2 – 6.6	6.5 ± 0.46(10)
Total Nitrogen	4.4 – 9.8	7.06 ± 1.6 (10)	5.1 – 9.6	7.3 ± 1.2 (10)
Total P	1.1 – 2.1	1.7 ± 0.33(10)	0.54 – 0.9	0.76 ± 0.12(10)
PO_4-P	1 – 2.2	1.5 ± 0.37(10)	0.5 – 0.88	0.7 ± 0.12(10)

Figures 2a, 2b, and 2c show COD, nitrogen, and phosphorus removal efficiency of the CFBBR, observed during operational modes I and II. The system achieved 94%, 80%, and 65% of COD, nitrogen removal, and phosphorus removal respectively during Mode I, whereas, 91% COD, 78% nitrogen, and 83% phosphorus removal during Mode II. Average effluent total COD concentrations were 18±3.2 and 16±4.8 $mgCOD.L^{-1}$ at organic loading rates of 3.43 and 3.3 kg COD $m^{-3}.d^{-1}$ during modes I and II respectively. The system attained average effluent

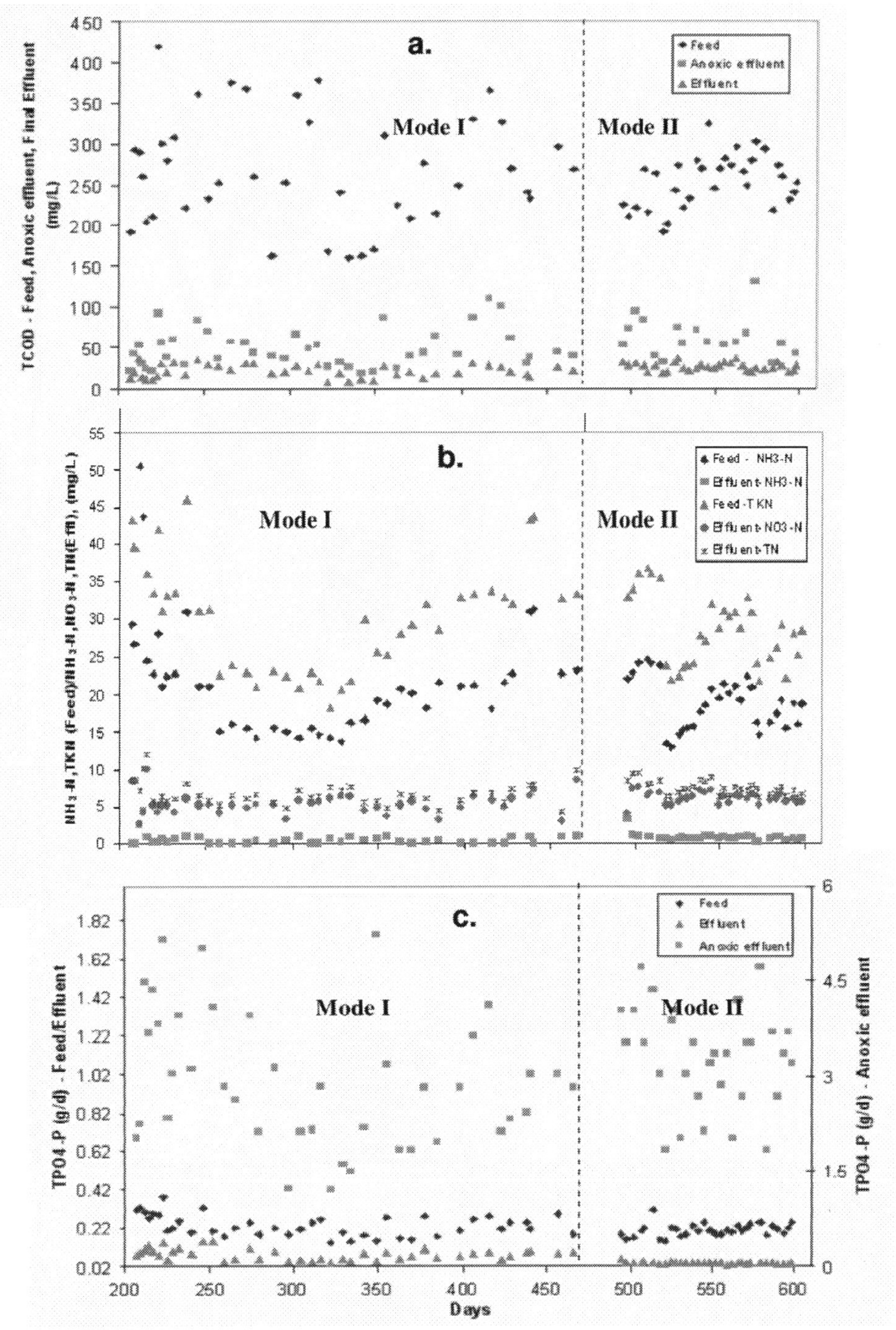

Figure 2: (a) organic removal (b) nitrogen removal and (c) phosphorus removal using CFBBR

TSS and VSS of 4-6 and 3-5 $mg.L^{-1}$ with the average influent TSS and VSS concentrations of 123-144 and 106-118 $mg.L^{-1}$ respectively. These results indicate that the CFBBR can generate tertiary effluent quality, which also diminishes the importance of the final clarifier. During Mode I, with average influent TKN, and NH_4-N concentrations were 35.5 and 23.6 $mg.L^{-1}$, the system achieved effluent TKN, NH_4-N, NO_3-N and total nitrogen (TN) concentrations of 1.7, 0.35, 5.8, and 7.5 $mg.L^{-1}$, respectively. Whereas, during Mode II, the system achieved effluent TKN, NH_4-N, NO_3-N and TN concentrations of 1.7, 0.67, 6.2 and 7.9 $mg.L^{-1}$ respectively with an average influent TKN and NH_4-N concentrations of 28.2 and 19.0 $mg.L^{-1}$. Tertiary standards typically stipulate total nitrogen (TN) of 10 $mg.L^{-1}$ [14]. As reported in Table 1 effluent PO_4-P concentrations were 1.5±0.37 and 0.7±0.12 $mg.L^{-1}$ in modes I and II respectively. Comparatively a higher phosphorus removal was observed when particles were recirculated between the downer and riser.

Recent optimization work has shown that the CFBBR can be designed at organic loading (OLR) of 5.3 $kgCOD.m^{-3}.d^{-1}$, nitrogen loading of 0.51 $kgN.m^{-3}.d^{-1}$, and phosphorus loading 0.1 $kgP.m^{-3}.d^{-1}$ with effluent characteristics of soluble biological oxygen demand ($SBOD_5$) ≤10 $mg.L^{-1}$, total nitrogen (TN) <10 $mg.L^{-1}$. Whereas, Rouse et al., (2007), and Wang et al., (2007) have reported 80-90% COD removal at an HRT of 0.5-4.0 h using moving bed and fluidized bed bioreactors respectively to treat wastewater characterized by influent COD of 275-367 $mg.L^{-1}$. However, these systems are not capable of biologically removing nitrogen and phosphorus. Approximately 70-80% nitrogen removal was observed by using FBBR for wastewater characterized by 17-35 $mg.N.L^{-1}$ [8,15]. Wang [16] have reported 90% total phosphorus removal using chemical coagulation and biological degradation in a pilot-scale fluidized bed pellet bed reactor as compared to, the CFBBR which can efficiently achieve 85% removal phosphorus biologically using particles recirculation [9].

4. Concluding remarks

It is evident from the comparative performance of the integrated CFBBR technology with FBBR individually dedicated for organics removal, nitrification, and denitrification that the technology outperforms conventional FBBRs in terms of efficiencies and effluent quality. The CFBBR will replace the complete wastewater treatment by a single compact unit process capable of economically achieving tertiary effluent quality. This technology has an excellent capability to entrain particulate organics and volatile suspended solids, facilitating hydrolysis and subsequent biodegradation. Furthermore, due to enhanced mass transfer processes, the CFBBR can biologically nitrify (convert ammonia and organic nitrogen to nitrates), denitrify (convert nitrates into insoluble nitrogen gas); thus effectively removing nitrogen as well as biologically removing phosphorus within a fraction of the contact time required in activated-sludge type BNR processes or multi-stage individual FBBR systems.

5. Nomenclatures

BNR biological nutrient removal
BOD biological oxygen demand
COD chemical oxygen demand
EBCT empty bed contact time
FBBR fluidized bed bioreactor
HRT hydraulic retention time
PAO phosphorus accumulating organism
SRT solid retention time
TKN total kjeldhal nitrogen
TN total nitrogen
TSS total suspended solids
VSS volatile suspended solids

6. References

1. Rittmann, B.E. 2007. "Where are we with biofilms now? Where are we going?" Water Sci. Technol., 559(8-9), 1-7.
2. Nicolella, C., van, Loosdrecht, M. C. M., and Heijnen, J. J. (2000) "Wastewater treatment with particulate biofilm reactors". *J. Biotechnol.*, 80, 1-33.
3. Ekama, G.A., Wentzel, M.C., 1999. Difficulties and developments in biological nutrient removal technologies and modeling. Wat. Sci. Technol. 39(6), 1-11.
4. Bolton, J., Tummala, A., Kapadia, C., Dandamudi, M., and Belovich, M. J. (2006). "Procedure to quantify biofilm activity on carrier used in wastewater treatment systems." *J. Environ. Eng.*, 132(11), 1422-1430.
5. Shieh, W. K., and Keenan D. K. (1986). "Fluidized bed biofilm reactor for wastewater treatment." In: Fiechter, A. (Ed.), Advances in Biochemical Engineering/ Biotechnology. Springer-Verlag, Berlin, 132-168.
6. Payraudeau, M., Paffoni, C., Gousailles, M., 2000. Tertiary nitrification in an up flow biofilter on floating media: influence of temperature and COD load. Wat. Sci. Technol. 41(4-5), 21-27.
7. Borregaard, V. R. (1997) "Experience with nutrient removal in a fixed-film system at full-scale wastewater treatment plants". *Water Sci. Technol.*, 36, 129-137.
8. Sen, P., and Dentel, S. K. (1998) "Simultaneous nitrification-denitrification in a fluidized bed reactor". Water Sci. Technol., 38(1), 247-254.
9. Patel, A., Zhu, J., Nakhla, G., 2006. Simultaneous carbon, nitrogen and phosphorus removal from municipal wastewater in a circulating fluidized bed bioreactor. Chemosphere 65, 1103-1112.
10. Nakhla, G., Zhu, J., Cui, Y., 2004. Liquid-solid circulating fluidized bed wastewater treatment system for simultaneous removal of carbon, nitrogen, and phosphorus. US patent no. 6,716,244, Int'l PCT patent awarded Aug. 2005.
11. Zhu J., Zheng Y., Karmanev D., Bassi A. "(Gas-) liquid–solid circulating fluidized beds and their potential applications to bioreactor engineering." Can. J. Chem. Eng, 2000; 78: 82-94.
12. Cui Y., Nakhla G., Zhu J., Patel A. "Simultaneous carbon and nitrogen removal in anoxic-aerobic circulating fluidized bed biological reactor (CFBBR)." Environ. Technol., 2004; 25: 699-712.
13. APHA, AWWA, WEF, 1992. Standard Methods for the Examination of Water and Wastewater, 18th Ed. American Public Health Association, American Water Works Association, and Water Environment Federation, Washington, DC.
14. U. S. EPA., 2004. Guidelines for water reuse. EPA-625/R-04-92-004, U.S. Environment Protection Agency and U.S. Agency for International Development, Washington, DC.
15. Fdez-Polanco, F., Real, F.J., Garcia, P.A., 1994. Behavior of an anaerobic/aerobic pilot scale fluidized bed for the simultaneous removal of carbon and nitrogen. Water Sci. Technol. 29(10-11), 339-346.
16. Wang, X.C., Yuan, H.L., Liu, Y.J., Jin, P.K. 2007. Fluidised pellet bed bioreactor: a promising technology for onsite wastewater treatment and reuse. Water Sci. Technol. 55(1-2), 59-67.

Describing Temperature Increases in Plasmon-Resonant Nanoparticle Systems

Michael P. Hoepfner, University of Utah, Salt Lake City, UT
D. Keith Roper, University of Utah, Salt Lake City, UT

Abstract
Plasmon-resonant nanoparticles are being integrated into a variety of actuators and sensors due to their extraordinary optical capabilities. We show a continuum energy balance accurately describes thermal dynamics and equilibrium temperatures in plasmon-resonant nanoparticle systems. Analysis of 18 data sets in which temperature increased ≤10.6 °C yielded a mean value of R^2 >0.99. The largest single relative temperature error was 1.11 %. A characteristic temperature was introduced into a linear driving force approximation for radiative heat transfer in the continuum energy description to simplify parameter estimation. The maximum percent error of the linearized description rose to 1.5 % for the 18 sets. Comparing the two descriptions at simulated temperature increases up to 76.6 °C gave maximum relative errors ≤7.16 %. These results indicate the energy balance and its linearized approximation are applicable to characterize dynamic and equilibrium temperatures for sensors and actuators in microfluidic and lab-on-chip systems over a broad range of heat-transfer lengths, power inputs and corresponding temperature increases.

1. Introduction

Applications of inorganic colloidal nanocrystals in sensing and analysis are expanding rapidly due to new methods that enable synthesis of nanocrystals at the scale of biomolecules and allow derivatization to use nanocrystals as biosensors and molecular probes. Gold (Au) nanoparticles have been used for DNA identification [1], to discriminate polynucleotide base-pair changes by colorimetry [2] and to enhance Raman spectroscopy [3-4]. Pump-pulse laser spectroscopy has been applied to examine transient electron dynamics of Au nanoparticles [5-7]. This body of work showed incident resonant photons can be dissipated as heat by way of surface plasmons on noble metal surfaces of suspended nanoparticles or thin films. Recently, the efficiency of converting incident laser light to heat that was subsequently dissipated by nanoparticle suspensions was measured using photocalorimetry [8].

The photocalorimetry data was analyzed by comparing conduction, convection and radiation losses with resonant photon energy used to irradiate thermally-isolated, colloidal-gold nanoparticles in order to calculate the efficiency of converting laser light to heat via surface plasmons. Determining the value of conversion efficiency allowed prediction of transient thermal dynamics and steady-state equilibrium temperature increases ≤10.6 °C for effective laser powers between 1.0 to 16.3 mW and for characteristic heat transfer lengths from 4.1 to 66.8 mm. Radiation, the predominant mode of heat transfer at these conditions and a quartic function of temperature [9], was linearized to estimate heat transfer coefficients from the data. But the accuracy of these descriptions were not evaluated and their applicability to a broader range of conditions was not studied.

This study examines the suitability of applying a continuum energy balance to analyze thermal dynamics and equilibrium temperatures obtained from heat dissipated by surface plasmons in aqueous suspensions of irradiated nanoparticles across a larger range of length scales, power inputs and temperature increases that includes the previous data. We evaluate the accuracy of employing a linearized radiation term to characterize heat transfer for effective power inputs ranging from 11.4 to 102.1 mW, and for characteristic heat transfer lengths ranging from 1.5 to 393.6 mm. These expanded parameter values produce temperature increases from 3.0 to 76.6 °C. We introduce a characteristic temperature into the linearized description that results in a maximum temperature error ≤7.16 %. These expressions appear useful to analyze and predict local thermal dynamics and equilibrium temperatures in a wide range of laser-induced surface plasmon resonant heating structures in lab-on-chips [10] and microfluidics. We demonstrate the utility of these expressions by analyzing a sensor in which double-stranded nucleic acid is cyclically denatured at 95°C and hybridized (annealed) at 55°C using precise, local control of temperature induced by plasmon-resonant particles [11].

2. Experiment

2.1 Experimental Procedure

Photocalorimetric data was obtained by dissipating laser light as heat using an aqueous suspension of nanoparticles as previously described [8]. A schematic of the sample cell can be found in Fig. 1, and the experimental setup is found in Fig. 2. The temperature profile was then analyzed by applying an overall energy balance to evaluate heat transfer from the thermally isolated system.

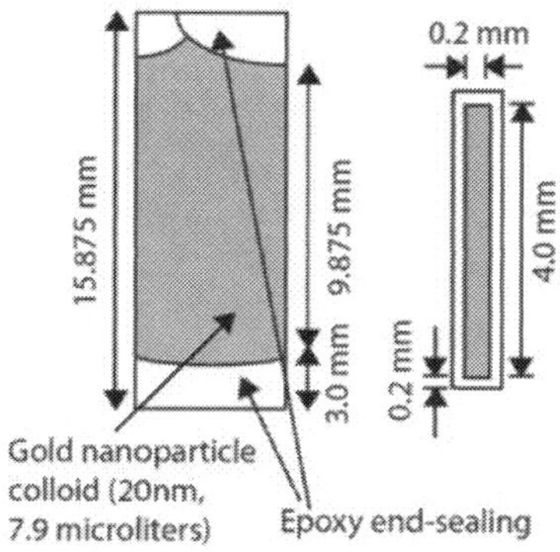

1.

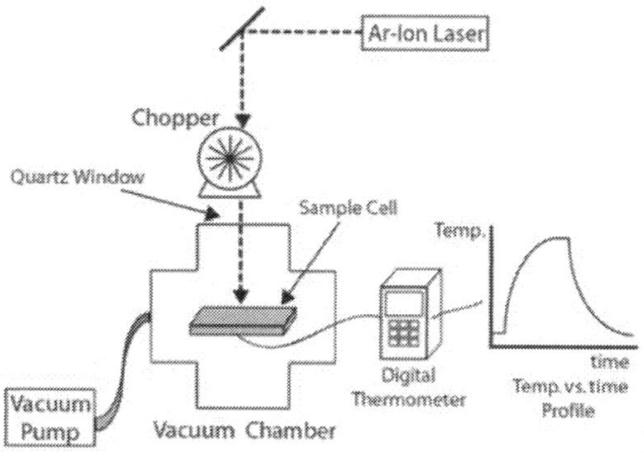

2.

Fig. 1. A schematic and dimensions of the sample cell filled with gold nanoparticle colloid. **Fig. 2.** A schematic of the experimental setup. The energy was provided by the Ar^+ -Ion laser. Temperature measured by a thermocouple was recorded by a digital thermometer.

2.2 Overall Energy Balance and Linearization of Radiation

The purpose of this study was to examine use of an overall continuum energy balance to describe thermal dynamics and temperature equilibria in plasmon resonant nanoparticle systems. A primary aim was to quantify the error associated with linearizing the radiative heat transfer term in the overall energy balance. Convection was reduced to a neglible amount by suspending the cell in a vacuum ≤1 Torr, leaving only conductive and radiative heat transfer terms in the energy balance. Rapid thermal equilibrium within the nanofluid in the sample cell [8] yields the following complete differential energy balance written in terms of temperature:

$$\frac{dT}{dt} = \frac{1}{\sum mC_P}\left[\eta Q_I - \frac{k_{Air}A_{Samp}}{L_{Air}}(T - T_{Amb}) - \frac{k_{TC}A_{TC}}{L_{TC}}(T - T_{Amb}) - A_{Samp}\varepsilon\sigma(T^4 - T_{Amb}^4)\right] \quad \textbf{(1)}$$

where ΣmC_P is the sum of the products of mass and specific heat of glass, epoxy and aqueous suspension in the sample cell, k_i are thermal conductivities, A_i are surface areas normal to heat transfer, L_i are characteristic heat transfer lengths between the sample and surrounding surfaces at T_{Amb}, A_{Samp} is the surface area of the sample cell, ε is the surface emissivity, σ is the Stefan-Boltzman constant, T is sample cell temperature, and the term ηQ_I represents the fractional value of incident laser light dissipated as heat in the cell. Subscripts i = Air and TC denoted parameters for conduction via air molecules that remained in the vacuum cell and via thermocouple wires, respectively. Eq. (1) was linearized using the form:

$$\frac{dT}{dt} = \frac{1}{\sum mC_P}\left[\eta Q_I - U(T - T_{Amb})\right] \quad \textbf{(2)}$$

where U is an overall heat transfer coefficient defined as:

$$U \equiv \sum_i \frac{k_i A_i}{L_i} + A_{Samp}\sigma\varepsilon\left(T'^2 + T_{Amb}^2\right)\left(T' + T_{Amb}\right) \quad \textbf{(3)}$$

We introduce here a characteristic temperature, T', which is intermediate between T_{Amb} and T_{Equil} and is defined by:

$$T' \equiv f_{\Delta T}T_{Equil} + \left(1 - f_{\Delta T}\right)T_{Amb} \quad \textbf{(4)}$$

where T_{Equil} is the temperature at which the system equilibrates for given values of ηQ_I and L_{Air}, and $f_{\Delta T}$ is a fractional relative difference between T_{Amb} and T_{Equil} which minimizes the maximum relative error between Eq. (1) and Eq. (2).

2.3 Criteria for Evaluating Error of Linearized Description

The first criterion for evaluating errors between measurements, Eq. (1), and Eq. (2) was the sum of squares of the errors, σ_i, given by [12]:

$$\sigma_i = \sum (T_i - T_{Ref})^2 \quad \textbf{(5)}$$

where T_{Ref} is a reference temperature, subscript i denotes the temperature predicted by either the complete model, i=2, or the linearized model, i=3, and the sum is taken over a complete data set that consists of one temperature point per second. The sum squared error, σ_i, has units of K^2.

The second criterion for evaluating the error between Eq. (1) and Eq. (2) was relative percent error, χ, given by:

$$\chi_i = \frac{T_i - T_{\text{Ref}}}{T_{\text{Ref}}} \times 100 \qquad \textbf{(6)}$$

where subscript i and Ref are defined the same as in the sum squared of errors.

A third criteria was used to evaluate the overall correspondence between a description and a reference data sets. It was the coefficient of determination, $R_i^2 = 1 - \sigma_i \Big/ \sum_j \left(T_j - \overline{T}_{\text{Ref}}\right)^2$, where $\overline{T}_{\text{Ref}}$ is the mean of all the values of the reference temperature, and T_j represents the individual predicted temperatures of the model.

A minimum value obtained for the sum of squares of the temperature errors, σ_2, was taken as the best fit when parameters were estimated by fitting Eq. (1) to measured data. For example, the value of characteristic heat transfer length, L_{Air}, for experimental data was determined by fitting Eq. (1) to measured photocalorimetric data using the cooling portion of the temperature profile (where $\eta Q_I=0$) in order to minimize σ_2. The effective power input to the cell, ηQ_I, was then determined using the value of L_{Air} derived from the cooling curve by minimizing σ_2 for the heating portion of the curve. When comparing Eq. (2) and Eq. (1), the same values for ηQ_I, L_{Air}, and other physical parameters were used. Differences in thermal dynamics and equilibrium temperatures predicted by Eq. (1) and Eq. (2) arose solely from the right-hand term in Eq. (3) into which a characteristic value of T' was inserted. T' was introduced in Eq. (4) and defined in terms of an optimum $f_{\Delta T}$. The sum-squared temperature error, σ_i, in Eq. (5) was used as the primary estimator for fit because it allowed for more precise comparison.

3. Results

3.1 Complete Energy Balance Describes Measured Data

To evaluate application of the energy balance in Eq. (1) to the photocalorimetry data, predicted temperatures were compared to temperatures measured in the vacuum apparatus. Fig. 3 compares photocalorimetric data with temperatures predicted by fitting Eq. (1) using L_{Air} in the cooling portion of the curve and ηQ_I in the heating portion of the curve. Predicted values deviate positively from experimental temperatures toward the tail of the cooling curve due to a small decrease in the ambient temperature, T_{Amb}, of the room induced by the automatic operation of the HVAC system at t>500 s and accounted for ~30% of the total error. The sum-squared temperature error, σ_2, between Eq. (1) and the experimentally measured data was 1.26 K^2, and the coefficient of determination, R_2^2, was 0.998. Values of characteristic heat transfer length, L_{Air}, and effective power input, ηQ_I, that minimized the sum of squares of the errors, σ_2, were 24.6 mm and 2.53 mW, respectively. Fig. 4 shows the relative percent error, χ_2, between measured temperature and temperatures predicted by Eq. (1) for the data in Fig. 3.

Eighteen temperature profiles similar to Fig 3. from five different photocalorimeter sample sets with nanoparticle concentrations ranging from 46 g Au/m^3 to 920 g Au/m^3 and characteristic heat transfer lengths from 4.1 to 66.8 mm were employed to examine applicability of Eq. (1). The heat transfer length, L_{Air}, in Eq. (1) that represents the characteristic distance for conduction by air, was adjusted to fit Eq. (1) to the cooling data. This accounted for deviations in sample cell size and orientation between the five sample sets. Standard deviations calculated from all replicates within each set relative to the mean value of L_{Air} were ≤30 %. By setting the fitted value of L_{Air} constant during the heating curve, values of effective power input, ηQ_I, for the eighteen profiles were obtained, ranging from 1.0 to 16.3 mW. Minimum and maximum values of sum-squared error, σ_2, for comparing predicted vs. actual temperatures across the entire data set were 1.1 and 7.9 K^2, respectively, with an average of 2.90 ± 2.48 K^2 (one standard deviation). These correspond to coefficients of determination, R_2^2, from 0.982 to 0.998, with an average of 0.992 ± 0.007. Minimum and maximum values of relative percent error, χ_2, in predicted temperature vs. actual temperature over the entire set were -0.67 and 1.11 %, respectively, with an average of 0.75 ± 0.22 % for $|\chi_2|$. These criteria suggest Eq. (1) adequately describes transient thermal dynamics and steady-state temperature equilibria for the photocalorimetry data.

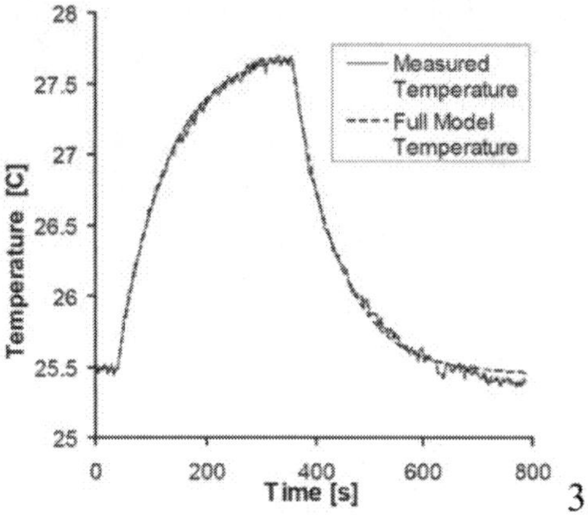

3.

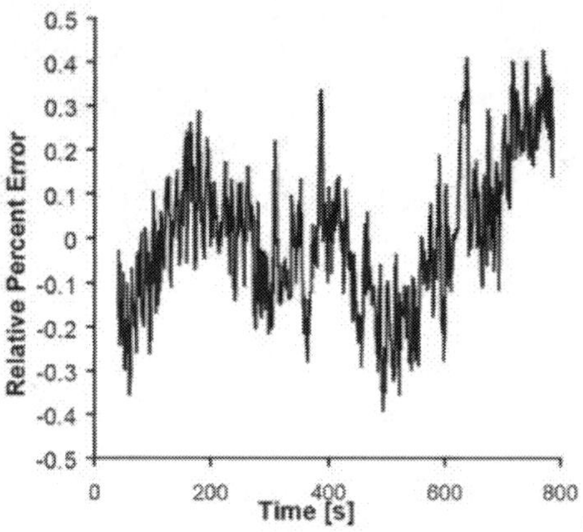

4.

Fig. 3. Plasmon-resonant heating in a nanoparticle suspension due to laser irradiation beginning at 30 seconds, followed by cooling when irradiation is terminated at 350 seconds. Eq. (1) was fit to experimental data using the method described in section 2.3. **Fig. 4.** A plot of the relative percent error, χ_2, between the full model, Eq. (1), and the measured data from the photocalorimetry experiment as seen in Fig. 3.

3.2 Linearized Model Approximates Complete Energy Balance

We next assessed the degree to which the linear approximation accurately describes the complete energy balance as a function of the physical dimensions of the system, L_{Air}, and the effective energy input to the system, ηQ_I. Comparing the linearized model, Eq. (2), with the profile in Fig. 3 resulted in a sum-squared temperature error, σ_3, of 1.27 K^2, and a coefficient of determination, R_3^2, of 0.998. The sum of squares of the error is slightly higher than the value of σ_2 = 1.26 K^2 obtained using Eq. (1) and suggests Eq. (2) describes the profile in Fig. 3 with comparable accuracy. The profile generated using Eq. (2) employs the optimum $f_{\Delta T}$ of 0.8.

The optimum $f_{\Delta T}$ value was determined by performing 378 simulations of both Eq. (1) and Eq. (2) for characteristic heat transfer lengths, L_{Air}, from 1.5 to 393.6 mm and effective power inputs, ηQ_I, from 11.4 to 102.1 mW. These values produced temperature increases, ΔT, ranging from 3 to 77 °C. It was shown in the previous section that at temperature increases <3 °C, as in Fig. 3 and 4, the error between the Eq. (1) and Eq. (2) is negligible.

We examined values of $f_{\Delta T}$ from 0.5 to 1.0 to find an optimum value that minimized χ_3. Fig. 5 is a plot of the absolute value of the maximum percent relative error, χ_3, as a function of effective laser-power input, ηQ_I, for $f_{\Delta T}$ values increasing from 0.5 to 1.0. The characteristic length, L_{Air}, in this calculation was 24.6 mm. The lowest values of the absolute value of the maximum relative error, $|\chi_3|$, resulted at $f_{\Delta T} = 0.8$ as shown by the solid triangles and was set as the optimum. Although introducing T_{Equil} into Eq. (3) by defining T' as in Eq. (4) requires specifying T_{Equil} in advance (from data or simulation), the resulting linearized form in Eq. (2) simplifies estimation of parameters from experimental data and gives a minimum-error linear driving force approximation of heat transfer in plasmon-resonant heating systems.

Next we assessed the accuracy of the linear model over the entire of range of conduction lengths, L_{Air} (1.5 to 393.6 mm), and transduced power inputs, ηQ_I (11.4 to 102.1 mW) using $f_{\Delta T} = 0.8$. Fig 6 shows the absolute value of the maximum relative error, $|\chi_3|$ over the entire range. The range of parameters used in photocalorimetry experiments ($4.1 \leq L_{Air} \leq 66.8$ mm, $1.0 \leq \eta Q_I \leq 16.3$) had a $|\chi_3| \leq 1.5$ %. As air conduction length, L_{Air}, decreases below 3 mm, the error decreases monotonically to a negligible amount as conduction approaches and ultimately exceeds radiation as the predominant heat transfer mode, and linearly approximating radiation results in negligible error. For $3 < L_{Air} \leq 200$ mm, the error of the approximation is affected by both heat transfer length and transduced power input. For $L_{Air} > 200$ mm, the maximum relative error, χ_3, was essentially a function of only ηQ_I.

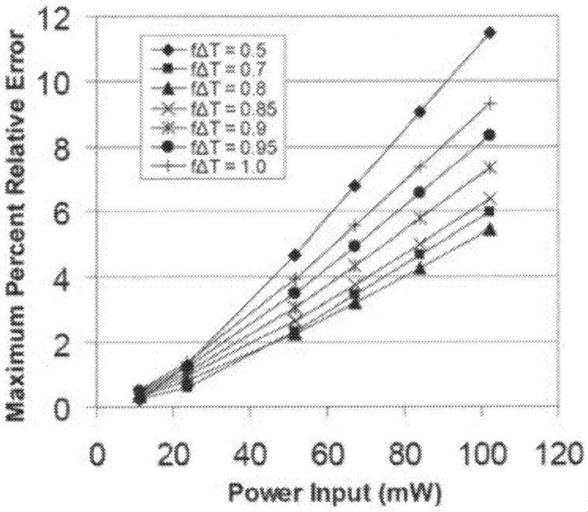

5.

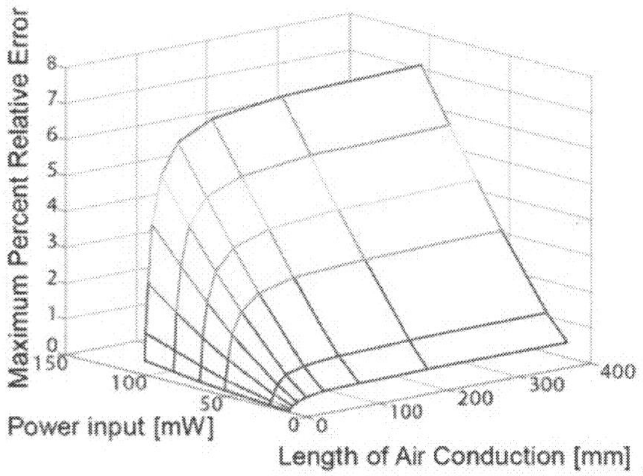

6.

Fig. 5. A plot used to determine the optimum $f_{\Delta T}$, described in section 3.2. **Fig. 6.** Maximum percent relative error, $|\chi_3|$, over full range of parameters as described in section 3.2.

3.4 Applying Linear Description to DNA Hybridization

One application of generating heat by transducing laser light via surface plasmons might be to thermally cycle nucleic acid hybridization on a DNA sensor. Complementary DNA strands are completely denatured at 95 °C, and anneal without producing mismatch errors at approximately 55 °C [11]. Temperature increases of 40 and 70 °C from ambient conditions are required for these steps and the error in the linear approximation was evaluated. From Eq. (1) we found that to attain these equilibrium temperature changes required effective power inputs of 51.6 and 102.1 mW for the 40 and 70 °C increases, respectively, at a characteristic heat transfer length of 2.5 cm (about 1 inch).

Fig. 7 shows heating/cooling curves obtained using Eq. (1) and Eq. (2) along with corresponding values of relative percent error for $\Delta T = 40$ °C (Fig. 7.a and 7.b) and $\Delta T = 70$ °C (Fig. 7.c and 7.d). Effective power input was initiated at t = 0 seconds and terminated at t = 550 seconds. Maximum values of relative percent error were -2.26 % and -5.44 % for $\Delta T = 40$ °C and 70 °C, respectively. The sum of squares of the errors, σ_3, for the 40 and 70 °C increases were 13.08 and 180.36 K^2, and coefficients of determination, R_3^2, were 0.9999 and 0.9993, respectively. The linear approximation underpredicts temperatures from the complete energy balance as heating begins and shortly after cooling begins when the characteristic temperature T', Eq. (4), inserted into the radiation term in Eq. (3) exceeds the actual temperature and overemphasizes effects of radiative heat loss. Similarly, the linear approximation overpredicts equilibrium temperature during heating because the characteristic temperature driving radiative heat loss is about 20 % smaller than the presumed equilibrium, T_{Equil}. Therefore at $L_{Air} = 24.6$ mm and $\eta Q_I = 51.6$ and 102.1 mW, Eq. (2) could be successfully applied to DNA hybridization-based sensors with relatively modest error. Eqs. (1) through (6) can be applied to quantitate the accuracy of linearizing the overall energy balance in order to simplify data interpretation and analysis in a variety of microfluidic or lab-on-chip devices.

4. Conclusion

A continuum energy balance accurately describes thermal dynamics of plasmon resonant nanoparticle suspensions with heat transfer lengths from 4.1 to 66.8 mm that are irradiated by effective resonant light intensities between 1.0 and 16.3 mW to yield steady-state equilibrium temperature increases up to 10.6 °C. Sum-squared temperature errors in this range vary from 1.1 to 7.9 K^2 with an average value of 2.90 ± 2.48 K^2 (N=18). Coefficients of determination, R_2^2, vary from 0.982 to 0.988 with an average value of 0.992 ± 0.007. Relative percent errors in temperature vary across the same range from -0.67 to 1.11 %, with the absolute values averaging 0.75 ± 0.22 % (N=18). Parametric analysis of experimental data using the energy balance is simplified by introducing a characteristic temperature into a linear driving force approximation for radiative heat transfer. The characteristic temperature was defined in terms of a fractional approach to equilibrium. An optimum value of 0.8 for the fractional approach minimized maximum percent relative error to values ≤ 7.16 % over a range of heat transfer lengths from 1.5 to 393.6 mm and a range of effective laser intensities from 11.4 to 102.1 mW that gave temperature increases from 3.0 to 76.6 °C.

The continuum energy balance and its linearized approximation were used to describe heating and cooling curves in plasmon-resonant DNA-hybridization sensors for temperature increases of 40 °C and 70 °C, respectively. Comparing the linearized and complete experessions gave maximum percent relative errors of -2.26 % for the 40 °C increase and -5.44 % for the 70 °C increase. Based on an analysis of photocalorimetric data it appears these expressions will be useful to describe and predict thermal dynamics and steady-state equilibrium temperatures in plasmon-resonant sensors and actuators across a wide range of geometries and operating conditions and will have application to microfluidic devices and lab-on-chips.

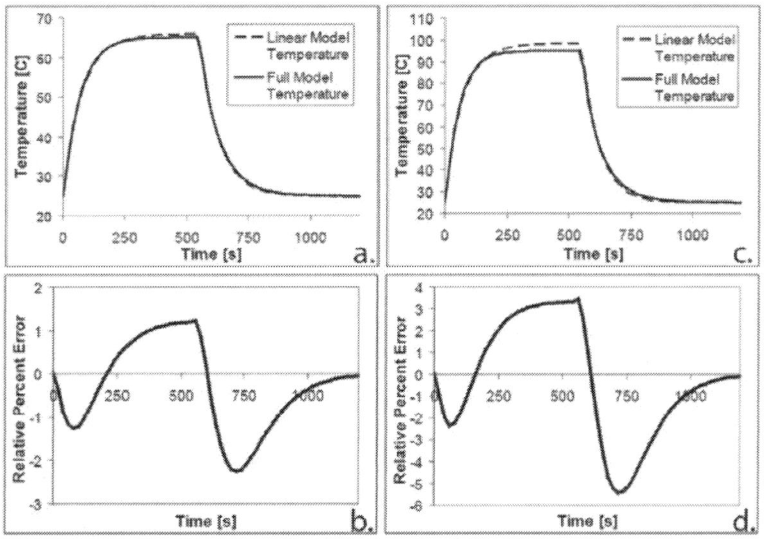

Fig. 7. A plot of the temperature profiles used to compare the linear, Eq. (2), to the full model, Eq. (1), and their respective relative percent differences, Eq. (6). Plots 7.a and 7.b are for the 40 °C temperature increase. Plots 7.c and 7.d are for the 70 °C temperature increase. Refer to section 3.4 for details.

References

[1] M. Xue, J. Li, W. Xu, Z. Lu, K. L. Wang, P. K. Ko, M. Chan, A Self-Assembly Conductive Device for Direct DNA Identification in Integrated Microarray Based System, IDEM '02 Digest, IEEE Int., Electron Devices Meet., San Fransisco, CA, USA, Dec. 8-11, 2002, p. 207-210.

[2] J. J. Storhoff, R. Elghanian, R. C. Mucic, C. A. Mirkin, R. L. Letsinger, One-Pot Colorimetric Differentiation of Polynucleotides with Single Base Imperfections Using Gold Nanoparticle Probes, J. Am. Chem. Soc., 120(9) (1998), p. 1959-1964.

[3] K. Kneipp, A. S. Haka, H. Kneipp, K. Badizadegan, N. Yoshizawa, C. Boone, K. E. Shafer-Peltier, et. al., Surface-Enhanced Raman Spectroscopy in Single Living Cells Using Gold Nanoparticles, Applied Spectroscopy, 56(2) (2002), p. 150-154.

[4] K. Kniepp, H. Kniepp, R. Manoharan, E. B. Hanlon, I. Itzkan, R. R. Dasari, M. S. Feld, Extremely Large Enhancement Factors in Surface-Enhanced Raman Scattering for Molecules on Colloidal Gold Clusters, Applied Spectroscopy, 52(12) (1998), p. 1493-1497.

[5] T. S. Ahmadi, S. L. Logunov, M. A. El-Sayed, Picosecond Dynamics of Colloidal Gold Nanoparticles, J. Phys. Chem, 100(20) (1996), p. 8053-8056.

[6] S. L. Logunov, T. S. Ahmadi, M. A. El-Sayed, J. T. Khoury, R. L. Whetten, Electron Dynamics of Passivated Gold Nanocrystals Probed by Subpicosecond Transient Absorption Spectroscopy, J. Phys. Chem., B, 101(19) (1997), p. 3713-3719.

[7] S. Link, M. A. El-Sayed, Optical Properties and Ultrafast Dynamics of Metallic Nanocrystals, Annual Rev. of Phys. Chem., 54(1) (2003), p. 331-366.

[8] D. K. Roper, W. Ahn, M. Hoepfner, Microscale Heat Transfer Transduced by Surface Plasmon Resonant Gold Nanoparticles, J. Phys. Chem., C 111(9) (2007), p. 3636-3641.

[9] F. Incropera and D. DeWitt, Fundamentals of Heat and Mass Transfer, 5th Ed., John Wiley and Sons, Hoboken, 2002, p. 699-756, 905-932 (Chapter 12, Appendix A).

[10] M. Roper, C. Easley, L. Legendre, J. Humphrey, J. Landers, Infrared Temperature Control System for a Completely Noncontact Polymerase Chain Reaction in Microfluidic Chips, Anal. Chem., 79(4) (2007), p. 1294-1300.

[11] R. Eeles and A. Stamps, Polymerase Chain Reaction (PCR) The Technique and Its Applications, CRC Press, Boca Raton, 1993, p. 4-11 (Chapter 2).

[12] G. Lindfield and J. Penny, Numerical Methods Using Matlab, 2nd Ed., Prentice Hall, Upper Saddle River, 2000, p. 200-249, 305.

542b Thermodynamic Analysis of the Gas Expanded Liquid Nanoparticle Size Separation Process

Madhu Anand[1], Seong-Sik You[1], Kendall M. Hurst[1], Steven R. Saunders[1], Christopher L. Kitchens[2], W. Robert Ashurst[3], and **Christopher B. Roberts**[1]. (1) Department of Chemical Engineering, Auburn University, 212 Ross Hall, Department of Chemical Engineering, Auburn, AL 36849, (2) Department of Chemical and Biomolecular Engineering, Clemson University, 127 Earle Hall, Clemson, SC 29634-0909, (3) Chemical Engineering, Auburn University, 212 Ross Hall, Auburn, AL 36849

A thermodynamic model and related equations were developed to improve our understanding of the size selective fractionation of ligand stabilized nanoparticles using a CO_2 gas expanded liquid precipitation process. This size selective precipitation process is based on controlled reduction of the solvent strength of the organic solution through increases in the antisolvent CO_2 concentration via pressurization. A decrease in the solvent strength reduces the repulsive Osmotic contributions to the total interaction energy of the dispersion by weakening the interactions between the solvent and the ligand tails. Eventually, these repulsive forces in the dispersion are lowered to the point where they can no longer overcome the van der Waals forces of attraction between particles, thereby causing the particles to precipitate. This results in an ability to tune the size of the nanoparticles that are dispersed or precipitated simply by adjusting the CO_2 pressure. In this chapter, Osmotic energy and Hamaker constant expressions were developed for CO_2 gas expanded solvent mixtures consisting of an organic solvent (e.g. hexane) and antisolvent (e.g. CO_2). These expressions, coupled with appropriate thermodynamic data, were used to estimate the maximum size of dodecanethiol capped Ag nanoparticles that could be dispersed at a given CO_2 pressure by equating the total interaction energy to the Boltzman threshold stabilization energy (-3/2 k_BT). A parameter sensitivity analysis was performed to determine which parameters most affect the predicted maximum particle size that could be dispersed at a given CO_2 pressure. It was found that the length of the ligand tail that interacts with the solvent medium had the most influence on these total interaction energy calculations. Assuming complete solvation of the entire ligand tail, the model always overpredicted the size of the particles that would be precipitated at a given CO_2 pressure compared to the experimental results. Consequently, we have proposed that the entire length of the dodecane chain is not accessible to the solvent medium. As such, three phenomenological model variations were developed that provide for differing abilities of the ligand tails to interact with the solvent including an Extended Ligand Length Solvation Model, a Condensed Phase Model, and a Limited Ligand Length Solvation Model with the best fit of the experimental size separation data coming from the Limited Ligand Length Solvation Model. Equations for a surface fraction model based on UNIQUAC parameters and its viability for this process is also discussed.

542c Formulation of Effective Mechanistic Aggregation Kernel/s for the Flocculation Process with Concurrent Mechanisms Using Inverse Problem Solution

Ingmar Nopens[1], Nandkishor K. Nere[2], Peter Vanrolleghem[1], and Doraiswami Ramkrishna[3]. (1) Department of Applied Mathematics, Biometrics and Process Control, Ghent University, Coupure Links 653, Gent, Belgium, (2) Purdue University, 1283, FRNY, 480 Stadium Mall Dr., West Lafayette, IN 47907, (3) School of Chemical Engineering, Purdue University, West Lafayette, IN 47907-1283

The activated sludge process is widely used for wastewater treatment. Before discharging the cleaned water to the receiving water body, the sludge needs to be separated from the sludge. This is traditionally done gravitationally, exploiting the higher density of the sludge. However, the flocculation state of the sludge plays an important role in this settling process. Modeling of activated sludge flocculation using a PBM framework with existing kernels from literature has proven to be a tedious task (Nopens et al., 2005). An alternative way of retrieving the kernels based on experimental data is to solve the inverse problem hereby exploiting self-similarity present in the data (Wright and Ramkrishna, 1992; Sathyagal et al., 1995).

We present here the application of the inverse problem approach to extract the aggregation kernel for the flocculation process. The computational intricacies associated with the inverse problem of interest are discussed. The quality of the inversion was adjudged through the comparison of the forward solution to the population balance equation with the experimental data. Results showed good description for the entire distribution dynamics, including the peaks and tails. The inverse problem approach results in an empirical equation for the aggregation kernel, which hardly conveys any physical significance. This limits the applicability range of the model to optimize and control the process in general. To overcome this, the empirical kernel can be confronted with mechanistic kernels. In the activated sludge flocculation application under study, these kernels might correspond to the shear aggregation, differential settling, charge effects, etc. The objective of the presentation is to discuss the formulation of the effective aggregation kernel that considers two or more concurrent and /or competitive mechanisms in the flocculation process.

References

Nopens, I., Koegst, T., Mahieu, K., Vanrolleghem, P., 2005. Population balance model and activated sludge flocculation: from experimental data to a calibrated model. AIChE J. 51 (5), 1548–1557.

Sathyagal, A., Ramkrishna, D., Narsimhan, G., 1995. Solution of inverse problems in population balances — ii particle break-up. Comp. Chem. Eng. 19 (4), 437–451.

Wright, H., Ramkrishna, D., 1992. Solutions of inverse problems in population balances — i aggregation kinetics. Comp. Chem. Eng. 16 (12), 1019–1038.

542d Study of Calcium Alginate Sol-Gel Transition by Population Balance Model

Yinyan Zhao, You-Yeon Won, and Michael T. Harris. School of Chemical Engineering, Purdue University, 480 Stadium Mall Drive, West Lafayette, IN 47907

Experimental and computational studies are presented to understand the Ca-alginate sol-gel transition mechanism. Ca-alginate microgel beads are prepared by using electrodispersion process. This method uses electric fields to atomize aqueous mixtures of alginate and $CaCO_3$ nanoparticles from a nozzle into an organic continuous phase containing a soluble organic acid. A critical concentration of alginate is observed for sol-gel transition as a function of the concentration of alginate and $CaCO_3$ nanoparticles. The polymerization kinetic model assumes an initial population of mondispersed alginate precursors, a spatially uniform distributed reactants in the system, an irreversible gelation process and no cyclization to simplify the problem. The model captures the effect of the reactant concentration, the rate constant ratio of the first step reaction (K_1, free Ca^{2+} reaction rate) to the second step reaction (K_2, cluster-cluster junction formation rate), and the functionality f on the sol-gel transition boundary. At high calcium ion concentrations, the formation of small alginate clusters is due to the saturation of Ca^{2+} along the polymer subchains. Typically, a large alginate concentration, large f and small K_1/K_2 result in fast gel formation. It is concluded that the model results are in reasonable agreement with experimental data and is capable of predicting the sol-gel transition mechanism of the Ca-alginate system.

Aggregation Modeling Using Sinc Methods of Solution to the Steiltjes Formulation of the Population Balance

Terry A. Ring

Department of Chemical Engineering, University of Utah, *50 S. Central Campus Drive, Salt Lake City, UT 84112*

Abstract

Aggregation and breakage processes are modeled the population balance equation, which accounts for the particles in the system. The population balance is an integro-partial differential equation that is coupled to the mass, heat and momentum balance equations that are simply partial differential equations. The population balance can be written in various forms, e.g. discrete or cumulative number as well as discrete or cumulative mass, with various bases, e.g. particle size or particle volume, as the internal coordinate. The population balance in all of these forms and bases is still an integro-partial differential equation. This paper solves the cumulative mass population balance in Steiltjes integral form using Sinc collocation, a finite element method, for several problems including deaggregation of fractal aggregates, aggregation of polystyrene latex particles and the combination of aggregation and breakage of NaCl crystals.

Introduction

The Stieltjes Integral[1] formulation of the population balance is often more useful to engineers than the more traditional number density based population balance [2,3] since it is 1) formulated on a mass (or volume) of particles basis instead of a number of particles basis, 2) assures conservation of both particle number and particle mass (or volume), 3) is a cumulative particle size distribution instead of discrete size distribution and 4) is functionally simpler than the number density based population balance. The Stieltjes integral formulation of the population balance is a relative new approach [2] in the field of population balances, however, this mathematical approach has been used for many years in traditional integral calculus [1]. Scarlet [4] has criticized the traditional number density based population balance as it 1) provides information (e.g. the particle number density within a given particle size range) that is less relevant to the process engineer who is interested in assuring mass is conserved in his process and 2) uses a particle length dimension that can not be used to determine the volume of the particle without a shape factor which is often different for each particle. His article goes on to requests that all further work on the population

---

[1] Taylor, A.E., (1952) *Advanced Calculus*, Ginn, Boston, p. 532.

[2] Ramkrishna, D.,(2000) *Population Balances*, Academic Press, New York, Chapter 2.

[3] Randolph, A.D. and Larson, M. A., (1971) *The Theory of Particulate Processes*, Academic Press, New York, Chapter 1.

[4] Verkoeijen, D.; Pouw, G. A.; Meesters, G. M.H.; Scarlett, B., (2002) Population balances for particulate processes—a volume approach Chem. Eng. Sci., 57[12], 2287-302.

balances be formulated on a mass based and the internal particle size coordinate be in terms of particle volume. The Stieltjes integral formulation of the population balance does just that - a population balance on a cumulative mass basis where the size coordinate is the particle volume.

There are two major reasons that the Stieltjes integral formulation of the population balance is not frequently used. The first reason is the various terms in the population balance for nucleation, growth, agglomeration on a number basis need to be reformulated in terms of a mass/volume based population and there are few numerical methods available for the Stieltjes integral formulation of the population balance since it is an integral differential equation of the Markov type[5]

.

This paper will focus on the later limitations by using the Stieltjes integral formulation of the population balance. We will present a Sinc numerical solution to this form of the population balance and compare it with experimental results.

Formulation

The Stieltjes integral formulation of the population balance has been derived by Bharucha-Reid[[2]] in terms of a cumulative mass based population, $F_m(x,t)$, where the particle size variable is the mass of a particle, x:

$$\frac{\partial F_m(x,t)}{\partial t}+\int_0^x x'\frac{\partial}{\partial x'}\left[\frac{g_M(x',t)}{x'}\frac{\partial F_m(x,t)}{\partial x'}\right]dx'=\hat{B}-\hat{D}=$$

$$\frac{\partial F_m(x,t)}{\partial t}+g_M(x',t)\frac{\partial F_m(x,t)}{\partial x'}-\int_0^x\frac{g_M(x',t)}{x'}\partial_x F_m(x',t)=\hat{B}-\hat{D}=$$

$$\int_x^\infty b(x')P_m(x|x')\partial_x F(x',t) \quad Breakage$$

$$-\mu_1(t)\int_{s=0}^x\int_{u=x-s}^\infty k_a\frac{a(s,u)}{u}\partial_u F(u,t)\,\partial_s F(s,t) \quad Agglomeration$$

with boundary conditions [1]

$$g_M(x,t)F_m(x,t)=\frac{\partial F_m(x,t)}{\partial t} \quad at \quad x=x_o \quad Nucleation\ at\ size,\ x_o$$

$$F_m(x,t)=0 \quad at \quad x=\infty$$

and initial condition

$$F_m(x,t)=F_{m,o}(x) \quad at \quad t=0$$

[5] Bharucha-Reid, A.T. (1960), Elements of the Theroy of Markov Processes and Their Applications, McGraw-Hill, New York.

where the growth term consists [6,7] of the cumulative particle growth rate on a mass basis, the hat's on B and D indicate that they are the cumulative mass based birth and death functions, b(x') is the breakage rate, $\mu_1(t)$ is mass density of particles, the $P_m(x|x')$ is the cumulative mass fraction of broken fragments due to the breakage of x', and a(x,x') is the agglomeration rate. This formulation of the population balance has fewer terms than the number based discrete form of the population balance because the breakage and agglomeration terms are written as single integrals with the Stieltjes formulation of the population balance.

The number based nucleation rate, a boundary condition for eq. [1], also needs to be reformulated in terms of the cumulative mass basis giving:

$$\frac{\partial F_m(x_o,t)}{\partial t} = x_o \frac{\partial f_N(x_o,t)}{\partial t} = \beta_v r_o^3 \frac{\partial f_N(\beta_v r_o^3,t)}{\partial t}. \qquad [2]$$

Where r_o is the radius of a particle with mass x_o and β_v is the volume shape factor (4/3π for spheres).

Numerical Solution

Sinc collocation methods [8] will be used for the solution to Stieltjes integral formulation of the population balance equation, eq. [1]. The function $F_m(x,t)$ is approximated by an interpolation function that consists of a summation of time based and position based Sinc basis function weights and values of the function, $F_m(x_j,t_k)$ itself at the Sinc points, x_j's on the x interval and t_k's on the t interval as follows:

$$F_m(x,t) \approx \sum_{j=-N}^{N} \sum_{k=-M}^{M} \mathrm{sinc}(\frac{\varphi(x)}{h_x} - j)\,\mathrm{sinc}(\frac{\phi(t)}{h_t} - k) F_m(x_j,t_k) \qquad [3]$$

where the function sinc(z) has the normal definition,

$$\mathrm{sinc}(z) = \frac{\sin(\pi z)}{\pi z}. \qquad [4]$$

The Sinc points are mapped across the x domain from zero to infinity using

[6] Ramkrishna, D. (2003) Private communication.
[7] Ramkrishna, D., A. Sathyagal and G. Narsimhan, (1995) Analysis of dispersed-phase systems: fresh perspective, AIChE J. 41, 35-44.
[8] Stenger, F., (1993) *Numerical Methods Based on Sinc and Analytic Functions*, Springer-Verlag, Berlin.

$$\varphi(x) = \ln(x)$$
$$x_j = \varphi^{-1}(jh_x) = \exp(jh_x) \qquad [5]$$
$$h_x = \frac{c_1}{\sqrt{N}}$$

and mapped across the t domain from zero to time T using

$$\phi(t) = \ln[t/(T-t)]$$
$$t_k = \phi^{-1}(kh_t) = \frac{T\exp(kh_t)}{1+\exp(kh_t)} \qquad [6]$$
$$h_t = \frac{c_2}{\sqrt{M}}$$

where c_1 and c_2 are $O(1)$ constants. T may be considered to be either the maximum time for the batch crystallization or the time for a piece of the calculation from zero to T. Convergence will always occur for T sufficiently small using this time map [[8]]. If a small time for T must be used for convergence then the calculation can be pieced together to form a complete solution for the total batch crystallization time. The x based Sinc points, x_j, for a vector, Zx, and the t based Sinc points, t_k, form a vector Zt.

To solve eq. [1], we need to integrate the time derivative giving:

$$\begin{aligned}
F_m(x,t) &= F_{m,o}(x,t=0) + \\
&- \int_0^t g_M(x',t')\frac{\partial F_m(x',t')}{\partial x'}dt' + \int_0^t\int_0^x \frac{g_M(x',t')}{x'}\partial_x F_m(x',t')dt' \quad \textit{growth} \\
&+ \int_0^t\int_x^\infty b(x')P_m(x|x')\partial_x F(x',t')dt' \quad \textit{Breakage} \qquad [7]\\
&- \int_0^t \mu_1(t')\int_{s=0}^x\int_{u=x-s}^\infty k_a\frac{a(s,u)}{u}\partial_u F(u,t')\,\partial_s F(s,t')dt' \quad \textit{Agglomeration}
\end{aligned}$$

Then each of the time integrals is formulated with the integration operator, $\overline{\overline{At_{I+}}}$, using the matrix manipulations described above. The resulting matrix formulation of the problem is given by:

$$\begin{aligned}
\overline{\overline{F_m}} &= \overline{\overline{F_{m,o}}} + \\
&- \overline{\overline{At_{I+}}}\bullet[\overline{\overline{g_M}}]\bullet[\overline{\overline{Ax_D}}\bullet\overline{\overline{F_m}}] + \overline{\overline{At_{I+}}}\bullet\overline{\overline{Ax_{I+}}}\bullet[\overline{\overline{\frac{g_M}{Zx}}}]\bullet[\overline{\overline{Ax_D}}\bullet\overline{\overline{F_m}}] \quad \textit{Growth} \qquad [8]\\
&+ \overline{\overline{At_{I+}}}\bullet\overline{\overline{Ax_{I+}}}\bullet\left\{[\overline{\overline{bP_m}}]\bullet[\overline{\overline{Ax_D}}\bullet\overline{\overline{F_m}}]\right\} \quad \textit{Breakage} \\
&- \overline{\overline{At_{I+}}}\bullet\left[diag[\mu_1(\overline{Zt})]\ \overline{\overline{X}}\ \tilde{L}\right] \quad \textit{Agglomeration}
\end{aligned}$$

This matrix equation is solved for $\overline{\overline{F_m}}$ at its sinc points in both x and t, iteratively until the norm of the matrix formed by subtracting the rhs from the lhs of eq. [8] is less than 0.001. A_{xI+} is a matrix for x integration, A_{tI+} is a matrix for t integration,

A_{xD} is a matrix for the x partial derivative and X is the eigen vector of A_{xI+}, g_M is a matrix of growth rates, bP_m is a matrix of breakage rates composed by multiplying the breakage rate b(x) and daughter distribution functions $P_m(x|x')$, $\mu_1(Z_t)$ is the specific mass of particles as a function of time and L is the Laplace transform matrix of the integrand of the agglomeration rate term.

The solution to eq. [8] for breakage and agglomeration only for a batch unit is shown in Figure 1a, 1b and 1c. For the breakage rate used in Figure 1a, we use a daughter function, $P_M(x|x')$, that consists of a power law function, seen in the figure legend and a breakage rate, b(x'), that is size independent with a value of 1.0. For the agglomeration rate we use an agglomeration kernel, a(x,y) = x+y and a particle mass density, $\mu_1(t)$, of 0.1. For the size and time intervals we use N=M=8 and an initial condition of a lognormal particle size distribution with a geometric mean size of 1.0 and a geometric standard deviation of 2.0. In Figures 1a, 1b and 1c, we see the change of the cumulative particle size distribution from that of the initial size distribution to that of the product as time in the batch unit increases. The effects of breakage alone are shown in Figure 1a, agglomeration alone in Figure 1b and their combination in Figure 1c where compensation between breakage and agglomeration is observed.

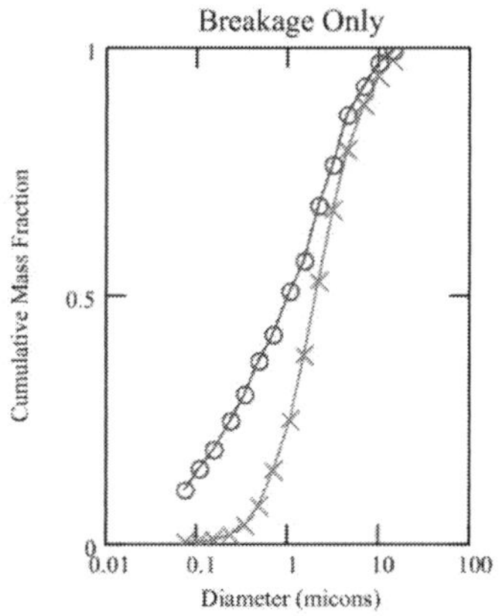

Figure 1a Cumulative size distribution for breakage only as time in the batch apparatus increases from 0 to 30 min. Data : $b(x')t=0.1$, $P_M(x|x') = (x/x')^{1.2}$.

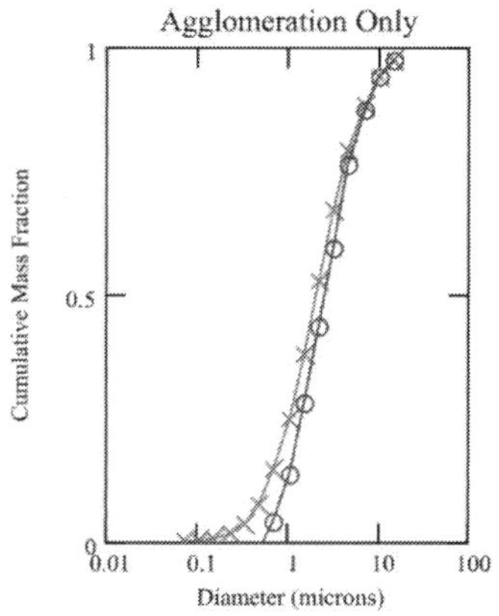

Figure 1b Cumulative size distribution for agglomeration only as time in the batch apparatus increases from 0 to 30 min. Data: $k_a t=0.5$, $a(x,y)=x+y$, $\mu_1(t) = 0.1$.

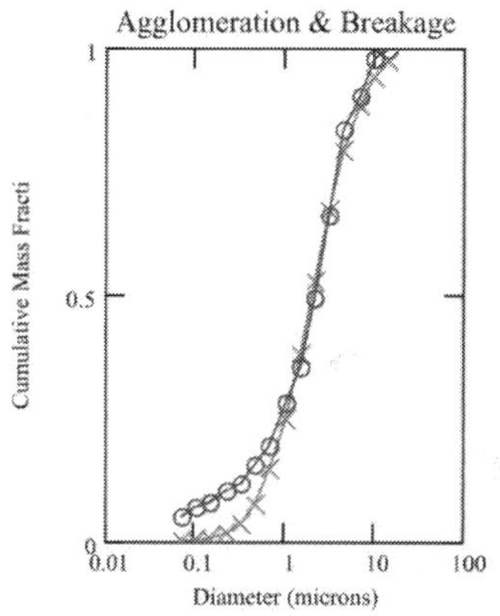

Figure 1c Cumulative size distribution for breakage and agglomeration as time in the batch apparatus increases from 0 to 30 min. Data : for breakage $b(x')t=0.1$, $P_M(x|x') = (x/x')^{1.2}$, and for agglomeration $k_a t=0.5$, $a(x,y)=x+y$, $\mu_1(t) = 0.1$.

Attrition experiments were performed with a powdered diatomite suspension by subjecting it to mixing in a round-bottom 1.4 L stirred tank with a Rushton impeller and 4 baffles of standard CSTR design. The particle size distribution was sampled every hour and measured using a Beckmann-Coulter LS230. The results are shown in Figure 2a. Here we see that the initial population decreases in size as time in the reactor increases. Simulations were performed using an exponential progeny function and a size dependent breakage rate[9] based upon turbulent collisions with the wall, impeller and baffles and are shown in Figure 2b. The fit is excellent.

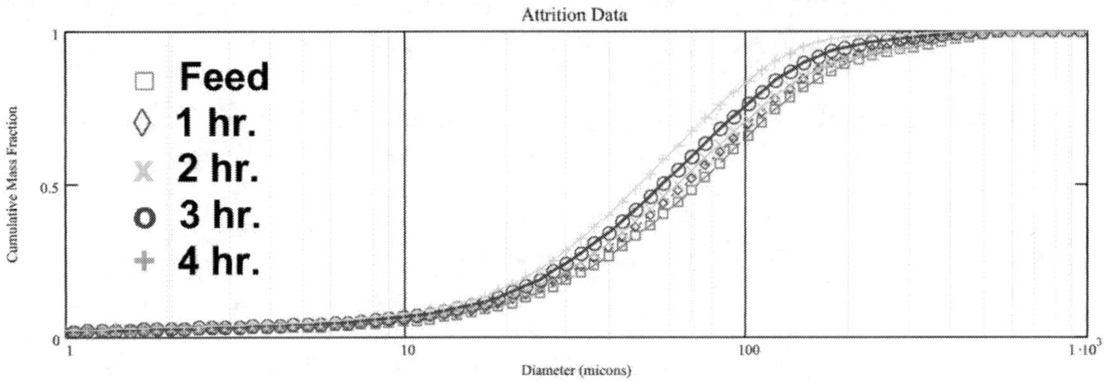

Figure 2a Experimental Data

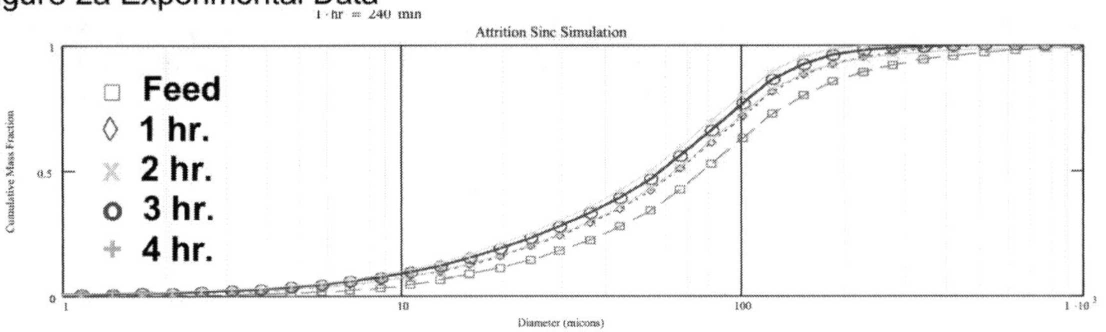

Figure 2b Sinc Simulation

Figure 2 Comparison of Diatomite Breakage Experimental Data with Sinc Simulation.

Aggregation experiments were performed with a 1 micron polystyrene latex sample by subjecting it to mixing in the same round bottomed 1.4 L stirred tank. A colloidally stable suspension was prepared initially and an increase in salt solution to 10^{-3} M NaCl was added to induce aggregation at time zero. The results of samples pulled at 20 min. intervals are shown in Figure 3. Here we see that the initial population decreases in size as time in the reactor increases. Simulations, also shown in Figure 3, were performed using both aggregation and breakage simultaneously with an exponential progeny function and a size dependent breakage rate dependent on turbulent collisions with walls, baffles and the impeller and a size dependent agglomeration rate based upon on turbulent collisions with walls, baffles and the impeller. The fit is again excellent.

9 Kusters, K.A. "The Influence of Turbulence on Aggregation of Small Particles in Agitated Vessels," Technical University Eindhoven, The Netherlands. P. 19.

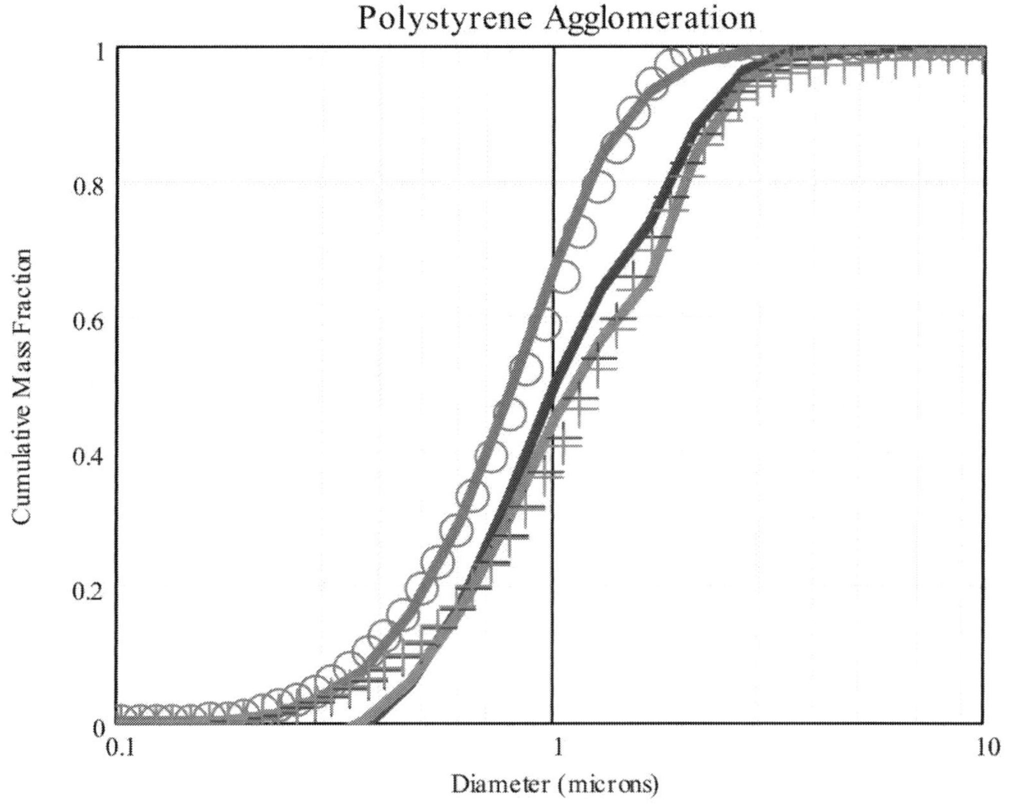

Figure 3 Comparison of 1 micron Polystyrene Latex Aggregation Experiments (data points, red o – initial distribution, blue + 20 min. and magenta + 40 min) with Sinc Simulations (red line - initial distribution, blue line 20 min. and magenta line 40 min).

Conclusions

Sinc finite element collocation methods are well suited to solving the Steiltjes integral formulation of the population balance. To get accurate simulations of experimental data for particle attrition and aggregation, the turbulent energy dissipation rate is needed in conjunction with collision models to provide accurate aggregation and breakage rate constants. Fits of the data with only 8 size intervals and 8 time intervals were of high fidelity making computation time short.

544a Advanced Materials And Nano-Energetic Development At Us Army, Rdecom-Ardec

Paul Redner, Energetics, Warheads and Environmental Technology Division, US Army, RDECOM-ARDEC, Bldg 3022, AMSRD-AAR-AEE-W, Picatinny, NJ 07806-5000

The US Army, at RDECOM-ARDEC, has been exploring various technologies to produce both energetic and non-energetic nanomaterials and advanced materials. ARDEC has established partnerships with industry and academia to synthesize and modify materials to meet identified needs and technology gaps as identified by DoD and Army leadership. These technologies are either being developed at ARDEC or will be transferred to the ARDEC campus so that a wide array of prototyping capabilities will be available to transition these technologies to core programs.

This briefing will discuss the various partnerships and technologies ARDEC is developing. It will bring to light the progress that has been made over the past few years as well as the challenges that have been overcome and the ones that the Army still faces as these materials and technologies are transferred to the Warfighter.

544b Heterogeneous Mixtures of Boron Compounds with Metals and Water for Hydrogen Generation

Moiz Diwan, Victor Diakov, Evgeny Shafirovich, and Arvind Varma. School of Chemical Engineering, Purdue University, 480 Stadium Mall Drive, West Lafayette, IN 47907

Sodium borohydride (SBH) and ammonia borane (AB) are promising hydrogen-storing compounds to feed fuel cells for portable electronic devices. To release hydrogen from these compounds, pyrotechnic mixtures of SBH and AB with solid oxidizers were considered by previous researchers. Alternatively, Hydrolysis of these compounds has also been studied extensively. To obtain high hydrogen yield without any adverse environmental effects, we use heterogeneous mixtures of SBH or AB with gelled water and nanoaluminum or magnesium. Due to the highly exothermic metal-water reaction, such mixtures, upon ignition, exhibit self-sustained propagation of combustion wave with simultaneous release of hydrogen from the boron compounds and water. Mechanistic studies of this process were conducted using digital video recording, pressure monitoring, thermocouple measurements, gas chromatography, mass spectrometry, TGA, and powder XRD analysis. Isotopic tests using D2O were performed to characterize roles of thermolysis and hydrolysis.

Fully dense, aluminum-rich Al-CuO nanocomposite powders for energetic formulations

Demitrios Stamatis, Vern K. Hoffmann, Mirko Schoenitz, and Edward L. Dreizin,
New Jersey Institute of Technology, Newark, NJ, USA

Abstract

The thermite reaction between Al and CuO is well known and highly exothermic. For a conventional thermite mixture comprising mixed metal and oxide powders, the reaction is rate limited by a slow heterogeneous mass transfer at the metal and oxide interface. The relatively low reaction rate and a difficult ignition have restricted practical applications for this reaction. For newly developed, nano-composed thermites, the interface area can be substantially increased resulting in a much higher reaction rate and a new range of possible applications. Recently, Al-CuO nanocomposite materials were produced by magnetron sputtering and by mixing of Al and CuO nanopowders. Such techniques realize the bottom-up approach, when the nano-domains are built from individual atoms or molecules. Respective materials are generally expensive and difficult to handle. An alternative, top-down approach is discussed in this project. Nanocomposite Al-CuO materials are produced using a technique referred to as arrested reactive milling. Regular metal and oxide powders are blended and ball milled at room temperature resulting in a fully dense and reactive nanocomposite powder. The milling is stopped (or arrested) before a self-sustaining exothermic reaction is triggered. The powder particles are in the 10-100 µm size range. Each particle has an aluminum matrix with copper oxide inclusions in the 20-200 nm size range, depending on milling parameters. This paper focuses on the compositions with excess of Al, useful for applications in propellants, explosives, pyrotechnics, as well as for joining small parts. The paper discusses the preparation and characterization of such Al-rich thermite compositions and suggests an approach for optimizing their manufacturing for the improved performance.

Introduction

Metal-based energetic materials are finding applications in propellants, explosives, and pyrotechnics[1-2]. The main advantage of using these materials is their high reaction enthalpy exceeding that of hydrocarbon based energetics. The limiting factors of using these energetic materials are their low reaction rates, long ignition delays, and incomplete burning. For example, long ignition delays cause agglomeration of molten aluminum particles in solid propellants[3]. New developments have led to reactive metal-based nanomaterials[4-10]. The main idea is to greatly increase the surface area leading to an increase in the reaction rate and decrease in metal ignition delays. There are two production approaches used to create the nanocomposite materials. A more common, bottom-up approach is based on nano-particles or nano-foils produced from elemental atoms or molecules. The related materials include mixed metal-oxide nanoparticles (termed metastable intermolecular composites or MIC)[4-5], sol-gel produced nanocomposites[7], and nanofoils[8-9]. The top-down approach used in this research is based on refining the bulk materials to achieve nano-scaled mixing between components. A related practical technique is called arrested reactive milling (ARM)[10-15]. ARM is based on high energy mechanical milling whereby regular micron-sized powders capable of highly exothermic reaction are ball milled together. The mechanically initiated exothermic reaction is prevented by arresting (or stopping) the milling when the desired powder refinement is achieved. The product is a metastable, fully dense, reactive nanocomposite powder. The mixing takes place at the nanoscale level, while the particle sizes are in the micron-range. Therefore, each micron-sized particle represents a nanocomposite structure of two or more reactive components. Systems of interest include metal-metal composites such as Al-Ni and Al-Ti, metal-metalloid such as Ti-B, Zr-B, and metal-oxide, such as Al-MoO_3, Al-CuO, Al-Bi_2O_3, etc. This research focuses on the particular system of Al-CuO for potential applications in explosives, propellant additives, and for joining/welding small components. Metal-rich compositions are being explored to take advantage of aluminum combustion with external oxidizers, such as air for explosives or ammonium perchlorate for propellants. Metal-rich composites can also provide more metal to form filler necessary to achieve better structural or strength properties for joining applications.

The focus of this project is to investigate whether shifting composition of nanocomposite Al-CuO thermite system towards Al-rich end would allow one to maintain the high reaction rates, similar to those observed for stoichiometric compositions[13]. The synthesis conditions appropriate for metal-rich materials are to be determined. Processing technology needs to be considered that is necessary to obtain powders with the most refined and uniform microstructure and at the same time, with the minimized amount of material that already reacted in the process of the nanocomposite synthesis. The project is aimed to synthesize and characterize a range of related nanocomposite powders and establish an approach for selecting the synthesis conditions leading to the maximum performance of the produced powders as energetic material additives.

Materials

The nanocomposite powders used in all the experiments were prepared by ARM. The initial mixture is simply a blend of regular aluminum and copper oxide powders. Aluminum used was provided by Atlantic Equipment Engineers, -325 mesh (> 45 μm), 99.5% pure. Several copper oxide powders were considered as a starting material, including a powder by Sigma-Aldrich (25 μm), 99 +% pure and a nanopowder by Alfa Aesar with nominal particle sizes of 30-50 nm. In addition, we used the Sigma Aldrich powder, which was initially pre-milled without Al, to reduce the CuO particle sizes. The pre-milling step used the same milling parameters and ball load as in the normal milling experiment (see below), used no liquid process control agent, and continued for 30 min. The measured size distributions of the starting powders (except for the nanopowder) are shown in Fig. 1. Both commercial Al and micron-sized CuO powders are clearly size classified based on a sharp cut-off in the size distribution on the side of coarse particles. The pre-milling step results in a remarkable particle size reduction of the commercial CuO powders.

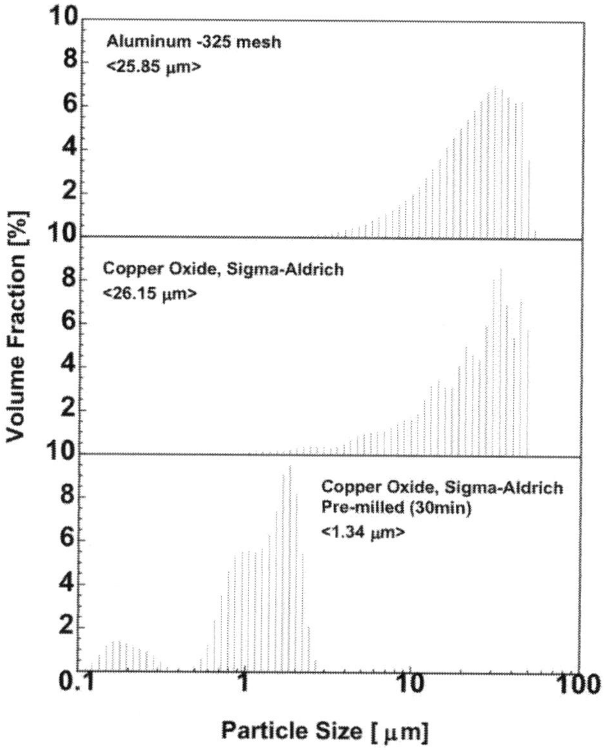

Figure 1. Particle Size Distribution of Starting Powders

The stoichiometric composition is $2Al+3CuO$ so that fuel-rich compositions can be identified as $xAl+3CuO$ with $x>2$. In this study 8, 10, and 12 were used for x. A parametric study of milling conditions was performed for the fixed composition of $8Al+3CuO$ and then two other compositions were prepared using the already optimized synthesis conditions.

ARM Parameters

Milling was performed using a Retsch 400PM planetary mill. The rotation speed was set to 350 RPM. The mill was set to reverse direction of rotation every 15 minutes. The vials used were custom made hardened steel jars designed for high pressure applications. The vials had internal diameter of 64 mm, height of 52 mm, and wall thickness of 18 mm. The milling compartment of the planetary mill was air-conditioned to maintain it at the temperature of 19 °C. The temperatures of the vials were monitored using wireless temperature sensors (Point Temperature Sensors including 418 MHz transmitters by Point Six Wireless, Inc.).

Milling medium was 10 mm balls made of AISI/SAE 1013 low-carbon steel. The ball to powder mass ratio was fixed at 3, and the powder mass load per milling vial was fixed at 30 g. Hexane was added as a process control agent (PCA) to prevent cold welding and minimize partial reaction of the components during the milling.

Several milling parameters were varied to determine their effect on the properties of the produced nanocomposite material. In this project, the variable parameters included the particle size of the starting CuO powder, milling time, and PCA amount. Further in the paper, each sample is identified by its composition (first number represents the number of moles of Al per three moles of CuO), following by a code for milling conditions (mc-0, mc-1, or mc-2). Each milling condition represents a specific combination of the amount of PCA and milling time. The last letter in the sample ID represents a type of the starting CuO powder. The powder by Sigma Aldrich is labeled with letter s, the pre-milled Sigma Aldrich CuO powder is identified by the last letter p, and the nano-powder by Alfa Aesar is labeled by letter a. Table 1 lists different samples prepared in the parametric milling study, showing their sample ID's and respective preparation parameters.

Table 1. Samples prepared in this study

Sample ID	CuO type	PCA, ml of Hexane	Milling Time, min
8mc-0s	Sigma-Aldrich 25 μm	15	60
8mc-1s	Sigma-Aldrich 25 μm	24	60
8mc-2s	Sigma-Aldrich 25 μm	24	120
8mc-0p	Pre-milled CuO	15	60
8mc-1p	Pre-milled CuO	24	60
8mc-1a	nano-CuO AlfaAesar	24	60
8mc-3a	nano-CuO AlfaAesar	48	60
10mc-1s	Sigma-Aldrich 25 μm	24	60
12mc-1s	Sigma-Aldrich 25 μm	24	60

Characterization of Nanocomposite Powders

Particle size distributions (PSD) for the prepared nanocomposite powders were measured using a Coulter LS230 particle size analyzer. The powders were also characterized using scanning electron microscopy (SEM). For SEM analyses, a LEO 1530 Field Emission Scanning Electron Microscope operating at 10 kV was employed. The particles were embedded in epoxy, cross-sectioned, and polished for the SEM analyses. Back-scattered electrons were used to obtain images showing the phase contrast between the materials.

X-ray diffraction was used to determine phase composition for each sample. The XRD was performed on a Philips X'pert MRD powder diffractometer operating at 45 kV and 40 mA using Cu Kα radiation (λ = 1.5438 Å). Scan angle was in the range of 30-50 degrees for as milled powders.

The powders were ignited in air on the surface of a metal wire which was electrically heated. Details of this experimental technique are available elsewhere[16]. A 0.5 mm diameter Chromel wire mounted in the experimental chamber was used as the filament. A small amount of powder was mixed with hexane to form slurry and this slurry was applied to the filament using a soft brush. Ignition was observed using a silicon photodiode (DET110 by Thorlabs, Inc.) equipped with an iris aimed at the powder coating at a distance of 4-5 cm. The temperature history of the heated filament was measured using a high-speed infrared pyrometer (DP1581 by Omega Engineering, Inc.). The pyrometer was focused on the uncoated surface of the filament adjacent to the powder coating. The experiments were performed with varied filament heating rates and in the air and argon environments.

Finally, the combustion behavior of the produced powders was evaluated using a constant volume explosion (CVE) experimental technique described in detail elsewhere[17-21]. The powders were introduced into a nearly spherical vessel, 9.2 l vessel as an aerosol and ignited at the center of the vessel. The value and the rate of pressure rise produced by combustion were compared for different powders. The CVE experiments were performed in air.

Results

Particle Size Distributions

Figure 2 shows particle size distributions of the 8Al+3CuO nanocomposite powders prepared with different milling conditions. It is clearly seen that for most samples, the size distributions extend to larger particle sizes, as compared to the size distributions of the starting powders shown in Fig. 1. The formation of larger particles can be seen as an indicator of the formation of composite materials. From that standpoint, the size distribution for the 8mc-1p sample is somewhat alarming. It ends sharply at around 40 - 50 μm, as does the original Al powder size distribution. This sample used pre-milled CuO powder and the shape of the size distribution function for the product powder indicates that formation of composite particles was limited. For a sample with the same starting materials but milled with less PCA (8mc-0p) the formation of larger particles is already observed as for other samples.

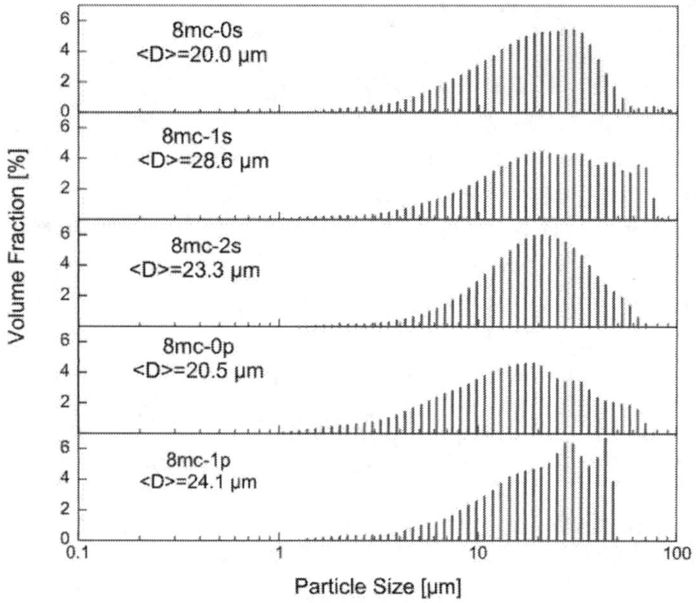

Figure 2. PSD of 8Al3CuO for different milling conditions

A longer milling time (sample 8mc-2s) clearly results in the formation of a PSD function that approaches the lognormal distribution. This shape of PSD is normally expected to be maintained at even longer milling times. The PSD of the powder milled for a shorter time but with less PCA (8mc-0s) is also fairly close to the lognormal shape, indicating a greater rate of particle size modification as compared to the similar conditions, but with more PCA (8mc-1s).

In general, the differences in the PSD shapes corresponding to different milling conditions are minor.

Nanocomposite Morphology

Typical shapes of the cross-sectioned 8Al3+CuO nanocomposite particles are shown in a low-magnification SEM image in Fig. 3. The observed particle sizes are well correlated with the measured PSD shown in Fig. 2. The particle shapes are irregular, as expected for the composites formed by mechanical milling. Some of the particles clearly formed by pressing powder between the elements of milling media, e.g., a large, flat-ended particles shown in Fig. 3. The particle interiors are rather homogeneous at the magnification shown in Fig. 3, but the nanocomposite morphology becomes apparent at a higher magnification.

A characteristic high magnification image is shown in Fig. 4. A few very dark spots represent voids filled with epoxy while nearly the entire image represents a cross-section of only one nanocomposite particle. The light colored areas indicate the denser copper oxide inclusions and the gray areas show aluminum matrix. The characteristic dimensions of the copper oxide inclusions are typically well less than 1 μm. Several images, similar to that shown in Fig. 4, were processed to determine the average Feret diameter of the oxide particles. It was found to vary between 100 and 200 nm for different samples. Copper oxide is nearly uniformly mixed into aluminum and very few voids are observed indicating that the density of the nanocomposite is close to its theoretical maximum density.

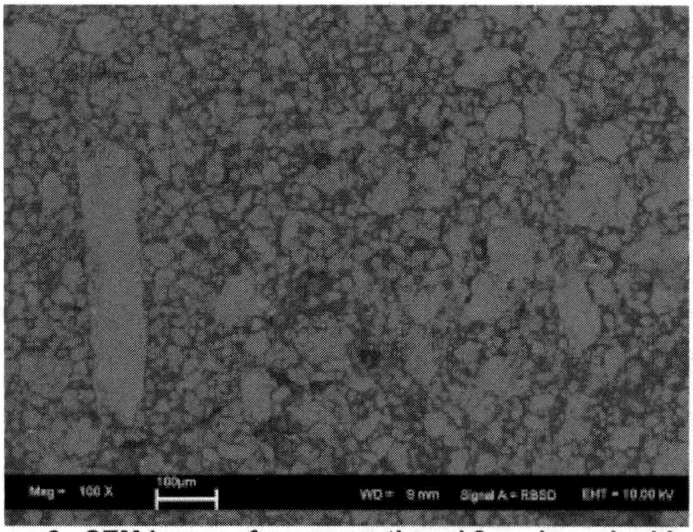

Figure 3. SEM image of cross-sectioned 8mc-1p embedded in epoxy

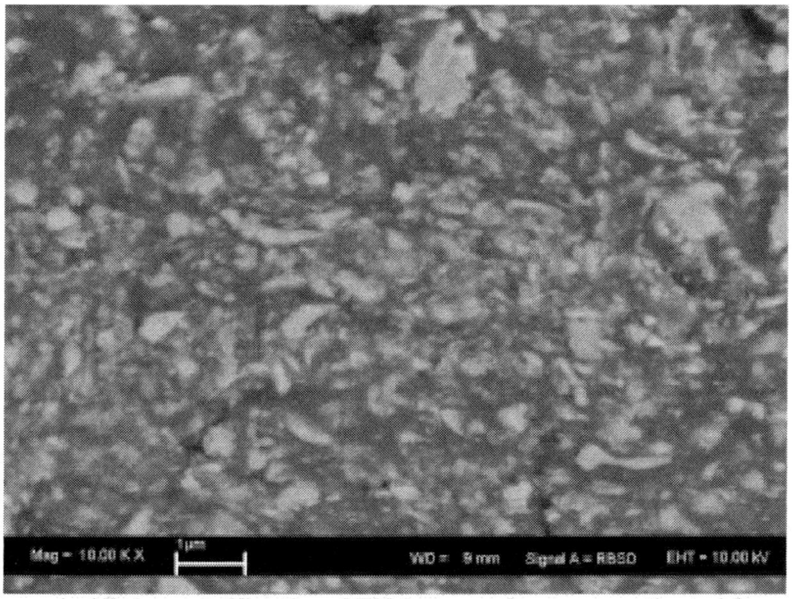

Figure 4. High magnification SEM image of cross-sectioned 8mc-1p

Generally, the shapes and sizes of the oxide inclusions were not observed to change significantly for the samples with the same compositions and prepared with different milling conditions or compositions. The only significantly altered morphology occurred in the nanocomposites that were made using nano-size CuO. Figure 5 shows cross section of 8mc-1a sample prepared using the nano-sized CuO. The copper oxide particles seen to form very thin layers, agglomerated into relatively large formations. At the same time, relatively large aluminum flakes appear to be pressed together with the layers of CuO but not include any CuO in the aluminum interior. It is hypothesized that agglomeration of the nano-sized oxide particles during milling prevented them from a more uniform mixing in aluminum. Thus, the reactivity of such powder is expected to be negatively affected so that the benefits of using the initially nano-sized CuO powder are limited.

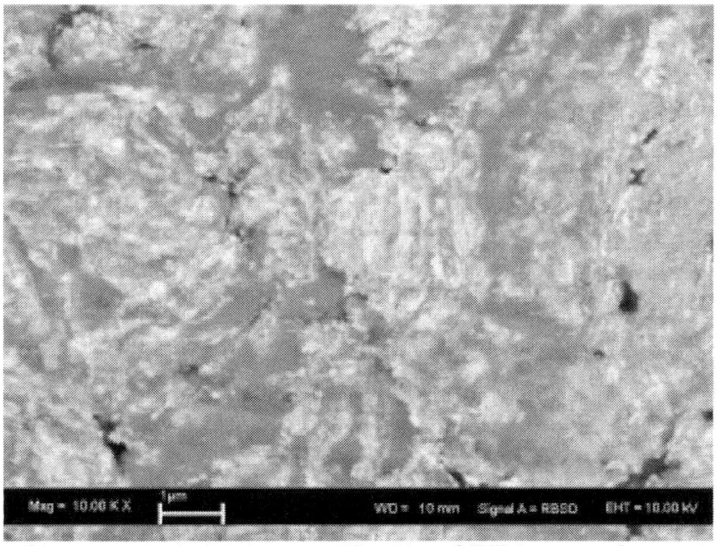

Figure 5. Characteristic high magnification SEM image of sample 8mc-0a

Material Phase Composition

X-ray diffraction (XRD) was used to find any evidence of partial reaction during the milling process. As a result of such reaction, formation of Cu_2O and Cu is expected that should be possible to detect from the XRD patterns. In addition, formation of Al-Cu alloys is possible if free Cu is produced as a result of partial reaction. Three patterns collected for the sample prepared using the same starting powders but with varying milling conditions are

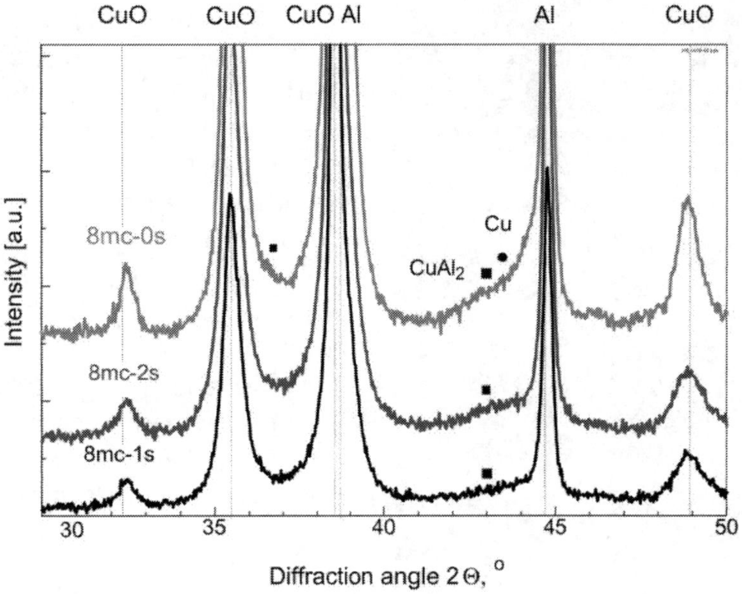

Figure 6. XRD traces of several 8Al+3CuO nanocomposite powders prepared with different milling conditions

shown in Fig. 6. The patterns are generally similar, with main peaks of Al and CuO as expected. The scale in Fig. 6 is selected to expand the trace background where signatures from the minor phases can be observed. The background is most prominent for the 8mc-0s sample prepared with smaller amount of PCA. Based on the locations of the raised background, signatures of Cu and $CuAl_2$ can be detected. Noticeable background indicative of presence of $CuAl_2$ is also visible for sample 8mc-2s, milled for a longer time. Based on this comparison, sample 8mc-1s appears to contain the smallest amount of the partially reacted material and thus it is expected to be most reactive in the combustion tests.

A similar comparison of the XRD traces is shown in Fig. 7, however, the milling conditions were maintained the same while the starting CuO powder varied. The sample prepared using regular micron-sized powder appears to contain the smallest amount of the products of partial reaction.

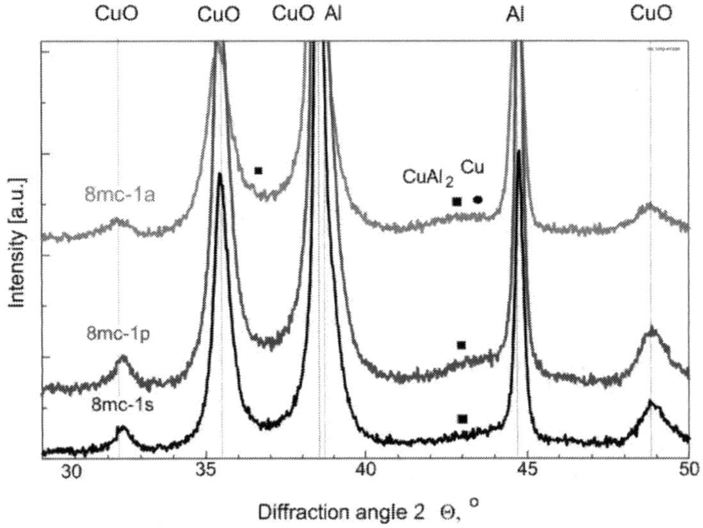

Figure 7. XRD traces of several 8Al+3CuO nanocomposite powders prepared with different starting CuO powders and the same milling conditions

Filament Ignition

A summary of results of the ignition temperature measurements is presented in Fig. 8. Despite the significant variation in the heating rate from about 2500 to almost 150000 K/s, the ignition temperatures measured in air remain nearly constant for all the powders tested. The ignition temperature is also not affected significantly by the milling conditions or specific composition of the metal-rich nanocomposite *x*Al+3Cuo powders with *x* varied from 8 to 12.

The experiments in argon were performed only with one powder and the measured ignition temperature (shown in Fig. 8) was found to be a bit lower than that in air. Once again, no effect of heating rate was observed.

Note that the measured temperature is that of a filament adjacent to the powder coated on its surface. The temperature of the igniting powder is typically somewhat higher, as explained in detail elsewhere[22]. Thus, the difference between the ignition temperatures measured in argon and air is most likely explained by a lower thermal conductivity of argon which results in smaller heat losses by convection, and in smaller temperature difference between the filament and the powder[22].

The results of measurements presented in Fig. 8 can be compared to the ignition temperatures measured using the same experimental technique for stoichiometric nanocomposite powders 2Al+3CuO prepared by ARM[13]. Several powders were prepared in Ref. 13 and the powder with the optimized milling conditions was observed to ignite in air at temperatures 650 – 750 K, lower than the ignition temperatures shown in Fig. 8. Furthermore, the ignition temperature for the optimized stoichiometric powder was affected by the heatingrate. However, other stoichiometric powders for which the milling conditions were not optimized, were observed to ignite in air at temperatures varied around 850 – 900 K which did not appear to be affected by the heating rate. These latter results directly compared to the current measurements for the metal-rich nanocomposite powders.

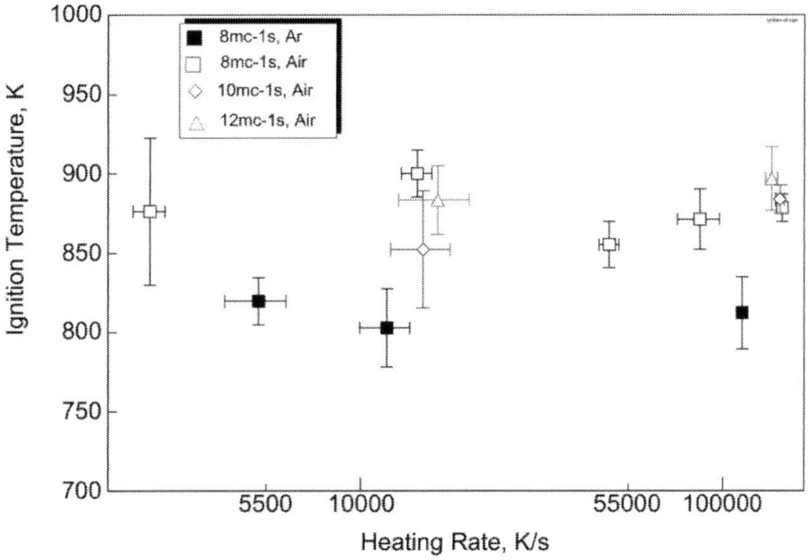

Figure 8. Ignition temperatures as a function of the heating rate measured fordifferent prepared samples in air and argon.

The independence of the measured ignition temperature on the heating rate indicates that ignition could be triggered by an invariant phase change occurring at a specific temperature. The specific ignition mechanism needs to be investigated further.

Constant Volume Explosion Experiments

The results of the constant volume explosion experiments are summarized in Fig. 9. For all tests, the volume of the loaded powder remained constant and equal to that of the volume of an aluminum powder load necessary for a stoichiometric flame in the constant volume experiment (2.89 g of Al). The load for 8Al+3CuO composition was 4.1 g and the loads for 10Al+3CuO and 12Al+3CuO were respectively 3.95 and 3.80 g.

The top plot shows the ratios of maximum explosion pressure, P_{max}, over the initial pressure in the vessel prior to ignition, P_{ini}. This ratio is generally proportional to the flame temperature. Comparison of the values of P_{max}/P_{ini} for the samples with the same bulk composition gives one a measure of the combustion efficiency. The results suggest that sample 8mc-1s demonstrates the highest efficiency among all similar samples. The plot at the bottom shows the maximum rate of pressure rise obtained from the CVE experiments. While no direct correlation between the P_{max}/P_{ini} ratio and dP/dt_{max} can be observed, the same sample that exhibited the highest P_{max}/P_{ini} ratio also shows the highest dP/dt_{max}. This same sample was also shown to contain the smallest amounts of partial reaction products (cf. Figs. 6, 7). Therefore, based on the CVE experiments and based on the XRD analysis of the materials phase compositions, the milling conditions corresponding to the sample 8mc-1s are selected as optimized. Samples with increased concentration of Al (10mc-1s and 12mc-1s) were prepared with the same milling conditions and the results of CVE experiments for those samples are also shown in Fig. 9. As expected, both the rate of pressure rise and the maximum pressure decrease as the amount of Al in the metal-rich nanocomposite powder increases.

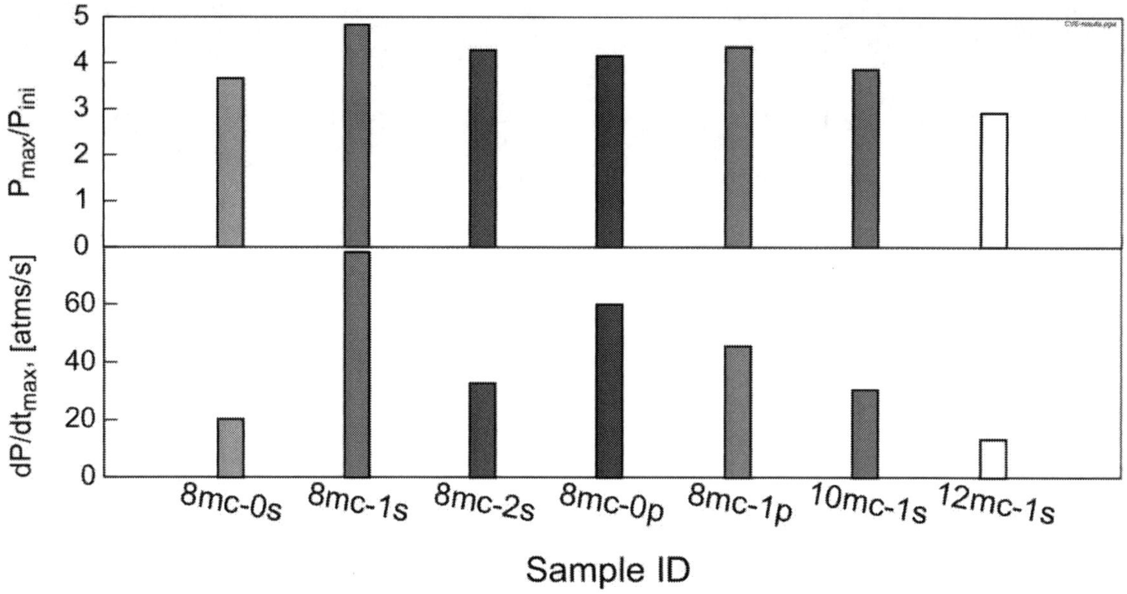

Figure 9. Results of constant volume explosion experiments with different nanocomposite powders

Conclusion

Different metal-rich nanocomposites in the thermite system Al-CuO were prepared by arrested reactive milling, characterized, and compared to one another. Based on the x-ray diffraction phase analysis and constant volume explosion experiment results, the optimized milling conditions for preparation of such powders were selected. It was also shown that the powder particle size distributions and ignition temperature are less sensitive to the specific milling conditions than the results of constant volume explosion experiments. Ignition temperature of metal-rich xAl+3CuO nanocomposites with $x \geq 8$ does not depend on the heating rate indicating that the ignition is likely triggered by an invariant phase transition. Using nano-sized or finely divided micron-sized CuO powder as a starting material did not result in a more refined or more reactive nanocomposite powder. Instead, the concentrations of undesirable products of partial reaction (Cu, $CuAl_2$ alloy) during the ball milling increased, reducing the quality of the produced nanocomposite powder.

Acknowledgements

This work was supported in parts by TACOM-ARDEC (Picatinny) and by the US Defense Threat Reduction Agency.

References

1. Palaszewski, B., and Zakany, J.S. (1995), "Metallized Gelled Propellants: Oxygen/RP-1/Aluminum Rocket Combustion Experiments"AIAA-95-2435
2. Gany, A., Netzer, D.W. (1985), "Fuel Performance Evaluation for the Solid-Fueled Ramjet" International Journal of Turbo and Jet Engines 2: 157–168
3. Beckstead, M.W. (2006) "Recent progress in modeling solid propellant combustion" Combustion, Explosion and Shock Waves 42 (6), pp. 623-641
4. Moorer, D.S., Son, S.F., Asay, B.W. (2004), "Time-resolved spectral emission of deflagrating nano-Al and nano-MoO3 metastable interstitial composites" Propellants, Explosives, Pyrotechnics 29 (2), pp. 106-111
5. Bockmon, B.S., Pantoya, M.L., Son, S.F., Assay, B.W., Mang, J.T. (2005), "Combustion velocities and propagation mechanisms of metastable interstitial composites" Journal of Applied Physics 98 (6), pp. 1-7
6. Plantier, K.B., Pantoya, M.L., Gash, A.E. (2005), "Combustion wave speeds of nanocomposite Al/Fe2O3: The effects of Fe2O3 particle synthesis technique" Combustion and Flame 140 (4), pp. 299-209
7. Prentice, D., Pantoya, M.L., Gash, A.E. (2006), "Combustion wave speeds of sol-gel-synthesized tungsten trioxide and nano-aluminum: The effect of impurities on flame propagation" Energy and Fuels 20 (6), pp. 2370-2376
8. Duckham, A., Brown, M., Besoin, E., Van Heerden, D., Knio, O.M., Weihs, T.P. (2004), "Metallic bonding of ceramic armor using reactive multilayer foils" Ceramic Engineering and Science Proceedings 25 (3), pp. 597-603
9. Mann, A.B., Gavens, A.J., Reiss, M.E., Van Heerden, D., Bao, G., Weihs, T.P. (1997), "Modeling and characterizing the propagating velocity of exothermic reactions in multilayer foils" Journal of Applied Physics 82 (3), pp. 1178-1188
10. Schoenitz, M., Ward, T.S., Dreizin, E.L. (2005) , "Fully Dense Nano-Composite Energetic Powders Prepared By Arrested Reactive Milling" Proceedings of The Combustion Institute 30 pp. 2071-2078
11. Dreizin, E.L., Schoenitz, M., US Patent Application for "Nano-composite energetic powders prepared by arrested reactive milling" 20060053970, filed 12 Nov. 2004.
12. Umbrajkar, S.M., Schoenitz, M., Dreizin, E.L. (2006), "Control of Structural Refinement and Composition in Al-MoO3 Nanocomposites Prepared by Arrested Reactive Milling" Propellants Explosives and Pyrotechnics, 31(5), pp. 382-389
13. Umbrajkar, S.M., Schoenitz, M., Dreizin E.L. (2006), "Exothermic reactions in Al-CuO nanocomposites" Thermochimica Acta 451, pp., 34-43
14. Takacs, L. (2002) "Self-sustaining reactions induced by ball milling" Progress in Materials Science 47 (4), pp. 355-414
15. Ward, T.S., Chen, W., Schoenitz, M., Dave, R.N., Dreizin, E.L. (2005), "A study of mechanical alloying processes using reactive milling and discrete element modeling" Acta Materialia 53 (10), pp. 2909-2918
16. Y.L. Shoshin, M.A. Trunov, M. Schoenitz, X. Zhu, E.L. Dreizin, (2006), "Ignition of aluminum-rich Al-Ti mechanical alloys in air", Combustion and Flame 144 688-697

17. Hertzberg, M., Cashdollar, K.L., and Opferman, J.J. (1979), “The Flammability of Coal Dust-Air Mixtures,” U.S. Bureau of Mines, Rept. RI8360,
18. Cashdollar, K.L., and Hertzberg, M., “20-L Explosibility Test Chamber for Dusts and Gases,” Review of Scientific Instruments, Vol. 56, No.4, April 1985, pp. 596-602
19. Cashdollar, K.L., and Chatrathi, K., “Minimum Explosible Dust Concentrations Measured in 20-L and 1m3 Chambers.” Combustion Science and Technology, Vol. 87, 1992, p. 157
20. Cashdollar, K.L., “Flammability of Metals and Other Elemental Dust Clouds,” Process Safety Progress, Vol. 13, 1994 pp. 139-145
21. Eapen, B.Z., Hoffman, V.K., Schoenitz, M., Dreizin, E.L., “Combustion of Aerosolized Spherical Aluminum Powders and Flakes in Air”, Combustion Science and Technology 176 (7), pp. 1055-1069
22. Umbrajkar, S., Mikhaylo, T.A., Schoenitz, M., Dreizin, E.L., (2007), “Arrested Reactive Milling Synthesis and Characterization of Sodium-Nitrate Based Reactive Composites

MODELING OF REDUCTION-OXIDATION REACTION IN Al-MoO_3 NANOCOMPOSITE POWDERS

Mikhaylo A. Trunov, Mirko Schoenitz, and Edward L. Dreizin
New Jersey Institute of Technology, Newark, NJ, USA

Introduction

Recently, reactive Al-MoO_3 nanocomposites were prepared by arrested reactive milling (ARM) for potential applications in various energetic formulations, including pyrotechnics, explosives, and propellants [1-2]. Other manufacturing approaches, including nanopowders mixing and sol-gel processing have also been recently explored to produce similar nanocomposite thermite compositions [3-5].

The reactions in nanocomposite powders produced by ARM were investigated by differential scanning calorimetry (DSC). DSC measurements were correlated with the powder ignition kinetics quantified from heated filament ignition experiments [6]. It was observed that many overlapping processes control the reaction rate and ignition so that a simplified model based on several independent processes is inadequate for practically useful predictions [6].

The objective of this work is to develop a model of oxidation processes in the Al-MoO_3 nanocomposites that enables one to describe the experimental data and account for differences in ignition kinetics as a function of the scale and morphology (e.g., spherical *vs.* flat) of nano-domains.

Model Development Roadmap

The model of the reduction-oxidation reaction in Al-MoO_3 nanocomposites is built upon the approach used previously to describe oxidation and ignition of aluminum particles in air [7]. The model accounts for simultaneous growth of different aluminum oxide polymorphs and for polymorphic phase transformations occurring within the aluminum oxide layer. In preliminary estimates, the oxygen generation term caused by decomposition of MoO_3 was simply added to the model of aluminum oxidation controlled by diffusion through various alumina polymorphs. It was observed that the phase transformations in alumina scale could not be used to adequately interpret the DSC measurements. Specifically, the shapes of the DSC peaks and the number of the peaks observed could not be reproduced in that simplified model. This negative conclusion was reached despite the model's capability to account for the size distribution and morphology of nano-domains. The model was, therefore, developed further to consider phase evolution in the molybdenum oxide occurring upon its heating.

Presently, the model accounts for reduction reactions in following molybdenum oxide phases: MoO_3, Mo_9O_{26}, Mo_4O_{11}, and MoO_2. The model monitors the balance between the phases listed above within the nanocomposite particles. Oxygen atoms produced as a result of reduction of various molybdenum oxides diffuse to aluminum and react with it forming aluminum oxide. As described in the earlier aluminum oxidation models, the growing aluminum oxide layer exists as a combination of different Al_2O_3 polymorphs, depending on its temperature and thickness.

The assumptions regarding the shapes and sizes of the multiple oxide layers present within the composite particles were made based on the analysis of the scanning electron microscope (SEM) images of such particles. Compared to the model of Al oxidation in air [7], the Al-MoO_3 nanocomposite reaction model includes new assumptions about diffusion characteristics of the growing alumina polymorph films sandwiched between molybdenum oxides and aluminum. The specific kinetic parameters are determined from processing the scanning calorimetry experiments similarly to our earlier work [7].

Predictions of the current model are compared with experimental DSC data. The experimental DSC peaks are interpreted as a combination of processes in aluminum and molybdenum oxide scales. The approach is promising and is capable of explaining multiple experimental DSC peaks and the observed peak shapes. Further work will focus on identification and more detailed description of specific processes included in the model.

Experimental Background

DSC experiments considered in this work were reported elsewhere [6]. Only a short description of the experimental background is provided.

Materials

Nanocomposite thermite powders of $2Al+MoO_3$ composition were prepared by ARM [2] in 50-ml steel vials using a SPEX 8000D shaker mill. The component powders were reagent-grade materials of –325 mesh nominal size (Al, 99.8 %, by Atlantic Equipment Engineers; MoO_3, 99.5 %, by Alfa Aesar). Steel balls of 5 mm diameter were used as milling media, the ratio of ball to sample mass was equal to 5. Milling for 60 minutes was carried out with 8 mL of hexane as a process control agent. The resulting nanocomposite powders were characterized by scanning electron microscopy. Figure 1 shows a representative backscattered electron image of an Al-MoO_3 nanothermite. The sample was embedded into epoxy and cross-sectioned. The phase contrast within the particles shows the formation of fully dense composites of Al and MoO_3 mixed on the scale of about 100 nm.

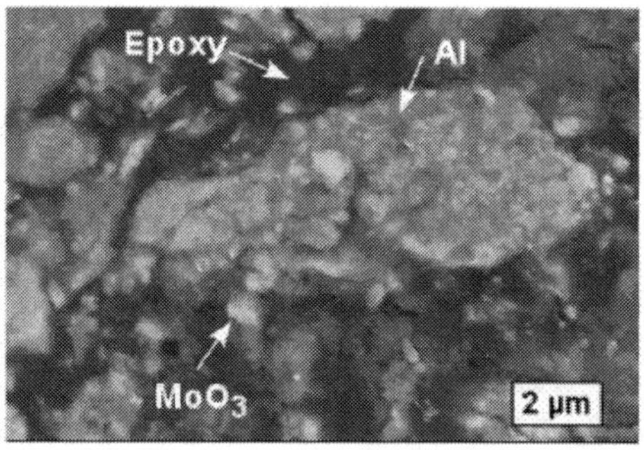

Figure 1. Backscattered electron image of an Al-MoO_3 nanothermite prepared by arrested reactive milling. Bright areas correspond to MoO_3, medium gray areas are Al.

Differential Scanning Calorimetry

Reactions occurring in the nanothermites during slow heating were characterized by DSC using a Netzsch STA409PC thermal analyzer. Platinum sample pans covered with lids were used, and the furnace was flushed with argon at approximately 10 ml/min. The temperature range was limited to a maximum of 1013 K (740 °C) in order to avoid melting and decomposition of MoO_3 above 1068 K (795 °C). After the initial heating cycle, the same sample was held at 1013 K for 30 min, cooled to room temperature at 40 K/min, and then heated again at the same heating rate in order to obtain a baseline for the measurement. An example of DSC traces obtained at 1 K/min heating rate is presented in Figure 2.

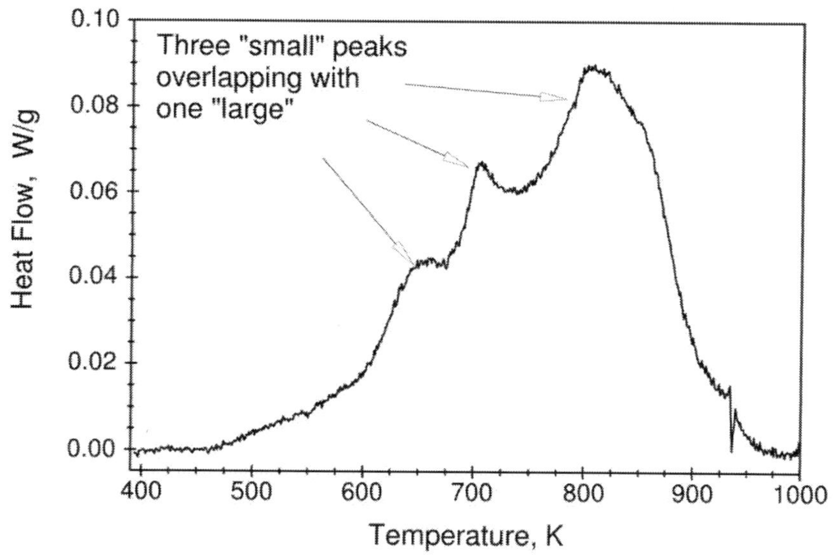

Figure 2. Experimental DSC trace of reaction in Al-MoO3 system obtained at slow heating rate of 1 K/min.

General Model Description

Structure of the nanocomposite materials

Electron microscopy images of the cross-sections of nanocomposite particles (see Figure 1) indicate that the inclusions of molybdenum oxide have characteristic dimensions of the order of several hundred nanometers. Such inclusions are fully embedded into aluminum matrix. The proposed model describes each inclusion as a single sphere of molybdenum oxide surrounded by a spherical layer of aluminum oxide, which is, in turn, surrounded by a spherical layer of aluminum. That structure forms an elemental domain. All elemental domains are assumed to be equivalent. Therefore, the redox reaction within a micron-sized particle of the nanocomposite material is modeled as a sum of reactions in multiple elemental domains. Note that different types of molybdenum oxides can co-exist simultaneously within the central sphere. The specific composition of the central sphere is determined using a model for the MoO_3 reduction, as described below. At the same time, the aluminum oxide layer can include two sub-layers with different alumina polymorphs.

In this presentation only a spherical symmetry case is presented, while the model can also easily describe a flat-layered aluminum/molybdenum oxide structure, e.g., as produced by magnetron sputtering. Both idealized geometric structures (spherical and flat-layered) are approximations for the real structure of nanocomposites produced by ARM.

Selection of the material structure affects the shapes of the predicted DSC peaks. The number of peaks and their characteristic temperatures depend only on the number and type of the specific reactions included in the model.

General Equation

The oxygen diffusion coefficients in molybdenum oxides are significantly higher than in different alumina polymorphs [8-9]. Therefore the concentration of oxygen within the molybdenum core (central sphere) was assumed to be uniform.

The general model formulation can be obtained considering mass conservation of oxygen whereas oxygen can be produced as a result of molybdenum oxide reduction and consumed as a result of aluminum oxidation:

$$V_{core}\dot{C}^o + \dot{V}_{core}C^o = -\sum_j R_j^o \dot{m}_j - \sum_i \mu_i^o \dot{m}_i \tag{1}$$

where V_{core} is the volume of the central molybdenum oxide sphere, C^o is the oxygen concentration in the central sphere, indices j and i indicate summation over all available types of molybdenum oxides and alumina polymorphs, respectively, R_j^o is the oxygen mass produced as result of reduction of a unit mass of a specific molybdenum oxide j, μ_i^o is the oxygen mass needed for formation of a unit mass of an aluminum polymorph i), m_i and m_j are the masses of alumina polymorph i and molybdenum oxide, j, respectively, and dots signify time derivatives (and show respective rates of change).

The rates of heterogeneous aluminum oxidation are introduced as in the earlier model describing aluminum oxidation in air [7]. However, now the effect of oxygen concentration C^o on the oxidation rates becomes significant. The rate of change for the oxygen concentration as a function of temperature is calculated from the reported P-T diagram for the Mo-O binary system (see below). Therefore, Equation 1 combined with a set of additional equations considering the system structure in order to obtain the masses of specific species available allows one to determine the rate of molybdenum oxide reduction. As noted above, in the preliminary calculations, only one molybdenum oxide, MoO_3 was considered. However, it was found out that inclusion of several different molybdenum oxide types is necessary to describe the experimental DSC curves.

Aluminum Oxidation Description

The mechanism of aluminum oxidation is fully adopted from the previous studies [7, 10], where it was discussed in detail. Here, it is important to note that aluminum oxidation is described as a combination of simultaneous growth of alumina polymorphs and phase transformations among them leading to changes in the crystal structure and density of the oxide scale. The oxidation rates are diffusion controlled. Phase transformations lead to formation of new polymorphs and, respectively, to changes in kinetic parameters of diffusion and oxidation. Values of kinetic parameters for both direct oxidative growth of alumina and for the occurring polymorphic transformations were determined from thermal analysis [7]. For modeling of oxidation reactions in nanocomposites, the activations energies found from the experiments on aluminum oxidation in air are expected to remain adequate. However, all the pre-exponents (in the Arrhenius type functions for diffusion coefficients) need to be adjusted. Currently, all the pre-exponents are considered as adjustable parameters.

Molybdenum Oxide Reduction Description

Equilibrium partial pressures for various types of gaseous oxides above solid molybdenum oxide surface are proportional to the exponent of inverse temperature [11]. In the model it was assumed that a similar Arrhenius type expression describes the concentration of oxygen within the central sphere:

$$C^o \sim \exp\left(-\frac{E_j^{eq}}{RT}\right). \tag{2}$$

The values of activation energies describing the reduction of specific types (j) of molybdenum oxide, E_j^{eq}, were derived from the reported equilibrium P-T diagram [11] for the Mo-O system.

Preliminary Results and Implications

In preliminary calculations, the types of introduced specific reactions were limited by those reported earlier to occur within the developing alumina scale [7]. These processes were expected to describe three "small" peaks in the experimental DSC signal (see Figure 1), while the availability of oxygen due to reduction of molybdenum oxide was expected to produce the broad (or "large") peak observed experimentally.

The results of calculations showed that the above approach does not allow one to match the calculated and experimental DSC traces. The main problem was observed to be caused by significant changes in the diffusivity of oxygen when transformations between alumina polymorphs occurred. The substantial changes (typically reductions) in diffusion coefficient caused by formation of the increasingly denser polymorphs were predicted to result in sharp drops in the DSC signal (because of the very low instant rates of oxidation). However, no such sharp drops in the signal were observed; instead, the typical DSC trace presented in Fig. 2 shows a continuously increasing baseline between the three "small" peaks.

Thus, it was suggested that a more sophisticated model is required that would consider a sequence of processes occurring during the molybdenum oxide decomposition including formation of different types of molybdenum oxides accompanied by staged release of oxygen. Such scenario should not result in a sharp reduction in the rate of aluminum oxidation, unlike the formation of higher density alumina polymorphs [7].

Model Development

Reduction Reaction Sequence

In the enhanced model, the reaction in the Al-MoO_3 nanocomposite is described as a sequence of several consecutive polymorphic transformations: $MoO_3 \rightarrow Mo_9O_{26} \rightarrow Mo_4O_{11} \rightarrow MoO_2$.

These transformations could be presented by a set of simplified equations (all equations are normalized by one molybdenum atom):

$$MoO_3 \rightarrow \frac{1}{9}Mo_9O_{26} + \frac{1}{9}O \tag{3}$$

$$\frac{1}{9}Mo_9O_{26} \rightarrow \frac{1}{4}Mo_4O_{11} + \frac{5}{36}O \tag{4}$$

$$\frac{1}{4}Mo_4O_{11} \rightarrow MoO_2 + \frac{3}{4}O \tag{5}$$

$$MoO_2 \rightarrow Mo + 2O \quad (6)$$

The first two transformations steps release nearly similar amounts of oxygen. The third step (Eq. (5)), decomposition of Mo_4O_{11}, releases about three times more oxygen. The majority of oxygen (2/3 of all the available oxygen in the initial MoO_3), is suggested to be released during the last step of MoO_2 reduction to metallic Mo.

Oxygen Concentration History: Qualitative Description

Calculations of the uniform oxygen concentration within the molybdenum oxide core based on the equilibrium phase diagram implies that at any given temperature, only two different molybdenum oxide types can coexist. One of these co-existing oxides, "parent" phase is reduced to the second, "child" phase. The model assumes that the oxygen concentration in the central sphere (or core) is always defined by the equilibrium partial pressure of oxygen for the "parent" phase. At the instant when the "parent" phase is fully reduced, the core includes only the "child" phase, while the instantaneous value of oxygen concentration remains defined by the "parent" phase that just disappeared. Immediately following that instant, the oxygen concentration starts to decrease due to the oxygen consumption in the aluminum oxidation reaction. For some time, the oxygen concentration can remain higher than that implied by the equilibrium partial pressure of oxygen for the "child" phase, but eventually it reaches that new equilibrium value. At that point, the old "child" phase becomes new "parent" and a new "child" phase is produced. Until that instant, however, the old "child" polymorph remains stable. The aluminum oxidation is still diffusion rate controlled but the concentration of oxygen within the core decreases, resulting in a gradual decrease on the overall rate of oxidation.

Current Results

The proposed model allows one to match the shapes of peaks in the modeled and experimental DSC signal. According to the model, the large peak (from 450 to 950 K) is due to the phase transformation of alumina from amorphous to gamma-polymorph. Initial shoulder of the large peak is due to an increase of diffusivity in amorphous alumina with increase in temperature. The decreasing part of the large peak is due to formation of the gamma alumina polymorph with a lower diffusivity coefficient. Thus, a substantial decrease in the DSC signal at about 950 K is not caused by oxygen deficit but is a result of the formation of protective gamma alumina film. The raising parts of the "small" peaks correspond to production of different molybdenum oxide phases, resulting in step-wise generation of oxygen as the temperature increases. The decreasing part of each "small" peak is descried by the reduction in the equilibrium oxygen concentration within the central sphere after the "parent" phase has been consumed while the "child" phase remained stable.

The model predicts that at 950 K the molybdenum oxide core is composed primarily of MoO_2, which needs to be verified experimentally.

Conclusions

The observed multiple experimental DSC peaks for the redox reaction in the Al-MoO_3 nanocomposite material cannot be explained only by phase transformations in the alumina scale, similarly to the approach successfully used to interpret aluminum oxidation in air [7].

It was proposed to further develop the reaction model for the Al-MoO_3 nanocomposite by inclusion of consecutive reduction reactions occurring in molybdenum oxide: $MoO_3 \rightarrow Mo_9O_{26} \rightarrow Mo_4O_{11} \rightarrow MoO_2$. The current model allows one to vary the amount of oxygen available for aluminum oxidation at different temperatures. The proposed approach allows reproducing the peak structure of the experimental DSC signals computationally. The model development continues with an effort to determine specific kinetic parameters of different reactions.

Acknowledgement

This work was supported by the US Defense Threat Reduction Agency.

References

1. Schoenitz, M., Ward, T. S., and Dreizin, E. L. (2005), "Fully Dense Nano-Composite Energetic Powders Prepared by Arrested Reactive Milling," *Proceedings of the Combustion Institute*, Vol. 30, pp. 2071-2078.
2. Dreizin, E. L., Schoenitz, M. (2006) "Nano-Composite Energetic Powders Prepared by Arrested Reactive Milling," *Patent Application 20060053970*, March.
3. Son, S.F., Asay, B.W., Busse, J.R., Jorgensen, B.S., Bockmon, B., Pantoya, M.L. (2001) "Reaction Propagation Physics of Al/MoO_3 Nanocomposite Thermites," *Proceedings of the International Pyrotechnics Seminar*, 28th, pp. 833-843.
4. Prakash, A., McCormick, A.V., Zachariah, M.R. (2004) "Aero-Sol-Gel Synthesis of Nanoporous Iron-Oxide Particles: A Potential Oxidizer for Nanoenergetic Materials," *Chemistry of Materials*, Vol. 16, No. 8, pp. 1466-1471.
5. Blobaum, K.J., Reiss, M. E., Plitzko Lawrence, J.M., Weihs, T. P. (2003) "Deposition and Characterization of a Self-Propagating CuO_x/Al Thermite Reaction in a Multilayer Foil Geometry," *Journal of Applied Physics*, Vol. 94, No. 5, pp. 2915-2922.
6. Schoenitz, M., Umbrajkar, S., Dreizin, E.L. (2007) "Kinetic Analysis of Thermite Reactions in Al-MoO_3 Nanocomposites," *Journal of Propulsion and Power*, Vol. 23, No. 4 , pp. 683-687.
7. Trunov, M.A., Schoenitz, M., Dreizin, E.L. (2006) "Effect of Polymorphic Phase Transformations in Alumina Layer on Ignition of Aluminum Particles," *Combustion Theory and Modeling*, Vol.10, No. 4, pp. 603-623.
8. Clair, T.P.St., Restad, J.M., Oyama, S.T. (1998) "Oxygen Diffusivity in MoO_3 as Determined by a Temperature Programmed Method," *Journal of Materials Research*, Vol. 13, No. 6, pp.1430-1433.
9. Riano, O.A, Wadsworth, J., Sherby, O.D. (2003) "Deformation of Fine-Grained Alumina by Grain Boundary Sliding Accommodated by Slip." *Acta Materialia*, Vol. 51, pp. 3617-3634.
10. Trunov, M.A., Schoenitz, M., Zhu, X., Dreizin, E.L. (2005) "Effect of Polymorphic Phase Transformations in Al_2O_3 Film on Oxidation Kinetics of Aluminum Powders," *Combustion and Flame*, Vol. 140, No. 4, pp. 310-318.
11. Levinskii, Yu.V. (1990) "P-T-X- Equilibrium Diagrams of Binary Metal Systems, Book 2," Moscow, USSR, Metallurgy Press.

Ignition Sensitivity of Nanoenergetics Produced by Various Processing Methods

Chris J. Bulian, Jan A. Puszynski, and Jacek J. Swiatkiewicz
Chemical and Biological Engineering Department
South Dakota School of Mines and Technology
Rapid City, SD 57701, USA

Abstract

Nanoenergetic materials are a class of materials that have been of much interest for over ten years now. Variation in methods of synthesis and processing have allowed for these materials to be tailored to many different applications (e.g., percussion primers, electric matches, low energy ignition devices, etc.). These applications require wide ranges of reactive energy outputs, reaction rates, and ignition sensitivities. Appropriate characterization methods have been developed for mechanical impact, thermal, and electrostatic discharge sensitivities. This study presents an explanation of ignition delay times of nanoenergetic mixtures during exposure to a laser pulse as a function of a sample density and laser power. It was shown that the porosity of the energetic material significantly affects the ignition delay time due to the dependence of thermal properties as the function of that parameter. The effect of porosity on ignition delay was also investigated using a mathematical modeling. The model verified that an increase in porosity significantly decreases ignition delay time due to a much lower thermal conductivity and heat capacity of the sample. Electrostatic discharge sensitivity of the nanoenergetic materials was also investigated. This study determined that individual particle size and its morphology (both fuel and oxidizer), degree of consolidation, and processing methods can all have a significant affect on the sensitivity of the nanoenergetic materials to ignition by ESD.

Introduction

Nanoenergetic materials, also referred to as nanothermites or metastable interstitial composites (MIC),[1-7] have become an area of growing interest and importance for the commercial and military as well as academic sectors. Generally comprised of a nanoscale fuel such as aluminum or zirconium and a nanoscale metal oxide, these nanoenergetic materials are often characterized by their very fast reaction rates and high heats of reaction.[8-11] Several variables have been investigated over the years to determine their effect of the reaction rates of these materials. Those variables include type of oxidizer, reactant particle size, processing methods, and initial sample density. While these variables have convoluted impact on the reaction rate and energy output of the nanothermites, another important factor, recently considered is ignition sensitivity to external stymuli. In order to properly characterize nanoenergetic materials and tune their reactivity and energy output to a desired level, safety and ignition sensitivity is of great concern. Despite high activation energies (> 150 kJ/mol), nanoenergetic materials are very susceptible to ignition by electrostatic discharge, impact, friction and thermal impulse due to small particle size and high specific surface area. Ultimately, any ignition is due to thermal effect of some sort but the mechanism by which the energetic material reaches its thermal threshold and the factors affecting it are different for each ignition stimulus. Calculated ignition energy thresholds, depending on type of stimulus, can vary from fractions of a microjoule to several millijoules. Factors affecting ignition sensitivity include particle size, specific surface area, initial sample density, sample configuration, and use of binder among others. All of these factors were considered and investigated in this study.

Previous studies have investigated the size, duration, and temperature of ESD spark kernels as a function of the discharge energy.[12] In the range of 0.03 to 1.0 mJ, testing showed that the gases present within the spark rose to temperatures of 850 to 1150 K within 3 μs from initial discharge and within volume in the range of 0.1 to 0.4 mm^3. These conditions should be taken into consideration when determining a nanoenergetic material's susceptibility to ESD ignition. Due to the small spark kernel volumes, only a small amount of energetic sample will actually interact with the discharge. The amount of material inside this volume, along with the individual particle size, specific surface area, and heat capacity must be considered. The objectives of this research study were twofold: i) investigation of the effect of particle size and morphology of reactants, mixing environment, and sample configuration on the susceptibility of nanothermites to ignition by electrostatic discharge and ii) experimental and theoretical studies of laser ignition of nanothermite systems.

Experimental

ESD Sensitivity Testing

Sensitivity to ignition by electrostatic discharge (ESD) was investigated using a Model 930D-FTS Electrostatic Discharge Test Firing Set from Electro-Tech Systems, Inc. Interchangeable capacitors with values of 0.1, 0.2, 0.5, 1, 2, 5, 10, and 20 nF were used along with a high voltage variable power supply with a voltage range of 0 – 5000 V to deliver a specific amount of discharge energy to the sample. Energy is calculated using the equation $E = \frac{1}{2} CV^2$. Additional resistances of 1.5 and 5 kΩ can also be added to change discharge time. A needle shape discharge electrode was used to allow for precise placement in or near a sample as well as easy replacement once the electrode tip has become degraded due to repeated discharge or ignition of energetic material. The distance between the discharge tip and the base electrode is set at the maximum distance at which discharge will still occur with the sample in place. This setup allows for maximum spark discharge kernel size. Powdered material was placed inside a teflon washer, having dimensions of 6.5 mm inside diameter and 1.5 mm thickness, fixed to a metal base electrode to allow for partial confinement of the powder, preventing scattering of material during spark discharge. When energetic material was tested in granular form, several granules were placed in the vicinity between the discharge tip and base electrode to assure that at least one of the granules would interact with the spark discharge kernel. Tests were repeated at least 20 times at each energy level to assure no ignition occurred in accordance with military testing standards.

Multiple nanothermite systems, including Al-Bi_2O_3, Al-Fe_2O_3, and Al-MoO_3 were tested for ESD sensitivity. Aluminum nanopowder with average particle size of 80 nm was used as the fuel in each nanothermite system. Each system was subjected to ESD spark at increasing voltages with a 100 pF capacitor and no additional resistance. Effect of oxidizer particle size is another factor that was considered when determining what affects a materials susceptibility to ignition by ESD. Nanopowders increase in specific surface area as the individual particles size decreases so both properties must be taken into affect when analyzing results. Three different sizes of bismuth oxide powders were used to illustrate this effect. The first powder has a specific surface area of 0.28 m^2/g and an average particle diameter of 2.4 μm from Clark Manufacturing LLC, the second powder has the SSA of 1.62 m^2/g and an average equivalent particle diameter of 416 nm from Accumet Materials, and the third has the SSA of 17 m^2/g and equivalent particle diameter of 40 nm from Nanophase Materials. Each oxide was combined with 80 nm aluminum powder at a wt% ratio of 14.5/85.5 fuel to oxidizer and ultrasonically mixed in isopropyl alcohol for ten minutes and air dried. After drying, the material was brushed out of a drying pan to render loose powder. Another method by which sensitivity can be affected without increasing individual particle size or decreasing specific surface area is to consolidate the energetic material into a pellet or granular form. Although such consolidation will still result in a material that has boundaries between individual particles and possibly require the use of binders, the overall number of contact points and total contact area between particles will allow for better dissipation of heat, helping to prevent the formation of hot spots. Small diameter granules (~2 mm diameter) of Al-Bi_2O_3 energetic material comprised of 80 nm aluminum from and bismuth oxide with the average particle size of 416 nm were produced by water-based processing technique.

Laser-pulse Igntion

A instrument was designed and built for measurement of ignition energy of superthermite samples irradiated by a laser pulse. A cw diode laser (LIMO Micro-optics and Laser Systems) powered by laser diode driver (model LDD50, LIMO) is operated in controlled pulse-length mode. The pulse-length can be regulated from 0.5 to 500 ms in this mode of operation. Accordingly, the diode laser delivers light pulses of comparable width and variable intensity, depending on the pre-set current at the laser diode driver. The pulse timing can be monitored on an oscilloscope using a synchronous TTL output from the driver. Optical pulses are carried away from the diode laser assembly by a multimode fiber (400 μm core diameter) to the collimating aspheric lens and send toward the sample through a limiting aperture. Event of the ignition is observed using an optical multimode fiber placed at the specimen site and leading intercepted optical signal to a photodiode. The photodiode signal was recorded as single sweep in a digital oscilloscope. Ignition time is clearly marked with a fast increase of the photodiode signal. This time was measured from the onset of the optical pulse or reference signal to the moment of ignition. Calibration of the equipment indicated a delay of 0.36 ms from the trigger of the oscilloscope to the moment the laser pulse falls incident on the sample. This delay was taken into consideration for all tests.

The nanothermite systems examined using laser-pulsed ignition were Al-Fe_2O_3, Al-MoO_3, and Al-Bi_2O_3. All three systems had been investigated using differential scanning calorimetry (DSC) method to determine kinetic parameters of pre-exponential factor and activation energy. Each of the three nanothermite systems was prepared for testing by ultrasonic mixing of 80 nm aluminum and Accumet bismuth oxide in isopropyl alcohol. The mixture was then air dried in a grounded metal pan and brushed out to render a loose powder form of the material. A small amount of nanothermite material was placed inside a small washer, having dimensions of 5.5 mm diameter and 1.3 mm thickness that is affixed to a thin plastic base. Laser ignition for each nanothermite mixture was measured by subjecting the material to a 1.7 W laser beam with a beam diameter of approximately 3.3 mm and recording the light trace from the photodiode.

Results

ESD Ignition

Loose powder ignition sensitivities of nanothermite mixtures were determined to be very low. Ignition thresholds for the Al-Fe_2O_3, Al-MoO_3, and Al-Bi_2O_3 were experimentally measured at 113 µJ, 50 µJ, and <1 µJ respectively. It was also determined that the increase in specific surface area, and subsequent decrease in average particle size of bismuth oxide, resulted in a decrease in the minimum ignition energy threshold of greater than one order of magnitude. Table 1 illustrates the effect of the bismuth oxide average particle size on minimum ESD ignition energy of the $Al-Bi_2O_3$ nanothermite in a loose powder form tested using zero added resistance.

Table 1. Effect of the bismuth oxide average particle size on minimum ESD ignition energy of the $Al-Bi_2O_3$ nanothermite.

SSA [m^2/g]	Average size [nm]	ESD sensitivity [µJ]
0.28	2400	2.0
1.62	416	0.125
17.0	40	0.075

It was previously shown that the system comprised of 80 nm aluminum and Accumet bismuth oxide was very sensitive to ignition by electrostatic discharge, having ignition threshold energy of only 0.125 µJ. When the same system was tested, but with aluminum powder produced at SDSM&T used in place of the 80 nm aluminum, the ignition threshold energy was four orders of magnitude higher at ~2.0 mJ while still maintaining its reactivity and impact sensitivity. This indicates that larger individual particles are able to dissipate the absorbed energy better, resulting in reduced ESD sensitivity.

ESD testing of granular material again showed a significant decrease in the susceptibility of the material to ESD ignition, however, if the granules were crushed into small particles once again, the material regained its high sensitivity to ignition by ESD. Table 2 lists the ESD ignition thresholds of the different $Al-Bi_2O_3$ systems to illustrate the effect of individual reactant particle size, as well as consolidation of the energetic material.

Table 2: ESD ignition thresholds for various $Al-Bi_2O_3$ systems.

Sample Configuration	Aluminum Size (nm)	Bismuth Oxide Size (nm)	Ignition Threshold Energy (µJ)
Loose powder	80	416	0.125
Loose powder	80	2400	2
Loose powder	Flake	416	2000
Granular	80	416	1500-2000

Typically, it is desirable to have energetic materials with ESD sensitivities in the range of several millijoules to prevent accidental ignition.

Laser Ignition

Figure 1 shows the constant laser signal (first sharp impulse) and sharp impulse from the photodiode for Al-Fe2O3 in both loose powder form and granular form.

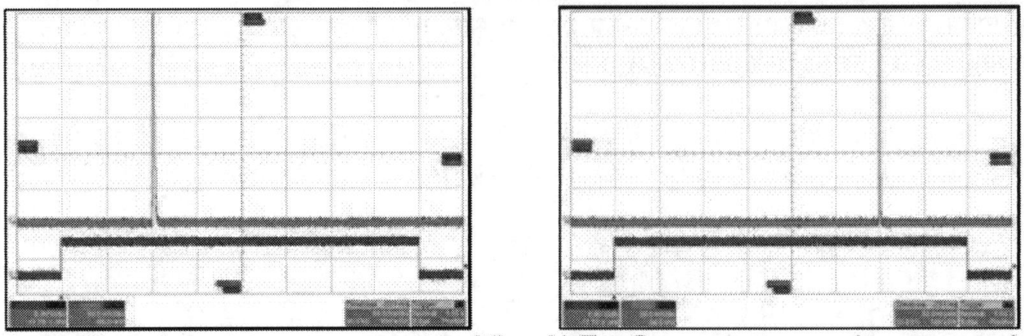

Figure 1: Ignition delay measurements of the Al-Fe_2O_3 system as a loose powder (left) and in granulated form (right) using a 1.7 W laser pulse. Time scale for the plots is 0.05 s per grid mark.

As with the ESD testing results, consolidation of the nanoenergetic material into granular form has a significant effect on the ignition delay and subsequently the amount of energy that needs to be absorbed for ignition to occur. In loose powder form, this ignition delay is 0.1 seconds, however, this value increases to 0.3 seconds when a granule of approximately 75% porosity is tested.

Conclusions

This research effort has concluded that sensitivity to ignition by electrostatic discharge is greatly affected by particle size, thermal properties of the nanothermite material, processing method and degree of consolidation. Initial laser ignition experiments confirm that consolidation has a significant effect on the amount of energy necessary to be absorbed in order for ignition to occur.

References

[1]Martin, J. A., Murray, A. S., Hollander, M. G., and Kung, H., "Metastable Interstitial Composites," *Warhead Technology*, Vol. 1, June 30, 1997, pp. 141-153.

[2]Son, S. F., Busse, J. R., Asay, B. W., Peterson, P. D., Mang, J. T., Bockmon, B. and Pantoya, M. L., "Propagation Studies of Metastable Intermolecular Composites (MIC)," *Proceedings of the 29th International Pyrotechnics Seminar*, Westminster, CO, July 2002, pp. 203-212.

[3]Bulian, C. J., Kerr, T. T., and Puszynski, J. A. "Ignition Studies of Aluminum and Metal Oxide Nanopowders," *Proceedings of the 31st International Pyrotechnics Seminar*, Ft. Collins, CO, July 2004, pp. 327-338.

[4]Son, S. F., Hiskey, H.L., Asay, B. W., Busse, J. R., Jorgensen, B. S., Bockmon, B., and Pantoya, M., "Reaction Propagation Physics of Al/MoO_3 Nanocomposite Thermites," The International Pyrotechnics Society, *The Twenty-Eighth International Pyrotechnics Seminar*, Adelaide, Australia, November 4-9, 2001, pp. 833-842.

[5]Somasundaram, V., Bulian, C. J., and Puszynski, J. A., "Investigation of Al-CuO Nanocomposite Reacting Systems," *Proceedings of EUROPYRO 2003*, Vol 2, 2003, pp. 536-541.

[6]Perry, W. L., Smith, B. L., Bulian, C. J., Busse, J. R., Macomber, C. S., Dye, R. C., and Son, S. F., "Nano-scale Tungsten Oxides for Metastable Intermolecular Composites," *Propellants, Explosves, Pyrotechnics*, Vol. 29, No. 2, 2004, pp. 106-111.

[7]Pusznski, J.A., Bulian, C.J., and Swiatkiewicz, J.J., "Processing and Ignition Characteristics of Aluminum-Bismuth Trioxide Nanothermite System," *Journal of Propulsion and Power.*

[8]Dorsett, H. E., Brousseau, P., and Cliff, H. E., "The Influence of Ultrafine Aluminum upon Explosives Detonation," *Proceedings of the 28th International Pyrotechnics Symposium*, Nov. 2001, pp. 239-244.

[9]Pivkina, A., Frolov, Y., Zavyalov, S., Ivanov, D., Schoonman, J., Streletskii, A., and Butyagin, P., "Ultra-Fine and Nanosized Oxidizers and Metals: Thermal Ignition Properties and Use in High Energetic Compositions," *Proceedings of the 31st International Pyrotechnics Seminar*, Ft. Collins, CO, 2004, pp. 285-298.

[10]Miziolek, A. W., "Nanoenergetics: An Emerging Technology Area of National Importance," *Amptiac Quarterly*, Vol. 6, No.1, Spring 2002, pp. 43-48.

[11]Mench, M. M., Kuo, K. K., Yeh C. L., and Lu, Y. C., "Comparison of Thermal Behavior of Regular and Ultra-Fine Aluminum Powders (Alex) Made From Plasma Explosion Process," *Combust. Sci. Technol.*, Vol 135, No. 1-6, 1998, pp. 269-292.

[12]Ono, R., Nifuku, M., Fujiwara, S., and Horiguchi, S., "Gas Temperature of Capacitance Spark Discharge in Air," *Journal of Applied Physics*, 97, 123307 (2005).

[13]Puszynski, J.A., Bulian, C.J., and Swiatkiewicz, J.J., "The Effect of Nanopowder Attributes on Reaction Mechanism and Ignition Sensitivity of Nanothemites," *Mater. Res. Soc. Symp. Proc.,* 2006, Volume 896, pp. 147-158.

ACKNOWLEDGEMENT

The authors gratefully acknowledge the financial support from the US Army Research Office, Grant # W911NF-05-1-0310 and the Naval Surface Warface Center, Indian Head, Contract # N00174-07-C-0013.

544f Characterization Of Shock Waves Produced By Nanothermites

Steve Apperson[1], Andrey Bezmelnitsyn[1], Rajagopalan Thiruvengadathan[1], Dan Tappmeyer[1], Zhen Chen[1], Keshab Gangopadhyay,[1], Shubhra Gangopadhyay[1], Paul Redner[2], Wendy Balas[2], Deepak Kapoor[2], and Steven Nicolich[2]. (1) University of Missouri Columbia, Columbia, MO 65211, (2) US Army ARDEC, Picatinny, NJ 07806

A bench-scale shock tube apparatus has been employed to measure the velocity and intensity of pressure waves produced by the combustion of nanothermite materials, synthesized in our laboratory. The combustion wave speed of these novel materials ranges from 1500 m/s to 2400 m/s for a combination of CuO nanorod and nano-Aluminum with optimized sizes and a proper equivalence ratio. The highest speed is achieved by the process of self-assembly between the fuel and oxidizer. The lowest speed corresponds to simple physical mixture of oxidizer and fuel. In both cases, we observed generation of shockwaves, the strongest one resulting in the self-assembled case. Small quantities of nanothermites (10-200mg) were loaded into a cylindrical volume, and placed against an air-filled tube. Pressure transducers were mounted along the length of the tube to record pressure profiles at various positions. Using time-of-arrival analysis the velocity of the pressure wave was calculated. Mach numbers ranging from 1.1-3.5 were measured for various compositions of CuO-Al nanothermite. This system has been used to study the effect of nanothermite mass on the shock waves and compare different variations of nanothermite composites. This technique allows rapid testing for tuning of nanothermite compositions for shock wave applications in the defense, energy and biomedical fields.

544g Reactivity of Conventional and Modified Nanothermites

Andrey Bezmelnitsyn[1], Rajagopalan Thiruvengadathan[1], Steve Apperson[1], Rajesh Shende[1], Keshab Gangopadhyay[1], **Shubhra Gangopadhyay**[1], Paul Redner[2], Wendy Balas[2], Deepak Kapoor[2], and Steven Nicolich[2]. (1) University of Missouri Columbia, Columbia, MO 65211, (2) US Army ARDEC, Picatinny, NJ 07806

Nanoengineered composites of metal oxide (oxidizer) and metal(fuel)have shown to exhibit enhanced combustion properties. More specifically, when the sizes of oxidizer and fuel particles are in nanoscale, the heat and the mass transport length scales are drastically reduced, leading to enhanced combustion speeds and reaction rates. Thus, it is quite possible to achieve tunable reaction rates by proper choice and combination of fuel, oxidizer and other chemicals, their dimensions in nanoscale and the mixing ratio. The present work is devoted to the study of reaction rates of conventional and modified nanothermites. We have used a variety of metal oxides, namely, Fe_2O_3, CuO and MoO_3 and Al as the fuel. The conventional nanothermites were also modified by mixing with ammonium nitrate (AN) nanoparticles. The reactivity was determined by monitoring the pressure as a function of time generated during the energetic reaction. The rate of increase in the pressure generated during combustion process is a measure of the reactivity of the material system. Reactivity rate tests were performed in a closed volume using a typical mass of 20 mg in a milli-well with a packing density of $0.33 g/cm^3$. Among the conventional nanothermites studied in this work, CuO-Al system possesses the highest reactivity. The rate of pressure increase is about 3.23MPa/μs and the peak pressure recorded is 40MPa. In comparison, it is 0.01MPa/μs for Fe_2O_3/Al nanocomposite. The reactivity of self-assembled CuO nanorod (NR) - Al nanoparticle composite (3.81 MPa/μs) is higher than that of the one prepared by random mixing of CuO NR and Al nanoparticle (3.23MPa/μs). The CuO NR – Al based conventional nanothermites were modified by mixing with low-grade explosives, such as ammonium nitrate (AN) nanoparticles. These nanoparticles were prepared by employing microemulsion route. The reactivity increases to about 250 MPa/μs and the peak pressure to 600 MPa for the modified thermite composition with 60 % AN nanoparticles. Mixing of micron size AN particles reduced the reactivity drastically. Our studies clearly show that nanoscale mixing of thermites with explosives is necessary to enhance the reactivity of the mixture.

544h Reduced Electrostatic Discharge Sensitivity Of Nanothermites

Rajagopalan Thiruvengadathan[1], Steve Apperson[1], Andrey Bezmelnitsyn[1], Keshab Gangopadhyay[1], **Shubhra Gangopadhyay**[1], Paul Redner[2], Wendy Balas[2], Deepak Kapoor[2], and Steven Nicolich[2]. (1) University of Missouri Columbia, Columbia, MO 65211, (2) US Army ARDEC, Picatinny, NJ 07806

Nanoenergetic materials with very high combustion speeds are ideal for use as green primers, kinetic rods, and reactive materials. However, their high sensitivity to ESD remains a major challenge for practical applications. These static sensitive materials may cause accidental ignition posing a major safety hazard. It is known that Al nanoparticles are typically used as the fuel component along with nano-sized metal oxides as oxidizers to prepare nanothermites. These Al nanoparticles are highly sensitive to the ESD, thereby making nanothermites vulnerable to ignition at extremely very low ESD energy. We explore the possibility of reducing the ESD sensitivity of Al nanoparticles through coating with fluoropolymers such as Teflon. Teflon being an energetic fluoropolymer is expected to sustain the reactivity of nanothermite while helping to reduce the ESD sensitivity. Well dispersed Al nanoparticles in Teflon were obtained by coating the surface of Al nanoparticles first with perfluorooctal mono-functional trimethoxy silane and then with teflon AF solution. Our measurements show that uncoated Al nanoparticles have a very low ESD energy of 0.98mJ. On the other hand, Al nanoparticles coated with 1 and 2 % of teflon have ESD energy 2.6 mJ and 4.7 mJ respectively. With increasing teflon weight to 4% and 10 %, the ESD energy increases to 25 mJ and 60mJ respectively. We have also studied the combustion properties and reactivity of the Teflon coated nanoparticles with copper oxide nanorods. Although there is a reduction in the burn rate due to Teflon coating, the peak pressure and reactivity of the material increases with the increase in percent of Teflon coating. In conclusion, our results clearly demonstrate that coating with fluoropolymers is very effective in reducing the ESD sensitivity of nanothermite materials.

544i Nanoparticulation Of CL20: A Feasibility Study To Produce Nanoparticles Of CL20 With Supercritical Fluids

Veera Boddu, Environmental Processes Branch, US Army Engineer Research and Development Center, 2902 Newmark Drive, PO Box 9005, Champaign, IL 61822, Rebecca K. Toghiani, Chemical Engineering, Dave C. Swalm School of Chemical Engineering, Mississippi State University, 330 Swalm, Mississippi State, MS 39762, and Reddy Damavarapu, Energetics and Warheads Division, U.S. Army - Armament Research, Development and Engineering Center, Picatinny Arsenal, NJ.

The Army is interested in developing green processes to support its industrial operations. Developing processes without the use of toxic solvents and chemicals is a major focus of the effort in developing the green processes. This presentation summarizes a recent study to develop nanoparticles of CL20 (2,4,6,8,10,12-Hexanitro-2,4,5,8,10,12-hexaazaisowurtzitane) using supercritical processes. In order to generate the desired nanocrystalline particles, the crude CL20 is dissolved into an environmentally benign solvent such as supercritical carbon dioxide and then condensed to ultrafine particles by reducing the pressure and temperature of the mixture. It is desired to have at least have estimates of supercritical solubility of CL20 in carbon dioxide, prior to the nanoparticulation experiments. Estimation of solubility of CL20 in supercritical carbon dioxide was carried out using the Peng-Robinson cubic equation of state. Solubility was predicted over the temperature range of 305.15 to 368.15 K and over the pressure range of 74 to 150 atm. In general, as the temperature increases, the solubility decreases, while as the pressure increases, the solubility increases. For CL20, the estimated vapor pressures are extremely small, on the order of 10-18 at ambient temperature, increasing to 10-13 at 368.15 K. Thus, the predicted solubilities are also small (range of 10-13 to 10-6), with the highest solubility predicted for 308.15 K (35,,aC) and 150 atm. Nanoparticulation experiments were conducted using a supercritical extraction apparatus supplied by Thar Technologies, Inc. Results of CL20 solubility predictions and SEM and XRD characterization of the nanoparticles will be presented.

Classification of Nano-powders in an Ultracentrifuge

Zac Doorenbos[1], Jan Puszynski[1], Deepak Kapoor[2], and Darold Martin[2]
1. Department Chemical and Biological Engineering Department
South Dakota School of Mines and Technology
2. Armament Research, Development and Engineering Center
Picatinny, NJ

Abstract

The classification of nano-powders is a large area of current research interest. This new wet classification of submicron powders is allowing for the classification of nano-powders for both reactive and structural applications. The focus of this research was on the classification of alumina and aluminum. Prepared samples of alumina and aluminum mixtures for two liquid classification were classified with repeatable results. A sample of aluminum produced in a plasma reactor was more difficult to classify due to shape effect. The important parameters for classification are: ultracentrifuge speed, run time, initial liquid, amount and type of dispersant, attrition milling of initial powders, and soak time. These and other aspects of the wet classification process are discussed in this research contribution.

Introduction

The use of nano-size materials in both energetic and structural materials is becoming wide spread due to the changes in the characteristics of the material. One use for nano-powders with a narrow particle size distribution is in the area of energetic materials specifically thermites. The smaller particles result in a much high surface area for reaction as well as shorter distances between fuel and oxidizer particles[1]. This new wet classification technique allows for the classification of reactive materials without the need for aerosol processing. This system will allow for the processing of nano-powders produced by plasma synthesis with a wide particle size distribution by classification of the powder into different fractions for use in different applications depending on the desired particle size distribution or average particle size.

Experimental

Materials

- A16sg alumina, purchased from Alcoa, average particle size 500 nm.
- Nano alumina purchased from Nanostructured and Amorphous Materials, Inc, primary particle size below 40 nm.

- 40 and 120 nm aluminum, purchased from Novacentrix.
- Plasma reactor aluminum, average particle size after attrition milling between 200 to 300 nm.

Material Pretreatment

The initial slurry for classification is prepared by using a predetermined particle loading, which is dependent on the material being classified and results from previous tests where the particle loading of the sample was varied. The proper amount of material is measured and added to the initial liquid along with the dispersant. The sample is sonicated using an ultrasonic probe and left to soak for at least six hours before classification. If the material is agglomerated it is processed with an attrition mill before being dispersed in the initial liquid.

Attrition Milling

Attrition milling was conducted using a lab scale attrition mill (Model 01HDDM, Union Process). For both the aluminum and alumina 0.4 mm yittria grinding media was used for the milling. Milling times were different based on the initial material and were as follows: nano alumina - 30 minutes, 40 nm aluminum - 15 minutes, and plasma reactor aluminum - 5 minutes. The milling time used was long enough to break up any agglomerates that may have formed during the drying process but not too long for the particles, specifically aluminum to form platelets.

Classification Setup

The classification of alumina and aluminum nano-powders has been conducted in a batch testing setup using two liquid setup as shown in Figure 1.

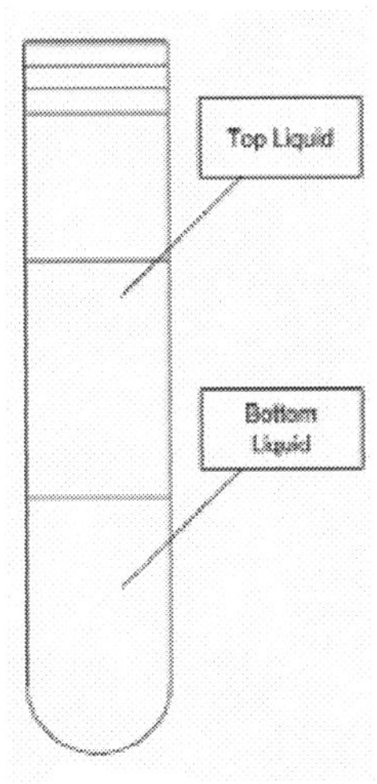

Figure 1: *Two liquid classification setup.*

In the two liquid setup the top and bottom liquids are immiscible to create an interface that the particles must pass through to settle to the bottom of the centrifuge tube. Several different fluids were tested as a bottom liquid; however FC-77 fluoroinert (3M) was found to be the best one as it is immiscible with a variety of liquids and the nano-particles do not disperses in the liquid. The centrifuge tube (50 mL) is made from polyallomer and is 28 mm wide and 105 mm long with a screw cap. A Beckman Coulter J-26 XPI ultracentrifuge was used with either a JA25.50 rotor (holds 8 centrifuge tubes at a 36 degree angle) or a continuous flow rotor, a diagram of each rotor is shown in Figure 2.

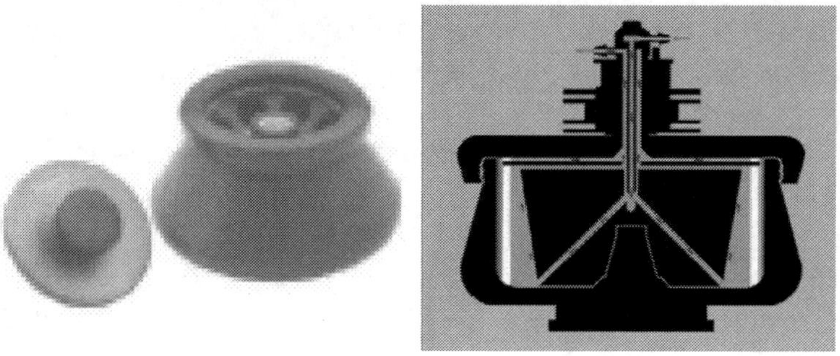

a b

Figure 2: *a) JA25.50 rotor[2], b) Continuous flow rotor with arrows indication direction of flow[2].*

Most classifications up to this point have been done using the batch rotor. The ultracentrifuge can handle speeds up to 24,000 RPM. Ultracentrifuge speeds for the classification experiments were varied between 500 to 15,000 RPM with run times between 2 to 30 minutes. The acceleration rate of the centrifuge between 0 to 500 RPM can be adjusted by setting the length of time from start up to 500 rpm. By increasing that time the classification results were more reproducible as the Coriolis effect is limited.

Analysis of Classification Results

Once the classification experiments are completed the centrifuge tube is removed from the centrifuge and the liquid is decanted off. The material left in the tube is considered the pellet (which contains the larger particles) and the material that did not cross the interface is considered the liquid sample (which contains the fines). Samples of each were then tested using a Nano-ZS zetasizer (from Malvern Instruments) that measures particle size between 6 nm to 6 μm. Results were verified using SEM analysis (Supra 40 VP, Zeiss).

Results

Alumina

Alumina classification was achieved using the two liquid setup with repeatable results. The initial mixture for two liquid alumina classification was 60 mg of a 50/50 wt% mixture of A16sg (Alcoa) and nano (Nanostructured) alumina dispersed in 10 mL of ethanol with 80 µL of Coulter Type 1C Nonionic dispersant. Before the two powders where mixed the nano alumina was attrition milled for 30 minutes to reduce the average agglomerate size from 1100 nm to 220 nm. The 10 mL initial sample is then placed on top of 20 mL of FC-77 fluoroinert in a centrifuge tube and classified. Selected results are shown in Table 1.

Table 1: *Alumina Two Liquid Classification Results*

RPM	Run Time (min)	Initial Average Particle Size (nm)	Liquid Average Particle Size (nm)	Pellet Average Particle Size (nm)
10000	10	303	199	317
10000	10	293	191	290
7500	14	292	150	293
7500	14	280	167	309

Table 1 shows that the classification of alumina using the two liquid classification setup is repeatable. The average particle size of the liquid slurry after classification can be adjusted by changing the ultracentrifuge speed and run time. This shows that the classification process is dependant of both: i) the ultracentrifuge speed, and ii) the run time.

Aluminum

Aluminum classification was done using two different aluminum nano-powders. The first was a mixture of attrition milled 40 nm and unmilled 120 nm aluminum and the second was aluminum that was produced in a plasma reactor with a wide particle size distribution. The nano-powders with average particle sizes of 40 and 120 nm were mixed into a 50/50 wt% mixture using the same procedure as the alumina but with a particle loading of 10 or 20 mg aluminum per 10 mL of initial liquid (IPA). The samples were prepared in the same manner as the alumina samples. Representative results from the classification of the aluminum mixture are shown in Figure 3. The particles in this mixture were all spherical in nature while the aluminum that was produced in the plasma reactor is not all spherical in nature as seen in Figure 4.

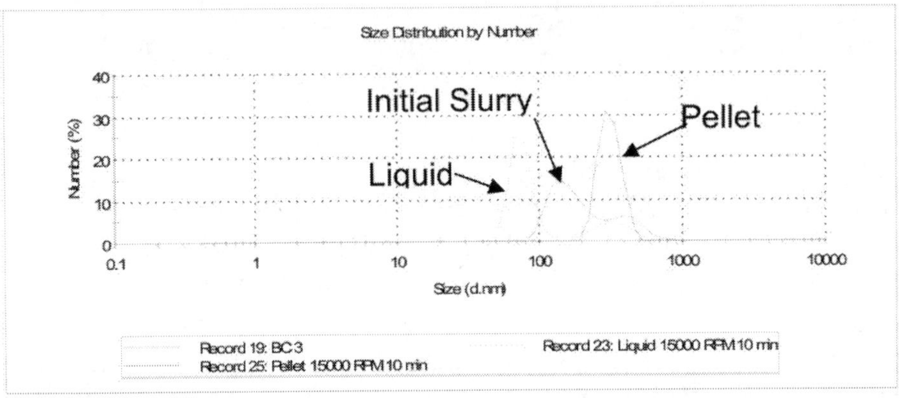

Figure 3: *PSD of two liquid classifications using 10 mg of a 50/50 wt% mixture of 40 nm and 120 nm Nanotechnologies Al dispersed using Coulter Type 1C nonionic dispersant in 10 mL of IPA and 20 mL of FC-77 bottom liquid with an ultracentrifuge speed of 3500 RPM and a run time of 5 minutes.*

Figure 4 shows that the aluminum made in the plasma reactor contains not only spherical particles but also cylindrical and platelet type particles which makes the classification process more difficult. The classification of this material was not as successful as the prepared spherical aluminum. Classification results of the plasma reactor aluminum are shown in Figure 5.

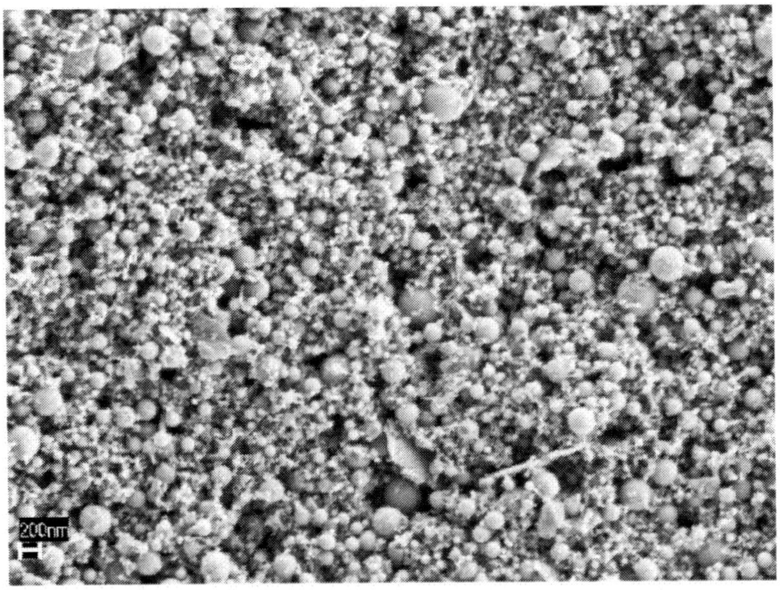

Figure 4: *Attrition milled plasma reactor aluminum*

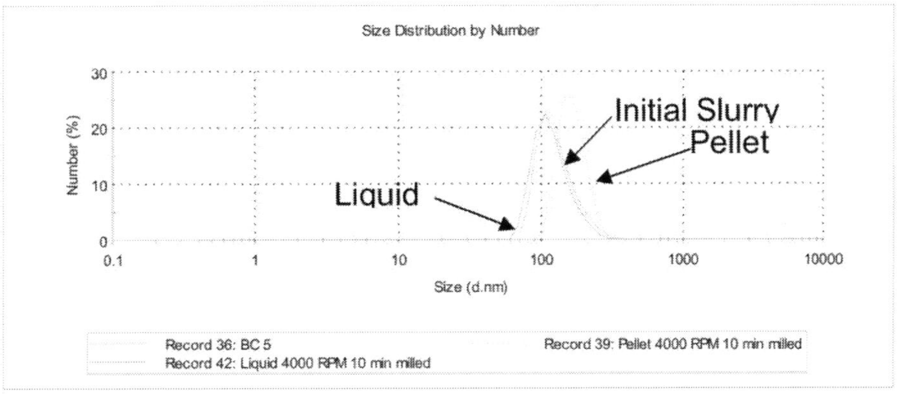

Figure 5: *PSD of two liquid classifications with Coulter Type 1C nonionic. 20 mg plasma reactor aluminum in 10 mL of IPA with a bottom liquid of 20 mL of FC-77 fluoroinert at an ultracentrifuge speed of 4000 RPM with a run time of 10 minutes.*

The most significant classification issue with the plasma reactor aluminum system is that it was not possible to remove all the large particles from the initial liquid slurry during the classification process. Even though the initial slurry had an average particle size of 220 nm classification resulted in insignificant size separation with an average particle size of 174 for the liquid and 194 nm for the pellet. It is possible that better classification of the plasma reactor aluminum is possible by variation of the top liquid or the amount and type of dispersant used.

Conclusions

Wet classification of nano-powders is possible with the proper pretreatment and operating conditions. This study has also shown that the classification of different materials requires different parameters to achieve the desired classification.

Acknowledgments

The authors gratefully acknowledge the financial support from ARDEC (Contract: W15QKN-06-D-0006)

References

1. C.E. Aumann, G.L. Skofronick, and J.A. Martin, *Journal of Vacuum Science & Technology B* 13(2): 1178-1183 (1995).
2. https://www.beckmancoulter.com

563a Coal Injection For Pyrolysis Process In A 2MW Hydrogen Plasma For Acetylene

Yi Cheng, Yulong Ding, Jiaqi Chen, and Yong Jin. Department of Chemical Engineering, Tsinghua University, Beijing, 100084, China

Coal pyrolysis in hydrogen plasma opens up a direct means for producing acetylene, together with the valuable carbon materials in solid state and some ethylene, methane, hydrogen and carbon monoxide as the secondary products. The process essentially consists of several steps: (1) the design of a hydrogen plasma torch, (2) fast heating of coal powders in milliseconds to release the volatile matter in coal in a downer reactor, (3) fast quenching of the reacting stream to prevent the decomposition of acetylene into soot and hydrogen. Since the complex reaction process is operated at very high severity conditions, e.g., reaction temperature at 1500K to ~3000K, high gas velocity and residence time in milliseconds, the nozzle design for coal injection into the hot hydrogen stream becomes the first critical aspect in the reactor design to guarantee the rapid heating of coal powders, and accordingly the acetylene yield. A 2MW thermal plasma torch was built using hydrogen as the heat carrier. To work in a novel V-shaped plasma torch, coal powders must be injected to the hot stream around the downer reactor (with inside diameter of 100 mm) instead of using a center jet. Due to the high gas velocity inside the column, it is challenging to inject the coal efficiently into the hot medium with great stability to well control the reaction progress. The different nozzle designs are compared in terms of the visualizing results on the coal distribution across the injection plane in the pilot-plant reactor with gas volume flow rates of >2000 m3/hr at room temperature. Computational fluid dynamics (CFD) model for multiphase flow simulation is conducted to assist the design of nozzle shape and placement. Flat nozzle shape is adopted with comparison to the conventional round one. By tuning the angles of the flat nozzles around the downer wall, swirling flow during the gas-solid contact is implemented, which in turn strengthens the heat exchange between phases and therefore improves the coal conversion and acetylene yield.

563b Drying Of Moist Solid Particulate In A Bubbling Fluidised Bed

Y. Makkawi[1], **R. Ocone**[2], Jamie Duncan[2], and Marc McAndrew[2]. (1) Heriot-Watt University, School of Engineering and Physical Sciences, Edinburgh, EH14 4AS, UK, (2) Chemical Engineering, Heriot-Watt University, School of Engineering and Physical Sciences,, Edinburgh, EH14 4AS, United Kingdom

The process of removing water from moist solid particles in a conventional bubbling fluidised bed has been experimentally studied. Air at ambient conditions was used to dry the moist particles at various fluidisation velocities. Throughout the drying process, images of the material distribution were recorded using Electrical Capacitance Tomography (ECT). Simultaneously, the outlet air moisture content was measured using temperature/humidity probe. The ECT data were used to quantify the bubbles characteristics while the air moisture content was used to calculate the drying rate. From the available experimental data, a simplified method, based on bubble-dense phase interaction, was used to calculate the overall mass transfer coefficient. To avoid the complexity of the process, the fluidised bed was operated at single bubbling regime. The experimental results are discussed in relation to the different drying mechanisms at the falling rate and constant rate.

MEASUREMENT OF HOLDUP AND PROPAGATION VELOCITIES IN A LIQUID-SOLID CIRCULATING FLUIDIZED BED USING ELECTRICAL RESISTANCE TOMOGRAPHY

Shaikh Razzak, Shahzad Barghi and Jesse Zhu
Department of Chemical and Biochemical Engineering,
The University of Western Ontario, London, ON N6A 5B9, Canada

Abstract

The application of liquid-solid circulating fluidized beds (LSCFB) is steadily increasing in the chemical, biochemical and other industries. Electrical Resistance Tomography (ERT) as an imaging technique was employed for the study of flow parameters such as phase holdup, velocity distribution of individual phases in LSCFB. ERT is a non-invasive technique based on conductivity measurement of the continuous phase, which also provides color-coded cross sectional view of phases with a frequency of up to 250 images per second. The local conductivity measured by a number of electrodes located at the periphery of the plane was then further converted into a local phase concentration distribution based on Maxwell's relation. Local solids holdup also measured by using optical fibre probe to verify and compare result produced by ERT. Both give good experimental agreement each other. By cross-correlation between the data obtained from a pair of planes of ERT, one upstream and one downstream, the phase propagation velocity was determined. Water was used as the continuous and conductive phase and glass beads of 500 micron size used as the solid and non-conductive phase in this investigation. Radial non-uniformity of phase holdups was observed at different superficial liquid velocities. Solids holdup was higher in regions close to the wall and low in the central area. Such non-uniformity in phase holdup decreased with increasing liquid velocity. Both solids holdup and radial non-uniformity increased with solids circulation rate. The propagation velocity did not change radially at low liquid velocities, but increased in the central region at higher superficial liquid velocities. The propagation velocities were lower than superficial liquid velocity due to local slip velocity and drag force. A good correlation between the ERT and Optical Fibre probe methods was verified.

1. Introduction

Liquid-Solid Circulating Fluidized Bed (LSCFB) now gets immense attention for its potential application in the large industrial fields such as: Waste water treatment, Protein separation, Pharmaceutical and Petrochemical. Good mixing and efficient heat and mass transfer have made fluidized bed reactors a unique choice in many industries. LSCFB emerged as lucrative equipment not only for its good advantages but also controllable residence time and operating conditions.

Most of the measuring techniques used in LSCFB systems are invasive or intrusive. There is an increasing interest in application of tomography techniques in the study of multiphase flow systems due to its non-invasive nature. The ERT method was applied to measure the local and cross sectional average of solid hold-up using single plane 16-electrode

ERT sensors. The radial non-uniformity in LSCFB was reported using conductivity probe, (Liang et al. [1]) and optical fiber probe (Zheng et al. [4]). It was claimed that radial flow structure is affected significantly by operating conditions and particle properties [3]. Zheng et al. [3] showed radial distribution of the solids holdup under wide range of operating conditions and tested the effect of particle density on the flow structure [3]. Radial distribution of the liquid velocity in LSCFB riser was studied by Zheng et al. [3] using dual conductivity probe. Radial solids velocity distributions still not studied and have scope to work in this field. In this study

In this study, ERT and optical fiber probe both used to measure radial and local solids holdup distributions. Optical fiber probe produces time based average of solids holdup, however the ERT is capable of zone based average and time based average of phase holdup. The phase propagation velocity was determined using cross-correlation between the data obtained from a pair of planes of ERT, one upstream and one downstream (Figure 4). Radial distribution of solids velocity was also determined for the first time using ERT method. All experiments were carried out at ambient temperature where water was used as the continuous and conductive phase and glass beads of 500 micron size used as the solid and non-conductive phase in this investigation.

2 Experimental Set-ups

Schematic diagram of the experimental set-up of LSCFB has shown in Figure 1. LSCFB consists of two main sections, riser and downer made of Plexiglas. All the measurements are carried out in the riser which is 7.62 cm ID and about 7 m high. The primary liquid distributor was made of seven stainless steel tubes extending 0.2 m into the riser column and auxiliary liquid distributor was a pours plate at the base of the riser. Riser is connected to the downer through the solid returning pipe at the top and solids feeding pipe at the bottom of the riser close to the distributor. Liquid was pumped from the reservoir and fed to the riser by two streams. One was the primary or main flow, which was introduced at the primary liquid distributor and the secondary or auxiliary liquid flow was introduced to the auxiliary liquid distributor.

Solid particles were circulated between the riser and downer in the LSCFB by ppropriate pressure balance. Auxiliary flow was adjusted at a threshold value to facilitate the solid particles flow. The main purpose of the auxiliary flow is to control the solid circulation rate and acts as a non-mechanical valve. The combined effects of both primary and auxiliary liquid flow produced the total liquid flow which carries the solid particles up in the riser. Entrained particles, collected in the liquid-solid separator at the top of the riser, were returned back to the downer after passing through the solid circulation rate measuring device located at the top of the downer.

2.1 The Optical Fiber Probe System

Optical fiber probe model PV-5, produced by Institute of Process Engineering, Chinese Academy of Sciences, was used in the system to measure local solids holdup. Schematic diagram of this system is shown in Figure 2. Probe consists of both light emitting and receiving quartz fibres, arranged in an alternating array, corresponding to emitting and receiving layers of fibres. The diameter of the fibers is 25 μm. The received light reflected by the particles was multiplied by the photo-multiplier and converted into voltage signals. The voltage signals were

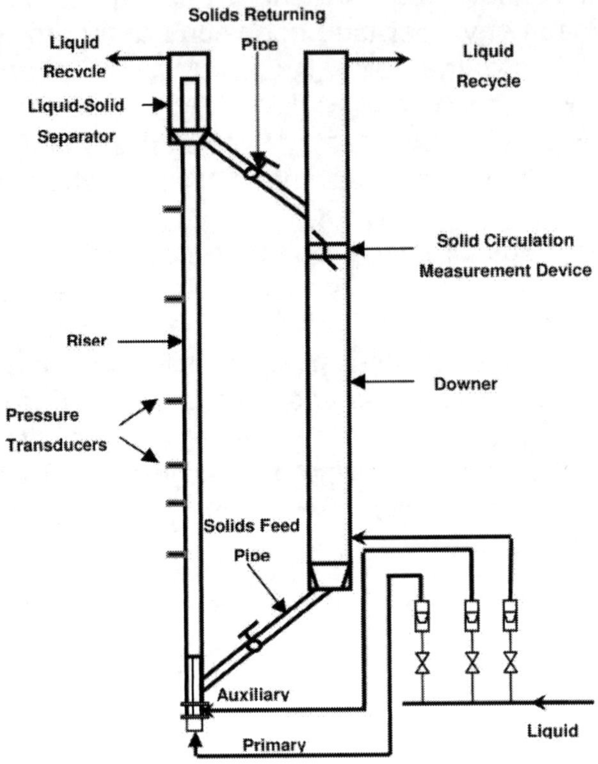

Figure1: Schematic Diagram of Experimental Setup of the LSCFB system.

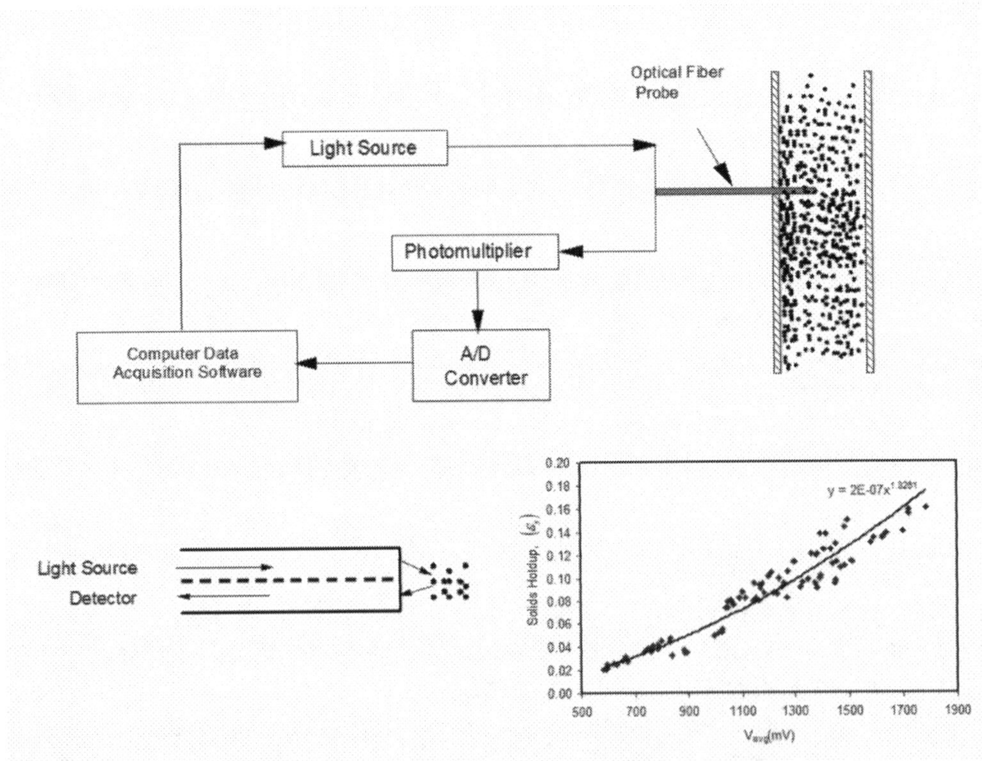

Figure 2: The Optical Fibre Probe System for Solids Holdup Measurement.

further amplified and fed into a PC via A/D converter. Because the relationship between the output signals of the optical fibre probe and the solids concentrations is nonlinear, a reliable calibration is required to ensure that the output correctly represents the measurement. The calibration was carried out in a conventional liquid solid system, which was assumed homogeneous and average solids holdup measured against the data produced by differential pressure transducers.

2.2 The ERT Measurement System

The local solids holdup and propagation velocity was measured using ERT system (EnERT) manufactured by En'Urga Inc. The ERT consists of a primary sensor, electronic circuit and a PC-based data acquisition system. The setup of the ERT system is shown schematically in figure 3.

The inner diameter of the primary sensor is built equal to the inner diameter of the riser or test section so that the sensor can be lined up with it. The liner of the primary sensor supports three planes of electrodes. Sixteen electrodes equally spaced on the first plane can provide voltages for reconstructing fine phase distributions. Each of the two planes contains eight electrodes and can provide voltages for reconstructing coarse phase distributions.

For each driving current, the ERT measures the electrical potential distribution through electrodes which are flush mounted on the pipe wall. With input values of electrical potentials and currents, the local conductivity (or resistivity) of the mixture can be reconstructed through a state-of-the-art optimization algorithm. The conductivity distribution is then further converted into a local phase concentration distribution based on Maxwell's relation. By cross-correlating two distributions from both upstream and downstream planes, the phase propagation velocity

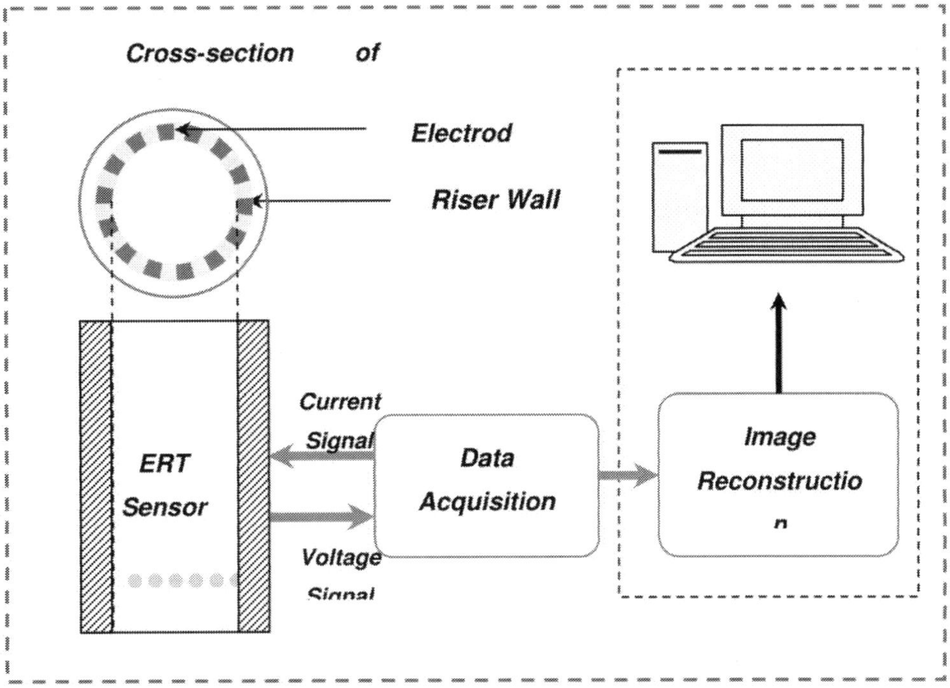

Figure 3: The Schematic Diagram of the Measurement Principle of ERT.

can be obtained. Figure 4 shows Cross-correlation techniques used for obtaining propagation velocity profile in LSCFB system. The ERT system obtains data at 250 (500 optional) frames per second. For a steady state condition, the data is collected for a certain period of time and averaged.

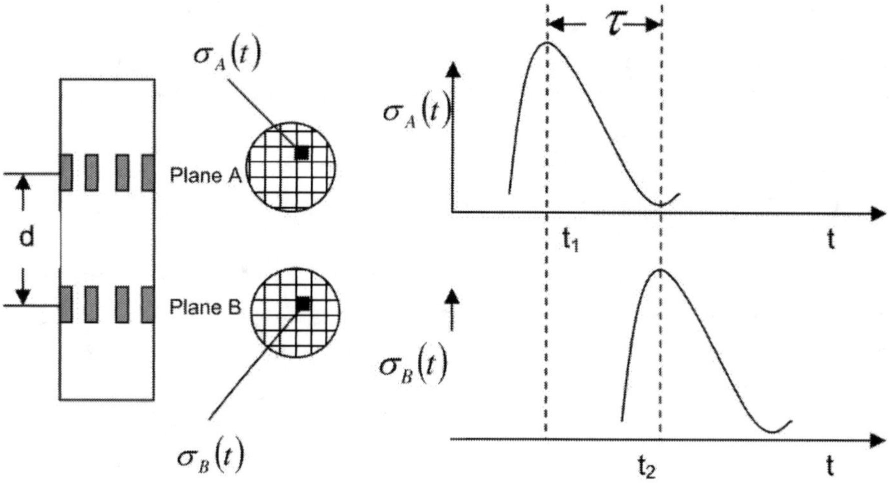

Figure 4: Cross-correlation technique used for obtaining propagation velocity profile in the LSCFB system.

3. Results & Discussion

3.1 Solids Holdup

Radial distribution in LSCFB riser was found to be non-uniform. Data produced using both ERT and Optical Fiber Probe systems. Figure 5 shows radial solids holdup against superficial liquid velocity at auxiliary liquid flow of U_a=1.4 cm/s. For any superficial liquid velocity solids holdup was lower at the center of the riser compared to the near wall regions. Solids holdup increased sharply at the wall region. Similar results were obtained in both systems. Liang et al. [1] and Zheng et al [3] also reported similar results. Figure 5 also shows that radial non-uniformity reduced with the increase of superficial liquid velocity. At the lower velocity, U_l=8.4 cm/s, solids holdup and radial non-uniformity was maximum whereas it decreases with the increase of superficial liquid velocity.

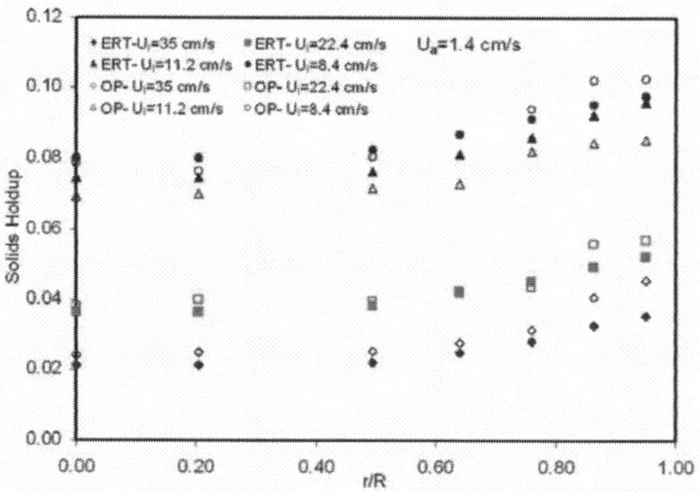

Figure 5: Solids holdup measurements using both ERT and Optical Fibre Probe system for different superficial liquid flow at auxiliary liquid velocity, U_a=1.4 cm/s.

Radial distribution of solids holdup depicted for different solids circulation rate at the same superficial liquid velocity is shown in Figure 6. The solids holdup increased with increasing solids circulation rate Figure 7 shows ERT and optical fiber results in a 3-D fashion. It was shown clearly that radial non-uniformity increase with the increase of solids circulation rate. It was also shown that for all five different solid circulation rates solids holdup was lower at the core region compared to the wall regions.

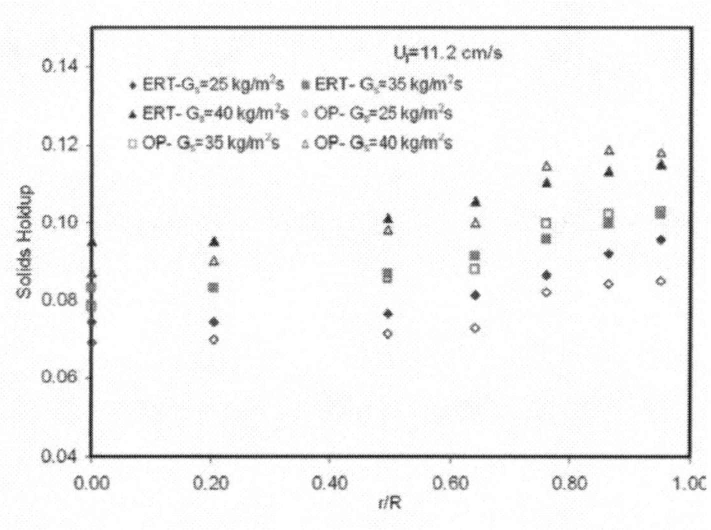

Figure 6: Radial solids holdup measurements for different solids circulation rates at superficial liquid velocity, U_a=1.4 cm/s.

3.1 Solids Propagation Velocity

ERT uses dual plane sensors as shown in Figure 4. By applying cross-correlation technique between the data obtained from the two distributions, upstream and downstream planes, the phase propagation velocity was estimated. Time taken for transit phase signal from upstream plane to downstream plane τ and the corresponding phase velocity v can be obtained by cross-correlating from pixel to pixel in both planes, $v = \tau / L$, where, L is the distance between two planes.

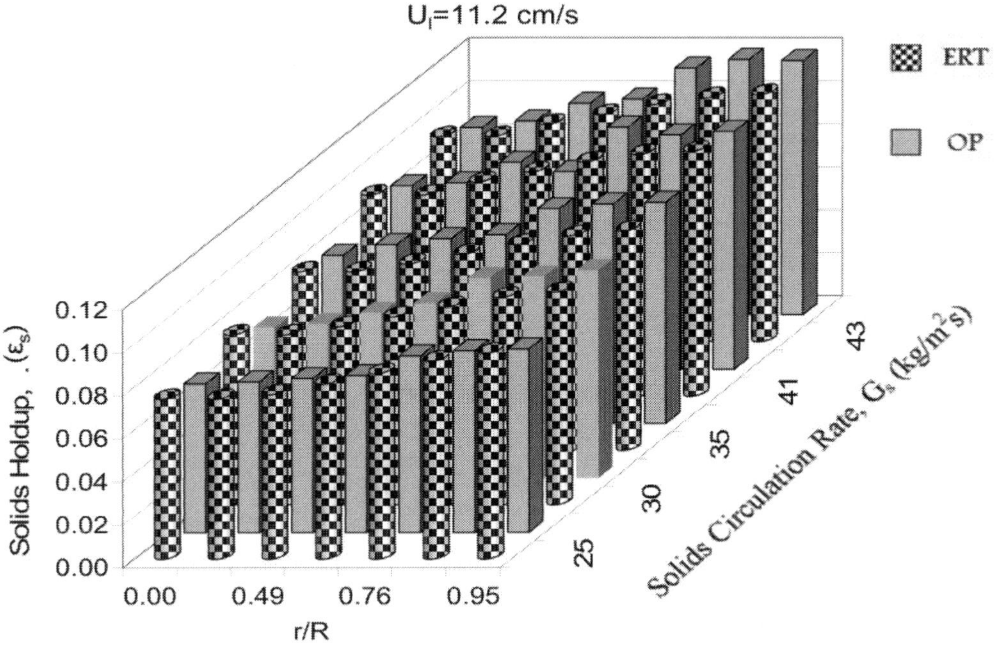

Figure 7: 3-D radial solids holdup measurements for different solids circulation rates at liquid superficial velocity, U_l=11.2 cm/s.

Figure 8 depicted the solids propagation radial velocity distributions. Figure 8(a) explained the velocity distribution for different solids circulation rate at the superficial liquid velocity U_l=11.2 cm/s. For all four different solids circulation rates the radial distribution of the solids velocity profile was almost parabolic in shape. The solids velocity at the center was maximum and decreased towards the wall regions due to the slip velocity and shear stress effect, which are dominant in near wall regions. Similar trend was observed for the liquid phase the velocity magnitude decreased with the increase of solids circulation rate due to the increase of solids holdup which farther increases the slip velocities thus reduced the local solids velocity.

With the increase of superficial liquid velocity solids velocity also increased. But the rate of increase was much higher at the central region than the wall region as shown in Figure 8(b).

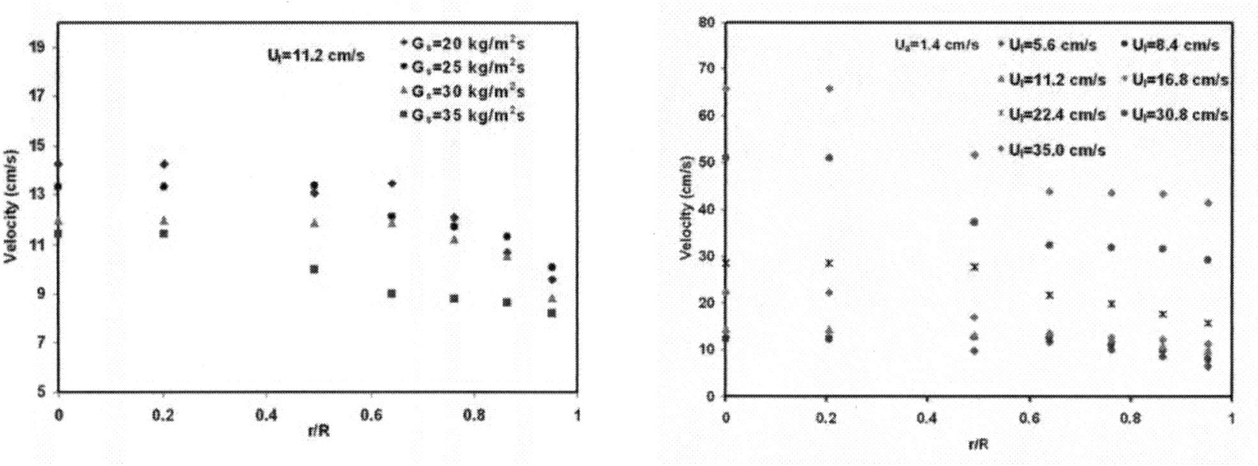

(a) (b)

Figure 8: radial velocity distribution for (a) different solids circulation rate at superficial liquid velocity, U_l=11.2 cm/s. (b)different superficial liquid velocity at auxiliary liquid velocity U_a=1.4 cm/s.

4. Conclusions

Radial distributions of solids holdup measured using both ERT and fiber optics probes. Solids hold up was found to be lower in the central regions of the riser and higher in regions close to the wall. Phase propagation velocity measured by ERT using cross-correlation techniques was also higher at the central regions compared to the wall regions. The results obtained in this study were in good agreement with literatures.

5. References

[1] W.-G. Liang, J.-X. Zhu, Y. Jin, Z.-Q. Yu, Z.-W. Wang and J. Zhou, Radial Nonuniformity of Flow Structure in a Liquid-Solid Circulating Fluidized Bed, Chemical Engineering Science, 51(1996), 2001-2010.

[2] W.-G. Liang, S.: Zhang, J.-X. Zhu, J. Yong, Z.-Q. Yu and Z.-W. Wang, Flow Characteristics of the Liquid-Solid circulating fluidized bed, Powder Technology, 90 (1997), 95-102.
[3] Y. Zheng, J.-X. Zhu, N. Marwaha and A. Bassi, Radial Flow structure in a Liquid-Solid Circulating Fluidized Bed, Chemical Engineering Journal, 88(2002), 141-150.
[4] Y. Zheng and J.-X. Zhu, Radial Distribution of Liquid Velocity in a Liquid-Solids Circulating Fluidized Bed, International Journal of Chemical Reactor Engineering, 1(2003), 1-8.
[5] Y. Zheng, Flow structure in a Liquid-Solids Circulating Fluidized Bed, PhD Thesis, Department of Chemical and Biochemical Engineering, The university of western Ontario, 1999.
[6] Y. Wu, H. Li, M. Wang, R.A. Williams, Characterization of Air-Water Two-phase Vertical Flow by Using Electrical Resistance Imaging, The Canadian Journal of Chemical Engineering, 83(2005), 37-41.

563d Particle Clump Formation and Its Effect on Entrainment

Roy C. Hays and S.B. Reddy Karri. Particulate Solid Research, Inc., 4201 W. 36th St., Chicago, IL 60632

Three cases are presented on the entrainment of various materials from fluidized beds in which particle clumps, resembling agglomeration, affected the measured entrainment rates. In the first case, direct evidence of "agglomeration" is observed in the freeboard of a fluidized bed of 20 μm particles in that large clusters are visible at sizes exceeding 1000 μm. In the second and third cases, the effects of agglomeration are seen indirectly by the measured entrainment fluxes. For the second case, fines from a secondary cyclone of a fluidized catalytic cracking unit were measured at different fluidized bed heights. Higher entrainment fluxes were observed with lower bed heights (and subsequently higher disengaging heights). The third case comes from replication of a phenomenon observed in an industrial scale process where loading into the solids recovery system was highly variable and led to flooding of the recovery system. It was found that with an initial high concentration of fines in the fluidized bed that the entrainment flux was very low, but that as the fines were gradually elutriated away, the entrainment flux increased dramatically. Following the dramatic increase, the entrainment flux then exhibited classical exponential decay as the fines were further elutriated from the fluidized bed.

Evidence suggest that fines of many materials may be clumping and resulting in a lower entrainment rate. These clumps appear to be formed in the fluidized bed and can be ejected into the freeboard intact. Thus, entrainment rates may be influenced by fines levels, bed heights, baffles, jet velocities, etc. Details of this mechanism and its implementation will be discussed.

Effect of Polydispersity, Particle Size and Compaction Pressure on the Tablet Properties

Alvaro Realpe, Teófilo Donaires and Carlos Velázquez. University of Puerto Rico, Mayaguez, Puerto Rico

Abstract

The particle size, polydispersity and binder viscosity are properties very important in the industry of pharmaceutical tablet manufacturing. Polydispersity affects the rearrangement of particles and it could lead to different compaction characteristics of powders that formed tablets. The aim of this work was analyze the effect of particle size, polydispersity, compaction pressure and binder properties on the dissolution time, hardness and friability of tablets manufactures by both direct compression and wet granulation. UV spectroscopy was used to determine the dissolution time of active ingredient. The dissolution time of tablets decreases with increasing binder viscosity. The friability test approved the FDA standards, with lost in mass smaller than 1%. The polydispersity and particle size will be studied to understand the effect on the dissolution time of tablets.

Introduction

Pharmaceutical tablets are the most common solid dosage forms and convenient methods for drug delivery. The tablet is mixture of active pharmaceutical ingredient (API) and excipients obtained by compaction of powders, which all contribute to its final properties. The properties of the tablets could be predicted based upon properties of the active drug and excipients.

The hardness is a property of interest for determining the compression process, as well as for the quality control of tablet manufacture. The hardness, dissolution, friability and density of a tablet depend of compaction pressure exerted on powder bed. Compaction of powders brings the particles into close proximity to each other. The reduced distance between the particles provide a larger total area to facilitate creation of bonds and makes the particles stick together to form a tablet.

Materials and Methods

Pharmaceutical powders such as lactose anhydrous supplied by Sheffield Products and supplied by *Mutchler Inc.* (Cayey, PR), magnesium stearate and Naproxen Sodium, were used in the experiments of wet granulation and direct compression to manufacture tablets.

Different particle size distribution (PSD) were prepared by screening lactose anhydrous using the following sieve trays Tyler's 270, 230, 200, 170, 140, 70, 50 and 40 mesh US Standard. The PSD of powders recollected in each sieve tray was measured by means of image processing and analysis. [1, 2]

Mixer/granulator was used to homogenize and granulate powder to manufacture the tablets. Cylindrical tablets were obtained under compaction pressures range between about 2000 and 20000 lbs, using a tabletting press

(Carver ™). Near infrared spectrometer was used to take spectrum of each tablet and determine dissolution profile by calibration with UV spectrometer.

Tables 1 and 2 depict the factorial design used for manufacturing tablets using wet granulation and direct compression, respectively, consisting of three factors, plus two replicates.

Table 1 Experimental Design for manufacturing tablets using wet granulation

Factors	**Granule size distribution**			**Amount of binder ($\% V_{Liquid}/W_{Solid}$)**		**Compaction pressure (lb)**			
Levels	PSD1	PSD2	PSD3	11.7	13.6	2000	4000	8000	12000

Table 2 Experimental Design for manufacturing tablets using direct compression

Factors	**Particle size distribution**			**Amount of binder (%W/W)**		**Compaction pressure (lb)**			
Levels	Fine	coarse	Fine-coarse	1	5.5	2000	4000	8000	12000

Results and discussion

Figure 1 depicts how tablet hardness increases with increasing of compaction pressure of both fine and coarse granules with 1% moisture content, while Figure 2 depicts that curve goes through a maximum for granules of both fine and coarse granules with 5.5% moisture content.

The tablet hardness is a function of the strength of the bond and the numbers of bonds [3]. Tablets manufactured with granules of 5.5% moisture content present high hardness in low compaction pressure due to the good water distribution that form strong bond. The high pressure on powder bed reduces volume of tablet, the particles get closer and facilitate the wetting of particles that debilitates the strength of the bond.

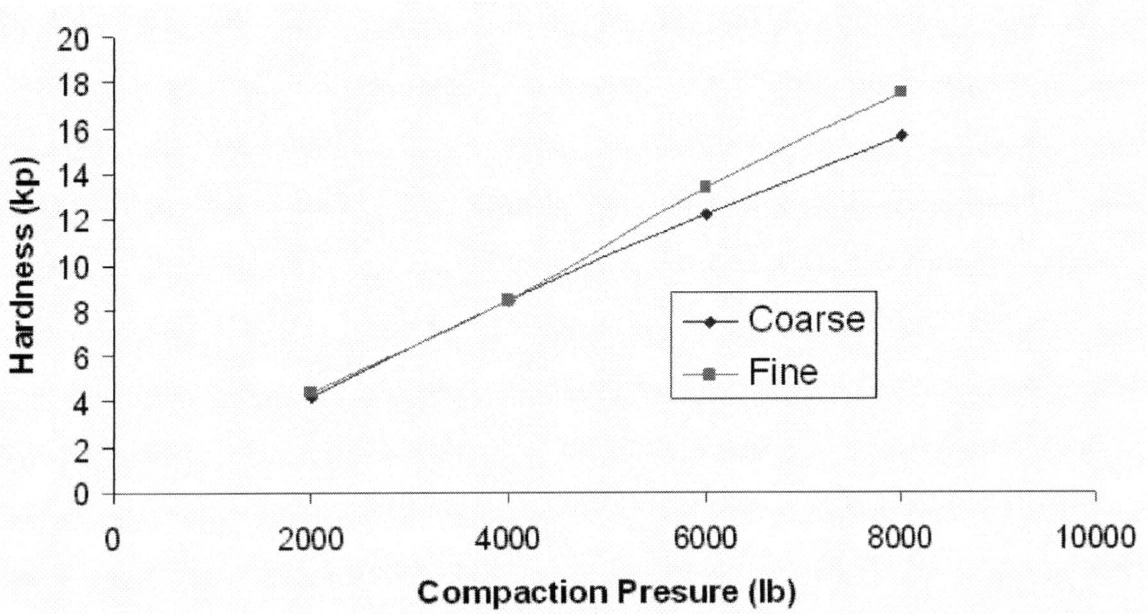

Figure 1. Tablet hardness versus applied compaction pressure at 1 % moisture content

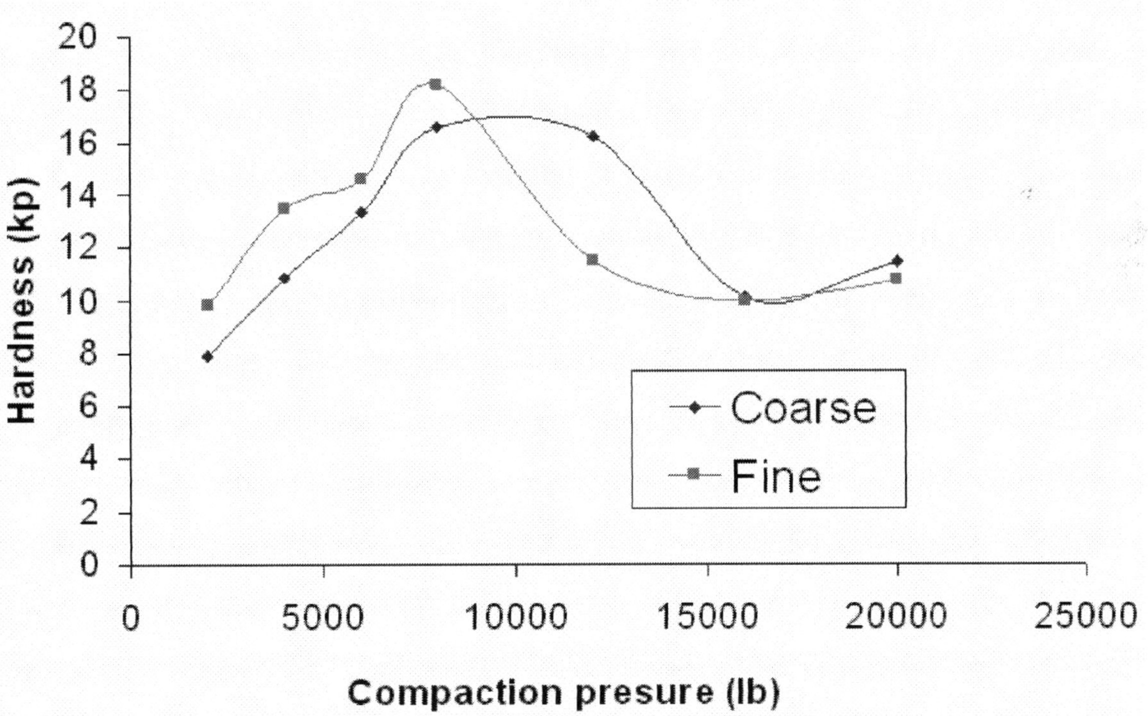

Figure 2. Tablet hardness versus applied compaction pressure at 5.5 % moisture content

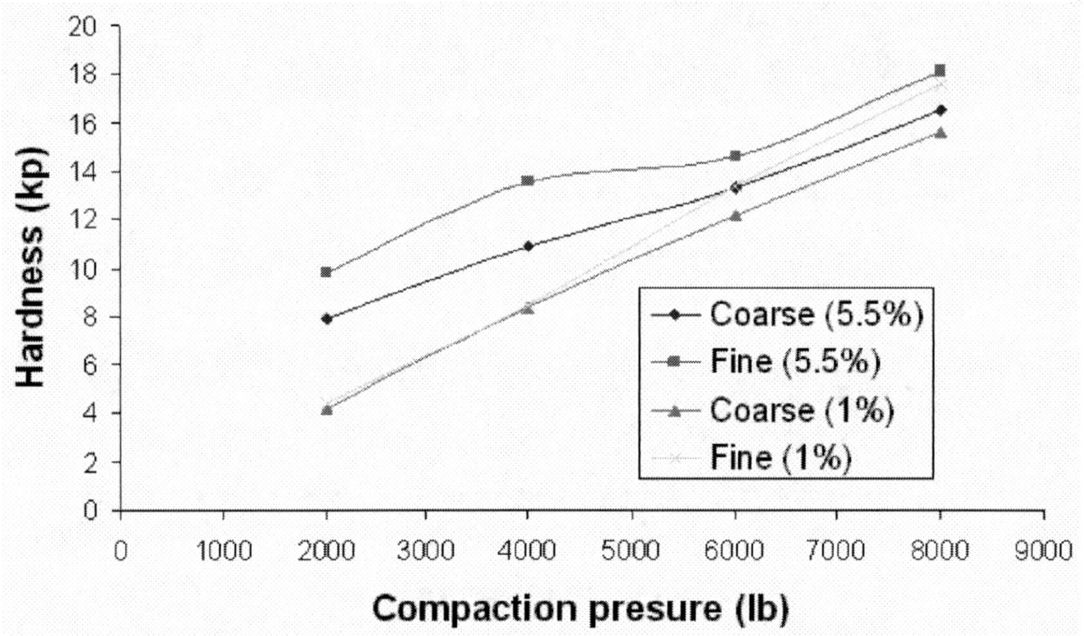

Figure 3. Tablet hardness versus applied compaction pressure

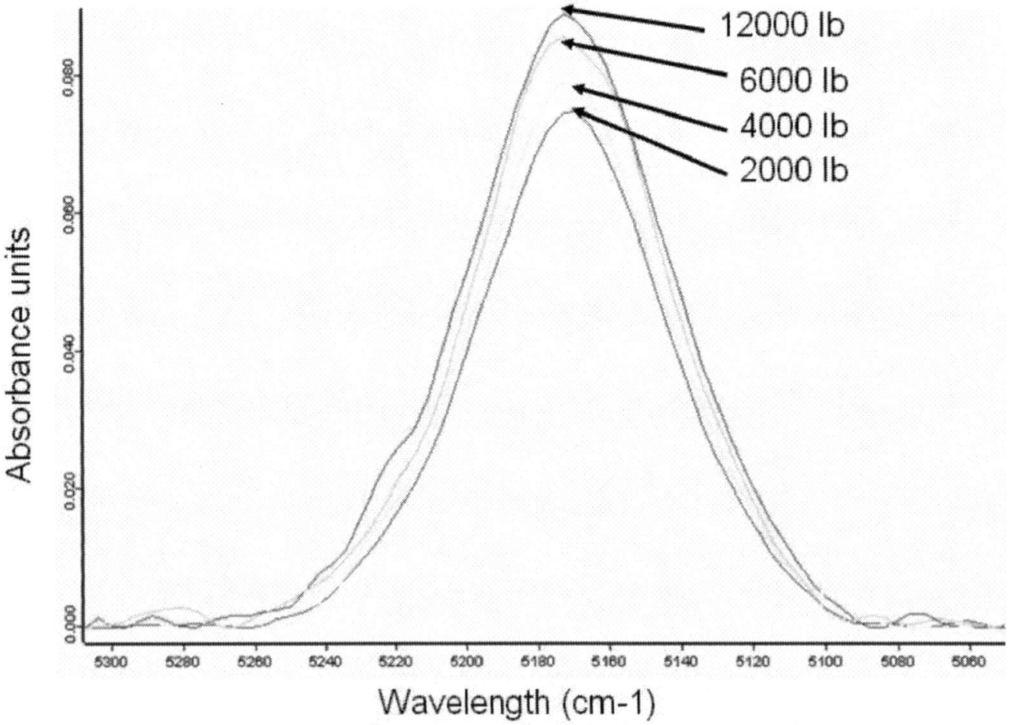

Figure 4. Absorbance versus compaction pressure for fine powders and 1 % moisture content

Future Works

- Make tablets by direct compression method
- Compare properties of tablets manufactured by wet granulation method and direct compression.
- Develop and validate NIR spectroscopy method for determination of hardness and dissolution profile of tablet
- Determine gap size between particles that formed tablet

References

1. Realpe A, Velázquez C. 2006. Pattern recognition for characterization of pharmaceutical powders. Powder Technology 169: 108-113.
2. Velázquez C, Realpe A, Obregón L. 2006. Image Analysis for Powder Concentration Estimation and Particle Shape and Size. Journal of Process Analytical Technology 3 (5): 20-24.
3. Carstensen Jens T. 1995. Drug stability. Principles and Practices. Markel Dekker. New York, United Stated.

Preparation and Performance of Organic Zeolites from a Diolefinic Monomer

Ozge Guney-Altay, Dmitry Pestov and Gary Tepper

Microsensor and Radiation Detector Research Laboratory
Department of Mechanical Engineering, Virginia Commonwealth University, Richmond, VA 23284

We report on submicron particles of a diethyl ester of *p*-phenylenediacrylic acid (EPA) whose molecular size selectivity can be tuned by an imprinting procedure applied during the solid-state photo-polymerization reaction. EPA monomer particles are produced by Rapid Expansion of Supercritical Solutions (RESS), and deposited directly on Surface Acoustic Wave (SAW) resonators as shown in Figure 1. Dynamic sorption isotherms of C_1- through C_9-alkanes were studied to characterize the particulate coatings on SAW sensors. The mass changes due to vapor uptake resulting from the selective sorption into the particulate coatings can be measured accurately due to the high stability of the coatings[1] and the nanogram sensitivity of the piezoelectric crystal. The selectivity and molecular porosity of the particulate coatings were studied *in situ* on the piezoelectric crystals by measuring sorption isotherms. Exclusion of progressively smaller molecules from the molecularly imprinted particulate coatings is observed as a function of the template molecule size[2]. The observed reversible adsorption of alkane analytes with different molecular sizes and shapes suggest that these imprinted polymers may be categorized as organic analogs of zeolites.

Keywords: organic molecular sieve, templated polymerization, RESS.

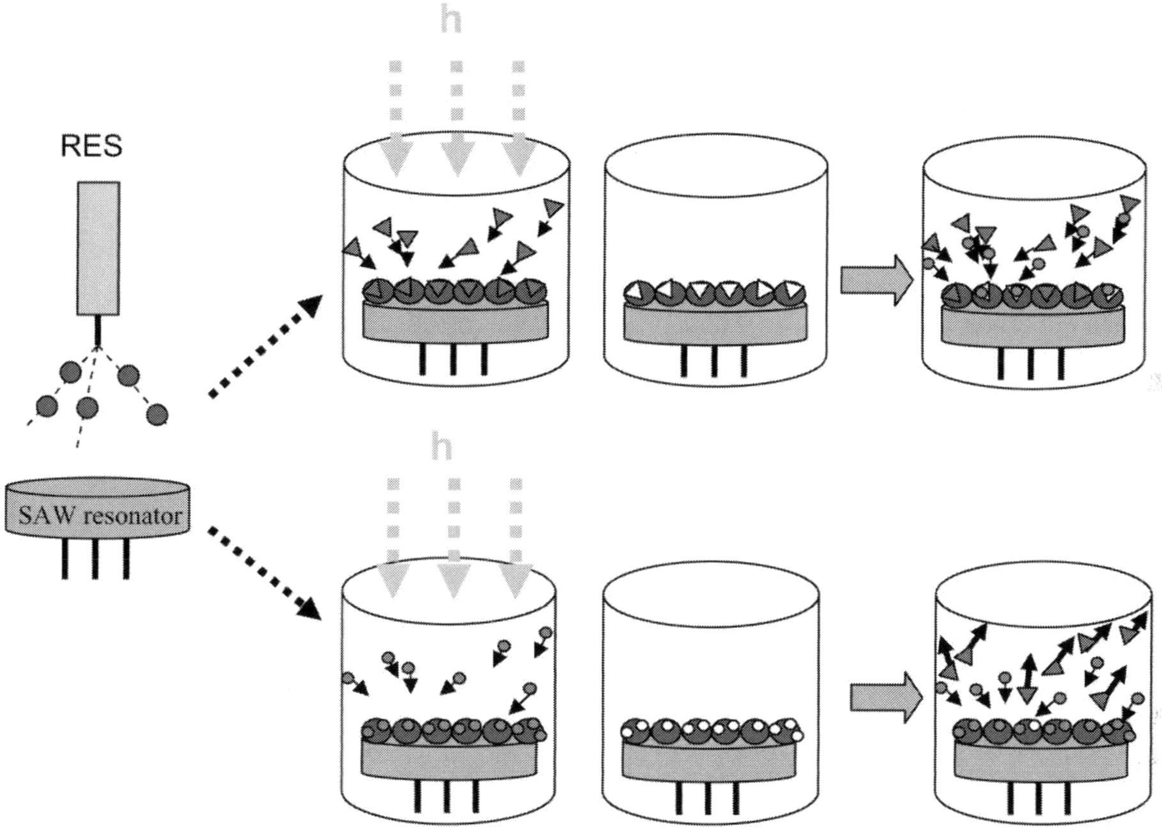

Figure 1. A schematic representation of the preparation of organic analogs of zeolites from EPA monomer.

References:

1. Pestov, D., Guney-Altay, O., and Tepper, G. "Improving the Stability of Surface Acoustic Wave (SAW) Chemical Sensor Coatings Using Photopolymerization", to appear in *Sensors and Actuators: B. Chemical.*

2. Guney-Altay, O., Pestov, D., and Tepper, G. "Organic Zeolites from a Diolefinic Monomer", to appear in *Journal of the American Chemical Society.*

571c Functionalized Single-Sheet Graphene by Oxidation and Thermal Expansion of Graphite: Exfoliation Mechanism and Characterization

Michael J. McAllister[1], Je-Luen Li[2], Douglas H. Adamson[3], Robert K. Prud'homme[1], and Ilhan A. Aksay[1]. (1) Chemical Engineering, Princeton University, A214 Engineering Quadrangle, Princeton, NJ 08544, (2) Institute of Atomic and Molecular Sciences, Academica Sinica, Taipei, Taiwan, (3) Princeton Institute for the Science and Technology of Materials (PRISM), Princeton Unversity, 70 Prospect Ave., Princeton, NJ 08540-0211

Individual layers of graphite, termed graphene, are composed of a honeycomb arrangement of carbon atoms and are the basis of the fullerene and carbon nanotube structures. Graphene sheets are expected to have tensile modulus and ultimate strength values similar to that of single wall carbon nanotubes (SWCNT). Recent studies of graphene produced by a scotch-tape method have been shown to possess interesting electrical properties.[1] Graphene sheets, much like SWCNTs, are expected to serve as fillers for the enhancement of mechanical and electrical properties in composite materials.[2] A recent study performed in our laboratories has demonstrated the formation of individual chemically modified graphene sheets by oxidation and thermal expansion of graphite.[3] However, the expansion mechanism and the characterization of the degree of exfoliation are not adequately understood in order to produce large quantities of this material consistently. In order to resolve these issues, we have studied the exfoliation mechanism and characterization of the material in greater detail. We conclude that this technique produces 80% single graphene sheets.

[1] Heersche, H.B.; Jarillo-Herrero, P.; Oostinga, J.B.; Vandersypen, L.M.K.; Morpurgo, A.F. Nature 446, 56-59 (2007).

[2] Ramanathan, T.; Liu, H; Brinson, L.C. J. Polym. Sci. B: Polym. Phys. 43, 2269 (2005).

[3] Schniepp, H.C.; Li, J.L.; McAllister, M.J. et al. J. Phys. Chem. B 110, 8535-8539 (2006).

571d Temperature Controlled Aggregation and Dispersion Behavior of Silica Clusters

Luciana Bava[1], Donald Feke[1], Ica Manas-Zloczower[2], and Stuart Rowan[2]. (1) Chemical Engineering, Case Western Reserve University, 10900 Euclid Avenue, Cleveland, OH 44106, (2) Macromolecular Science and Engineering, Case Western Reserve University, 10900 Euclid Avenue, Cleveland, OH 44106

Chemical binders or processing aids have been widely used to control processing conditions of fine powders by changing the intrinsic cohesivity of the solid material and/or the filler-media interactions. This study is aimed at achieving this control with polymers that are responsive to external stimuli, such as temperature. In this paper, we will examine aggregation and dispersion of silica clusters induced by the temperature dependent conformational changes in an adsorbed layer of poly(N-isopropyl acrylamide) (PNIPAM). Experimental studies of silica powder incorporating PNIPAM and water were performed at low and high temperatures. Results show that as the PNIPAM undergoes a coil to globule transition with an increase in temperature, modifications in the interparticle interactions of the PNIPAM-treated powders take place. This dependence of the cluster morphology and cohesivity on temperature was evidenced through changes in the aggregate particle size and in the dispersion mechanism and kinetics for the treated silica clusters.

571e Effect Of Dry Particle Coating On Packing Density Of Cohesive Powders Under Low Consolidation

Lauren Beach, Yuhua Chen, Miguel Angel Sanchez Quintanilla, Rajesh Dave, and Laila Jai Jallo. New Jersey Center for Engineered Particulates, New Jersey Institute of Technology, 138 Warren Street, YCEES 208, Newark, NJ 07102

Dry coating is an innovative technique in which nano particles are mechanically coated onto the surface of larger (from a micron to few microns) host particles to impart useful properties to the final product, which are engineered particles. Dry coated engineered particles can have improved flow and handling properties, and hence are very useful for a number of industrial applications. In this paper, we present preparation, characterization and applications of such engineered particles to pharmaceutical applications.

Cornstarch particles, with a mean size of around 15 microns are coated with 20nm hydrophobic fumed silica nanoparticles by dry particle coating method to reduce the cohesiveness of the original powder. Acetaminophen and tolmetin sodium particles were also used as host particles. The weight percentage of additive is varied between 0.025% and 5%, which influences the cohesion of the coated powder. We examine the influence of the cohesion on packing density.

Flowability of fine powders is strongly dependent on memory effects such as the effect of previous consolidation stresses on the packing density of the powder. In this work we present measurements of the particle packing density of fine cohesive powders as a function of the consolidation stress previously applied on the sample. Experiments are carried out using a novel device for measuring the tapped density as well as the Sevilla Powder Tester apparatus, which is based on the use of gas flow either to fluidize to erase memory of the powder or to compress the bed in order to subject it to a controllable consolidation stress. The particle volume fraction is derived from the height of the bed, which is measured by a high-accuracy sensor placed on top of the vessel. It is observed that powders with lower amounts of coated additive have larger cohesion and form larger agglomerate structures and hence show lower packing densities. As amount of coated additive increases, cohesion is reduced and smaller agglomerates are formed which pack better in both fluidized and non-fluidized states. The effect of low consolidation stresses on the packing density is also examined.

Lack of flowability is a continual problem in manufacturing processes of pharmaceuticals, for example. By altering the flow characteristics of the powder, these manufacturing processes can be improved and using one or both of these powder testing methods would help in further formulations by quantifying the improvement. Specific applications potentially include mixing and tableting processes.

571f Highly Sensitive Optical Detection Of Humidity On Polymer/metal Nanoparticle Hybrid Films

Norman A. Luechinger, Robert N. Grass, and Wendelin J. Stark. Chemistry and Applied Biosciences, Swiss Federal Institute of Technology (ETH Zurich), Wolfgang-Pauli-Str. 10, ETH Hönggerberg, Zurich, 8093, Switzerland

Control and detection of humidity is most crucial in process and packaging applications. At present, mainly sensors with an electronic read-out are in use. This contribution introduces an alternative, directly optically (visually) readable humidity detection method. The used advanced hybrid metal/polymer surfactant films allows precise detection of water and solvent vapor levels without requiring an electrical reactant or power supply. More specifically, the sensor film can be easily obtained from copper nanoparticles protected by a 2-3 nm carbon shell, a polymeric dispersant and a wetting agent. The films showed an exceptional sensitivity with optical shifts in the visible light range of up to 50 nm for a change of 1% in relative humidity. This coloration response of the nanocomposite film could be attributed to the formation of a water-surfactant interference film when brought in contact to air with a certain humidity level.

Figure: Deposition of C/Cu dispersions on a glass substrate results in metallic copper films (a). Exposure to a stream of ethanol vapor resulted in rapid and reversible, intense iridescent coloration (b). Exposure to saturated water vapor (c) at room temperature provoked an intense green coloration that gradually shifted back to pink (d), blue (e), and orange (f) under ambient humidity.

References:

Athanassiou, E. K., R. N. Grass, et al. **(2006)**. "Large-scale production of carbon-coated copper nanoparticles for sensor applications." *Nanotechnology* 17(6): 1668-1673.

Luechinger, N. A., S. Loher, et al. **(2007)**. "Highly sensitive optical detection of humidity on polymer/metal nanoparticle hybrid films." *Langmuir* 23(6): 3473-3477.

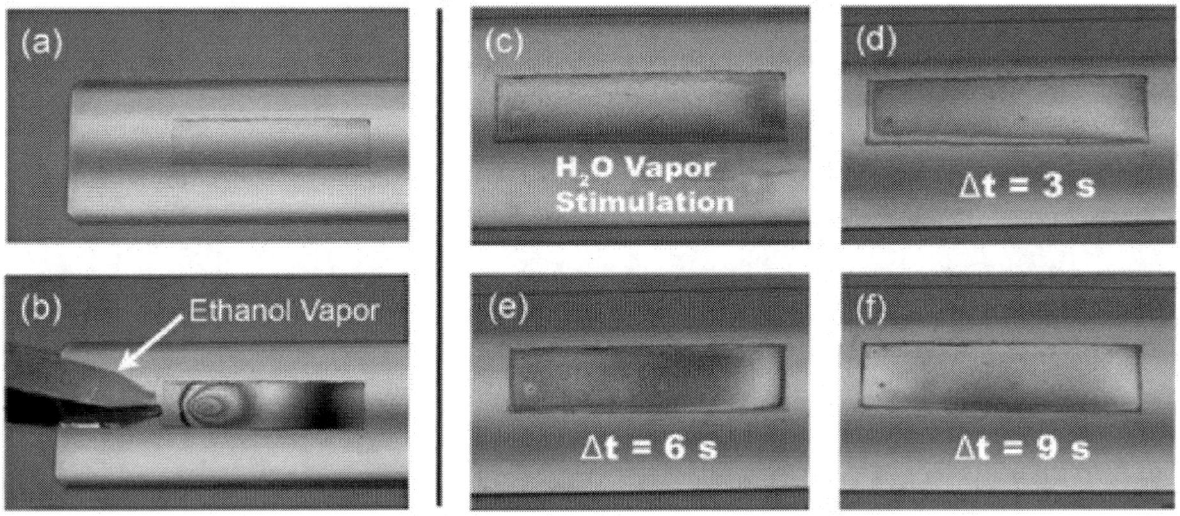
(a)
(b)
Ethanol Vapor
(c)
H2O Vapor
Stimulation
(d)
Δt = 3 s
(e)
Δt = 6 s
(f)
Δt = 9 s

571g Antibacterial Coatings On Glass And Plastic Materials Via Single-Dip Coating

Stoyan K. Smoukov and Bartosz A. Grzybowski. Chemical & Biological Engineering, Northwestern University, 2145 Sheridan Rd/TECH E136, Evanston, IL 60208

Self-organization of 5nm silver particles results in dense coatings on a variety of surfaces via a single dip solution procedure, thus conferring antibacterial properties to the surfaces.

585a Advanced Signal Analysis Methods For Selective Agglomeration Detection In Fluidized Bed Energy Conversion

Malte Bartels[1], John Nijenhuis[1], J. Ruud Van Ommen[1], Peter Verheijen[2], and Freek Kapteijn[1]. (1) DelftChemTech, Delft University of Technology, Julianalaan 136, Delft, 2628 BL, Netherlands, (2) Biotechnology, Delft University of Technology, Julianalaan 67, Delft, 2628 BC, Netherlands

The utilization of coal in fluidized bed conversion (combustion and gasification) is well researched and an established technology in industrial practice. Especially when utilizing biomass, however, agglomeration of bed/ash particles often forms a major operational problem. In industrial practice, typically measurements of average pressure drop over the bed and average temperatures are used to evaluate the fluidization quality. However, in many cases these measurements do not detect agglomeration problems in an early stage of the agglomeration process; they are therefore not suitable to indicate when corrective measures against poor fluidization have to be taken.

The attractor comparison method [AIChE Jour. 2000;46:2183-2197] has been developed in order to solve this problem. This method is based on high-frequent pressure fluctuation measurements and has been successfully applied on lab- and industrial scale bubbling fluidized beds. Yet, the source of the changed hydrodynamics cannot be unambiguously determined. We have therefore developed a new screening methodology to assess the capability of many other signal analysis methods in addition to attractor comparison. First, the raw pressure fluctuation data is pre-treated (filtered) with a method out of a pool of m methods. By filtering one can potentially extract specific signal properties relevant to agglomeration and therefore increase the sensitivity of an analysis method. Consecutively, the pre-treated data is analyzed with an analysis method out of a pool of n methods. For each of the (m x n) combinations of pre-treatment / analysis method, the screening methodology then calculates a quantitative measure for the sensitivity towards agglomeration as compared to other effects, such as changes in bed mass and fluidizing gas velocity. Results show that this methodology is indeed capable to reject inappropriate methods and to highlight methods specifically sensitive towards agglomeration as compared to other effects. Besides the application in bubbling fluidized beds, special attention is given to circulating fluidized beds in light of their frequent application in the energy sector.

In addition to the agglomeration detection, we are presenting some first results of counteracting agglomeration during biomass combustion on lab- and bench scale. Our monitoring approach is applied to determine the appropriate time to impose different strategies to counteract the agglomeration process. Hereby it is possible to successfully maintain stable fluidization.

585b Co-Ni/Al2O3 Oxygen Carrier For Fluidized Bed Chemical-Looping Combustion

Mohammad M. Hossain, Kelly E. Sedor, and **Hugo I. de Lasa**. Chemical and Biochemical Engineering, The University of Western Ontario, Biochemical Engineering Building, Room 85 (100), Faculty of Engineering, London, ON N6A 5B9, Canada

In recent years, chemical-looping combustion (CLC) has received growing interest as an efficient combustion technology for the capture of CO2 and the substantial reduction of NOx emissions from fossil fuel based power generation. This non-conventional technique involves the use of a solid oxygen carrier to supply oxygen, instead of air, during fuel combustion, thereby preventing the dilution of CO2 in the flue gas with N2 from air. Consequently, CO2 is delivered without any energy penalty for separation. Formation of NOx is also negligible because the fuel burns in absence of air without flame. This study deals with the development of a bimetallic Co-Ni/Al2O3 oxygen carrier suitable for fluidized bed CLC. Temperature programmed characterization shows that the addition of Co enhances the reducibility of the oxygen carrier by influencing the state of the surface minimizing the formation of nickel aluminate. Reactivity study of the prepared oxygen carriers in a CREC fluidized bed riser simulator, demonstrates that the Co-Ni/Al2O3 particles are highly reactive and stable over multiple reduction/oxidation cycles. The XRD analysis shows similar phases on both the fresh and used samples, confirming no phase transformation occurred during the CLC processes. The addition of Co also inhibits metal particle agglomeration by maintaining consistent metal dispersion during the cyclic reduction/oxidation processes, which is confirmed by pulse chemisorption analysis over repeated reduction/oxidation cycles. The SEM images of a fresh and a used carrier sample evidences similar distribution of the metal crystallites suggesting minimum agglomeration of the nickel species at high temperature CLC processes. The solid-state kinetics for both the reduction and oxidation cycles is established using a clarified Avrami-Erofeev model at non-isothermal conditions. This random nucleation model describes solid phase changes adequately. The activation energy for Co-Ni/Al2O3 reduction is found to be significantly lower than the activation energy for the unpromoted Ni/Al2O3 sample, indicating the positive effect of Co on the Co-Ni/Al2O3 oxygen carrier.

Keywords: CO2 capture, NOx, CLC, Co-Ni/Al2O3 oxygen carrier, reactivity, stability, solid-state kinetics.

585c Enhanced Hydrogen Production with Integrated Carbon Capture and Sequestration

Hyung Rae Kim[1], **Ah-Hyung Alissa Park**[2], Luis G. Velazquez-Vargas[3], Fanxing Li[3], and Liang-Shih Fan[1]. (1) Chemical and Biomolecular Engineering, The Ohio State University, 140 W. 19th Ave., Columbus, OH 43210, (2) Earth and Environmental Engineering & Chemical Engineering, Columbia University, 918 S. W. Mudd Hall, Mail Code: 4711, 500 West 120th Street, New York, NY 10027, (3) Ohio State University, 140 W 19th Ave, Columbus, OH 43210

A novel carbon mineral sequestration process developed by Park and Fan immobilized carbon dioxide into thermodynamically stable solid matrix of magnesium carbonate using serpentine, hydrous magnesium silicate mineral. Serpentine generally contains approximately 5 wt% of Fe. Thus, during this process, two other solid products are also generated: high surface area silica and iron oxide particles. By carefully controlling the pH of the system, these solids product can be produced with high purity. In order to ensure the economic feasibility as well as the environmental impact of the carbon mineral sequestration process, the development of various industrial pathways that may utilize these solid products, including iron oxide, is proposed. One of the potential applications of the iron-bearing solid product is the enhanced hydrogen production via the chemical looping process that produces hydrogen either directly from coal or syngas, while producing sequestration-ready CO2 stream. Iron oxide is one of key ingredients of chemical looping sorbent. In this study, the optimum characteristics of the solvent extracting iron from serpentine and the subsequent precipitation process of iron oxide that will benefit the chemical looping process are investigated. The prepared chemical looping sorbents are characterized and tested for the hydrogen production capability and recycleabiltiy.

Exergy Analysis on Chemical Looping Reforming Process

Fanxing Li; Luis G. Velazquez-Vargas; Liang Zeng; Joseph Chen; Liang-Shih Fan

Department of Chemical and Biomolecular Engineering
The Ohio State University
140 W 19th Avenue, Columbus OH, 43210

Abstract

The Chemical Looping Reforming (CLR) process developed at The Ohio State University (OSU) converts the coal into H_2. In the CLR process, highly reactive metal oxide particles, rather than air or oxygen, are reacted with coal to produce reduced metal particles and an exhaust gas stream containing mainly CO_2 and Steam. By condensing the steam, a sequester-ready CO_2 stream can be obtained. The reduced metal particles are subsequently regenerated with steam and/or air in a second reactor to produce hydrogen or syngas gas respectively. The reactions oxidize back the oxygen carrier particles to their original state. These particles are then cycled back to the first reactor to react with the coal.

This study presents the configuration of the CLR process for the production of H_2 from coal. ASPEN plus simulation shows that the energy conversion efficiency of CLR process can be as high as 80% (HHV), much higher than the 64% efficiency obtained by the conventional Coal Gasification-Water Gas Shift (CG-WGS) route. This increase in efficiency for the CLR process is due to the better energy management strategy. In the conventional CG-WGS process mixtures of CO_2 and H_2 are produced, from which H_2 is obtained after several separation and purification steps. Further, the CLR process offers better exergy recuperation compared to the conventional CG-WGS process. In this study, an exergy (availability) analysis on both processes is carried out. The analysis shows that the regenerated particle serves as an exergy carrier, thus improving the overall energy integration in the CLR process.

Introduction

Gasification is the process of converting solid carbonaceous fuels into gaseous fuels, such as hydrogen and syngas, via a partial oxidation reaction. Because it is cleaner and more flexible than direct combustion, gasification is a prime candidate for use in next-generation coal-based plants, especially in a carbon constrained scenario. Although gasification is considered to be more advantageous when compared to combustion from a carbon management viewpoint, carbon capture would nevertheless impose significant energy penalties towards gasification systems. Moreover, the efficiencies of solid fuel gasification technologies are low due mainly to the relatively complicated processing steps needed for solid fuel utilization. Therefore, future coal gasification processes should be simplified through process intensification and coupled with more efficient carbon capture schemes. The patented Chemical Looping Reforming process developed in the

Ohio State University is such a technology that has the potential to offer significantly more affordable products while effectively capturing CO_2 (Fan, et al., 2007).

Process Overview

Figure 1 shows a simplified flow diagram of a specific configuration of the CDCL process. This process has three reactors, i.e., the fuel reactor (unit 1), the hydrogen reactor (unit 2), and the combustion train (unit 3). The fuel reactor converts carbonaceous fuels such as coal into CO_2 while reducing Fe_2O_3 to a mixture of Fe and FeO, the hydrogen reactor oxidizes the reduced Fe/FeO particle to Fe_3O_4 using steam, producing a H_2 rich gas stream. The combustion train pneumatically transports the Fe_3O_4 particles from the H_2 reactor outlet to the fuel reactor inlet using air. The Fe_3O_4 particles are reoxidized to Fe_2O_3 during the conveying process.

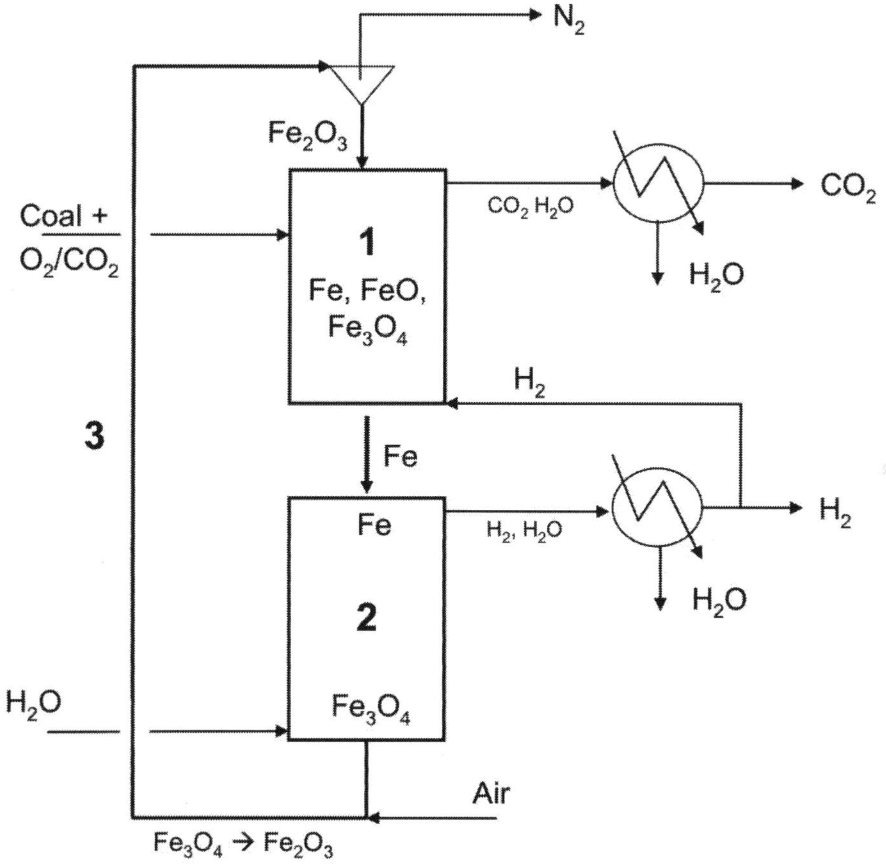

Figure 1. Simplified schematic of the Chemical Looping Reforming (CLR) system

Fuel Reactor

The fuel reactor is a moving bed reactor operating under 750 – 900 °C, 1 – 30 atm. The overall reaction in the fuel reactor is:

$$C_{11}H_{10}O + 6.44\ Fe_2O_3 + 3.34\ O_2 \rightarrow 11CO_2 + 5\ H_2O + 12.88\ Fe \quad \Delta H = 0\ @\ 900\ °C$$

Here the coal considered is Pittsburgh #8 and is represented as $C_{11}H_{10}O$ given the elemental composition (Stultz et al. 1992). The substoichmetric amount of O_2 (according to coal) sent into the fuel reactor is used to compensate the heat needed for the reduction reaction of Fe_2O_3 using coal.

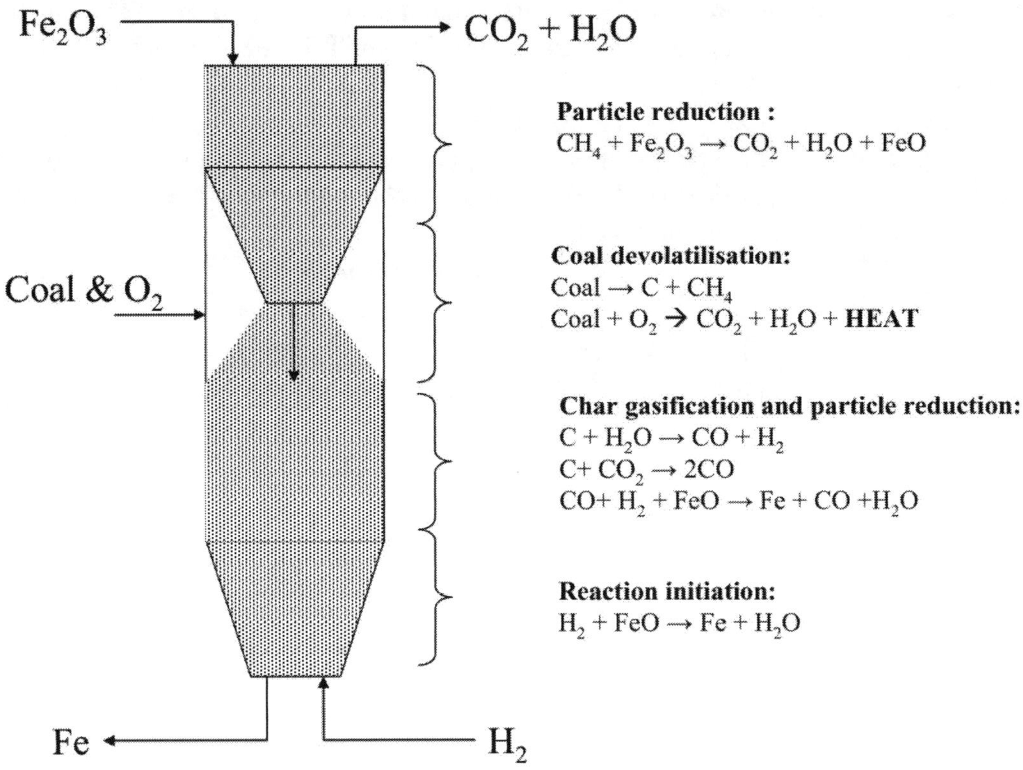

Figure 2 Gas-solid contacting pattern of the fuel reactor

In order to promote the reaction rate between the Fe_2O_3 and coal particles, a small portion of the hydrogen (< 5%) produced from the hydrogen reactor is introduced to the bottom of the fuel reactor. By doing this, the steam formed from the reaction between hydrogen and FeO at the bottom of the fuel reactor would help to gasify the char.

Figure 2 shows the gas-solid contacting pattern inside the fuel reactor. Fresh Fe_2O_3 composite particles are fed from the top of the fuel reactor while coal is introduced in the middle section of the fuel reactor. This configuration allows the volatiles from the coal to react with Fe_2O_3 particles to form CO_2 and H_2O as they move towards the gas outlet of the fuel reactor at the top. The devolatilized coal char, however, would move downwards along with the partially reduced Fe_2O_3 particles. The coal char would be gradually gasified by the CO_2 and H_2O formed at the lower part of the reactor. Provided enough residence time is (30 – 90) mins depending on the operating temperature and pressure) give to the solids, coal char can be fully converted before it reaches the solid outlet of the reactor at the bottom. Meanwhile, Fe_2O_3 can be reduced to a mixture of metallic iron and FeO. Coal ash will also come out along with the particles in solid form. Thus, the product from the fuel reactor is a solid stream with Fe, FeO, and coal ash and a gas stream with mainly CO_2 and H_2O. By condensing out the H_2O in the gas stream, a ready-to-sequester CO_2 stream is obtained from the fuel reactor without the inherent separation costs.

Hydrogen Reactor

The hydrogen reactor is a moving bed reactor operates under 700 – 900 °C, 1 – 30 atm. In the hydrogen reactor, the Fe and FeO mixture from the fuel reactor reacts with steam in a countercurrent mode to achieve maximum solid/gas conversion. The products of the reactor are Fe_3O_4 particles and H_2. The reaction in the hydrogen reactor is as follows,

$$3\,Fe + 4\,H_2O \rightarrow Fe_3O_4 + 4H_2$$

This reaction is slightly endothermic and therefore, steam with lower temperature is introduced into the reactor to moderate the temperature of the reactor.

Combustion Train

In order to fully convert the CO and H_2 from the char gasification into CO_2 and H_2O, Fe_2O_3 had to be used as the feedstock for the fuel reactor according to thermodynamics. Therefore, the Fe_3O_4 from the hydrogen reactor needs to be oxidized. This regeneration step is done in the combustion train.

The combustion train is a fluidized bed reactor operating at similar a pressure as the fuel reactor and the hydrogen reactor under adiabatic mode. Air is used to pneumatically convey the Fe_3O_4 particles from the outlet of the hydrogen reactor to the fuel reactor inlet. The Fe_3O_4 particles are oxidized to Fe_2O_3 in this step:

$$4Fe_3O_4 + O_2 \rightarrow 6Fe_2O_3$$

This highly exothermic reaction will heat up the solid as well as the gas. The hot solid would be subsequently fed into the fuel reactor to (partially) compensate the heat needed to convert the coal. The coal ash, with significantly smaller size than the Fe_2O_3 composite particles, will be separated out from the cyclone on the top of the fuel reactor, <1% makeup particles will also be introduced to the fuel reactor along with the regenerated particles.

Earlier studies (Gupta et al, 2006) showed that a maximum of about 86% coal to hydrogen conversion efficiency in term of HHV (or 0.18 kg H_2/ kg coal) can be achieved in the CDCL process.

Exergy Analysis

In order to better reveal the advantages of the chemical looping reforming process, an exergy analysis is performed over the traditional coal gasification – water gas shift process for hydrogen production and the chemical looping reforming process. Figure 3 illustrates the two schemes in a simplified manner.

The exergy (availability) of a system is defined as the maximum amount of usable work extractable from the system during a transformation that brings the system into equilibrium with a reference state (Kotas, 1985). The ambient environment is commonly used as the reference state. The exergy rate is the ratio between the available work and the total amount energy. The second law of thermodynamics indicates that energy degrades in all irreversible processes. Although it is not possible to eliminate the energy degradation, the exergy loss in a process can be minimized through strategic energy management.

The following assumptions are made in the calculation for results presented in the examples:

- The environmental temperature is T_0 = 273.15 K and the environmental pressure is P_0 = 101325 Pa;
- The reference substance for carbon and CO is 400 ppm CO_2, and the reference substance for H_2 is pure water, reference substance for Fe, FeO, Fe_3O_4 is pure Fe_2O_3;
- The upper limit for gas turbine operating temperature is 1800 K;
- Coal is considered as pure carbon;
- There is no heat loss and the heat can be integrated with a 100% efficiency whenever feasible;
- The enthalpy of devaluation is defined as the enthalpy change for a substance from current state to its reference state
- The feedstock in the outside of the reaction loop is at the environmental state

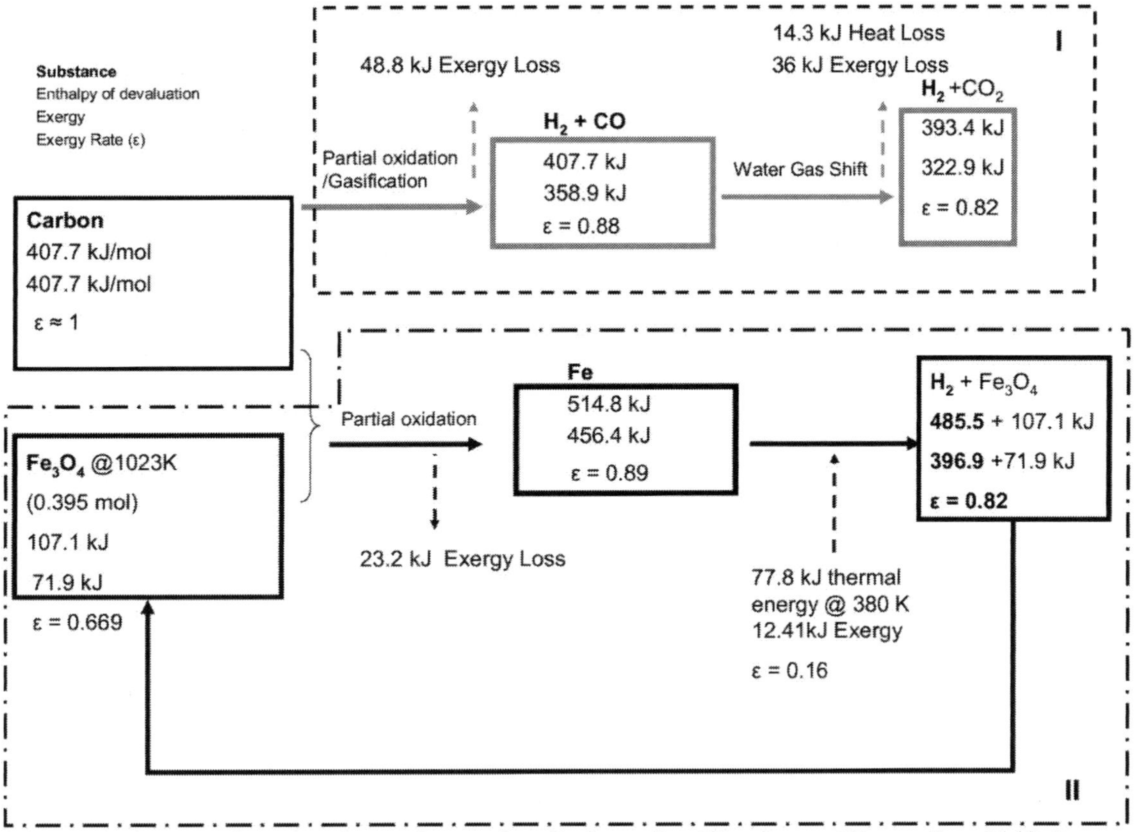

Figure 3. Exergy Recovering scheme for Carbon Gasification – Water Gas Shift Process (reaction temperature is assumed to be 1123K for scheme II)

Scheme I is the traditional route where carbon is converted into syngas followed by the water gas shift reaction to produce hydrogen. Scheme I consist in two steps. The first step is the gasification step, which incurs about 12% exergy loss due to partial oxidation reaction. The second step consists of the water gas shift reaction, which has another 8.8% exergy loss due to the conversion of CO into H_2. Hydrogen has both lower enthalpy of devaluation and lower exergy rate. Overall speaking, scheme I has an exergy loss of 20.8 % relative to the fuel input.

Scheme II offers an alternative approach using chemical looping concept. Here Fe is used as a chemical intermediate to convert carbon into hydrogen. There are also two steps involved in scheme II, the reactions are as follows:

$C + 0.395Fe_3O_4 + 0.21O_2 \rightarrow 1.185Fe + CO_2$ Step 1

$Fe + H_2O \rightarrow Fe_3O_4 + 4H_2$ Step 2

Small amount of O_2 is introduced in the first reaction to provide the heat needed to reduce Fe_3O_4 to Fe. Much less exergy loss occurs in step 1 (23.2 kJ or 4.8% according to total energy input) compared to traditional gasification step (48.8 kJ). This is due to the recovering of a low exergy rate chemical, Fe_3O_4 ($\varepsilon = 0.669$), with carbon ($\varepsilon = 1$) to form a chemical, Fe ($\varepsilon = 0.89$), with a medium exergy rate.

The second step in scheme II, steam-iron reaction, requires only a small amount of low grade heat to feed H_2O into the system under environmental conditions. This makes it much easier to balance both the heat and exergy needed in the second step. Zero exergy loss is achievable in the second step of scheme II.

As a result, the traditional gasification – water gas shift process would lead to at least 68.7 kJ (20.8%) exergy loss for every mole of H_2 generated; however, when the chemical looping process is used and when Fe – Fe_3O_4 are used as the looping medium, the exergy loss can be reduced to 14.7 kJ for each mole of H_2 generated. Thus, the exergy loss for H_2 production is reduced by more than four times for using the chemical looping process.

It is noted that the results presented in the examples are based on the conditions dictated by the thermodynamic principles. Thus, they represent only the upper bound of the energy conversion efficiencies in the conversion process. Nevertheless, due to the common assumptions employed in the evaluation of process options, the results obtained serve as a guide in comparing the two processes. To conduct the energy/exergy analysis of processes that are more comparable to practical situations, the analysis of the detailed process schemes is necessary.

The better performance of chemical looping reforming process lies in the use of suitable reaction intermediates with desirable chemical exergy rates. The integration of energy with low exergy rate to improve the system energy conversion efficiency has already been explored (Tsutsumi, 2006). The utilization of chemical intermediates, however, is an area of special interest to the development of next generation energy conversion systems. The exergy analysis, as it is illustrated above, would serve as a good indicator for the potential of a novel system.

Conclusion

The Chemical Looping Reforming technology is an attractive approach to produce hydrogen from coal. The process involves the use of Fe_2O_3 in cyclic reduction and oxidation reactions with coal and steam respectively. Exergy analysis reveals that this process is more thermodynamically advantageous than the state-of-the-art coal gasification – water gas shift process for H_2 production by reducing the exergy loss by more than 75%.

Acknowledgements

The authors would like to acknowledge the Ohio Coal Development Office of the Ohio Air Quality Development Authority and United States Air Force for financial support for this project.

References

Fan, L.S., Gupta, P., Velazquez Vargas, L. G., Li, F. Systems and methods of converting fuel. PCT Int. Appl. 2007:788506(2007)

Stultz S.C.; and Kitto J.B. "Steam – its generation and use" 40th ed. Barberton: Babcock and Wilcox Company (1992).

Gupta, P; Velazquez-Vargas L.G.; Li F.; Fan L.S. "Chemical Looping Reforming Process for the Production of Hydrogen from Coal". *Proc. 23th Annu. Int. Pittsburgh Coal Conference*, Pittsburgh, PA, USA, Sep (2006).

Kotas, T.J. The Exergy Method of Thermal Plant Analysis. Butterworths: London. (1985)

Tsutsumi, A. Exergy recuperative gasification technology for hydrogen and power co-production. Abstracts of Papers, 231st ACS National Meeting, Atlanta, GA, United States, March 26-30, 2006 (2006).

585e Solar Thermal Conversion of Biomass for Synfuel Production

Christopher Perkins[1], Paul Lichty[2], and Alan W. Weimer[2]. (1) University of Colorado, 1111 Engineering Drive, Boulder, CO 8030900424, (2) Chemical and Biological Engineering, University of Colorado, 1111 Engineering Drive, Boulder, CO 8030900424

Aerosol steam gasification and water-free pyrolysis of biomass components was conducted at temperatures achievable using solar thermal energy (>1000 °C). Gasification and pyrolysis of cellulose and lignin were performed using an electrically heated tube furnace to simulate the solar source. High conversions of biomass to syngas (>98%) were found for temperatures exceeding 1100 °C, while significantly lower conversions (< 65%) were obtained at temperatures less than 1000 °C. The low temperature experiments showed a large number of side products, leading to the conclusion that gasification should be conducted at as high a temperature as possible. These results were verified "on-sun" at the National Renewable Energy Laboratory, showing that high yields of syngas could be obtained in short residence times (<2 s) using a solar thermal source. Gasification of a naturally occurring biomass product (Poa pratensis, Kentucky bluegrass) showed similar results, with very high yields of syngas in the solar reactor. An economic analysis of hydrogen production by this method was conducted and yielded a selling price of $0.95 - $1.30 / kg H2. This is significantly lower than the conventional price of fossil fuel derived hydrogen.

IMPROVEMENTS IN THE SYNTHESIS OF GUANIDINIUM AZOTETRAZOLATE (GuZT)*[1]

Shannon M.Lenahan and Jerry S. Salan
Naval Surface Warefare Center, Indian Head Division
Indian Head, MD

ABSTRACT

This project involves optimizing the synthesis of guanidinium azotetrazolate (GuZT) for yield and purity. Experiments were first performed maximizing reactant content. Next, a design of experiments (DOE) was performed screening synthesis parameters for effect on product yield, purity, and particle characteristics. After the DOE, four more experiments were performed to investigate washing parameters. The project was undertaken to determine the effects of selected processing parameters on specific responses and to determine operating parameters for the synthesis of GuZT at the 100-gallon scale. Three of the four factors studied showed effects that were considered significant. Based on this Design of Experiments, using 20% excess guanidine hydrochloride (Gu-HCl) and cooling to 5°C before filtering maximize yield. The product GuZT should be washed with 2° – 4°C water to remove sodium and chloride ion impurities. Further work in this area might include investigations into further increasing the yield by increasing the reaction time or the excess Gu-HCl. Also, altering the crystal habit of GuZT may prove beneficial for formulation .

INTRODUCTION

Guanidinium azotetrazolate (GuZT) is a high-nitrogen material with potential application as a burning rate modifier in gun propellants or an ingredient for insensitive explosives or propellants. It appeals to the energetics industry because of its straightforward synthesis and relative insensitivity to friction and impact. The reactants and products are displayed in Figure 1. Properties of GuZT are given in Table I. The synthesis method used to make GuZT in these studies involves dissolving NaZT in water, heating to a desired reaction temperature, and then adding Gu-HCl in aqueous solution. GuZT precipitates, is cooled to a temperature prescribed by the DOE, and then filtered. A room temperature water wash was applied to each batch of GuZT on the filter. During synthesis, Lasentec Focused Beam Reflectance Measurement (FBRM) and Particle Vision Microscopy (PVM) probes were used in the reaction vessel. These instruments provided real-time information on GuZT particle size and shape.

---

*This effort was performed under the Navy's Manufacturing Technology Program

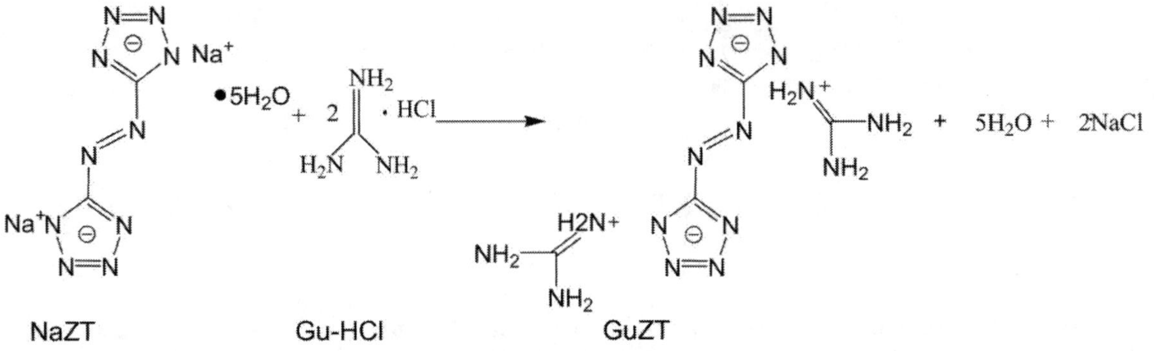

Figure 1. GuZT synthesis reaction scheme

Table I. Properties of Guanidinium Azotetrazolate (GuZT)[1]

Density (g/cm^3)	1.538
Heat of Formation (kcal/mol)	+98
Impact Sensitivity (cm, Type 12) (HMX=26)	>320
Vacuum Stability (ml/g for 48h at 100°C)	0.21
Friction Sensitivity (BAM, kg)	Not Sensitive at 36 kg
Detonation Velocity (km/s) (calc.)	7.1
Detonation Pressure (GPa) (calc.)	155

Indian Head has made approximately 12 kg of GuZT at the 50-gallon scale. With the goal of developing a more efficient process, the Chemical Development group at Indian Head contacted Dr. David Chavez of Los Alamos National Labs (LANL), who in collaboration with Indian Head engineers more than doubled the quantity of GuZT obtained per batch by increasing the concentration of reactants in the reaction solution. The work reported here is based on this increased-density synthesis and includes execution of a design of experiments (DOE) screening guanidinium azotetrazolate (GuZT) synthesis parameters for effect on product yield, purity, and particle characteristics. A short study of washing parameters was also performed. The results of these studies will be used to determine operating parameters for the synthesis of GuZT at the 100-gallon scale. Table II gives the factors and levels used for the 19 experiments performed. Measured responses were percent yield, the sum of sodium and chloride ion weight percentage in samples, average percent error in carbon, hydrogen, and nitrogen content, and FBRM statistics on particle size and shape. The primary responses were yield and purity; particle characteristics were evaluated because GuZT exhibits acicular crystal morphology similar to TAGZT, which has proved difficult in remote feeding operations and is not optimal for energetic material processing. Factor effects on particle size and shape will be used for future studies and will not be taken into account during 2007 GuZT manufacture at the 100-gallon scale.

Table II. Factors and Levels

Factor	Low Level	High Level	Description
Excess Gu-HCl	0%	20%	Weight % excess beyond stoichiometric quantity required
Cool Temperature	5°C	60°C	Final temperature of GuZT slurry before filtering
Reaction Temperature	60°C	80°C	Temperature during Gu-HCl dosing
Gu-HCl Concentration	0.4 g/ml	1.3 g/ml	Concentration of Gu-HCl in feed solution

RESULTS AND DISCUSSION

All 19 DOE experiments were performed using a RC1*e* Mettler Toledo reaction calorimeter in a 1.2-liter vessel and a glass upward impeller. Dosing of varying concentrations of Gu-HCl/water solution was performed at 5 grams per minute using a Mettler Toledo RD10 dosing controller and Ismatec ISM 829 peristaltic pump. Experimental design and analysis was done using MODDE 6.0.

The effects plots resulting from the DOE were analyzed to determine which factors or interactions between factors have a statistically significant effect on the measured responses. In Figures 2,3,5, and 7 through 9, the effect on the individual responses is either positive or negative when the factor moves from low to high levels as set up in the DOE matrix. These levels are provided in Table II. Error bars on the effects plots indicate that with a 95% confidence level, the actual magnitude of the effect could lie at any point along the error bar. For this reason, effects with error bars crossing the x-axis are generally not considered. This test of significance is not exhaustive and a bit of discrimination on the part of the experimenter must be exercised. Both response and interaction graphs were used to make educated decisions as to the significance of the factors.

Figure 2 shows factor effects on yield. As shown by the fact that the only error bars that do not cross the x-axis are those for excess Gu-HCl and filtration temperature, these are the only two factors taken to have significant effects on yield. The error bars on the plots indicate that with a 95% confidence level, the actual magnitude of the effect could lie at any point along the bar. Figure 2 shows that as the GuZT filtration temperature moves from the low (5°C) to high (60°C) level, yield decreases and as the weight percent of excess Gu-HCl increases, yield is positively affected. Adding Gu-HCl at the higher level (20% excess) increased the yield, as did cooling to the lower temperature of 5°C before filtration. Higher yields were effected by cooling to a lower temperature before filtering and by increasing the amount of excess Gu-HCl added.

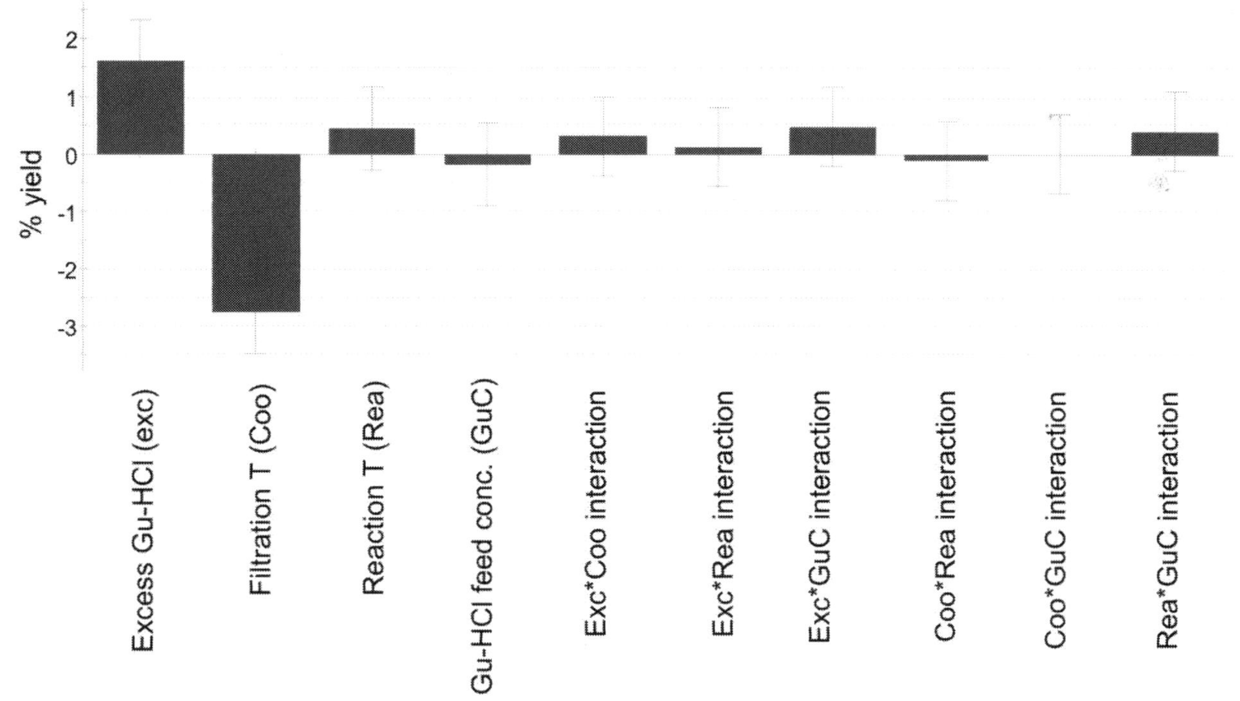

Figure 2. Factor and interaction effects on yield

Complicating the interpretation of this result is a varying amount of water present in each experiment. Total quantities of water present in the vessel after dosing of Gu-HCl varied between 486 and 591 ml. Obviously, the quantity of GuZT lost in this water varied as well, affecting the calculated yields. For this reason, an adjusted yield was calculated to compensate for the dissolved GuZT product lost in the water. The solubilities of GuZT used for this calculation are given in Table III. Experiment 1, for example, was filtered at 5°C and there were 568.1 ml of water present at the time of draining. If at 5°C there are 4.4 mg/ml GuZT in the water, 2.5 grams of GuZT were theoretically dissolved in the water present. Previously the yield was recorded as 78%, but with the 2.5 g dissolved GuZT added, the adjusted yield is 81%. Figure 3 shows the effect of excess Gu-HCl on yield and adjusted yield. This plot verifies that excess Gu-HCl is, indeed, a significant factor, and that the varying amounts of dissolved GuZT in each run had only a small effect on evaluation of this factor's significance.

Table III. GuZT Solubility in Water

Temperature, °C	GuZT Solubility, mg/ml
5	4.4
32.5	6.3
60	10.4

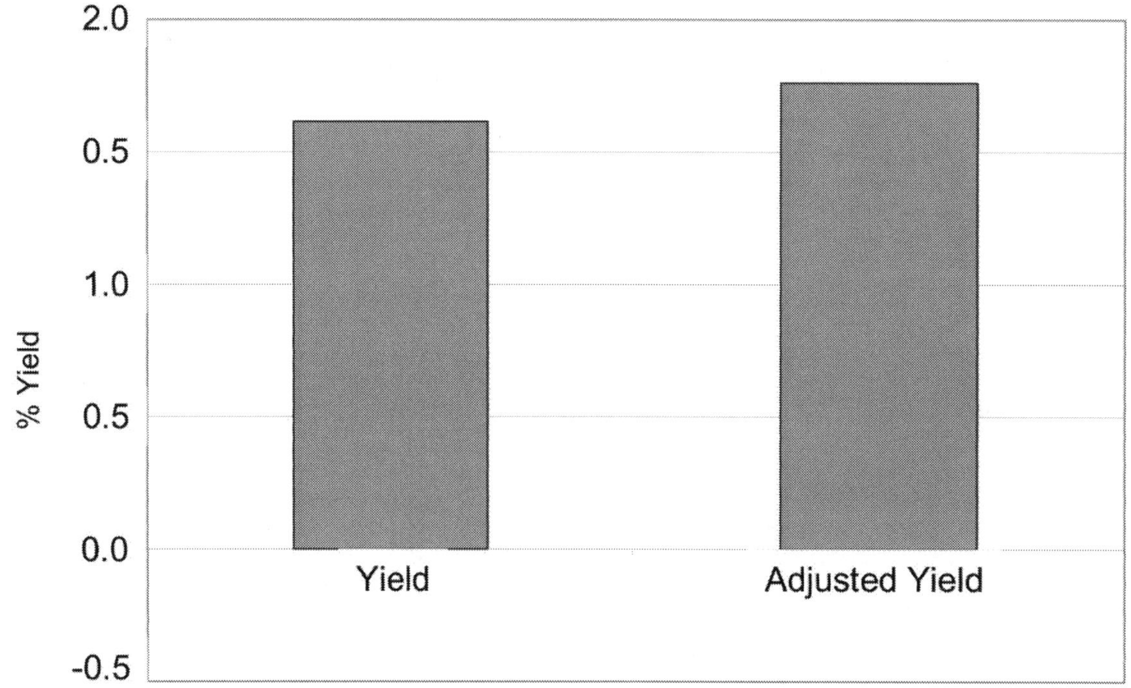

Figure 3. Effect of excess Gu-HCl on yield and adjusted yield

Figure 4 shows the prediction plot for the effect of excess Gu-HCl on yield. Percent yield increases linearly with increasing excess Gu-HCl percentage, and it is possible that by increasing the percent Gu-HCl beyond the maximum included in this study, further increases in yield may be obtained. A second DOE is proposed to investigate the possibility that additional excess GuZT may increase yields.

The yields in this study were consistently lower than those achieved by Dr. David Chavez (LANL) in his GuZT optimization work. One possibility is that the NaZT used for the experiments was not completely dry as supposed. NaZT is a pentahydrate and drying operations can remove one or two of the complexed water molecules from it. NaZT is a poor candidate for Karl Fischer moisture analysis because it will not completely dissolve in solvents other than water. So to evaluate the moisture in the NaZT used in these experiments, a sample of the material was weighed, dried under vacuum, and then placed in a container with a cup of water to reattach any complex waters that had been removed while drying. After this procedure, the NaZT sample exhibited no loss in weight and it was deemed dry (0% water). It is possible that the sample used for moisture analysis was not representative of the bulk NaZT used in the experiments. Also, the sample may not have been large enough to precisely measure mass loss due to drying using the balances available. These possibilities are being investigated.

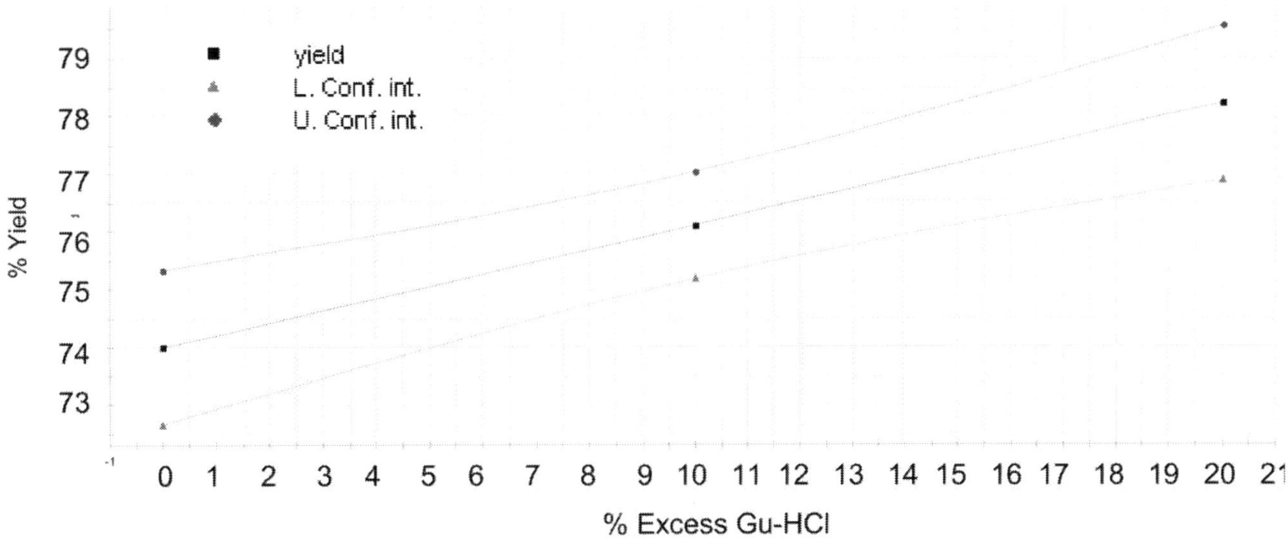

Figure 4. Prediction plot: effect of excess Gu-HCl on percent yield

On CHN purity there were no significant factor or interaction effects, but the quality of the model given by MODDE is statistically poor, suggesting that a factor not included in the DOE was responsible for variation in CHN composition. All samples were within 0.4% of theoretical value and acceptable per C and H requirements, but N measurements were consistently low. These low measurements most likely result from the nitrogen standard used for calibrating the elemental analyzer; the standard currently in use only contains approximately 20% nitrogen whereas a sample of GuZT contains nearly 80% nitrogen. Sodium and chloride ion contamination were determined using ion chromatography. Analysis of the effects plot for sodium and chloride ion contamination (Figure 5) shows that reaction temperature may have an impact on sodium and chloride ion content, but all 19 samples exhibited acceptable levels of contamination less than 1%. Further investigation into process effects on purity were deemed unnecessary since all samples analyzed were acceptably pure.

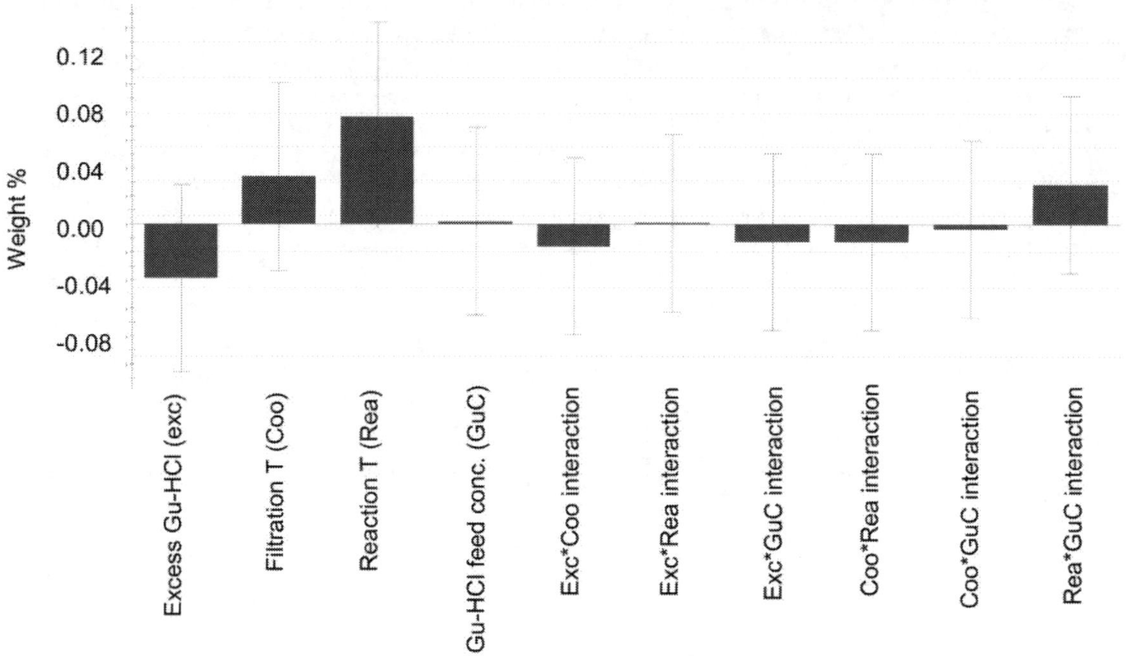

Figure 5. Factor and interaction effects on sodium and chloride ion contamination

FBRM and PVM were used to track particle properties of the GuZT crystals. The particles varied by experiment in needle size and number. Figure 6 shows PVM images taken during a few of the experiments.

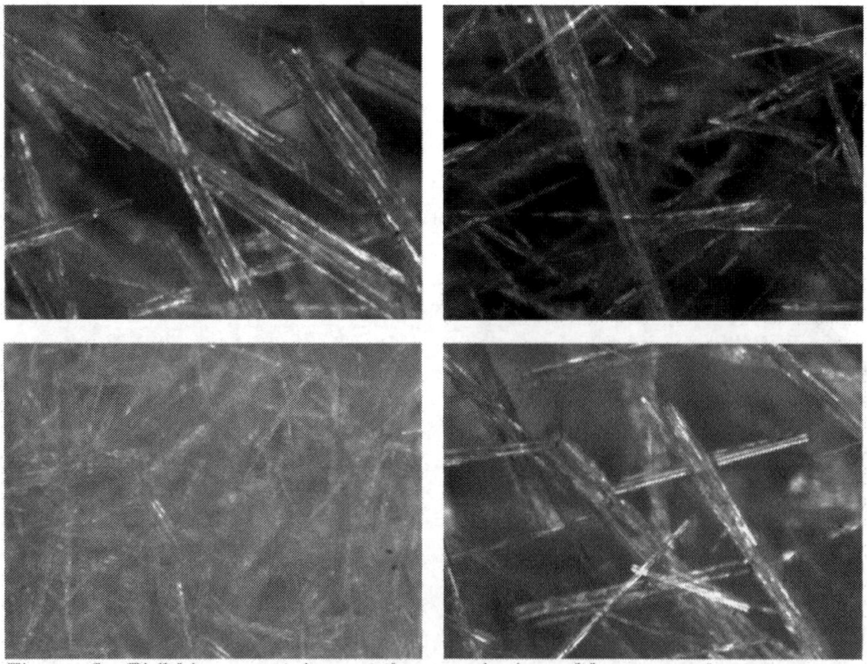

Figure 6. PVM images taken at the conclusion of four experiments

Using FBRM statistics to represent rough diameter and length measurements of the GuZT needles, an L/D ratio was calculated to estimate process parameter effects on needle aspect ratio. The L/D ratio is calculated by dividing the square-weighed mean statistic by the unweighted median statistic taken just before filtration. Square-weighting the data is a means of emphasizing the longer measurements. An L/D ratio approaching one indicates approximately spherical or cubic particles. Figure 7 demonstrates that a lower L/D is achieved using excess Gu-HCl. A significantly lower L/D is favorable for several reasons. First, as in the case of triaminoguanidinium azotetrazolate (TAGZT), a material very similar to GuZT, formulators experience trouble remotely feeding ingredient to mixing equipment with large L/D ratios. Crystals of this morphology may also inhibit good mixing during formulation. In addition to processing challenges, needle-shaped particles have been shown to increase the sensitivity of both the neat material and formulations[2]. The Navy's Insensitive Munitions efforts require that munitions under development be resistant to initiation upon unplanned stimuli. Controlling crystal morphology may be a way to achieve acceptable sensitivity for some materials and formulations.

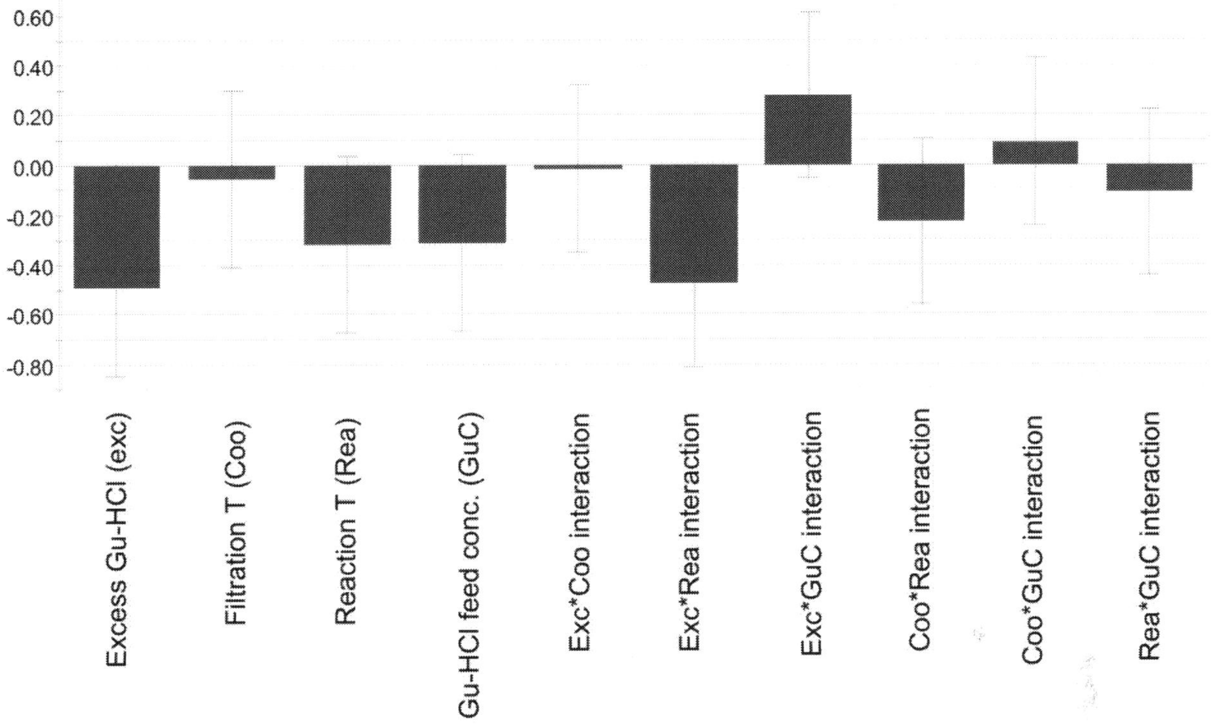

Figure 7. Factor and interaction effects on L/D

When the estimates of diameter (median, no wt) and length (mean, sqr wt) are considered separately instead of as a ratio, factor and interaction effects are pronounced and similar. Increasing reaction temperature increases average crystal diameter and length, and reaction temperature and Gu-HCl feed concentration interacted to exert a significant effect. The effects plot for particle diameter can be seen in Figure 8, and that for particle length is shown in Figure 9. This effect of reaction temperature is also clearly presented in Figure 10, where the unweighted FBRM chord length distributions for six experiments are compared. These distributions were recorded just before filtration. Reactions occurring at 60°C consistently produced more GuZT particles of smaller size than those occurring at 80°C. Slow filtration due to fine particles has not been a problem with GuZT synthesis, so these factor effects on particle size will not affect GuZT synthesis on the 100-gallon scale.

Particle size, in addition to particle morphology, can affect sensitivity and processability of materials. Particle size of ingredients can also affect sensitivity of formulations including that ingredient. To date, no study of the effects of GuZT particle size on sensitivity has been performed. In the event that such a study is performed in the future, data obtained here may be useful in adjusting GuZT synthesis parameters.

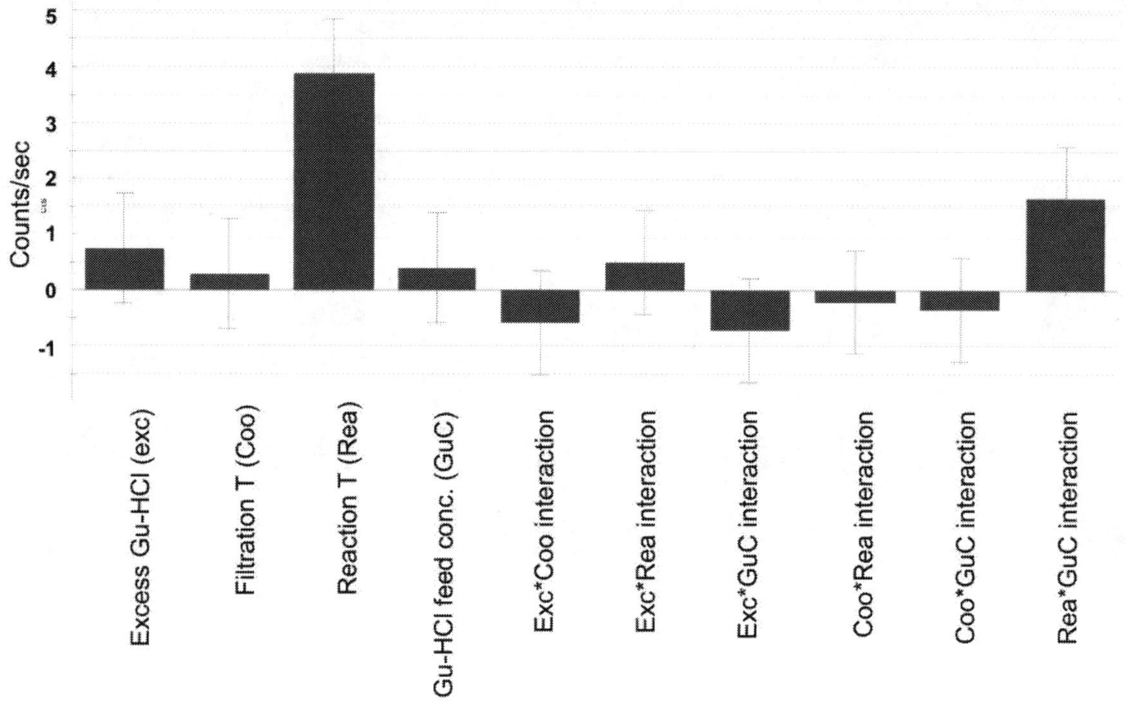

Figure 8. Factor and interaction effects on needle diameter

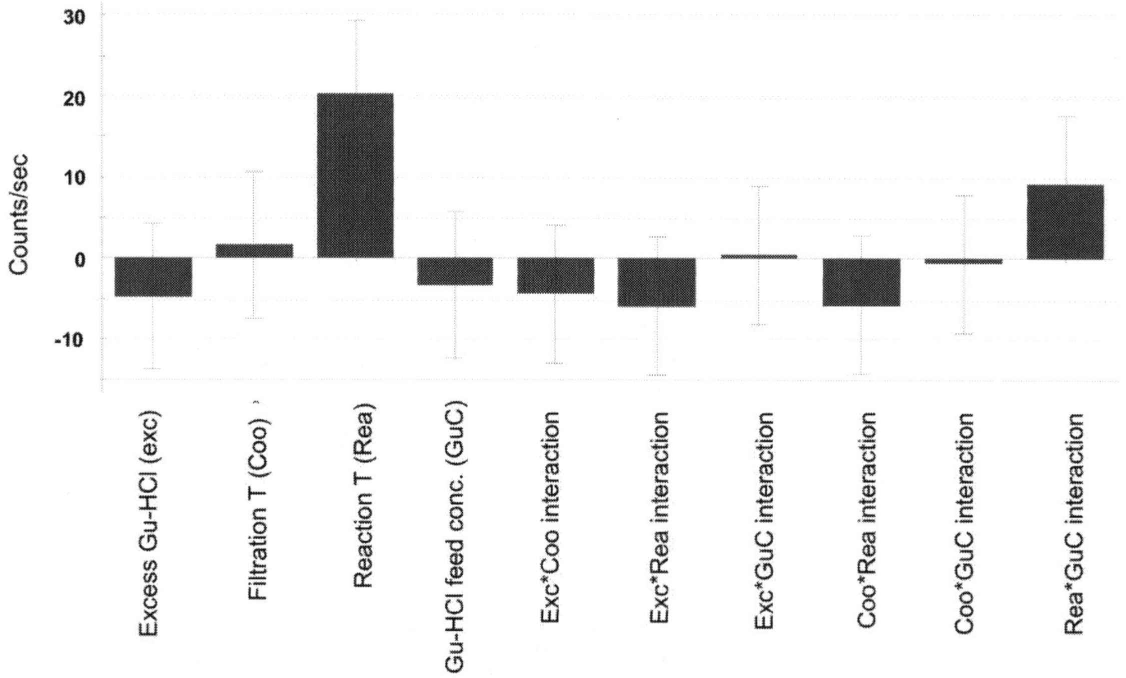

Figure 9 Factor and interaction effects on needle length

Further evidence of the effect of reaction temperature on GuZT particle size is shown in Figure 10. These FBRM distributions show six different batches of GuZT synthesized at the high and low reaction temperatures. The distributions were taken at the end of the 30-minute waiting time after cooling to the filtration temperature. Experiments 6, 13, and 16 took place at 80°C, and experiments 1, 2, and 12 reacted at 60°C.

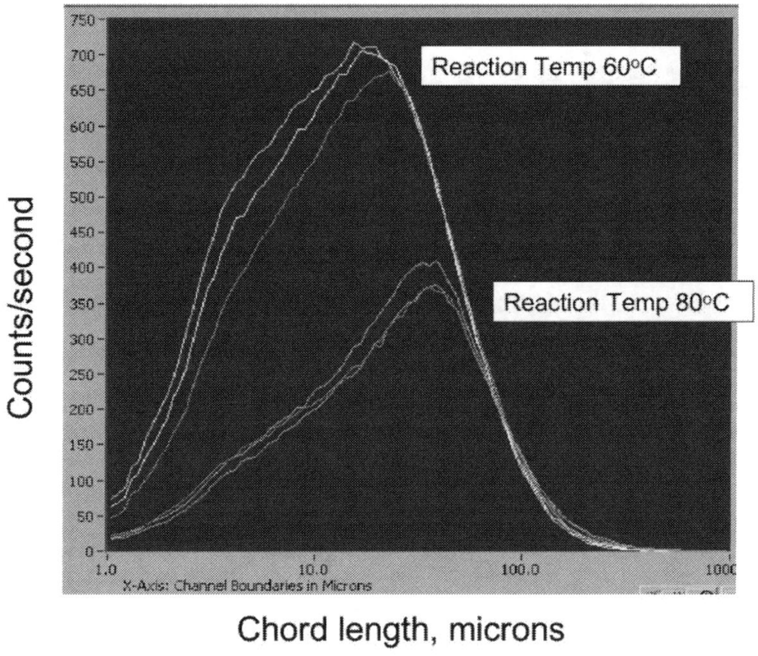

Figure 10. FBRM distribution, no weight

WASHING EXPERIMENTS

After completion of the DOE, two of the 19 experiments exhibiting the highest yield (not adjusted yield) were repeated twice: once with no wash, and once with 225 grams of a cold wash (2 to 4°C). Washes applied during the DOE consisted of 400 grams of room temperature water. These washing experiments and the yield results are described in Table IV.

A wash is required to ensure that sodium and chloride ion content remain under 1%. The temperature and quantity of the wash water appears to make little difference, at least at the 1.2-liter scale. It is possible that at the 100-gallon scale, a loss in yield not evident at the small scale will be apparent if room temperature wash water is used versus chilled water. The average temperature of the tap water used for the "room temperature" washes was 15°C. The difference in GuZT solubility between 4°C and 15°C is approximately 0.2 mg/ml. If 225 ml of wash water at 4°C were to become saturated with GuZT, it would contain approximately 1 gram of material. The loss of this quantity of material at the scale under study can equate to as much as 2% loss in yield. If the same solubility data is correlated to washes at the 100-gallon scale, the difference in GuZT lost to wash water at 15°C and 4°C assuming 114 liters of water is 13 grams, or 0.03% of the proposed yield.

Table IV. Washing experiments

Experiment No.	Wash	% Sodium	% Chloride	Yield
2	None	0.7	1.4	82+/-1
2	Room Temp	0.1	0.0	81+/-1
2	Cold	0.1	0.1	80+/-1
14	None	0.4	0.6	81+/-1
14	Room Temp	0.1	0.1	82+/-1
14	Cold	0.1	0.2	80+/-1

SUMMARY AND CONCLUSIONS

Based on this DOE, using at least 20% excess Gu-HCl and cooling to 5°C before filtering should maximize yield under the conditions included in this study. Purity of the product is expected to meet purity requirements under all conditions used in this DOE. Washing experiments performed at this scale demonstrated no clear requirement for chilled water washes, and calculations indicate that at the 100-gallon scale, losses of approximately 0.05% of predicted yield may be expected. Since preparing chilled versus room temperature water washes requires no additional processing time, labor, or equipment, it is likely that chilled water washes will be employed at the large scale. Higher reaction temperatures can be expected to produce slightly larger GuZT particles; however, the medians of the unweighted distributions and the means of the square weighted distributions are on the same order of magnitude for both the high and low reaction temperatures. Based on conservation of energy and time principles, it is recommended that the GuZT reaction be run at 60°C, and that Gu-HCl be fed at the higher concentration; however; the proposed second optimization study may contradict these recommendations.

FUTURE WORK

It is critical that the greater than 10% discrepancy between yields achieved at LANL and those at the 1-liter scale at Indian Head be explained. Using 0% excess Gu-HCl, Dr. Chavez achieved a higher yield than that exhibited by any of the experiments in this study. Once this phenomemon is explained, the positive effect of excess Gu-HCl on yield may be investigated further, since it may be possible to increase the yield even more by adding excess Gu-HCl. Another factor worthy of study is reaction time. The experiments in this DOE included a 30-minute stir time after addition of Gu-HCl, but perhaps a longer reaction time will increase yield, as well. A follow-on DOE to the first has been proposed to investigate the effects of increased excess Gu-HCl and reaction time on yield and should be executed before GuZT synthesis at the 100-gallon scale begins.

GuZT is relatively insensitive to impact and friction, but it may be worthwhile to study the effects of crystal morphology, quality, and size on the sensitivity of GuZT and its formulations. In addition to using crystallization to control the aspect ratio of GuZT particles, wet milling may be pursued to increase GuZT processability.

ACKNOWLEDGEMENTS

The authors would like to acknowledge Mr. Chuck Painter, Director, Navy Energetics Manufacturing Technology Center, David Chavez, for optimization work performed at LANL, and Claire Wells, for helping perform experiments and for data analysis.

REFERENCES

References
[1] Hiskey, M.A., Goldman N., and Stine J.R.. High-Nitrogen Energetic Materials Derived from Azotetrazolate, Journal of Energetic Materials 16 (1998): 119-127.
[2] Ulrich Teipel, ed. 2005: Energetic Materials (WILEY-VCH Verlag GmbH & Co. KgaA, Weinheim).

Ignition of metal powders by electric spark

Ervin Beloni, New Jersey Institute of Technology, Newark, NJ
Eric Whitman, Princeton University, Princeton, NJ
Mikhalylo A. Trunov, New Jersey Institute of Technology, Newark, NJ
Edward L. Dreizin, New Jersey Institute of Technology, Newark, NJ

Abstract

Energetic formulations with metal fuel additives are extensively used in propellants, explosives, pyrotechnics, and incendiaries. Currently, replacement of regular metal powders with nanocomposite metal-based materials is being considered. Such nanocomposite reactive materials are capable of very high reaction rates while maintaining the high combustion enthalpies characteristic of metals. However, such novel reactive nano-materials are often highly sensitive to impact, friction, and electrostatic discharge, making them difficult to handle. In particular, their high electrostatic discharge sensitivity (ESD) was reported. While ESD testing is common and standardized, the mechanisms of powder ignition by electric spark remain unclear. This project is aimed to establish the relationship between thermal ignition mechanism for metal powders and related nanocomposites and their ESD sensitivity. The project includes both experimental and modeling components. An explicit numerical model of the heat transfer within a powder bed subject to a pulsed electric discharge will be developed. The model will initially describe the behavior of metal powders for which the thermal ignition kinetics is well established and quantified, such as spherical magnesium. It will be further expanded for novel nanocomposite materials for which the kinetics of exothermic processes leading to ignition will be determined in separate experiments. The predictions will be validated experimentally using a standard ESD testing apparatus. This paper will present initial model formulation and experimental results.

Introduction

The current standards, such as MIL STD-1751A by which the ESD testing is performed today came about through an effort in the 1940s by the US Bureau of Mines to test the hazards of explosives by electrical sparks[1]. In a test apparatus, a moving electrode needle discharged a spark upon approaching the surface of the material. If some minimum threshold energy delivered to the powder caused a self-sustaining chemical reaction, ignition would occur. The potential energy based on the capacitance of stored charge delivered to the pin electrode and the potential difference between the electrodes offered a qualitative ranking of powder sensitivities. However the results of these tests may be misleading and a number of issues need to be addressed before a reliable and useful ranking for new reactive materials can be obtained.

There are two generic sets of issues that need to be addressed. The first set of issues has to do with the testing procedure itself. Related test procedures for the ESD sensitivity of gas mixtures have been well developed and documented[2]. It has been established that the material and shape of the electrodes, the discharge duration, the presence of moisture, and other parameters affect the ESD sensitivity substantially. However, no comparable studies are available for ESD sensitivity testing of powder-like materials. In the current testing standards, there is no clear guidance regarding the environmental factors that need to be considered, which could either accelerate or mitigate ignition. There is also no detailed description for the shape and material of the high-voltage electrode. The pin electrode used to transfer the stored charge from the capacitor to the test surface is important in how the discharge is delivered to the surface. The pin shape affects the point of contact where the discharge will strike and it is vital to have a consistent shape for repeatable results. Also pin material affects how the discharge energy is transferred through the pin. For instance it has been reported that for high energies for nickel coated brass pins which are typically used, the coating was vaporized which in turn reduced the discharge energy due to heat losses[1]. The current standards do not specify the discharge duration, which can make it difficult to make meaningful comparisons between different powders. In particular, the powder electric conductivity would affect the time constant of the capacitor discharge, so that different discharge durations can occur for different powders tested using the same experimental setup.

The second set of issues is unique to properties of powder-like energetic materials. For example, there is also little to no guidance regarding the experimental pressure or sample confinement. In testing of propellants, pressure can affect internal mechanical stresses which in turn affect ignition[3]. Confinement of the powder being tested can be crucial as well. For example, it was reported that metal powders exhibit a greater ESD ignition sensitivity being unconfined while black powder is more sensitive under confinement[1]. In addition to the issues common for all powders, reactive nanocomposite materials pose additional challenges. Their inherently heterogeneous structure with extremely developed reactive interface area can result in unusual thermal, mechanical, and electric properties, all of which affect their ESD sensitivity. Some of the related heterogeneous processes have recently been considered while modeling flame propagation for self-propagating high-temperature synthesis (SHS)[4] and similar considerations are necessary for analysis of the ESD sensitivity.

The goal of this work is therefore twofold. First, an understanding of ESD sensitivity as related to the ignition of powders is to be established. A numerical model will be developed for the heat transfer in a powder bed subject to ESD sparking. The numerical model will be developed and calibrated considering properties of spherical Mg powder. This powder is selected as a benchmark material because of its well established properties and thermal ignition kinetics. A connection will be made as to whether ESD is responsible for the thermal initiation of metal powders. The numerical model solutions will be compared to actual experimental results based on current testing methodology. The second goal of this work is to conduct experimental analysis of the ignition of powders using ESD as the ignition trigger. Parametric work will be conducted to determine the effect of electrode materials, capacitance and polarity and will provide a baseline to sample measurements. Optical measurements will be taken of the ignition of Mg from which temporal and energetic characteristics of ignition will be determined. This paper presents the initial effort aimed to achieve the goals stated above.

ESD Ignition Model

Formulation

A numerical heat transfer model was created for studying ignition by ESD. The powder is contained in a cylindrical cup electrode. The domain in which the heat transfer is analyzed is a disk with elements taking the shape of a series of inserted and stacked rings with central cylindrical elements, as illustrated in Fig. 1. The element's vertical dimension is z_m and the element radius is r_n. For each element, the heat transfer equation is considered as

$$\rho VC\frac{T(t+\Delta t)-T(t)}{\Delta t}=\dot{Q}_{Cond}+\dot{Q}_{Conv}+\dot{Q}_{Rad}+\dot{Q}_{Disch}+\dot{Q}_{Joule}+\dot{Q}_{Chem}$$

where ρ is density, V is volume, C is specific heat, T is temperature, t is time, Δt is the time step and $\dot{Q}_i$ are different heat transfer rate terms.

Conduction occurs between all neighboring elements in contact and is a major source of heat input for elements not receiving energy directly from the spark. The rate of conductive heat transfer is affected by the thermal conductivity, which is a function of temperature. The initial value of the thermal conductivity is determined by laser flash experiments[5]. Its change as a function of temperature is poorly known for the loosely packed powders. In particular, a step-wise increase in the thermal conductivity is expected upon melting and respective assumptions are being made in the current model subject to future verification.

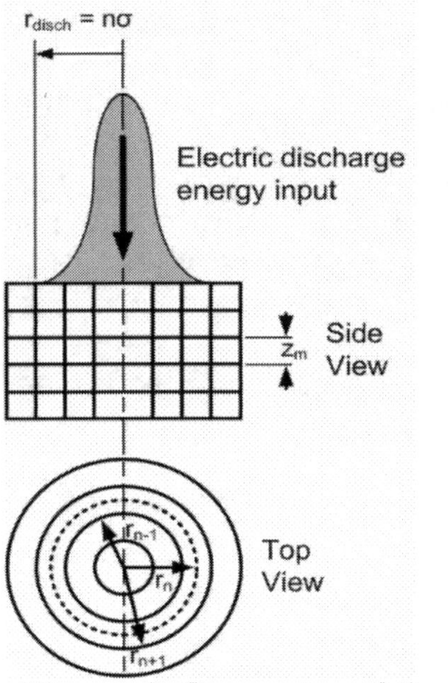

Figure 1. Geometry of the Computational Domain

Convection and radiation are assumed to be the only modes of heat transfer between the powder and ambient environment. These heat transfer terms are considered for the elements at the top of the mesh, which are exposed to air.

The discharge is the major source of energy input into the powder bed. The discharge energy is assumed to be spatially distributed by a revolved Gaussian distribution and exponentially decays in time. Two cases for the heat transfer between the discharge and powder were analyzed. In the first case, the discharge was simply assumed to represent an energy flux. Thus, the entire discharge energy was distributed only among the top elements of the mesh based on the assumed Gaussian profile of the discharge energy. In the second case, the current produced by the discharge in the powder bed was calculated and assumed to transfer the energy to the powder by Joule heating. Both the current distribution within the powder and the heat release were calculated considering the contact resistances between adjacent mesh elements (representing individual particles). In the current model, the changes in the electric conductivity of elements were assumed to follow the changes in the thermal conductivity.

Lastly, the chemical heat release within each element describes the reaction leading to ignition and was initially described by an Arrhenius type equation. The activation energy and pre-exponent used were taken from the existing literature reports on Mg ignition[6,7]. At the same time, the consumption of oxygen required to produce the heat release according to the Arrhenius equation was monitored. When significant consumption of oxygen occurred for a specific element (typically, one of the surface elements), the Arrhenius term was corrected and eventually replaced by a formula derived to describe the rate of oxygen diffusion towards the element. Once the reaction for the surface elements became diffusion limited, the elements located below received no oxygen and the chemical heat release became equal to zero.

Several boundary conditions were considered to close the model. It was assumed that the air and cup surfaces remain at a constant environmental temperature. The cup surfaces are assumed to be electrically grounded.

Calculation Results

Several numerical model runs were conducted to observe the temperature distribution and history throughout the powder prior to ignition. Examples of the obtained temperature distributions are shown in Fig. 2. Each element of the numerical mesh is selected to represent a 10 μm particle. When an element boils off, white color is assigned to indicate its location. Boiling of one or more of the mesh elements is expected to indicate an ignition event.

Assumption of the ESD energy transfer to the powder from the powder surface only (cf. left side of Fig. 2) results in extremely sharp temperature gradients within the powder. Such gradients are developed because the thermal conductivity of the powder is relatively low while the rate of energy transfer from the discharge to the exposed elements (particles) is very high. As a result, a very small powder volume is heated and boiled very rapidly.

When Joule heating distributed within the powder volume is considered (cf. right side of Fig. 2), the heating of the powder bed occurs much slower and the temperature is distributed over a relatively large domain. The temperature distributions are significantly affected by the assumed effect of temperature on the electric conductivity of the powder. Because a step-wise increase in the conductivity was assumed to occur upon melting, the electric field and current density around a molten element change substantially. These changes result in the continuous re-distribution of the current density in the sample once the first elements are predicted to melt. Such redistribution causes the formation of characteristic molten columns within the sample, as can be seen on the right side of Fig. 2.

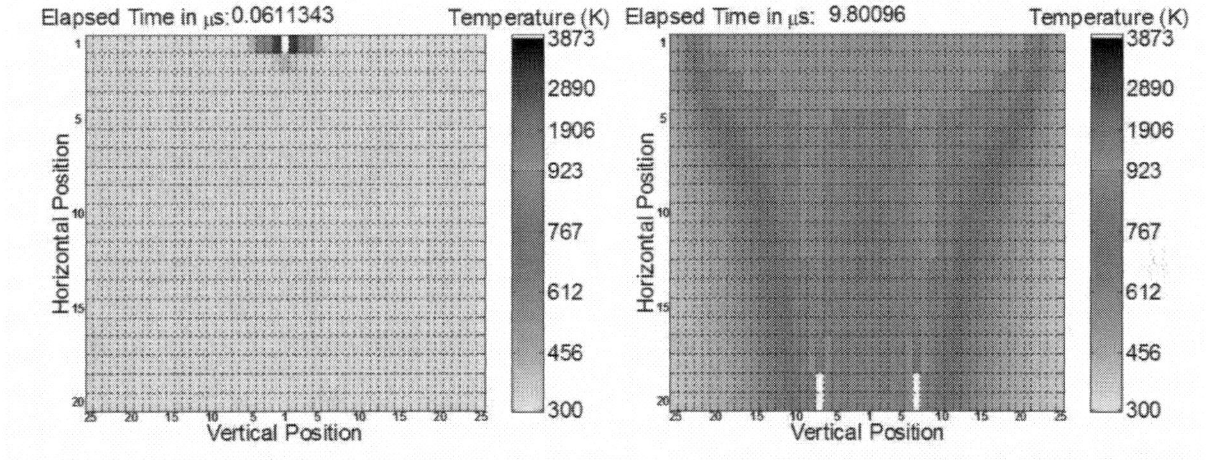

Figure 2. Calculation results showing the temperature distributions in the samples for the cases when the entire energy of the discharge is supplied to the top layer of the powder (left) and when the Joule is considered to distribute the discharge energy within the powder (right). For the case on the left the energy is 25mJ and the case on the right the energy is 50mJ.

Experimental

A schematic diagram of the experimental setup is shown in Fig. 3. ESD testing in accordance with the MIL-STD 1751A standard was performed using a Model 931 Firing Test System (FTS) by Electro-Tech Systems, Inc. Free metallic powder was placed into a stainless steel cup 15 mm in diameter and 3 mm in depth which was affixed to a grounded base. FTS includes a capacitor bank with capacitance varied in the range of 100 pF to 10000 pF which can be discharged through the spark gap directly or through an additional resistor varied from 500 to 5000 Ω. The discharge pulse voltage that can be produced by the apparatus varies from 100 V to 26 kV. A short Nichrome wire was used to ground the sample cup and the voltage drop measured across it was used to determine the spark current. A LeCroy WaveSurfer 64Xs Series oscilloscope was used to acquire and record signal traces.

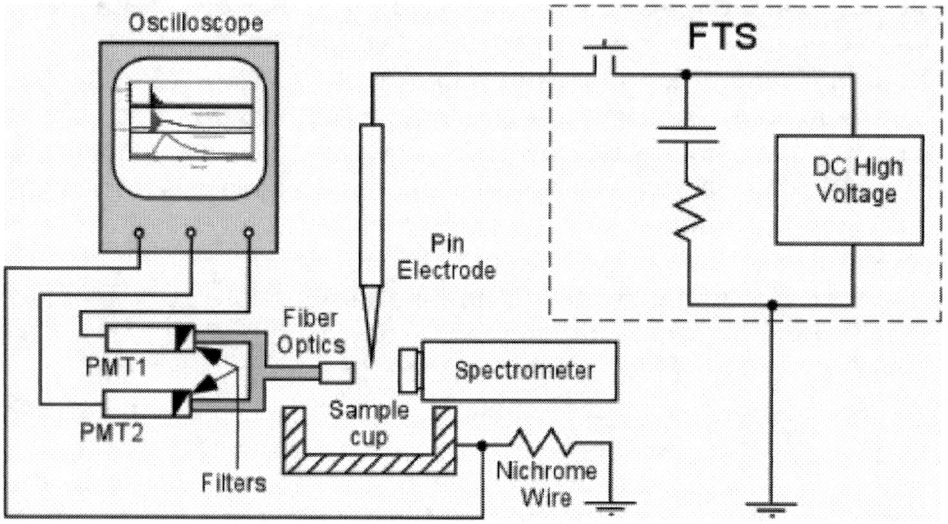

Figure 3. Schematic diagram of the experimental setup

Testing was performed in the room air environment. Optical traces of the spark emission were acquired at two selected wavelengths by two photomultiplier tubes by Hamamatsu Corp. and recorded by the oscilloscope. The choice of the bandpass interference filters for the optical traces was determined based on two initially measured optical spectra produced respectively by sparks between the pin electrode and an empty cup and between the pin electrode and igniting Mg powder. The spectra were recorded using an EPP2000 spectrometer by Stellarnet Inc. and are shown in Fig. 4. Main characteristic peaks are labeled based on the references [8] and [9]. It was observed that the spark produced between the pin electrode and an empty cup emitted primarily in the ultraviolet range and respectively a 280 nm filter by Cheshire Optics was used to detect the spark emission. Alternatively, the spectrum of the igniting Mg powder exhibited a strong and broad peak in the range of 500 – 600 nm with an additional characteristic MgO peak prominent at 500 nm. Therefore, a 500 nm filter by Edmund Industrial Optics was used to select the light produced by the igniting powder.

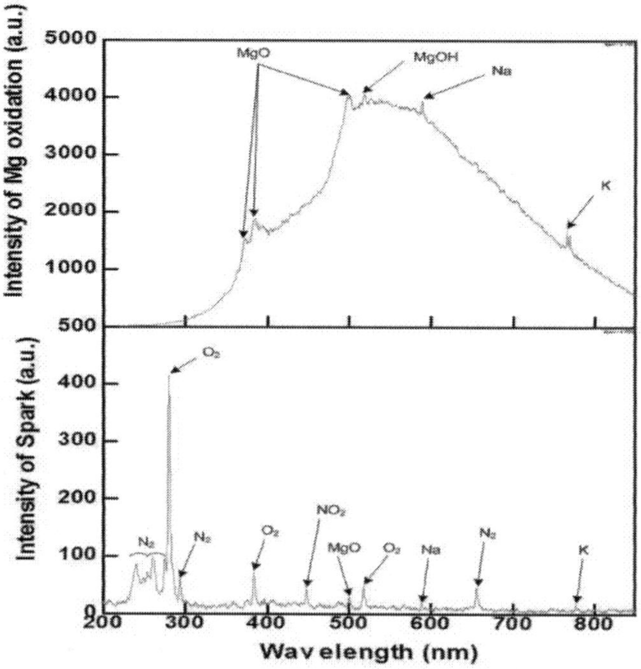

Figure 4. Emission spectra of the sparks between the pin electrode and empty sample cup (top) and between the pin electrode and igniting Mg powder (bottom)

A set of calibration measurements was conducted in order to determine the presence of any additional parasitic resistance and parasitic capacitance associated with each specific capacitor of the FTS capacitor bank and to calibrate the voltage measurements across the Nichrome wire used to ground the sample cup into specific current values.

Spark Current and Emission

A set of preliminary measurements was performed to establish the baseline spark current and emission signatures. In addition, the effects of the selected spark polarity and bulk electrode material on such signatures were investigated.

Characteristic measured spark current and radiation signals filtered at the wavelengths of 280 and 500 nm are shown in Fig. 5.

The current produced by the spark appears to comprise mostly of the AC component. There is no noticeable DC component affected by the selected discharge polarity. Note that in the calibration measurements, when the spark gap was shunted with a known resistor, the current trace looked distinctly different and clearly showed a DC component affected by the selected pulse polarity.

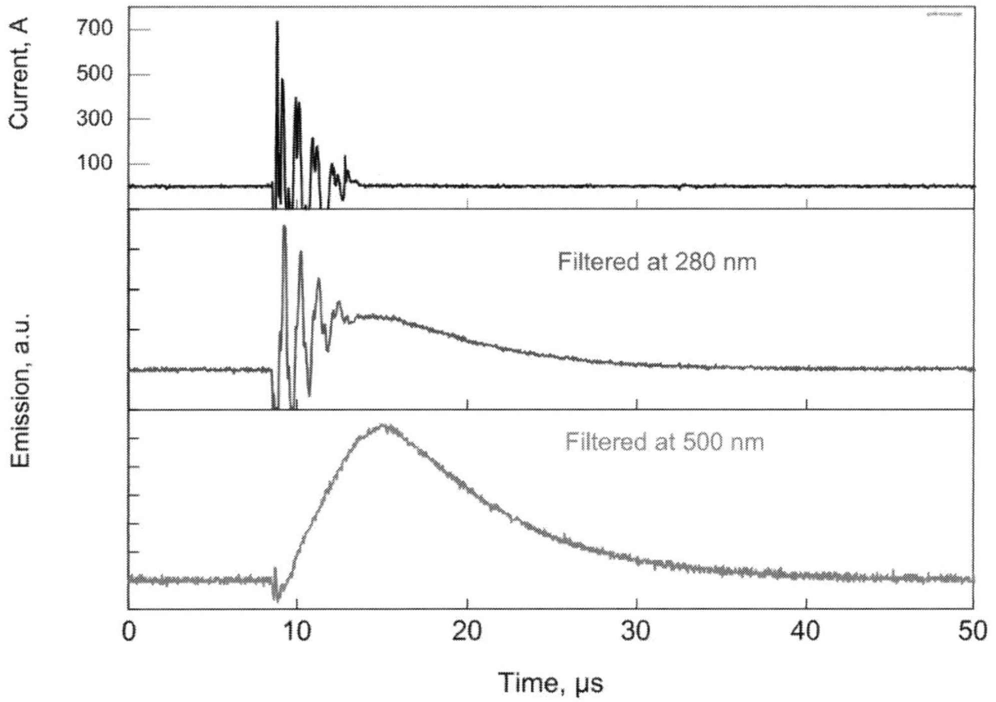

Figure 5. Voltage traces showing spark current and emission signatures at 280 and 500 nm.

Similar to the discharge current, the spark's UV emission measured through the 280 nm filter shows a decaying periodic pattern clearly correlating with the current trace. The UV emission continues to decay after the spark is over as suggested by the ending current trace. Unlike the UV emission generated by the spark itself, the green emission filtered at 500 nm shows an intensity increasing almost linearly during the entire spark duration. After the discharge is over, the green emission is decaying, qualitatively similar to the UV emission trace. This green emission signal is, most likely, generated by the heated metal and represents the respective grey body emission.

The discharge traces were studied for plates of aluminum and stainless steel serving as bottom electrodes. The period of oscillations as well as time of optical signal decay were considered as a function of the selected capacitor for both materials. It was observed that the period of oscillations is clearly affected by the capacitor but is not affected by the selected electrode material. On the other hand, the rate of decay of the green radiation after the end of the spark discharge was substantially higher for aluminum than for stainless steel electrode.

This observation supports the assumption that the green emission is indicative of the grey body emission generated by the heated electrode material.

Preliminary Experiments on Mg Powder Ignition

The first qualitative observation made when igniting Mg powder placed in the sample holder by an ESD was that substantial portion of the powder is ejected from the sample holder even if no ignition was observed. Typically, an indentation of about 3 mm diameter is produced within the powder bed under the pin electrode. Such powder ejection is likely a result of a shock typically produced by a spark discharge[10]. Therefore, an adequate description of the ESD ignition of an unconfined powder needs to take this phenomenon into account. For simplicity, initial experiments can employ powders which are either pressed or slightly bonded together to separate only the thermal effect of the ESD, as described in the model presented above.

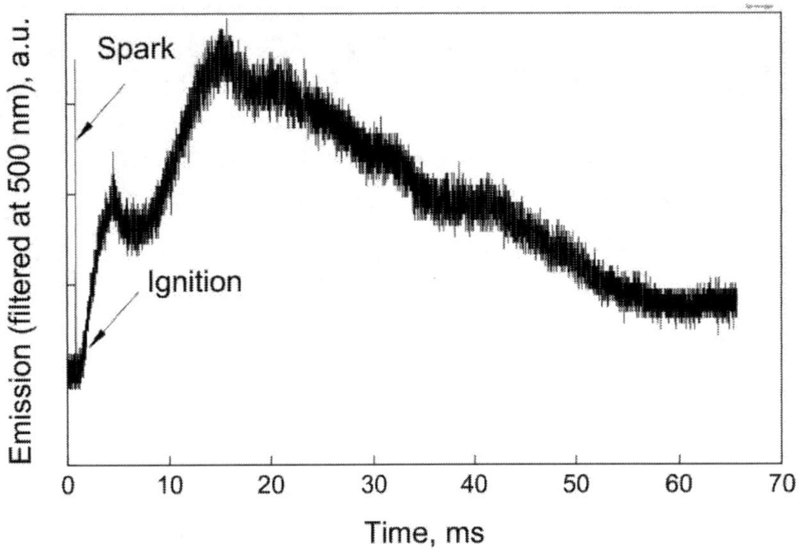

Figure 6. Characteristic emission trace filtered at 500 nm and recorded for ESD igniting Mg powder.

A characteristic emission trace filtered at 500 nm recorded for the igniting Mg powder is shown in Fig. 6. It is remarkable that while the spark duration is close to only 0.06 μs for non Joule heating and 9.8 μs for Joule heating (seen in Fig. 5), the ignition inferred by the sharply rising green radiation signal is observed after almost a 500 μs delay. The following radiation trace exhibits several peaks, most likely resulting from combustion of different portions of the powder bed. The significant ignition delay indicates that the ESD energy was well distributed within the powder as opposed to the predicted case of sharp temperature gradients occurring when the ESD energy is introduced only into the surface powder layer. Therefore, the ESD energy transfer to the powder due to Joule heating is a more realistic assumption that will need to be considered in more detail in future work.

Conclusions

A heat transfer model was developed to determine whether the thermal initiation is the main cause of powder ignition under the ESD stimulation. The model will be further developed and calibrated using a spherical Mg powder as a convenient benchmark. The model considers the conductive heat transfer in a powder bed, which is controlled by the thermal resistance of particle contacts. The model relies on an experimentally determined thermal diffusivity in the powder bed. The model considers Joule heating as a major heat component in the delivery of heat away from the discharge area. It also includes convection and radiation heat transfer terms. The chemical reaction is introduced and described by an Arrhenius function. The limitations on the reaction rate due to the finite rate of oxygen diffusion to the reacting surface are also considered.

The firing test system used in standardized ESD sensitivity tests was calibrated and prepared for detailed measurements. The effect of the cup electrode material, polarity, and capacitance on the spark current and emission traces was determined. Preliminary experiments on Mg ignition show a significant delay between the spark pulse and the rise in the radiation signal due to the powder combustion.

In the future, the model will take into account the changes of thermal conductivity and resistivity over a wider ranger of temperatures. The model will be validated for a wide range of powders and testing conditions and used to study and understand the mechanisms of ESD ignition sensitivity of reactive materials. Parametric studies, both computational and experimental, will be conducted to establish the effect of the discharge power, duration, electrode material and shape and other parameters on the ESD sensitivity of selected reactive powders. The effect of gas environment will be addressed in particular. The findings from the study will then be incorporated into the ignition model to further the understanding of ESD powder ignition processes.

Acknowledgement

This work was supported by Dr. Ralph Anthenien of the US Army Research Office.

References

1. Skinner, D., Olson, D. and A. Block-Bolton (1997), "Electrostatic Discharge Ignition of Energetic Materials." *Propellants, Explosives, Pyrotechnics* 23, pp. 34-42.

2. Mellor, A. M., and D. R. Stoops (1990), "Optimization of Spark and ESD Propellant Sensitivity Tests." *Propellants, Explosives, Pyrotechnics* 15, pp. 1-7.

3. R. L. Raun (2000), "Modeling of ESD-Induced Confined Ignition of Solid Rocket Propellants," *Combustion and Flame* 120, pp. 107-124.

4. A. Makino (2001), "Fundamental aspects of the heterogeneous flame in the self-propagating high-temperature synthesis (SHS) process," *Progress in Energy in Energy and Combustion Science* 27, pp. 1-74.

5. Parker, W.J., Jenkins, R.J., Butler, C.P., and G.L. Abbott (1961), "Flash method of determining thermal diffusivity, heat capacity, and thermal conductivity," *Journal of Applied Physics* 32 (9), pp. 1679-1684.

6. Roberts, T.A., Burton, R.L. and H. Krier (1993), "Ignition and combustion of aluminum/magnesium alloy particles in oxygen at high pressure," *Combustion and Flame* 92 (1-2), pp. 125-143.

7. Ward, T.S. Trunov, M.A., Schoenitz, M. and E.L. Dreizin (2006), "Experimental methodology and heat transfer model for the identification of ignition kinetics of powered fuels," *Heat and Mass Transfer* 49, pp. 4943-4954.

8. Pearse, R.W.B and A. G. Gaydon (1976), The Identification of Molecular Spectra. London, Fourth Edition Chapman and Hall.

9. NIST Atomic Spectra Database v3, http://physics.nist.gov/PhysRefData/ASD/index.html, NIST Physics Laboratory, 2007.

10. Borghese, A., D'Alessio, A., Diana, M., and C. Venitozzi (1988), "Development of hot nitrogen kernel, produced by a very fast spark discharge," *Twenty-Second Symposium (International) on Combustion/The Combustion Institute*, pp.1651-1659.

604c High Energy Chemistry And Its Chemical Engineering Challenges

Jerry Salan, Patrick B. Skahan, and **Kimberly Hanson**. Naval Surface Warfare Center Indian Head Division, 101 Strauss Ave., Indian Head, MD 20640

The majority of synthetic routes to energetic materials involves at least one step involving high energy chemistry. Although steps are typically taken early in the development phase to evaluate alternatives to high energy chemistry there will always be a need to address the process safety for these types of reactions when developing energetic materials. An appreciation of the risks and potentially catastrophic consequences involved with high energy chemistry and the products of such reactions must be maintained.

Indian Head Division, Naval Surface Warfare Center recently scaled up the nitration of 2,6 dimethoxydinitropyrazine. This energetic intermediate's chemical reaction was evaluated using reaction calorimetry and contiuous IR measurements to determine heat flow, reaction progression and identify factors of interest for the process. One key parameter observed was the reagent order of addition and its affect on heat flow. The process was also evaluated for processing continously in lieu of the traditional batch process. Various chemical engineering challenges including heat flow, product stability and isolation are evaluated. The results of these experiments and rationale for scaling high energy chemistry will be presented.

Web Page: www.ih.navy.mil

Laser Ignition of Aluminum Particles in Water

Salil Mohan, Luc Furet and Edward L. Dreizin
New Jersey Institute of Technology, Newark, NJ 07102, U.S.A.

Abstract

This study presents experimental results of ignition of Al particles with a CO_2 laser in water vapor environment. Al powder, with nominal particle sizes in the range of 4.5 μm – 7 μm, is aerosolized using a parallel plate capacitor by charging particles contacting the electrodes. A thin, laminar aerosol jet is carried out by nitrogen heated to ~150 °C through a small opening in the top electrode and is fed into a focused CO_2 laser beam. A shroud flow of superheated steam, also at ~150 °C, is maintained around the central aerosol jet. The velocities of particles in the jet can be varied in the range of 0.1 – 3 m/s. The particle velocity is controlled by the inert gas jet velocity while keeping the shroud steam flow velocity constant. Numerical simulation using Fluent CFD code was used to determine the gas composition at the laser focal spot. For each selected central flow velocity, the laser power was increased until the particles were observed to ignite. The ignition was detected optically using a photomultiplier. The ignition thresholds for spherical aluminum powder were measured at varied particle velocities resulting in varied heating rates and vapor concentration. Similar experiments were conducted by replacing superheated steam with hot air. Laser ignition threshold from air was found to be lower than that for superheated steam.

Introduction

Aluminum powder is widely used as an additive in various applications such as propulsion systems, explosives, and pyrotechnics. Aluminum powder, due to high energetic properties, is a preferred choice for propellant applications requiring high specific impulse. Ignition and combustion of aluminum particles has been studied for a long time and still generates a lot of interest [1-3]. Different mechanisms, based on disruption of protective Al_2O_3 layer on aluminum particle, have been proposed for ignition of aluminum powder in dry air [4-6]. However, practical applications involve ignition and combustion of aluminum particles in mixed environment involving CO_2 and H_2O species [7, 8]. The reported aluminum ignition and combustion in wet environment, has been qualitative in nature and do not take into account the very high heating rates experienced in practical application. This study employs a new experimental technique that enables one to quantify ignition kinetics of aluminum particles in wet environments with heating rates approaching to or exceeding 10^6 K/s. The experimental technique uses CO_2 laser heating of aerosolized aluminum powder, which is particle size sensitive. As a result, only particles in a narrow range of sizes are effectively heated and ignited. This allows interpreting the results using a detailed heat transfer model only for the specific, close to 3.4 μm, effectively heated particle size, even though a commonly available polydispersed powder is used in experiments. The laser is fired at a minimum possible power at which ignition starts to be observed, so that only the most effectively heated particles are ignited. The technique allows achieving different particle heating rates by varying the speed of particles fed into the laser beam. The aluminum powder can be aerosolized in dry air as well as wet environment of superheated steam and N_2 gas. Experiments were conducted in dry and wet environments and preliminary results are presented. The study shows ignition of aluminum particle is relatively difficult in the wet environment compared to dry air environment.

Experimental

The powder used in these experiments was spherical, 99% pure aluminum by Alfa Aesar with nominal average particle size of 4.5 – 7 μm. The experimental setup is shown in Fig. 1 and includes a generator of an aerosol jet, a 125 W CO_2 laser (by Synrad, Evolution 125 series) with a ZnSe convex lens (0.75″ aperture and 4″ focal length), and a modulated green laser (SUWTECH model DPGL-3000 by Photop Technologies, Inc) operated with a set of a semi-cylindrical and convex glass lenses to produce a laser sheet for the jet visualization. The setup also includes a superheated steam generator which supplies a superheated steam/nitrogen mixture as a shroud jet around the central nitrogen jet carrying aerosolized particles. Also, not shown in the Fig. 1 but employed in the experiments were a digital camera (Panasonic GS-35), used to obtain particle streaks, and a photomultiplier tube (PMT) used to measure emission traces of the heated and ignited particles. In addition, a power meter (Synrad, POWER WIZARD 250), was used to measure the laser beam energy and verify the accuracy of the pre-set laser power.

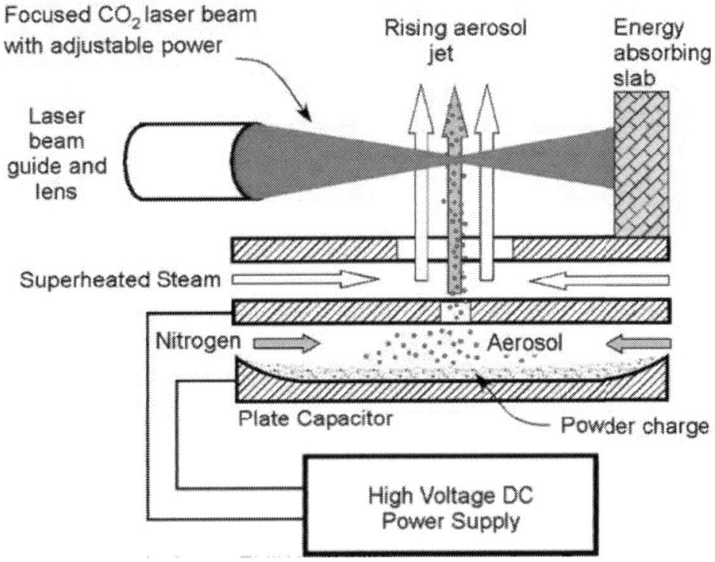

Figure 1. Experimental set-up showing heating of Al aerosol in a H_2O/N_2 gas mixture by a CO_2 laser.

The aerosol jet generator uses electrostatic aerosolization, described in detail elsewhere [9]. In this technique, a conductive (e.g., metal) powder is placed between the electrodes of a circular parallel plate capacitor. A high DC voltage in a range of 1-15 kV is applied and conductive particles acquire electric charge. They are repelled from the bottom electrode and attracted to the top electrode, at which they re-charge upon collision. The motion of the charging and re-charging particles continues so that an aerosol is produced in the space between the capacitor's electrodes. Because of the multiple collisions of the charged particles with electrodes, particle agglomerates are broken and the resulting jet contains single particles only, as was confirmed in the previous reports [10]. The applied voltage can be used to control the particle number density in the produced aerosol. In the device used in this project, a small opening (nozzle) was made in the center of the top electrode to allow formation of a thin aerosol jet escaping from the capacitor. The number density of the produced aerosol was maintained small so that the number of particles fed into the aerosol jet was of the order of 1000 per second. The space between the capacitor's electrodes was enclosed in a chamber. A controlled flow of nitrogen gas, measured by a gas flow meter (M-200SCCM-D by Alicat Scientific Inc.), was fed into the chamber to carry the aerosolized particles out of the nozzle. The nitrogen flow was used to control the aerosol jet speed which could be readily varied in the range of 0.1 - 3 m/s. The setup also includes a parallel enclosed heated flow system, as shown in Fig. 1, which was used to feed the superheated steam at 423 K (150 °C) as a shroud around the central nitrogen jet carrying metal particles. The top surface of the chamber housing the plate capacitor filled with nitrogen was also heated to 423 K to avoid significant temperature gradients between the central and shroud jets.

Note that in initial experiments, argon flow for the central jet was used. However, successful production of aerosol in argon with heated gas and heated elements of the experimental setup could not be achieved. Indeed, the breakdown potential of argon is lower than that for nitrogen; furthermore, the breakdown potential is substantially reduced at elevated temperatures. Therefore, electric field achievable without breakdown with argon between the plate capacitor electrodes is very low and is inadequate for aerosol jet production.

A superheated steam generator included a peristaltic pump (Variable Speed Pump by Control Company) supplying the metered amount of liquid water (typically, 0.1 ml/min) into a steel tube (3.1 mm diameter, 5 m overall length) coiled inside a tube furnace (Tube Furnace 21100 by ThermoLyne). A metered nitrogen flow was also fed through the coiled steel tube placed in the furnace. The furnace was pre-heated to 673 K to produce a superheated steam at 673 K (400 °C) mixed with nitrogen gas at a controlled ratio. The steam/nitrogen mixture was fed through a heated hose maintained at the temperature of 423 K, consistent with the temperature of the heated elements of the aerosol jet generator.

The aerosol jet was illuminated by a vertical green laser sheet. To enable particle image velocimetry, the laser sheet was modulated at 900 and 1500 Hz, with the higher frequency selected for higher jet speeds. Produced particle streaks were recorded using the digital camera and the streak lengths were measured to determine the jet velocity.

The CO_2 laser was focused about 2 cm above the jet nozzle using an auxiliary red laser aligned with the CO_2 laser beam and the ZnSe lens. Once a stable aerosol jet was established, the CO_2 laser was fired continuously for 5 seconds, at a preset power level. Visible radiation generated by heating and/or ignition of particles was monitored using a PMT connected to an oscilloscope (Wavesurfer 64Xs by LeCroy). The experiment was repeated with the laser powers gradually increased until ignition events were clearly detected. The ignition events were detected based on the shapes of particle emission pulses. Typical examples of the recorded PMT output pulses produced by particles crossing the laser beam are shown in Fig. 2. Figure 2 (a) shows a peak produced by a particle that ignited and burned. A sharp voltage rise is followed by small changes in the radiation signal occurring during the particle combustion. On the other hand, the peak shown in Fig. 2 (b) shows the heating and cooling of the particle that never ignited, and the voltage rise is immediately followed by the voltage decrease as the particle exits from the laser beam. The minimum ignition threshold was determined if at least one ignition event was detected during a 5-s period the laser was fired. A 5-s experiment was performed at least three times for each laser power setting. Both the jet velocity and the laser power measurements were repeated immediately before and after each ignition experiment.

Two sets of experiments were performed, each measuring laser ignition threshold power for different particle velocities. One set used dry air for both shroud as well as central jet gas while the other set used a mixture superheated steam and nitrogen for the shroud jet and nitrogen for the central jet. Similar temperatures (423 K) were maintained for the gas feeding tubes and enclosures in both sets of experiments. The experimental results obtained when heated dry air environment was used were compared with the results reported earlier for the room temperature dry air environment [11].

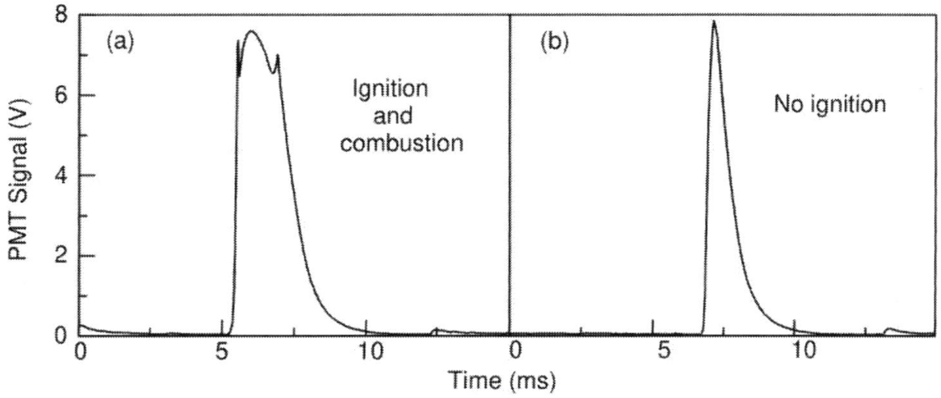

Figure 2. PMT signal from micron size Al particles crossing a CO_2 laser beam. (a) Signal corresponding to ignition and combustion of a particle; (b) Signal corresponding to heating and cooling of an unignited particle.

Gas flow calculations

The steam flow rate in the shroud jet and nitrogen flow rate in the central jet were considered in a numerical simulation to determine the gas composition at the location of the focal spot of the CO_2 laser, where the particle ignition events could occur. Gambit, a pre-processing software, was used to model the geometry of the aerosol generator outlet. Experimental geometry of the central jet nozzle and the shroud flow nozzle was directly introduced into the code. An axisymmetric cylindrical geometry for the jet was considered. For the calculations, the space was covered by 2 different meshes. Both meshes used quadrilateral elements. The first, internal mesh had a relatively small node-to-node distance of 0.1 mm. It was used in the central zone where jet was produced and where the concentration gradients were likely to be high. The second, external mesh had a distance between nodes of 5 mm, and was used for the ambient environment. Computational fluid dynamic (CFD) code Fluent was used for numerical calculations. Calculations assumed laminar gas flow and considered a gas jet neglecting the presence of particles.

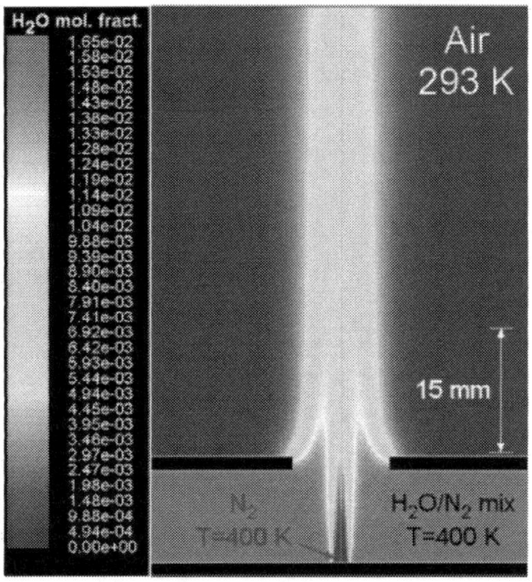

Figure 3. Mole fraction of water as a function of vertical and horizontal coordinates for the produced mixed jet. The specific conditions are: Flow rate of N_2 (central jet): 100 ml/min at 293 K; Flow rates of

H_2O/N_2 mixture represent 100 ml/min of N_2 at 293 K and 0.085 ml/min of liquid water. Temperature of all entering gases is 400 K. The final mixture consists of 57.5 mass % of N_2 and 42.5 mass % of H_2O. The environment is air at 293 K.

The gases were introduced into the system using mass flow inlets, corresponding to the mass flow rate measurements obtained in experiments. The heat transfer to the walls was neglected and walls were assumed to be at the same temperature as the adjacent gas elements.

The numerical model was validated by calculations for a simple expanding jet and comparing the calculation results with the available analytical expressions for the gas velocity profiles. The results of calculations for a specific case are presented in Fig. 3. The conditions used in this specific calculation were representative of those used in experiments.

Figure 3 shows that mixing of the starting H_2O/N_2 mix with the central nitrogen jet is nearly completed about 1 cm above the nozzle and that the water concentration across the jet is nearly uniform at the location of the laser beam (about 2 cm above the shroud nozzle). The specific water concentration present at that location is important for the further analysis of the experimental results.

Results

The first set of experiments was performed with dry air fed in both central jet and shroud flow. The main difference compared to the previous experiments [11] was that shroud flow was pre-heated to about 400 K. In addition, the top surface of the aerosol generator chamber was also pre-heated to the same temperature. The results of these experiments as well as the results of our previous measurements presented in ref. [11] are shown as open symbols in Fig. 4.

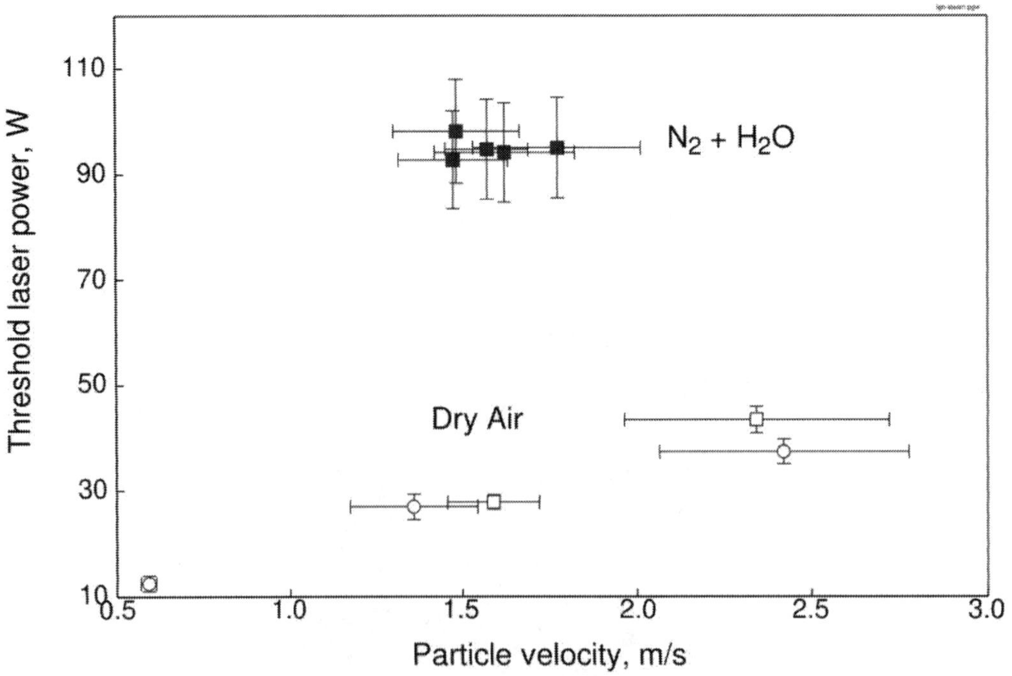

Figure 4. Threshold laser power required for ignition of aluminum particles crossing the laser beam in different gas environments. Open symbols: dry air environment. Circles: room temperature air, ref [11]. Squares: air heated to 423 K. Filled symbols: N_2/H_2O mixture at 423 K.

The results of these new measurements correlate very well with our previous measurements of the threshold laser power required to ignite aluminum particles in air. The pre-heating of air does not produce substantial difference in the measured laser threshold power and this result is consistent with the predictions using the particle heating model developed in ref. [12].

The results of measurements performed when air was replaced with nitrogen-water mixture are shown with filled symbols. The filled symbols lie noticeably higher than the open symbols suggesting that aluminum ignition in water is much more difficult than it is in oxygen. Additional measurement at different particle velocities and respectively at different heating rates will be performed and the results will be further analyzed.

Acknowledgement

This work was supported by the US Defense Threat Reduction Agency.

References

1. R. Friedman and A. Maček (1962), "Ignition and Combustion of Aluminium Particles in hot Ambient Gases," *Combustion and Flame,* 6, pp. 9-19.
2. Trunov, M.A., Schoenitz, M., and Dreizin, E.L. (2005), "Ignition of Aluminum Powders Under Different Experimental Conditions," *Propellants Explosives and Pyrotechnics*, 30 (1), pp. 36-43.
3. Andrzejak, T.A., Shafirovich, E. and Varma, A. (2007), "Ignition Mechanism of Nickel-coated Aluminum Particles," *Combustion and Flame*, 150(1-2), pp. 60-70.
4. Trunov, M.A., Schoenitz, M., and Dreizin, E.L., "Effect of Polymorphic Phase Transformations in Alumina Layer on Ignition of Aluminum Particles," *Combustion Theory and Modeling*, 10 (4), 2006, pp. 603-624.
5. Rozenband, V.I. and Vaganova, N.I. (1992), "A Strength Model of Heterogeneous Ignition of Metal Particles," *Combustion and Flame*, 88 (1), pp. 113-118
6. Rozenband, V.I. (2004), "Thermo-Mechanical Aspects of the Heterogeneous Ignition of Metals," *Combustion and Flame*, 137 (3), pp. 366-375.
7. Servaites, J., Krier, H., Melcher, J. C. and Burton, R. L. (2001), "Ignition and Combustion of Aluminum Particles in Shocked $H_2O/O_2/Ar$ and $CO_2/O_2/Ar$ Mixtures," *Combustion and Flame*, 125 (1-2), pp. 1040-1054.
8. V. Sarou-Kanian, J.C. Rifflet, F. Millot and I. Gökalp (2006), "Aluminum Combustion in Wet and dry CO_2: Consequences for Surface Reactions," *Combustion and Flame*, 145 (1-2), pp. 220-230.
9. Shoshin, Y., and Dreizin, E. (2002), "Production of Well-Controlled Laminar Aerosol Jets and Their Application for Studying Aerosol Combustion Processes," *Aerosol Science and Technology* 36(9), pp. 953-962.
10. Shoshin, Y.L. and Dreizin, E.L. (2004), "Laminar Lifted Flame Speed Measurements for Aerosols of Metals and Mechanical Alloys," *AIAA Journal*, 42 (7), pp.1416-1426.
11. Mohan S., Trunov M. A. and Dreizin, E.L (2007), "Heating and Ignition of Metallic Particles by a CO_2 Laser," *45th AIAA Aerospace Sciences Meeting Proceedings.*
12. Mohan, S., Shoshin, Y.L. and Dreizin, E.L. (2006), "Ignition of Aerosolized Reactive Particles at High Heating Rates," *Materials Research Society Symposium Proceedings*, 896 (Multifunctional Energetic Materials), pp 165-170.

604e Twin Screw Extrusion Of Nanoenergetics With Processing Geometries That Are Tunable To The Targeted Thermo-Mechanical History

Dilhan Kalyon[1], Seher Ozkan[1], Moinuddin Malik[2], James E. Kowalczyk[3], and Mark Mezger[4]. (1) Stevens Institute of Technology, Castle Point St., Hoboken, NJ 07030, (2) HfMI, Stevens Institute of Technology, Castle Point St., Hoboken, NJ 07030, (3) Material Processing & Research, Inc., 31 Mercer St. 2-E, Hackensack, NJ 07601, (4) US Army Research Development Engineering Command, US Army, Picatinny Arsenal, Dover, NJ 07806-5000

The dispersion of nanoparticles into energetic formulations requires relatively high shearing stresses to be applied to allow the deagglomeration of particle clusters and the encapsulation of the separated nanoparticles by the binder phase. For such dispersion the twin screw extrusion process provides multiple advantages over the conventional batch processing methods including a significantly higher surface to volume ratio for better process and product quality control and flexible and tunable geometries (combinations of reversing or forwarding fully-flighted or kneading disc type screw elements configured at differing pitches and stagger angles) to tailor the thermo-mechanical history that the energetic formulation is to be exposed to during the incorporation of the nanoparticles. Here the technology base will be introduced and results from processing studies obtained with a MPR mini twin screw extruder, in conjunction with 3-D FEM based mathematical models of the process and the detailed rheological characterization of formulations involving gel based binder systems and different types of nanoalumina, will be presented.

Experimental investigation of aluminum oxidation in water

Mirko Schoenitz, Mikhaylo A. Trunov, Chi-Mon Chen, and Edward L. Dreizin
New Jersey Institute of Technology, Newark, NJ 07102, USA

Abstract

The use of aluminum as an energetic component in pyrotechnics, solid fuel formulations, and propellants is motivated by its high gravimetric oxidation enthalpy. However, the rate of its oxidation is significantly lower than that of other, typically hydrocarbon-based, components of fuel formulations. As a result, aluminum combustion will take place in the combustion products of those components, i.e. H_2O and CO/CO_2. In order to design and optimize fuel formulations as well as the devices where these formulations are used, the thermodynamics and kinetics of the reaction of aluminum with the component gases need to be understood. This report shows results of oxidation of aluminum powders in water-argon gas mixtures at temperatures between 300 – 1100 °C using thermogravimetry. In this environment, aluminum oxidizes in several distinct stages. Initial slow oxidation between 300 – 500 °C is followed by stepwise oxidation of approximately 10 % near 550 °C. At low heating rates (<5 K/min), the next sharp oxidation step of about 10 % is observed at the aluminum melting point of 660 °C. This oxidation step shifts to higher temperatures at increased heating rates. The initial two oxidation steps are followed by one or more distinguishable spread-out oxidation reactions that terminate when the material is fully oxidized above 1000 °C. The temperatures and degrees of oxidation achieved in the individual steps depend on the particle size, the heating rate, and the water concentration in the atmosphere. The results are comparable to previously published results of oxidation in oxygen-argon mixtures, although there are significant differences. Most notably, the oxidation step associated with aluminum melting was not previously observed experimentally. Phase analysis will be conducted on fully and partially oxidized powders, and a mechanistic model for aluminum oxidation in water will be developed.

Background: state of the art metal ignition models discussed in the literature

Aluminum and other metal powders are widely used as fuels or fuel additives in explosives, propellants, and pyrotechnics. However, metal ignition processes crucial for accurate modeling of metallized energetic materials remain poorly understood. For example, aluminum, the most common reactive metal additive, is known to ignite after the integrity of a protective Al_2O_3 layer is disrupted. A quantitative description of the processes leading to such a disruption is lacking. In most practical models, it is simply assumed that the oxide coating stops being protective at a fixed temperature close to the melting point of Al_2O_3. Recent research involving ignition of powders with different size distributions and, in particular, nanopowders, showed that the value of the fixed ignition temperature must be adjusted as a strong function of particle size [1]. One hypothesis posed that the disruption of integrity of the Al_2O_3 layer occurs as a result of thermal expansion mismatch between Al and Al_2O_3, and led to the suggestion that ignition should be triggered by aluminum melting, when this mismatch is at its maximum [2, 3]. However, this hypothesis did not explain many reports of much higher ignition temperatures measured for micron-sized Al powders and somewhat lower than melting point ignition temperatures for nano-sized Al powders. Furthermore, multiple thermal analysis studies with micron-sized and nanosized Al powders [4 – 6] showed no accelerated Al oxidation during melting, contradicting the expected higher oxidation rate that would be caused a non-continuous Al_2O_3 layer.

A new model of heterogeneous oxidation of Al powder leading to ignition was recently developed and quantified for reaction of aluminum in environments containing oxygen as the only oxidizer [6 – 8]. This model considers the oxidation process that is limited by the rates of diffusion of aluminum and oxygen ions through the growing layer of aluminum oxide. Most importantly, this model considers the changing properties of the aluminum oxide layer at elevated temperatures, including polymorphic phase transitions accompanied by substantial changes in the oxide density. The model predicts a broad range of ignition temperatures for aluminum powders of different sizes, consistent with many available experimental reports. The quantitative description is currently available for ignition of aluminum particles in air [7] but not in the many practically interesting environments, in which H_2O and CO/CO_2 molecules serve as primary oxidizers.

Experimental

The existing Netzsch Simultaneous Thermal Analyzer STA409 PC was upgraded to install a new furnace enabling measurements with the controlled flow of water vapor. The water feed lines are pre-heated and the water flow rate is metered. In addition, dedicated sample holders for thermo-gravimetric (TG) measurements only were acquired. These sample holders are flat dishes of approximately 3/4 inch diameter, designed to increase the accuracy of the TG measurements by working with a larger sample and by exposing a larger surface area of the sample to the furnace environment. This minimizes problems with oxidizer gas diffusion within the powdered sample as well as sintering of the sample, although the change in sample geometry precludes simultaneous DSC or DTA measurements. The installation of the new furnace was completed and the furnace with new TG sample holders was calibrated. Temperature calibration was carried out with a set of certified pure metals, the melting point of which can be recovered from the recorded temperature signal.

Preliminary measurements

Spherical aluminum powder with nominal particle size of 3 – 4.5 μm, 97.5 % pure by Alfa Aesar was used. The TG balance was continuously flushed with ultra high purity argon provided by Matheson. The gas flowrate was maintained at 20 mL/min.

Preliminary measurements were performed to establish the qualitative behavior of aluminum oxidizing in water and to determine experimental conditions appropriate for more detailed analyses.

The first preliminary measurement was aimed to directly compare the kinetics of oxidation of aluminum in water and in oxygen using the same experimental conditions. The obtained oxidation traces for the same aluminum powder processed in the same TG sample holder with oxygen and water serving as oxidizers in different tests are shown in Fig. 1. The tests were designed to achieve similar oxidizer flow rates. In general, the rate of aluminum oxidation in water is noticeably greater than that in oxygen. A striking difference between the measured oxidation traces is observed in the vicinity of the aluminum melting point. Consistent with our earlier results, the oxidation trace measured in oxygen does not exhibit any substantial weight increase. On the contrary, a substantial and rapid weight increase is observed at the melting point of aluminum for the oxidation curve measured in water. The observed behavior suggests that in the presence of water, the oxidation rate increases step-wise upon aluminum melting, unlike for the reported earlier oxidation in oxygen. Such behavior can be attributed to the mechanical disruption of the continuity of the oxide coating and needs to be investigated further.

The effect of water flow rate was addressed in the following measurement. The result is presented in Fig. 2.

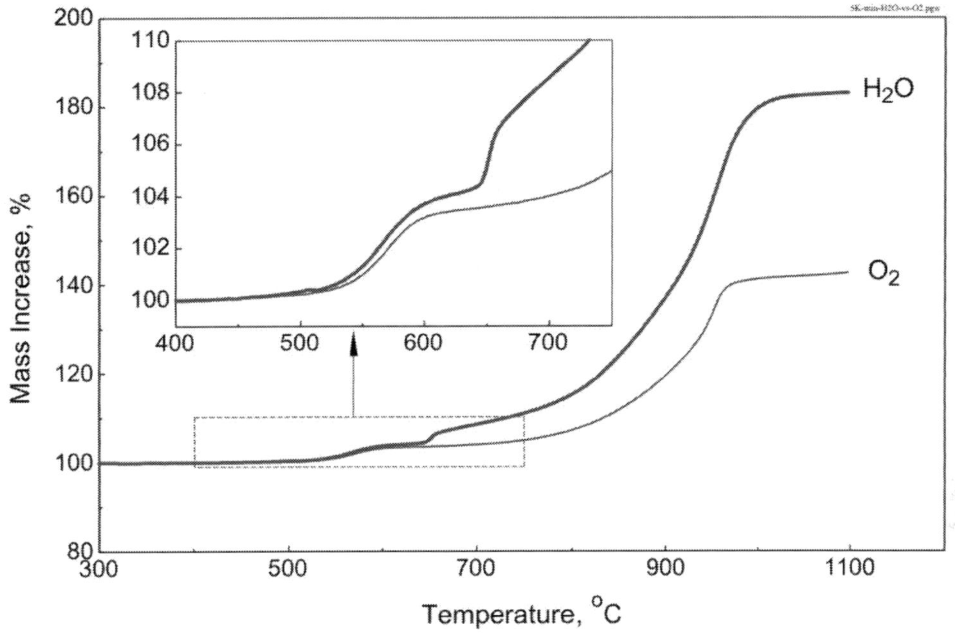

Fig. 1. Experimental TG traces for Al powder, 3 – 4.5 μm nominal size oxidized in H_2O/Ar and in O_2/Ar gas mixtures. Water and oxygen concentrations are fixed at 77 vol-% each. The heating rate is 5 K/min.

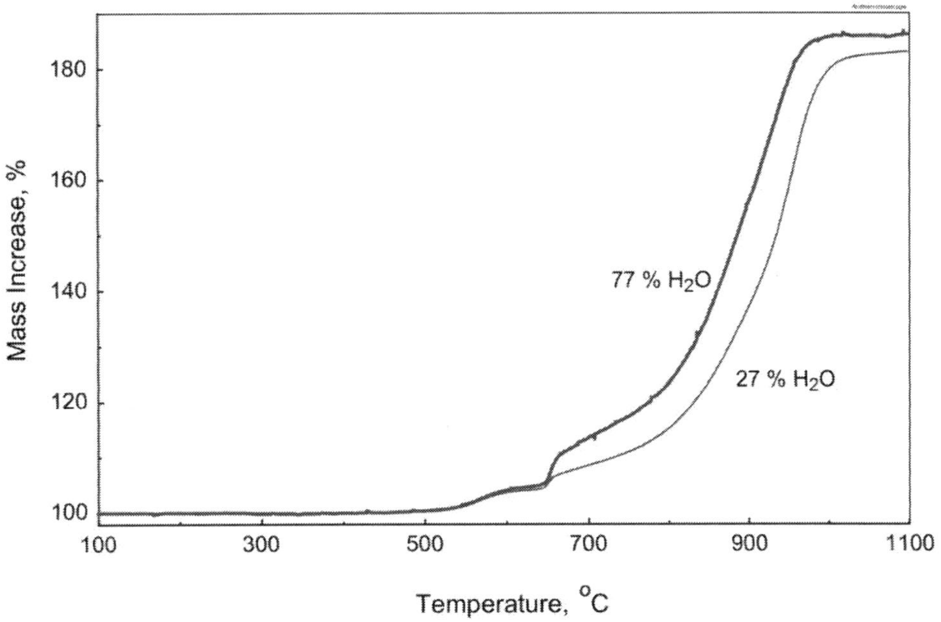

Fig. 2. Experimental TG traces for Al powder, 3 – 4.5 μm nominal size oxidized in H_2O/Ar gas mixtures with different water concentrations. The heating rate is 5 K/min.

It is noted from Fig. 2 that the water concentration affects the rate of oxidation as expected. As in Fig. 1, a clear stepwise increase in the oxidation rate is observed to occur upon aluminum melting.

The powders recovered from the TG sample holders were observed to be strongly agglomerated. The agglomeration could have occurred as a result of powder melting and sintering upon heating. Such processes reduce the available powder surface and can substantially skew the interpretation of the TG measurements. Therefore, several experiments were performed for Al powder was blended with nanosized alumina before being loaded into the TG sample holder. Thus, aluminum particles were separated from one another and reduction in their agglomeration upon heating was expected. The blending was performed using a dispersion of aluminum and alumina powders placed into an ultrasonic bath. The TG traces presented in Fig. 3 compare the experiments performed under identical conditions and using pure aluminum powder and aluminum powder blended with the nanosized alumina.

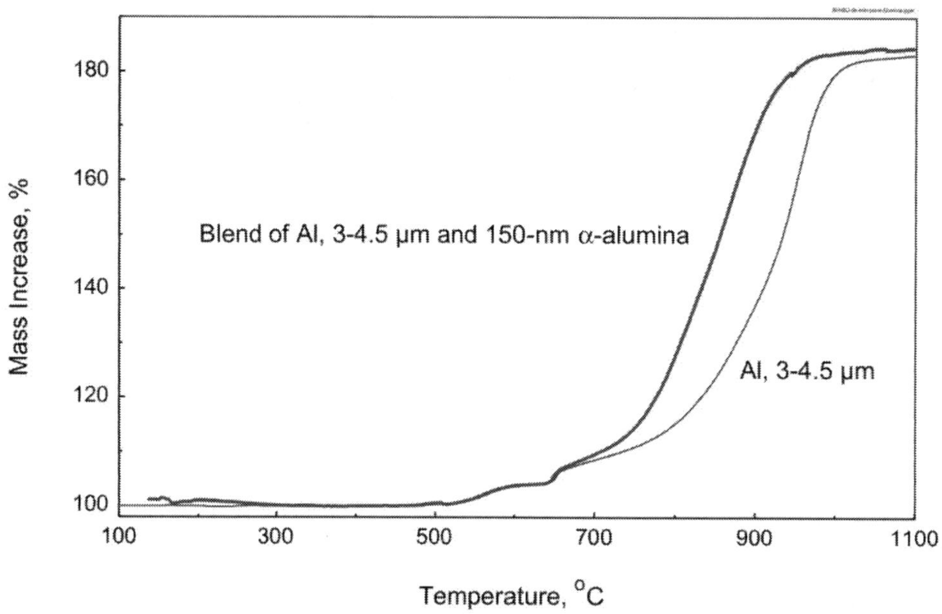

Fig. 3. Experimental TG traces for Al powder, 3 – 4.5 µm nominal size oxidized in H_2O/Ar gas mixtures. Water concentration is fixed at 27%. One of the samples is mixed with fine Al_2O_3 powder to minimize the effect of particle coalescence upon melting. The heating rate is 5 °C/min.

It is observed from Fig. 3 that coalescence of particles upon their heating and melting substantially reduces the measured rate of oxidation. For a similar experiment at a higher heating rate of 20 °C/min, it is also noted that the stepwise increase that apparently is associated with Al melting is slightly shifted to high temperatures for the pure Al powder sample. Note that the effect of coalescence may not have been fully removed by mixing aluminum and fine alumina powders.

Detailed kinetics measurements

Based on the preliminary results presented above, a set of measurements was performed, in which the heating rate was systematically varied. The measurements consistently used aluminum powder blended with nanosized alumina and used constant water flow rate resulting in its consistent concentration of 27%. The results of the measurements are presented in Fig. 1.4.

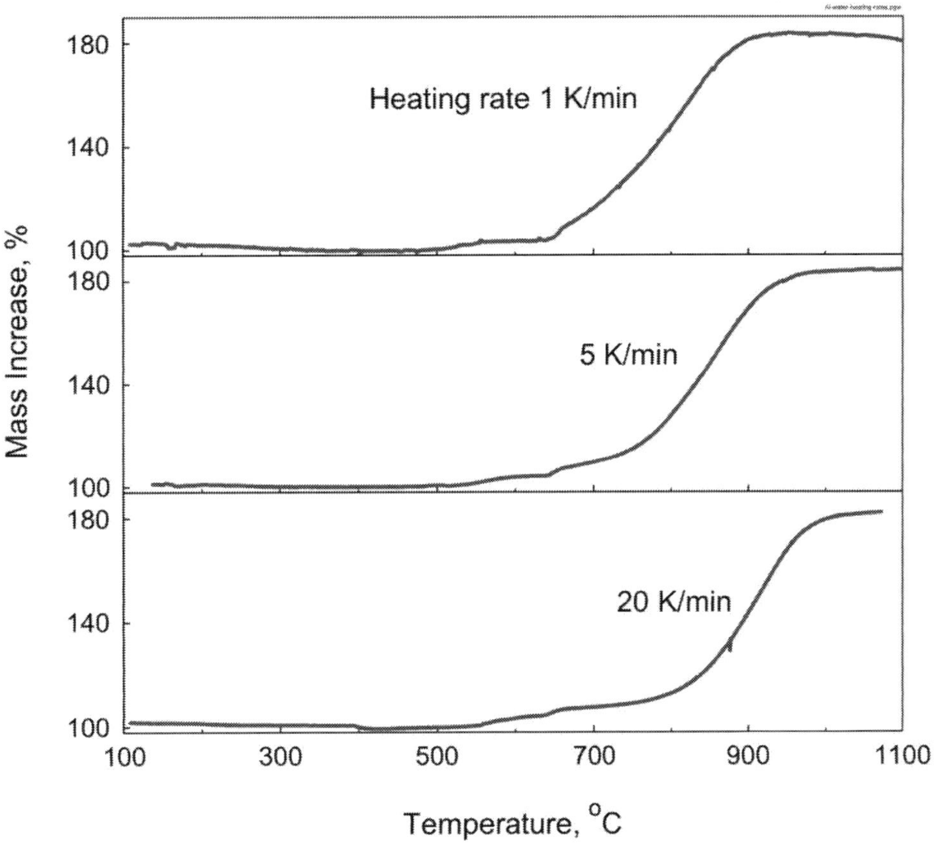

Fig. 4. Experimental TG traces for Al powder, 3 – 4.5 μm nominal size oxidized in H_2O/Ar gas mixtures. Water concentration is fixed at 27%. The samples are mixed with fine Al_2O_3 powder to minimize the effect of particle coalescence upon melting. The heating rates vary from 1 to 20 °C/min.

The data presented in Fig. 4 show that despite the change in the heating rate, one sharp, stepwise increase in the rate of Al oxidation always occurs at the same temperature closely matching the aluminum melting point. This further supports the idea that this oxidation step is caused by a mechanical disruption of the oxide coating as a result of aluminum melting, a phenomenon which was not observed in dry oxidation experiments [4-6]. Future work will focus on processing the experimental data presented in Fig. 4 and establishing the quantitative description of kinetics of aluminum oxidation in water. Further experiments will consider other oxidizing gases, such as CO and CO_2 and different gas mixtures.

Acknowledgements

This research was supported by the Defense Threat Reduction Agency.

References

1. Trunov, M.A., Schoenitz, M., and Dreizin, E.L. *Propellants Explosives and Pyrotechnics*, V. 40, No 1, pp. 36-43 (2005)
2. V.I. Rozenband, N.I.Vaganova, *Combustion and Flame*, 88 (1), 113-118 (1992)
3. V.I. Rozenband *Combustion and Flame*, 137, 366-375 (2004)
4. C. Johnson, T. Parr, D. Hanson-Parr, R. Hollins, S. Fallis, K. Higa, Proc. 37-th JANNAF Combust. Subcom. Meeting (November 2000), 2000, pp. 539-55
5. Bulian, C.J., Kerr, T.T., Puszynski, J.A. *Proceedings of the Thirty-First International Pyrotechnics Seminar*, F.J. Schelling - Editor, IPSUSA Seminars, Inc., Fort Collins, CO, pp. 327-338 (2004)
6. M.A. Trunov, M. Schoenitz, E.L. Dreizin *Combustion and Flame* V. 140 No 4, pp. 310-318 (2005)
7. Trunov, M.A., Schoenitz, M., Dreizin, E.L. *Combustion Theory and Modeling*, 10 (4) pp. 603-624 (2006)
8. Trunov, M.A., Umbrajkar, S.M., Schoenitz, M. Mang, J.T., Dreizin, E.L. *Journal of Physical Chemistry B*, 110(26), pp. 13094-13099 (2006)

Bis Triaminoguanidinium Azotetrazolate (TAGzT) Scale Up And Production

Chris Radack, Jerry Salan, and Shannon Lenahan
Naval Surface Warfare Center
Indian Head Division
Indian Head, MD

ABSTRACT

The manufacture of bis triaminoguanidinium azotetrazolate (TAGzT), along with its precursor sodium azotetrazolate pentahydrate (NaZT), has been well documented by employees at the Naval Surface Warfare Center, Indian Head Division (NSWC-IHD). In 2003 NSWC-IHD successfully validated both the NaZT and TAGzT 100-gallon batch scale processes working with senior synthesis chemists at Los Alamos National Laboratory (LANL). Recently in 2006 the TAGzT process was optimized to improve operational safety and to maximize the product yield. This report addresses the various process improvements made to the manufacture of the novel energetic ingredient TAGzT.

INTRODUCTION

TAGzT is a novel energetic ingredient that was synthesized and characterized by Hiskey et al at Los Alamos National Laboratory.[1] This high nitrogen molecule generates high volumes of gas with little or no smoke. It is currently being explored as an energetic ingredient in explosive and propellant applications.[2,3,4]

NSWC-IHD is optimizing the TAGzT process to reduce operating costs and increase the safety of personnel involved with these operations under the Navy's Manufacturing Technology Program (Navy MANTECH). In 2003, the manufacture of TAGzT was first demonstrated at the 100-gallon scale with the goal of producing 100 kilograms of material for evaluation in energetic formulations. During the execution of this original manufacturing campaign, the following process challenges were revealed:

1. Personnel had to remain present during the entirety of the NaZT and TAGzT processes since both reactions were controlled locally.

2. The hot filtration of the precipitated manganese dioxide (MnO_2) byproduct produced during the NaZT reaction was slow, which resulted in NaZT crystals forming in the MnO_2 cake.

3. The process involved the isolation, drying and handling of the NaZT (Class 1.1 explosive).

4. The NaZT and TAGzT reactions were not optimized, resulting in a higher cost per kilogram.

5. While executing both the NaZT and TAGzT processes, substantial quantities of process water were accumulated, incurring storage and disposal costs.

CHEMISTRY OVERVIEW

The NaZT synthesis reaction is shown below as Figure 1. Five water molecules are complexed to the azotetrazoleate ion. The NaZT synthesis is accomplished in a single-step oxidation. 5-aminotetrazole (5-AT) is added to a 2M aqueous solution of sodium hydroxide (NaOH) followed by the addition of potassium permanganate ($KMnO_4$), which produces the NaZT and the insoluble MnO_2 byproduct.

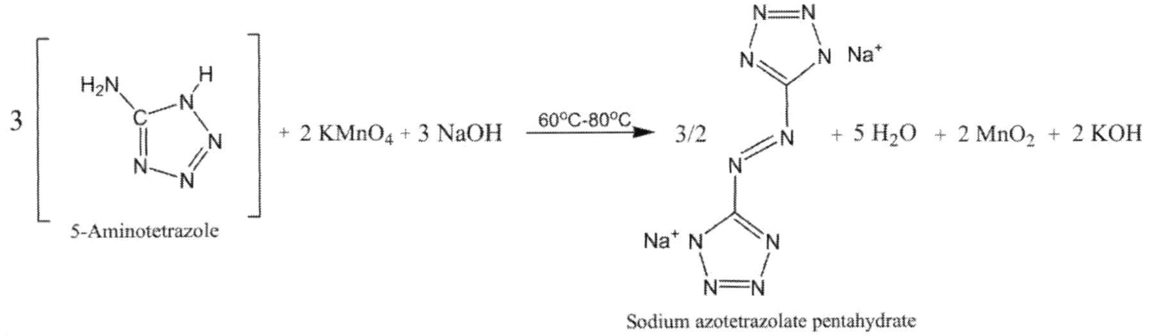

Sodium azotetrazolate pentahydrate

Figure 1. NaZT stoichiometrically balanced equation

The TAGzT synthesis is accomplished with the addition of triaminoguanidine hydrochloride (TAG-HCl) to a heated aqueous solution of NaZT. The sodium ion is exchanged for the triaminoguanidine ion. The TAGzT reaction is shown below as Figure 2. As seen below sodium chloride (NaCl) is formed as a byproduct of the reaction and the complexed waters on the azotetrazolate molecule are released.

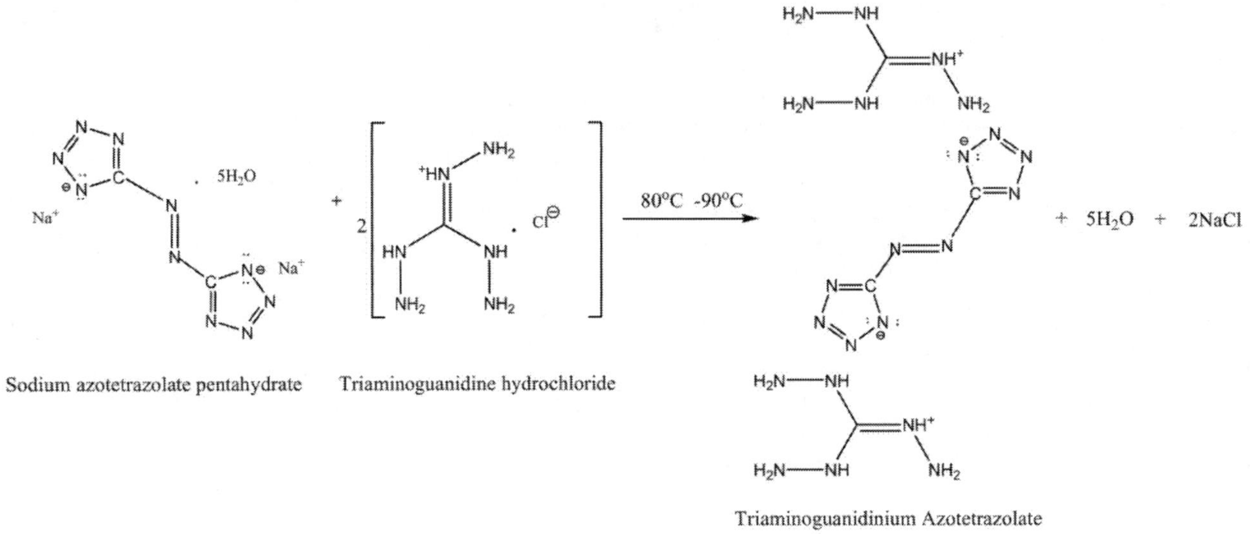

Figure 2. TAGzT stoichiometrically balanced equation

ORIGINAL NAZT AND TAGzT PROCESS DESCRIPTION

In 2003, the manufacture of TAGzT was first demonstrated at the 100-gallon scale with the goal of producing 100 kilograms of material for use in energetic formulations. The process requires three batches of NaZT to produce one batch of TAGzT in the 100-gallon reactor. Figure 3 depicts the equipment setup for the NaZT process.

The NaZT process started with the manual addition of water to a 100-gallon reactor using a hose and flow totalizer. Sodium hydroxide was added manually using a scoop. This exothermic reaction was monitored using local temperature readout. 5-AT was added to the caustic solution by means of a scoop. Once the solution reached the required reaction temperature of 60°C, the $KMnO_4$ was added using a pneumatic vibratory feeder. The operator was required to open and close a local valve on the feeder to adjust the feed rate as a function of temperature, since the oxidation produced a significant heat of reaction. The operator was present for all of the addition steps. As the NaZT is formed it remains in solution and the MnO_2 byproduct precipitates out of solution.

The MnO_2 byproduct is removed by hot filtration utilizing a Buchner filter and double diaphragm pump. The operator was required to open the drain valve to the reactor locally, which allowed the slurry to fill the Buchner filter. The operator would then close the drain valve as the filter approached capacity. The operator turned on the double diaphragm pump using a local valve. The solution was pumped to a receiving vessel of the same volume. The solution was cooled to an adequate temperature for the NaZT product to precipitate from solution.

After the crystallization of the NaZT, it was captured on a second Buchner filter by opening the vessel drain valve. Again, the operator controlled the capacity of the filter by manually opening and closing the drain using a local valve. The filtrate was pumped into drums using a manually operated double diaphragm pump. The NaZT was allowed to air dry on the filter, after which it was transferred to drying trays. The air-dried NaZT was weighed and packaged. Approximately 10 kilograms of dry NaZT were collected per batch.

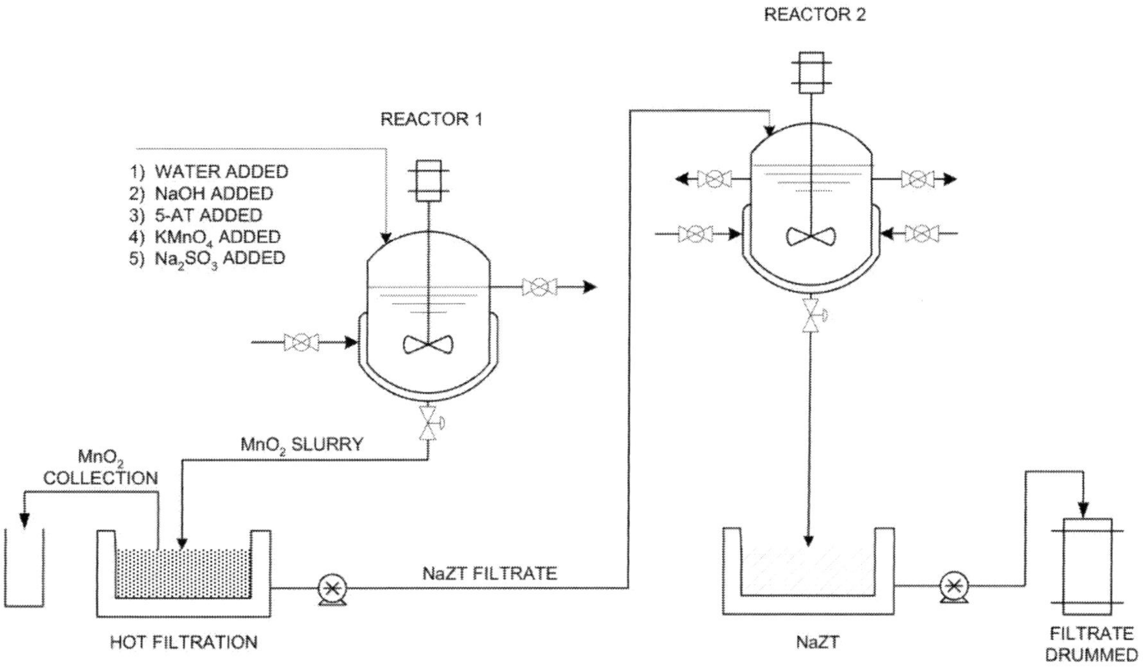

Figure 3. Original NaZT Process

The TAGzT process required the operator to weigh out a quantity of dry NaZT to be fed to a predetermined volume of warm water. The original process called for the addition of TAG-HCl (non-energetic material), but for three 100-gallon scale batches at Indian Head, Triaminoguanidine Nitrate (TAGN) (Class 1.1 explosive) was used instead because a supply existed. TAGN was metered into the vessel by the operator using a conductive scoop. This reactant was later replaced with TAG-HCl, which was metered to the vessel using a vibratory feeder. After the addition of the TAG-HCl, the TAGzT solution was cooled to ambient temperature. The TAGzT slurry was then drained onto a Buchner filter for collection. The TAGzT was allowed to air dry on the filter and then was transferred to drying trays. The product was dried at 60°C under vacuum, weighed and packaged. Approximately 30 kilograms of dry TAGzT were collected per process run.

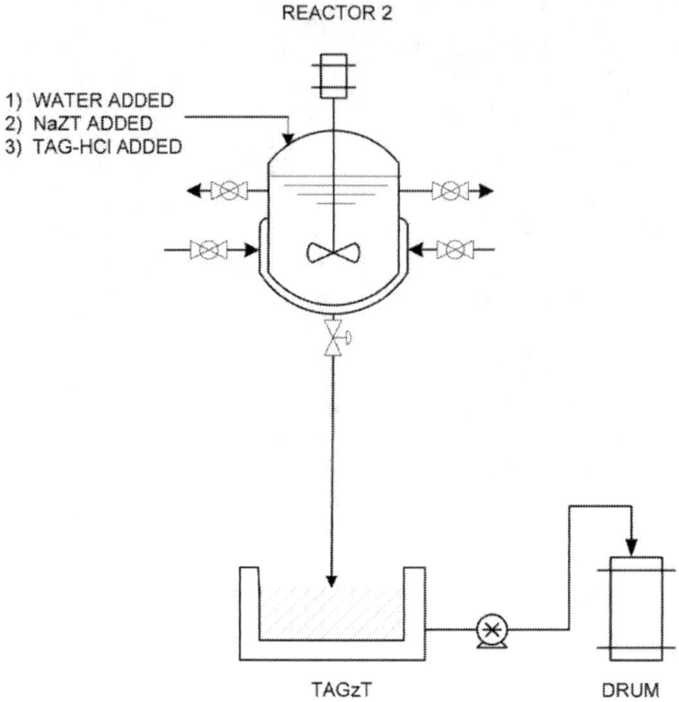

Figure 4. Original TAGzT Process

RESULTS AND DISCUSSION

Before beginning the 2006 TAGzT production campaign, each one of the aforementioned challenges were addressed:

1. PERSONNEL EXPOSURE

The original NaZT and TAGzT manufacturing processes were attended operations. Reactants were added manually and all process transfers were initiated and monitored locally. The implementation of automated process controls and a centralized monitoring system removed the operators from the majority of these hazards. This programmable logic control (PLC) system was designed to include safety alarms and interlocks to aid the operators for safe processing of materials. The potassium permanganate ($KMnO_4$) addition is now performed remotely. It is considered the most hazardous addition since the explosive NaZT is produced during this step in a caustic solution. During all filtrate transfers the operators have been removed from the hazardous environment by remotely operating valves and pumps. The installation of video cameras allows the operators to visually confirm the all of the material transfers. The addition of an intercom system enables operators to hear the pumps and communicate between the operating bay and the control room during troubleshooting activities.

2. HANDLING EXPLOSIVE INTERMEDIATES

The original TAGzT production campaign required the handling of TAGN and NaZT. The handling of explosive intermediates is typically avoided because it increases the hazards of the overall production operation. NaZT is a particularly dangerous material with a high sensitivity to electrostatic discharge and friction. It has been involved in one known incident where in a laboratory setting the dried material was unintentionally initiated, causing injury to personnel. The cause was thought to be an electrostatic discharge.

The original NaZT production process required three batches of NaZT to be air dried, weighed and packaged for later use in the TAGzT synthesis. The NaZT was later weighed and added to the reactor manually, exposing the operators to a dangerous explosive material. There were several solutions considered to eliminate the handling of this sensitive intermediate, including the use of a centrifuge with automated discharge, allowing the NaZT to remain wet after synthesis, searching for a new intermediate, and using automated feeders to introduce the material into the vessel. The search for a new intermediate was not thoroughly pursued and out of the scope of this work. The other options all demonstrated methods to eliminate exposure but none provided a total solution. The selected solution involves collecting two or three batches of wet NaZT in a Nutsche filter, dissolving it in hot water when it is required to make TAGzT, and pumping it in solution to the TAGzT synthesis reactor, entirely eliminating operator handling of NaZT. One additional challenge is introduced by using this method: determining NaZT yield. The NaZT is never dried and weighed, but the amount present must be determined so that the appropriate quantity of TAG-HCl is added to form TAGzT.

The first implemented solution to this challenge came in the form of introducing a new technology to the process. ReactIR had been useful in the past for monitoring reaction progression by identifying starting materials, intermediates, and products. This *in situ* technology views materials in solution using mid-infrared (IR) spectroscopy. The plan was to dissolve the NaZT collected on the Nutsche filter, collect it in the reactor used to make TAGzT, and use ReactIR to measure the concentration of NaZT in solution. This data would be used to calculate the total NaZT yield and the required amount of TAG-HCl reactant needed for a successful TAGzT reaction. A reactor sampling system was set up to withdraw a continuous sample from the reactor so that the ReactIR probe inserted in a custom-designed flow cell could analyze the sample stream. Figures 5 and 6 show the equipment setup for the optimized NaZT and TAGzT processes, respectively.

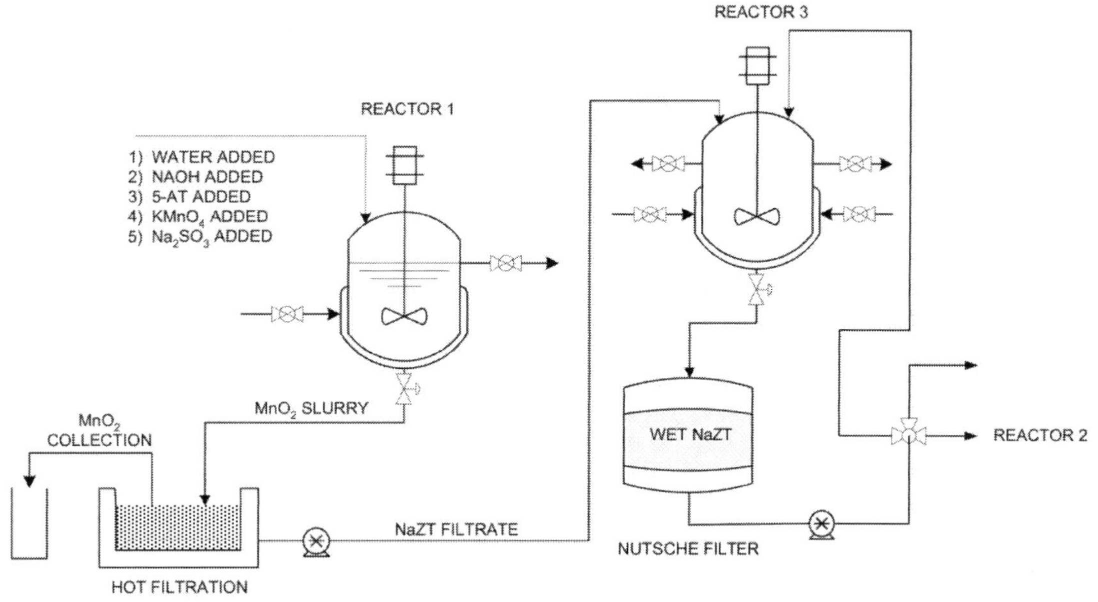

Figure 5. Optimized NaZT Process

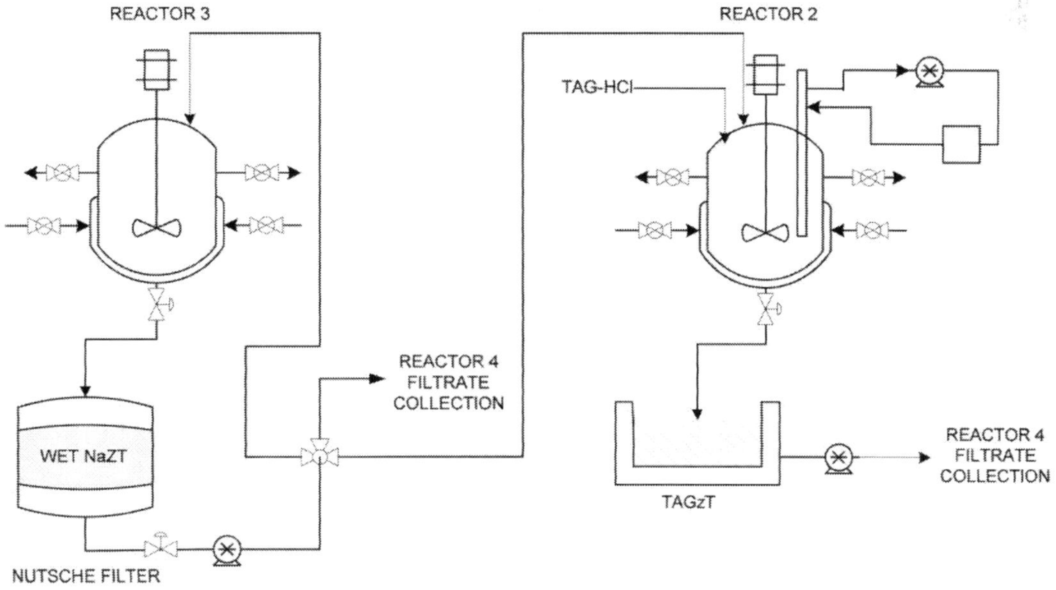

Figure 6. Optimized TAGzT Process

Although ReactIR technology had previously been used by NSWC-IHD personnel to obtain qualitative measurements of species in solution, quantitative measurements had not yet been attempted. NSWC-IHD personnel consulted with the equipment manufacturer and determined that it was possible with proper equipment calibration. Since the TAGzT reaction is completed in a temperature range of 70-80°C, the calibration data was done across the same temperature range. Calibration data was collected for multiple NaZT concentrations in the stated temperature range. The data was obtained by circulating known NaZT concentrations through the flow cell with a lab-scale setup, simulating the reactor sampling system.

The IR spectra obtained during the calibrations were incorporated into a quantitative program that concentrated on a peak or peaks in the full spectrum that grew proportionally with discrete increases in concentration. Figure 7 is the complete infrared spectrum for an aqueous solution of NaZT. Figure 8 shows the identified peak of interest that changes intensity with varying NaZT concentration. This peak was used to construct the NaZT quantitative program. In both plots the x-axis is the wavelength (cm^{-1}) and the y-axis is absorbance. Figure 9 below is a validation plot of the quantitative program. The target NaZT weight percent was assigned a value of 100 and the high and low concentration range was established after this target weight percent was assigned. The values on the y-axis are the predicted values produced by the quantitative program.

The validation of the quantitative program appeared to be promising; however, during the actual production process the ReactIR proved to be unreliable. Working with the manufacturer, a robust quantitative program could not be generated that was accurate for a range of concentrations in the given temperature range. It was determined that the equipment is too sensitive to small changes such as conduit alignment, probe fouling and changes in solution temperature. This lack of reliability resulted in abandoning the technology and pursuing other alternatives.

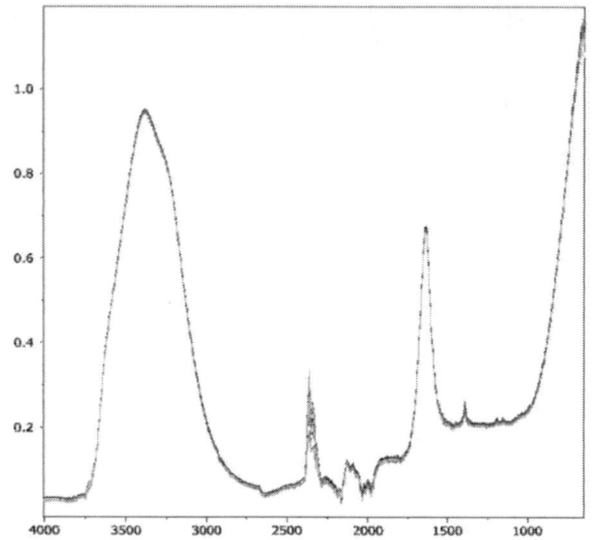

Figure 7. NaZT IR spectra

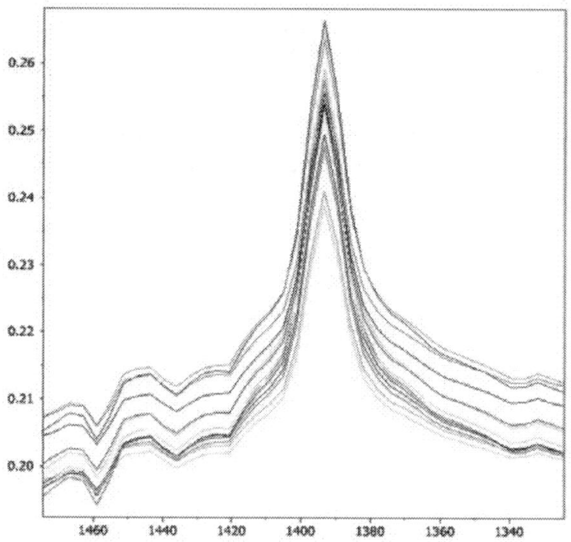

Figure 8. NaZT Identified Peak

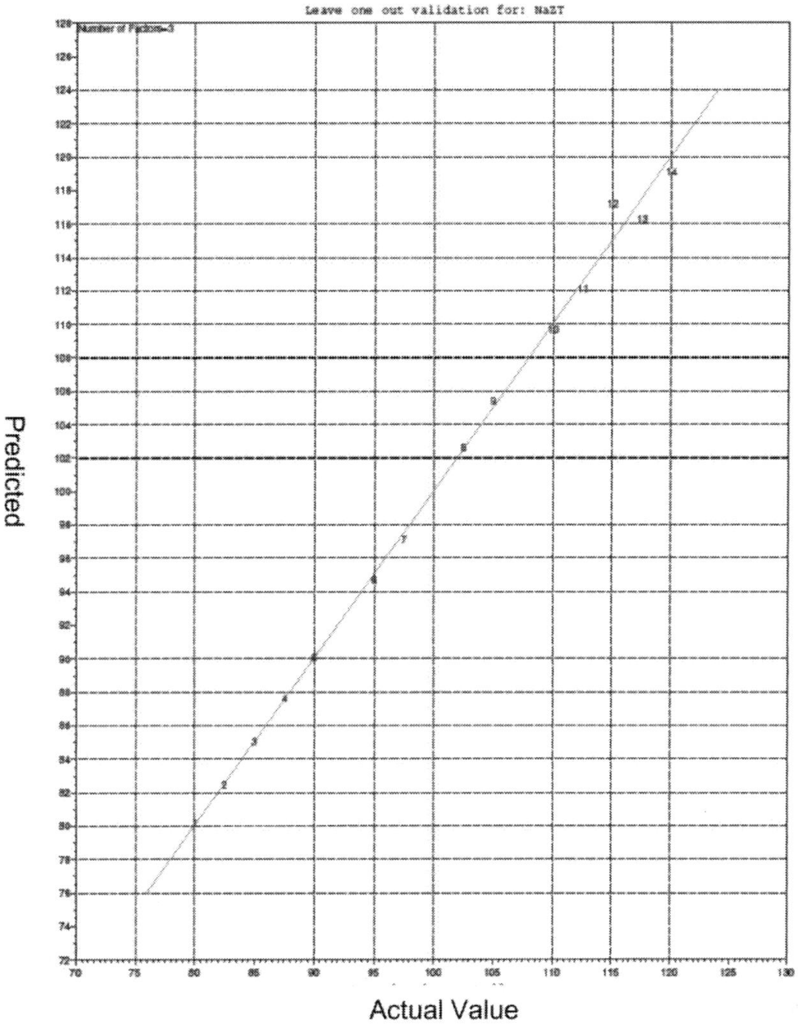

Figure 9. Validation plot

As an alternative, a Cary 50-UV/VIS spectrophotometer was calibrated using diluted NaZT solutions of varying concentrations. From the collected data, a calibration plot was created correlating concentration with absorbance. Figure 10 is the concentration calibration plot for NaZT in water.

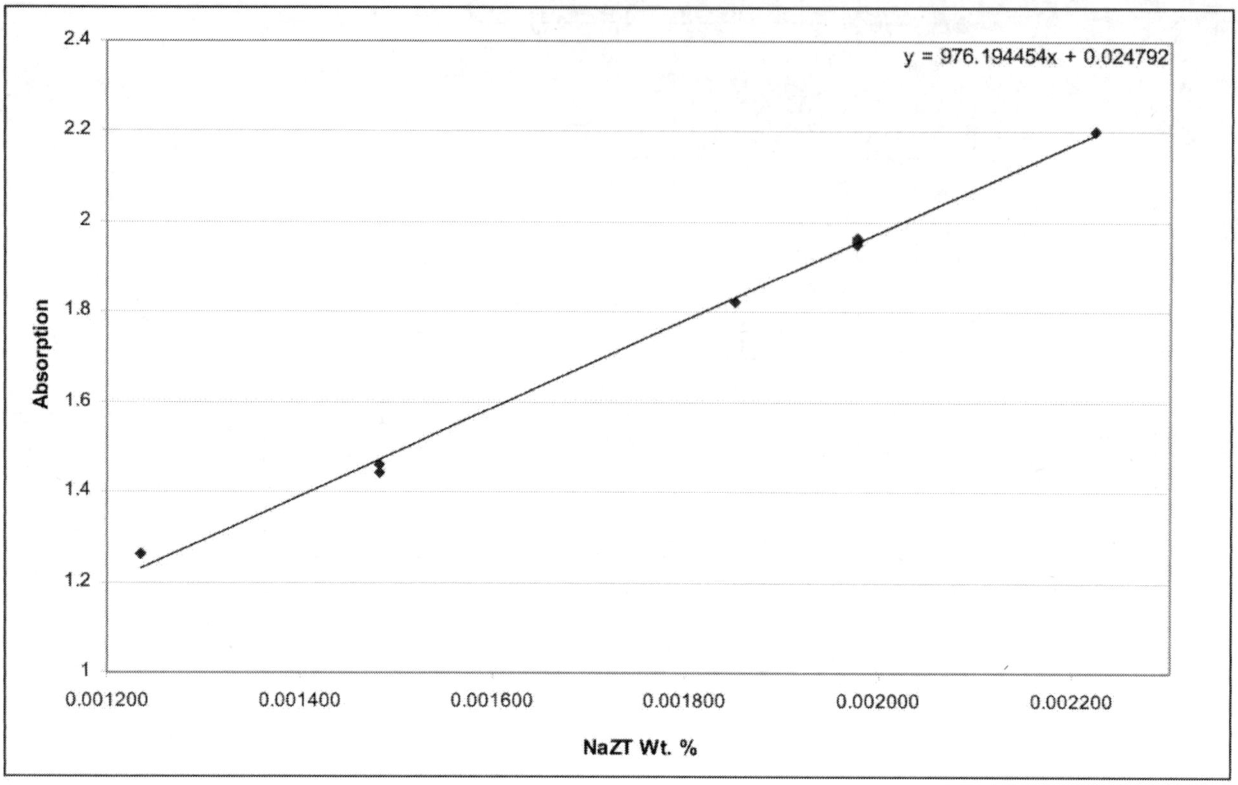

Figure 10. NaZT concentration calibration plot using Cary 50-UV/Vis Spectrophotometer

Once the NaZT is made and in solution, a sample will be retrieved from the reactor and diluted with a predetermined amount of water. The diluted sample is analyzed using the spectrophotometer to obtain the absorbance value, which corresponds to an NaZT concentration based on the calibration data. This procedure has been validated in the laboratory, but has not yet been performed during TAGzT production.

3. INCREASE QUANTITY PER BATCH

A collaborative effort between NSWC-IHD and LANL resulted in the optimization of the NaZT and TAGzT syntheses. The starting point for these optimization studies was taken from a research article by Hiskey, et. al.[1] The original goal was to improve the quantity per batch of both reactions twofold over the original procedure.

The NaZT reaction involves adding potassium permanganate ($KMnO_4$) to a solution of 5-amino-tetrazole in 2N sodium hydroxide (NaOH) solution. LANL effectively doubled the reactant density and yield possible from the same volume reactor. Table I shows a comparison between the original procedure and the optimized one. An example of the improved laboratory preparation is provided here: To 2N NaOH (150 ml) was added anhydrous 5-aminotetrazole (8.5 g, 100 mmol). The solution was brought to 60 °C and $KMnO_4$ (12 g, 76 mmol) was added portion wise over 1 hour while maintaining the temperature between 60 and 65 °C throughout the addition. The excess oxidant was treated with sodium sulfite and the reaction filtered at 60-65 °C through a bed of Celite. The filter cake was washed with hot water (90 °C, 150 ml) and the filtrate was concentrated to 175 ml and the product filtered. The filtrate was then neutralize with approximately 20 ml of concentrated HCL to neutralize to pH=7. The product yield of NaZT.5H_2O was 11.7 g.

Table I. Comparison of original and improved NaZT synthesis at the 100-gallon scale

	Original procedure, mol	Improved procedure, mol
5-Aminotetrazole monohydrate	110	227
Potassium permanganate	87	172
Projected NaZT yield (78%)	44	88

The quantity of TAGzT obtained per batch was also increased, though not doubled. The reaction concentration was increased in a manner similar to that used in optimizing the NaZT reaction, but the reaction was also cooled to 5 °C prior to isolation of the TAGzT, increasing the yield. The reaction concentration was increased to 0.5M in NaZT from 0.4M in the original process. The combination of the increase in the molarity of the reaction and the decrease in temperature prior to isolation provides a 70% increase in TAGzT isolated per batch. Table II compares the original TAGzT preparation with the improved one. A typical experimental lab procedure is provided here: To a solution of water (20 ml) is added NaZT.5H_2O (3 g, 10 mmol, 1 equiv.), the reaction was heated to 70 °C and triaminoguanidine hydrochloride (3.09 g, 22 mmol, 2.2 equiv.) was added in one portion. The reaction was allowed to stir at 70 °C for 10 minutes, then allowed to cool to room temperature and finally to 5°C. The product was then filtered and air-dried to provide 3.51 g (94%) of TAGzT.

Table II. Comparison of original and improved TAGzT synthesis at the 100-gallon scale

	Original procedure, mol	Improved procedure, mol *
NaZT	114	171
TAG-HCl	233	376
Projected TAGzT yield (94%)	95	161

*Also cooled to lower temperature than original procedure

When the improved process was scaled up to the 100-gallon scale, the quality of each TAGzT lot was evaluated using differential scanning calorimetry (DSC), elemental analysis (carbon, hydrogen and nitrogen weight percent), and ion chromatography to determine percent sodium and chloride ions. The NaZT yield per batch was approximately 70%. The TAGzT yields per batch varied from 65% to 88%. Product was lost during the crystallization of the TAGzT due to a significant quantity adhering to the reaction vessel wall, which was not easily recovered. In most cases this material was not recovered and was lost during the reactor cleaning process.

4. HOT FILTRATION OF MANGANESE DIOXIDE

The synthesis of NaZT includes the formation of the byproduct manganese dioxide (MnO_2). This byproduct forms as a fine solid and turns the reaction mixture into a slurry. When the slurry is filtered, the consolidated particulates form a sludge-like material that is difficult to filter. In previous campaigns, a double diaphragm pump with Teflon diaphragms was used to draw the NaZT solution through the manganese dioxide cake. Because of the limited ability of the pump to draw the filtrate through the forming MnO_2 cake, the filtrate would cool and the NaZT product would begin to crystallize out in the filter cake, resulting in the loss of a significant amount of NaZT product.

Several engineering challenges were addressed to alleviate the hot filtration issues. The Teflon diaphragm pump was replaced with a thermal elastic diaphragm pump that had approximately 4-times the dry vacuum capacity. The Buchner filter was insulated in an attempt to minimize the amount of heat loss from the solution to the surroundings. Running the optimized NaZT reaction 20°C higher than the original procedure ensured that more NaZT product would stay in solution as it was being drawn through the filter cake and pumped to the crystallization vessel.

5. PROCESS WATER DISPOSAL

The amount of wastewater generated during the TAGzT production created a disposal challenge. The NaZT and TAGzT processes produce approximately 250 gallons of wastewater each for a total of approximately 500 gallons of wastewater per batch of TAGzT. This is approximately 10-12 gallons of wastewater per kilogram of TAGzT produced. Drumming, storing and disposing of 500 gallons of wastewater for each lot of TAGzT is expensive, labor intensive, and requires large amounts of storage space. An investigation into alternatives yielded the possibility of discharging the wastewater to Indian Head's sanitary sewer under the condition it did not disturb sewage treatment operations.

NaZT and TAGzT process water were tested to insure they did not inhibit the respirometric activity of the wastewater treatment plant organisms. Lab-scale aerobic and anoxic chambers were set up to represent the treatment plant and NaZT or TAGzT waste water was dosed into half of these chambers at the concentration that the treatment plant is likely to see. The oxygen uptake and/or nitrogen generation was measured for all of the chambers and the results from the contaminated chambers were compared to the results from the uncontaminated chambers (baseline). The results indicated that NAZT and TAGzT do not inhibit the aerobic biomass activity with up to 0.1% weight percent explosive material, which is less than expected concentrations of the process water.

CONCLUSIONS

The process for manufacturing TAGzT was optimized to reduce operator exposure to a hazardous environment. This was accomplished by utilizing remote actuated valves, a PLC system, cameras, and an intercom system. The handling of explosive intermediates was eliminated by using analytical methods to determine the concentration of NaZT in solution prior to adding the reactant for manufacturing TAGzT. As a result, there is no longer a need to isolate the NaZT intermediate. The syntheses for NaZT and TAGzT were optimized to produce significantly increased quantities per batch when compared to the original syntheses. The crystallization of NaZT and subsequent loss of NaZT product during the hot filtration of MnO_2 was alleviated using higher reaction temperatures and an insulated Buchner filter. Finally, the process water generated is now disposed conveniently in the existing sanitary sewer with minimal costs.

Approximately 220 kilograms of TAGzT has been manufactured using the optimized NaZT and TAGzT production processes. The estimated cost per kilogram using the optimized process is calculated at $650/kilogram. The optimized process labor costs are one half that of the original process due mostly to increased quantities per batch and the elimination of handling the NaZT intermediate.

FUTURE WORK

NSWC-IHD plans to scale the NaZT and TAGzT processes to the 500-gallon scale. At the 500-gallon scale, the expected quantity per batch is approximately 200 kilograms. The cost per kilogram will be further reduced by increasing the NaZT and TAGzT quantities per batch. It is anticipated that the cost will be reduced to approximately $200-300/kilogram.

There are several engineering challenges remaining in both the NaZT and TAGzT processes. These include the ability to remove larger quantities of MnO_2 at the 500-gallon reactor scale. A candle type filter may be a solution to this filtration. This jacketed filter can back pulse the MnO_2 from the filter providing uninterrupted service throughout the filtration. The filter will also provide a method to capture the MnO_2 for subsequent disposal. The existing process requires the operator to manually scoop the material from the filter into disposal cans. The candle filter discharge can automatically place the material into disposal receptacles. Other challenges include eliminating the caking of the product on the reactor walls and the ability to dry the TAGzT in a more efficient manner. The existing process utilizes a vacuum shelf dryer. A tumble type vacuum dryer would aid in the quality of the bulk material without the need to manually manipulate the material on drying trays to avoid encrustation of the product.

ACKNOWLEDGEMENTS

The authors would like to acknowledge Mr. Chuck Painter, Director, Navy Energetics Manufacturing Technology Center for his continuous support, Wes Walker of Mettler Toledo Autochem for tireless aid with ReactIR, and Dr. David Chavez of Los Alamos National Laboratory for his work on the optimization of these processes.

REFERENCES

1) Hiskey, M. A., Goldman, N. and Stine, J.R. High-Nitrogen Energetic Materials Derived from Azotetrazolate, Journal of Energetic Materials. 1998, Vol. 16, 119-127.

2) Bryce C. Tappan, Arif N. Ali, Steven F. Son, Thomas B. Brill. Decomposition and Ignition of the High-Nitrogen Compound Triaminoguanidinium Azotetrazolate (TAGzT), Propellants, Explosives, Pyrotechnics 31, No. 3 (2006).

3) Walsh, C. and Knott, C., "Advanced Gun Propellant Formulations", JANNAF PDCS Meeting 2002.

4) Walsh, C. and Knott, C., "Gun Propellant Formulations with High Nitrogen Modifiers", JANNAF PDCS Meeting 2003

5) Wilden Pump & Engineering, LLC. EOP-P2RP. Section 6, April 2005.

6) Respirtek Inc. "Toxicity analysis of sodium azotetrazolate and triaminoguanidinium azotetrazolate streams from Navy facility to the biomass in their wastewater treatment facility." Biloxi, MS. 31 March 2005.

608a Environmentally Benign Nanomixing by Sonication in Supercritical CO2

Ganesh Sanganwar[1], Alexandre Ermoline[2], James Scicolone[2], Rajesh Dave[2], and Ram B. Gupta[1]. (1) Chemical Engineering, Auburn University, Auburn, AL 36849-5127, (2) New Jersey Center for Engineered Particulates, New Jersey Institute of Technology, Newark, NJ 07102

Due to increased use of nanocomposites and nanomaterials, mixing at nanoscale has become important. Conventional dry mixing is not effective in achieving mixing at nanoscale, and wet mixing suffers from different disadvantages including the need for a non-solubilizing and wetting liquid, and the removal of the liquid after mixing. To overcome the disadvantages, we propose to replace the liquids by supercritical carbon dioxide, which is environmentally benign, non flammable, non toxic, low viscous and inexpensive fluid. In the proposed method, ultrasound is applied to the suspension of agglomerated nanoparticles in supercritical carbon dioxide. Due to the high enough molar density and low viscosity of carbon dioxide, high impact collisions are created during sonication. These collisions are strong enough to provide a good mixing, and the final mixture is obtained by simple depressurization. The method was tested for binary mixtures of alumina/silica, silica/titania, MWCNT (multiwalled carbon nanotubes)/ silica, and MWCNT/titania. The effects of sonication intensity and pressure (e.g., molar density of carbon dioxide) on the degree of mixing are studied. Comparative study was also done with liquid hexane as a media for mixing. Quantitative characterization (e.g., composition standard deviation, degree of homogeneity) of mixing of alumina/silica and silica/titania was done with energy dispersive x-ray spectroscopy. The qualitative characterization of binary mixing of MWCNT/silica and MWCNT/ titania was done using field emission scanning electron microscope and day light illumination spectrophotometer. Results show that mixing in carbon dioxide and at higher ultrasound amplitude was as good as in liquid hexane. And the final mixed product does not contain any residual media as in the case of liquid hexane in conventional method.

608b Aggregate Breakage Under Laminar Conditions

Miroslav Soos, Lyonel Ehrl, and Massimo Morbidelli. Department of Chemistry and Applied Biosciences, Institute for Chemical and Bioengineering, ETH Zurich, Zurich, 8093, Switzerland

Design and usage of appropriate units for handling of fragile materials requires the knowledge of conditions leading to the breakage of such materials. Unfortunately, this information is commonly obtained using stirring devices with a broad shear rate distribution, and often under conditions where, besides breakage, aggregation is also present. An alternative approach to measure the breakage phenomena is to use aggregates produced under static and dilute conditions and break them in various types of flow fields. In this way, although no aggregation is present, obtained results will be affected by the restructuring of the particles within the aggregates. Therefore, in order to avoid possible artifacts from both aggregation and restructuring, we choose the following approach. In a first step, we prepare compact aggregates using a stirred vessel operated under turbulent conditions. In a second step, such aggregates, diluted to a very low concentration where aggregation is negligible, are broken either in extensional flow using geometry with a sudden contraction or in simple shear flow using a rheometer. The evolution of the aggregates size and structure over the course of the breakage process is measured using a combination of small angle static light scattering and image analysis. The steady state values obtained for the two types of flow are compared to the steady state aggregates sizes collected in various stirring devices operated under turbulent conditions, always using the same size of primary particles.

608c Deagglomeration of Nanoparticle Agglomerates by Rapid Expansion of Supercritical Carbon Dioxide through a Nozzle

Daniel To[1], Alexandre Ermoline[1], Rajesh Dave[1], Xiaolong Yin[2], and Sankaran Sundaresan[2]. (1) New Jersey Center for Engineered Particulates, New Jersey Institute of Technology, 138 Warren Street, YCEES 208, Newark, NJ 07102, (2) Chemical Engineering, Princeton University, A-315 Engineering Quadrangle, Princeton, NJ 08544

A major challenge in making and utilizing materials based on nanoparticles and nanocomposites is the tendency for individual nanoparticle constituents to aggregate due to Van der Waal forces and form large fractal structures on the order of tens of microns. Intimately mixing two or more constituents at the nanoscale requires one to first break the original agglomerates and then mix the individual nanoparticle constituents. To that extent, the focus of this study is to investigate the deagglomeration mechanism in a novel mixing method that utilizes the rapid expansion of supercritical suspensions (RESS). The expansion results in the simultaneous deagglomeration and mixing of the nanoparticle constituents. The focus of the present study is on the deagglomeration process where suspension containing agglomerates of nanoparticles in supercritical CO2 expands to atmospheric conditions.

Agglomerates of a single nanoparticle constituent, alumina (dp = 13 nm) was stirred in a high pressure vessel with CO2 at pressures in the range of 550 to 1700 psi and at a temperature of 45oC. The suspension was then rapidly expanded through a variety of sub-millimeter nozzles into a small expansion chamber. The deagglomerated nanoparticles, which were in an aerosol form, were sized by an in-situ method based on laser diffraction, and were examined by two ex-situ methods: differential mobility analysis and electron microscopy imaging of agglomerates collected by inertial impaction. Various pre-expansion pressures were considered and their influence on the size reduction of alumina agglomerates was studied. This work also investigated deagglomeration of other types of nanoparticles.

To better understand the deagglomeration mechanisms in RESS, a theoretical analysis based on one-dimensional mass, momentum, and energy conservation principles was conducted to determine the changes in flow and thermodynamic properties in the RESS process. The calculations indicated that the flow in a thin capillary nozzle was governed by friction and was always subsonic. When CO2 exited the nozzle, it expanded very rapidly and the flow became supersonic. After the pressure dropped to a level that matched the ambient condition, this supersonic flow experienced a sudden deceleration as it traveled through a normal shock, which is a step change in pressure, temperature, and density and became subsonic again. Thus, we believe that deagglomeration in our experiments was due to two independent mechanisms: first the strong velocity gradient in the thin nozzle, and then the impact with the normal shock.

608d Interparticle Forces in a Supercritical Fluid: A Simulation Study

Aleksey Vishnyakov, Department of Chemical and Biochemical Engineering, Rutgers, the State University of New Jersey, 98 Brett rd, Piscataway, NJ 08854, Yangyang Shen, Rutgers University, 98 Brett Road, Piscataway, NJ 08854, and **M. Silvina Tomassone**, Chemical and Biochemical Engineering, Rutgers University, 98 Brett Road, Piscataway, NJ 08854.

Rapid expansion of supercritical solutions (RESS) is a promising technique for separation and mixing of nanopartricles. Further development and tuning of RESS require fundamental understanding of interparticle forces in supercritical fluids. We present a molecular simulation study of interactions of silica nanoparticles of different sizes in supercritical carbon dioxide ($T = 318K$, $60\ atm < P < 200\ atm$), focusing on the dependence of the interparticle force on particle size, carbon dioxide pressure, and particle surface modification.

The potentials for fluid-fluid and solid-fluid interactions are fitted to experimental data on (a) PVT data of bulk CO_2 at similar temperatures and pressures (b) experimental sorption isotherms at cryogenic and critical conditions. We employ two different approaches to the modeling of interparticle forces. The forces between very small particles (2.2nm) are determined directly: two nanoparticles are placed in a bath of supercritical CO_2. To calculate the forces between the larger particles (20-60nm in diameter), we first study the disjoining pressure between planar silica surfaces, which form a classical slit-shaped pore. Then the disjoining pressure between similar spherical surfaces is derived from the planar disjoining pressure via the Derjaguin approximation.

Our simulations show that the interactions of particles with supercritical CO_2 result in notable effective attraction between the particles. The attraction is most prominent for strongly hydroxylated particles at the lower limit of the considered pressure range. The effective attraction is substantially weaker for dehydroxylated particles. The interparticle attraction also weakens as the pressure increases from 68 to 100 atm, but barely changes when the pressure of CO_2 increases further to 200atm.

PHYSICOCHEMICAL MEASUREMENTS ON INSENSITIVE MUNITIONS COMPOUND, N-METHYL-4-NITROANILINE (MNA) FOR ENVIRONMENTAL APPLICATIONS

Veera M. Boddu, Krishnaiah Abburi (Speaker) and Stephen W. Maloney
Environmental Processes Branch, US Army Engineer Research and Development Center, 2902 Newark Drive, PO Box 9005, Champaign, IL 61822-9005.

Reddy Damavarapu
Energetics and Warheads Division, U.S. Army - Armament Research, Development and Engineering Center, Picatinny Arsenal, NJ 07806-5000.

Abstract

Physicochemical properties such as solubility, octanol-water partition coefficient and Henry's law constant of energetic compounds provide invaluable information for the overall understanding of environmental distribution, biotransformation, and potential treatment processes. The solubility, octanol-water partition coefficient, and Henry's law constant were measured for N-methyl-4-nitroaniline (MNA), an insensitive munitions compound at 25, 35, and 45°C. The effect of ionic concentration on solubility, using electrolytes such as NaCl and $CaCl_2$, was also studied. The aqueous solubility of MNA was measured by adding an excess amount of the solid compound (0.2 g) to a flask containing deionized water (100 mL). The equilibrium concentration of MNA was determined using UV spectroscopy. The octanol-water partition coefficients (log K_{OW}) values were determined directly with the shake-flask procedure. A modified gas-purging technique was used for the determination of Henry's law constant (K_H) of MNA. The data on the physico-chemical parameters were correlated using the standard van't Hoff equation. The enthalpy and entropy of phase transfer were derived from the experimental data. Details of the experimental studies and results will be presented.

Introduction

Insensitive Munitions (IM) program is to develop new explosive materials and formulations which are more stable than the materials currently used. These new munitions will be incorporated into the Department of Defense (DoD) inventory and will be packed at current Load, Assemble and Pack (LAP) facilities. Production and loading, assembling and packing of these munitions will generate new waste streams at Army industrial facilities. The fate and distribution of an insensitive munitions compound released into the environment are primarily controlled by; 1) prevailing environmental conditions at the point(s) of discharge and 2) the physicochemical properties of the compound. Thus, accurate determination of physico-chemical properties is critical to develop valid environmental models and impact assessments. The most important physico-chemical properties relating to the environmental behavior of hydrophobic organic compounds are aqueous solubility (S_w), octanol-water partition coefficient (k_{ow}) and Henry's law constant (K_H). The aqueous solubility of organic compounds usually decreases in the presence of inorganic salts, which is known as the salting-out effect. Solubility, octanol-water partition coefficient and Henry's law coefficient quantify the extent to which a chemical partitions between solid-liquid, liquid-liquid and air-water, respectively.

Although the importance of insensitive munitions compounds has been long recognized, experimental data of their aqueous solubility, octanol-water partition coefficient and Henry's law coefficient are scarce. We present here new experimental data on aqueous solubility, partition coefficient (octanol-water) and Henry's law constant for MNA at 25, 35, and 45°C. The salting-out effect on aqueous solubility of MNA was investigated. The effect of temperature on S_w, k_{ow} and K_H was used to evaluate enthalpy and entropy of phase transfer.

Experimental methods

The aqueous solubility of MNA was measured at temperatures of 298.15, 308.15 and 318.15 K by adding an excess amount of the solid compound (0.2 g) to glass flasks containing deionized water (100 mL). Experiments were conducted in a shaker water bath and the temperature was controlled within $\pm 0.5^0$C. The flasks were stirred at 150 rpm for about 36 hours. Once thermal equilibrium was established at a given temperature, three samples (2 mL each) were withdrawn, centrifuged for 10 minutes at 2000 rpm and immediately diluted with water to avoid precipitation. The concentration of MNA was determined spectrophotometrically.

A stock solution of approximately 1 g/L of MNA in 1-octanol was prepared and the concentration of the solution was determined. The experiments were conducted in 40 ml vials with silicone septa. The two solvents (octanol and water) were mutually saturated at the temperature of the experiment before adding MNA dissolved in 1-octanol. The vials were shaken in the shaker water bath at a desired temperature controlled to ±0.5 K. The contents were then allowed to equilibrate at the set temperature for about 36 hours. Later the vials were centrifuged at 2000 rpm for about 10 minutes at the experimental temperature in order to achieve complete separation of the two phases. Both the phases were analyzed for MNA concentration. The octanol-water partition coefficient was obtained as the ratio of concentration of MNA in organic and aqueous phases.

The gas-purging bubble column system used for the experimental determination of Henry's law constant was similar to that of Mackay et al. [1]. High purity compressed air was bubbled through the aqueous solution containing MNA, which was stripped from the solution into the gas phase. The Henry's law coefficient was calculated from the rate of decrease of aqueous concentration of MNA. Experiments were conducted at 10 K intervals between 298 and 318 K. The gas flow was controlled by calibrated mass flow meter. The gas exited from the upper end of the column through a three-way stopcock and was discharged into cooled acetonitrile through Teflon tubing. One mL of the sample was withdrawn at one hour time intervals and the concentration in the liquid phase was measured using UV absorption spectroscopy. Recovery of MNA in the gas purge experiment was calculated from the mass balance between the initial aqueous concentration and the total amount of MNA collected in the cooled acetonitrile plus the remainder in the water phase after the experiment. This recovery value was used to estimate the uncontrolled losses of MNA and to validate the measured K_H. The values of K_H were obtained following the procedure reported by Fu et al., [2].

Results and discussion

Solubility

The solubility of MNA in water and salt solutions of NaCl and $CaCl_2$ (1, 5 and 10%) at 298.15, 308.15 and 318.15 K are shown in Table 1. The variation of solubility of MNA with reciprocal temperatures at different salt concentrations is graphically represented in Figs. 1 and 2 respectively. The increase in temperature results in an increase in solubility in pure water as well as in solutions of electrolytes. Assuming that the saturated solution of MNA has an ideal behavior (which is justified in solutions of limited solubility of hydrophobic materials, [3]), the influence of temperature on the solubility of MNA can be quantitatively described by the Van't Hoff equation as;

$$Log(S_w) = \frac{-\Delta_{sol}H^0}{RT} + b \qquad (1)$$

The enthalpy of MNA dissolution in pure water has been determined by plotting log (S_w) versus 1/T and found to be 20.02 $kJ.mol^{-1}$. The solubility of MNA decreases gradually with increase in concentration of sodium chloride and calcium chloride at all the three temperatures studied. The decrease in solubility of a solute in the presence of an electrolyte is correlated by the classical Setschenow equation [4] as given below;

$$Log\left(\frac{S_o}{S}\right) = K_s C \qquad (2)$$

where S_o is the solubility of MNA in pure water; S is the solubility in presence of a salt; K_s is the Setschenow constant, and C is the concentration of the salt in mol/L. The values of Setschenow constants (Table 2), obtained from experimental data, are positive K_s ($S_o > S_s$), which indicate the salting-out effect.

Table 1. Aqueous solubility of MNA in presence of NaCl and $CaCl_2$ salts at different temperatures.

Medium	Solubility (mg/L)		
	298.15 K	308.15 K	318.15 K
Water	85.42 ± 0.85	111.51 ± 1.12	141.93 ± 1.42
1% NaCl	79.49 ± 0.80	96.65 ± 0.97	128.17 ± 1.28
5% NaCl	63.69 ± 0.64	75.51 ± 0.76	113.96 ± 1.14
10% NaCl	49.56 ± 0.50	58.22 ± 0.58	76.28 ± 0.76
1% $CaCl_2$	82.82 ± 0.83	104.99 ± 1.05	134.81 ± 1.35
5% $CaCl_2$	72.30 ± 0.72	95.03 ± 0.95	117.29 ± 1.17
10% $CaCl_2$	60.62 ± 0.61	85.84 ± 0.86	101.28 ± 1.01

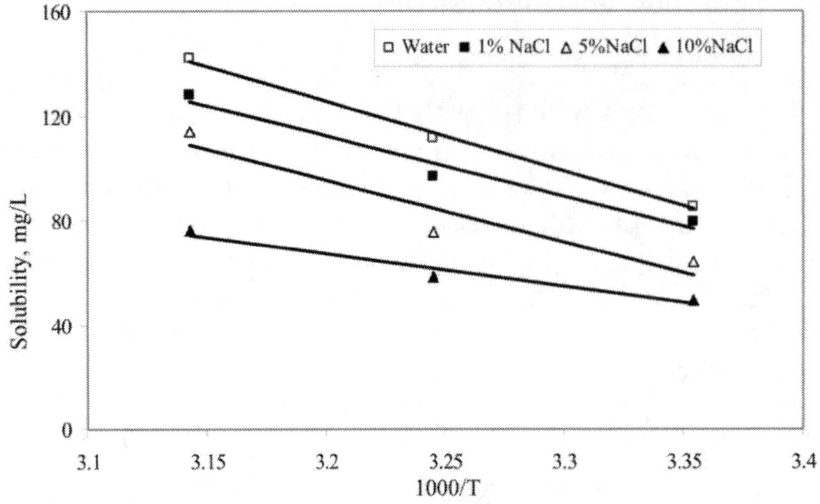

Fig. 1. Solubility of MNA versus reciprocal temperature in the presence of NaCl.

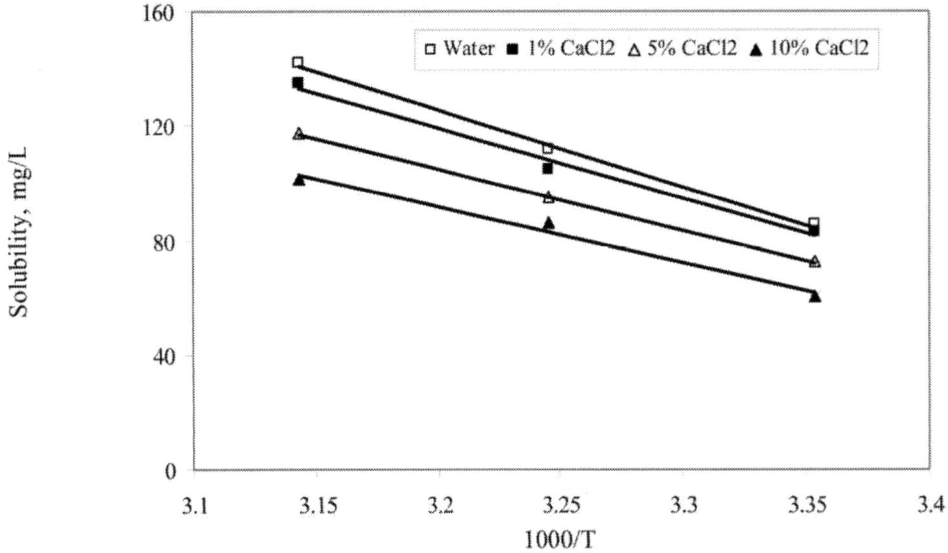

Fig. 2. Solubility of MNA versus reciprocal temperature in the presence of $CaCl_2$.

Table 2. Constants of Setschenow equation for MNA solubility at different temperatures.

Temperature, K	Sodium chloride		Calcium chloride	
	K_s	R^2	K_s	R^2
298.15	0.1368	0.9980	0.2195	0.9998
308.15	0.1580	0.9807	0.1884	0.9733
318.15	0.1490	0.9617	0.2121	0.9945

Octanol-water partition coefficient

The experimental values of octanol-water partition coefficient (Log k_{ow}) of MNA at 298.15, 308.15 and 328.15 are given in Table 2 and variation of Log k_{ow} with reciprocal temperature is shown in Fig. 3. From the Van't Hoff's isotherm, the free energy of transfer from water to octanol is given by,

$$\Delta G = -2.303RT \log k_{ow} \tag{3}$$

We have used the Van't Hoff's approach, which involves plotting the log k_{ow} of a substance versus reciprocal temperatures, and obtaining from the Van't Hoff isochore [equation (4)] the enthalpy and entropy of partitioning.

$$Logk_{ow} = -\Delta H/2.303RT + \Delta S/2.303R \tag{4}$$

The Gibbs free energy (at 298.15 K), enthalpy and entropy of partitioning of MNA in the water-octanol are -11.95 $kJ.mol^{-1}$, -15.06 $kJ.mol^{-1}$ and -10.44 $J.mol^{-1}.K^{-1}$, respectively.

Table 3. Octanol-water partition coefficients (Log k_{ow}) for MNA at different temperatures

Temperature (K)	Log k_{ow}
298.15	2.1028 ± 0.0209
308.15	1.9846 ± 0.0305
318.15	1.9377 ± 0.0146

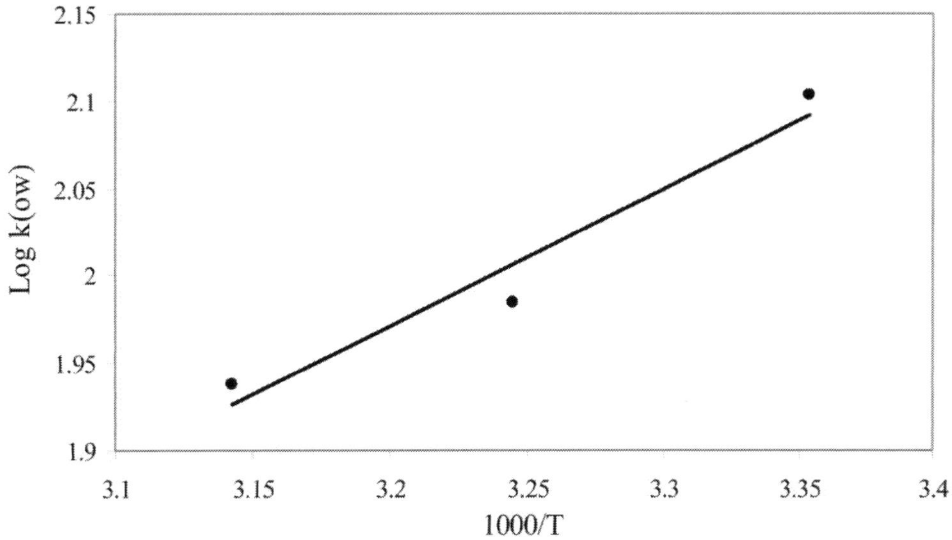

Fig. 3. Octanol-water partition coefficient (Log k_{ow}) of MNA versus reciprocal temperature

Henry law constant

The Henry's law constant (K_H) is also often expressed as a dimensional quantity with units of (pressure. volume)/mass,

$$K_H = p_g / C_w \tag{5}$$

where p_g is the gas-phase partial pressure and C_{ow} is the aqueous concentration in units of mass of MNA per volume of water, or as the dimensionless Henry's law constant (K_{H}'),

$$K_{H'} = C_g / C_w \tag{6}$$

where the gas-phase concentration (C_g) is expressed in units of mass of chemical per volume of air. The partial pressure can be expressed in moles per cubic meter of air assuming ideal behavior, [$p_i = (n_i/V)RT$]. The relationship between K_H and $K_{H'}$ is as follows;

$$K_{H'} = K_H / RT \quad (7)$$

where R is the gas constant (8.314 $J.K^{-1}.mol^{-1}$) and T is the absolute temperature. The experimental values of Henry's law constants for MNA are given in Table 4. The temperature dependence of K_H is described by the equation,

$$-RTLnK_{H'} = \Delta H_H - T\Delta S_H \quad (8)$$

where ΔH_H and ΔS_H, the enthalpy and entropy of the phase transfer from the dissolved phase to the gas phase, were obtained from a plot of Log$K_{H'}$ versus reciprocal temperature (Figure 4).

Table 4. Henry law constants for MNA at different temperatures

Temperature (K)	K_H (M^3 Pa mol^{-1})	$K_{H'}$ (dim.-less g/aq)
298.15	0.6100	$2.46x10^{-4}$
308.15	0.5566	$2.25x10^{-4}$
318.15	0.5160	$2.08x10^{-4}$

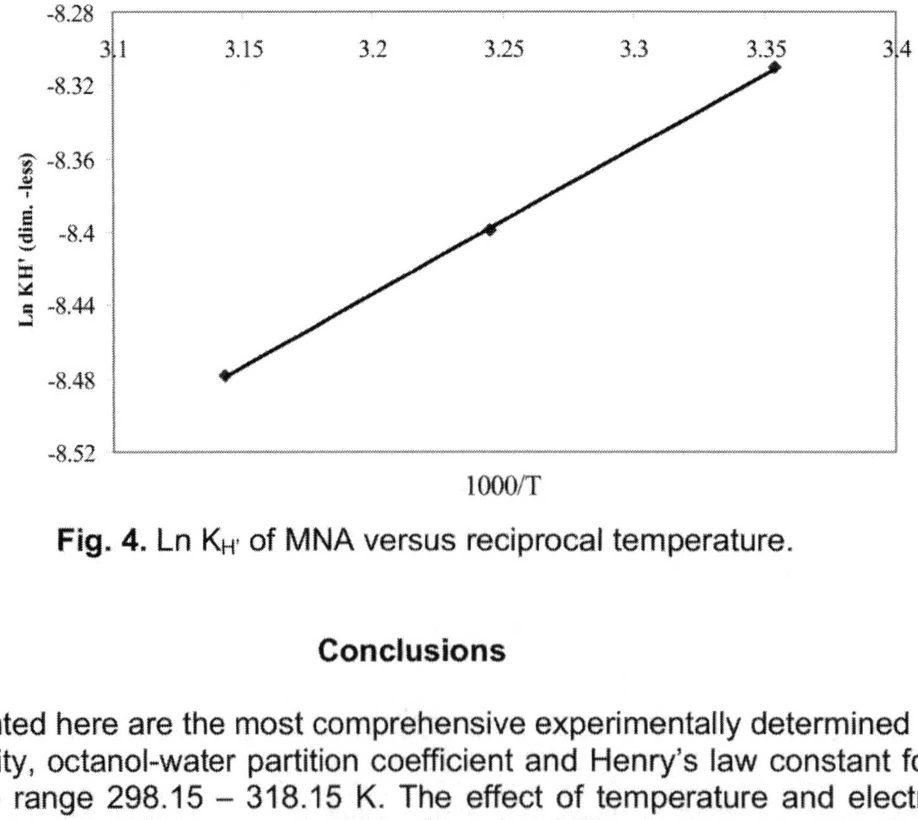

Fig. 4. Ln $K_{H'}$ of MNA versus reciprocal temperature.

Conclusions

The data presented here are the most comprehensive experimentally determined values for aqueous solubility, octanol-water partition coefficient and Henry's law constant for NMA in the temperature range 298.15 – 318.15 K. The effect of temperature and electrolytes on aqueous solubility of MNA was studied. The solubility increased with increase of temperature and decreased with electrolyte concentration. Enthalpies of phase change from solid to liquid, liquid to liquid and liquid to vapor were obtained by fitting the experimental data on S_w, k_{ow} and K_H respectively to Van't Hoff equation. These data enable prediction of the concentrations of MNA to be expected in wastewater and air from munitions facilities. Projectiles will normally be loaded by melt pour techniques, and the facilities are cleaned with hot water or steam. Thus MNA would be at a higher concentration than expected at ambient temperatures due to the temperature dependence of the physical parameters studied.

References

[1] D. Mackay, W.Y. Shiu, R.P. Sutherland, 1979. Determination of air–water Henry's law constants for hydrophobic pollutants. Environ. Sci. Technol. 13 (1979), 333-337.

[2] F. Fu, C. Shaogang, H. Chia-Swee, Air-water Henry's law constants for PCB congeners: Experimental determination and modeling of structure-property relationship, Anal. Chem. 78 (2006), 5412-5418.

[3] N. Adel, A. Manef, P. Elise, Temperature and salt addition effects on the solubility behavior of some phenolic compounds in water, J. Chem. Thermodyn. 39 (2007), 297–303.

[4] J. Setschenow, Uber die konstitution der salzlosungen auf grund ihres verhaltens zu kohlensaure. Z. Phys.Chem. Vierter Band. 1 (1889) 117-125.

628b Thermophysical Property Prediction of Energetic Materials with Atomistic Computer Simulations

Nandhini Sokkalingam[1], MaryBeth Helen Ketko[2], Aishuang Xiang[2], and Jeffrey J. Potoff[2]. (1) Department of Chemical Engineering, Wayne State University, 5050 Anthony Wayne Dr, Detroit, MI 48202, (2) Department of Chemical Engineering, Wayne State University, 5050 Anthony Wayne Dr., Detroit, MI 48202

The majority of DOD generated hazardous waste streams can be traced to the production, maintenance and disposal of weapons systems. In this work, we discuss the use of molecular simulation as a predictive tool for the determination of bioaccumulation potential of a particular energetic material.

Newly developed molecular models, or "force fields" for dinitroanisole (DNAN) and n-methyl-p-nitroaniline (MNA) are presented. Monte Carlo simulations are used to predict the vapor-liquid coexistence curves, vapor pressures and critical points for each compound. Molecular dynamic simulations coupled with thermodynamic integration are used to determine octanol-water partition coefficients, key predictors in the bioaccumulation potential for each material.

628c Green Methodologies for Separation of Explosive Components Rdx and Hmx

Veera Boddu[1], Reddy Damavarapu[2], Ann J. Randolph[3], and Jessie Creamean[1]. (1) Environmental Processes Branch, US Army Engineer Research and Development Center, 2902 Newmark Drive, PO Box 9005, Champaign, IL 61822, (2) Energetics and Warheads Division, U.S. Army - Armament Research, Development and Engineering Center, Picatinny Arsenal, NJ, (3) Environmental Processes Pranch, US Army Engineer Research and Development Center, 2902 Newmark Drive, PO Box 9005, Champaign, IL 61822

Explosive compounds with Reduced Sensitivity (RS) or Insensitive Munitions (IM) are desired for reasons of safe loading, pack and assemble operations and to meet the improved performance requirements of the energetics. The two most widely used dense energetic materials are the monocyclic nitramines RDX (1,3,5-trinitro-1,3,5-hexahydrotriazine,) and HMX (1,3,5,7-tetranitro-1,3,5,7-tetraazacyclooctane). The manufacturing process for these explosives yield a mixture of both RDX and HMX. The sensitivity of the final explosive depends on the composition of RDX and HMX and increases with the composition of HMX. There is a need to selectively separate HMX from the mixture in order to develop RS and IM munitions. This presentation summarizes our research on environmentally benign processes to separate HMX from RDX. The approach was to dissolve the crude RDX and HMX mixture into a suitable solvent and then separate the components. Our approach focused on identifying a suitable solvent to solubilize RDX and HMX and evaluating membrane and chromatographic separation processes.

628d Assessment Of Waste Treatment Technologies For Energetic Materials

Rebecca K. Toghiani[1], Hossein Toghiani[1], Stephen W. Maloney[2], and Veera Boddu[2]. (1) Chemical Engineering, Dave C. Swalm School of Chemical Engineering, Mississippi State University, 330 Swalm, Mississippi State, MS 39762, (2) Environmental Processes Branch, US Army Engineer Research and Development Center, 2902 Newmark Drive, PO Box 9005, Champaign, IL 61822

One important criterion in the development of an energetic material is an assessment of its potential environmental impact. During production, waste streams generated may contain small amounts of the energetic material. Knowing how this material may partition in the environment is beneficial to development of waste treatment technologies. Physicochemical parameters define how this partitioning will occur. While these parameters can be experimentally determined once suitable amounts of the energetic material are available for testing, the ability to predict these important properties using the structure of the energetic material will provide a means for assessing their environmental impact during the preliminary development stage. In this work, the chemical structures of six energetic materials are used to estimate their physicochemical parameters. These energetic materials have been identified by researchers at CERL as materials under consideration as energetics in short-term development or longer-term development, and include dinitroanisole (DNAN), n-methyl-p-nitroaniline (MNA), nitro-triazolene (NTO), triaminotrinitrobenzene (TATB), dinitro pyrazole (DNP) and m-trinitro imidazole (MTNI). Aqueous solubility as a function of temperature, octanol-water partition coefficient and Henry's constant are predicted and used to assess the potential viability of commonly employed wastewater treatment technologies. Development of a QSPR/QSAR relating the structure of the energetic material to potentially effective treatment technologies is discussed.

628e Navy Demilitarization R&d Efforts

Joshua R. Geary, NSWC Crane, 300 Highway 361, Attn: Code 4073, Building 2540, Crane, IN 47522

The primary objective of the Demilitarization Technology Project at Naval Surface Warfare Center, Crane Division (NSWC Crane) are to pursue the development of safe and environmentally acceptable demil processes that will remove, reclaim, and reuse the energetic material found in munition items; for those items where it is unfeasible to reclaim the energetic, contained detonation technology is being pursued. The Demil Technology Project increases the joint service activities of Team Crane by strengthening the bond between NSWC Crane and Crane Army Ammunition Activity (CAAA) by providing new demil capabilities to CAAA for the benefit of its demil execution mission. Some of the technologies being explored and developed by NSWC Crane include: Abrasive Waterjet Cutting/ Washout, Cryofracture of Submunitions, Induction Heating Melt-Out, Manufacturing Commercial Products from Large-Grain Gun Propellants, HMX Recovery and Requalification, Magnesium Recovery, Contained Detonation Chambers, and Molten Salt Oxidation.

AUTHOR INDEX

AUTHOR INDEX

AUTHOR INDEX

AUTHOR INDEX

AUTHOR INDEX

AUTHOR INDEX

AUTHOR INDEX